SEAWEEDS AND THEIR ROLE IN GLOBALLY CHANGING ENVIRONMENTS

Cellular Origin, Life in Extreme Habitats and Astrobiology

Volume 15

Series Editor:

Joseph Seckbach
The Hebrew University of Jerusalem, Israel

For other titles published in this series, go to
www.springer.com/series/5775

Seaweeds and Their Role in Globally Changing Environments

Edited by

Alvaro Israel
Israel Oceanographic and Limnological Research, Ltd.
The National Institute of Oceanography, P.O. Box 8030,
Tel Shikmona 31080. Haifa, Israel

Rachel Einav
Blue Ecosystems, 26 Hagat St., Zichron Yaakov, Israel

and

Joseph Seckbach
The Hebrew University of Jerusalem, Israel

Editors
Alvaro Israel
Israel Oceanographic and Limnological
Research, Ltd.
The National Institute of Oceanography
P.O. Box 8030, Tel Shikmona 31080. Haifa
Israel
alvaro@ocean.org.il

Rachel Einav
Blue Ecosystems
26 Hagat Street
Zichron Yaakov
Israel
einavr@blue-ecosystems.com

Joseph Seckbach
The Hebrew University of Jerusalem
Israel
seckbach@huji.ac.il

ISBN 978-90-481-8568-9   e-ISBN 978-90-481-8569-6
DOI 10.1007/978-90-481-8569-6
Springer Dordrecht Heidelberg London New York

Library of Congress Control Number: 2010925024

© Springer Science+Business Media B.V. 2010
No part of this work may be reproduced, stored in a retrieval system, or transmitted in any form or by any means, electronic, mechanical, photocopying, microfilming, recording or otherwise, without written permission from the Publisher, with the exception of any material supplied specifically for the purpose of being entered and executed on a computer system, for exclusive use by the purchaser of the work.

Printed on acid-free paper

Springer is part of Springer Science+Business Media (www.springer.com)

TABLE OF CONTENTS

Preface/**Alvaro Israel and Rachel Einav**.. ix

Acknowledgements xiii

Introduction/**Joseph Seckbach** xv

List of Authors and Their Addresses xxi

PART 1:
CHANGES IN THE MARINE ENVIRONMENT

Sea-Level Changes in the Mediterranean: Past, Present,
and Future – A Review **[Lichter, M. et al.]**............................... 3
Global Climate Change and Marine Conservation
[Olsvig-Whittaker, L.].. 19

PART 2:
BIODIVERSITY IN MARINE ECOSYSTEMS
IN THE GLOBALLY CHANGING ERA

Is Global Warming Involved in the Success of Seaweed
Introductions in the Mediterranean Sea?
[Boudouresque, C.F. and Verlaque, M.]................................... 31
Climate Change Effects on Marine Ecological Communities
[Rilov, G. and Treves, H.] 51
Fucoid Flora of the Rocky Intertidal of the Canadian Maritimes:
Implications for the Future with Rapid Climate Change
[Ugarte, R.A. et al.] 69

PART 3:
ECOPHYSIOLOGICAL RESPONSES OF SEAWEEDS

GIS-Based Environmental Analysis, Remote Sensing,
and Niche Modeling of Seaweed Communities
[Pauly, K. and De Clerck, O.] 93

Physiological Responses of Seaweeds to Elevated
  Atmospheric CO_2 Concentrations [Zou, D. and Gao, K.] 115
The Role of Rhodolith Beds in the Recruitment of Invertebrate
  Species from the Southwestern Gulf of California, México
  [Riosmena-Rodriguez, R. and Medina-López, M.A.] 127
The Potential Impact of Climate Change on Endophyte
  Infections in Kelp Sporophytes [Eggert, A. et al.] 139

PART 4:
THE EFFECTS OF UV RADIATION ON SEAWEEDS

Interactive Effects of UV Radiation and Nutrients
  on Ecophysiology: Vulnerability and Adaptation
  to Climate Change [Figueroa, F.L. and Korbee, N.] 157
Ecological and Physiological Responses of Macroalgae
  to Solar and UV Radiation [Gao, K. and Xu, J.] 183
Ultraviolet Radiation Effects on Macroalgae from Patagonia,
  Argentina [Helbling, E.W. et al.] 199

PART 5:
BIOFUEL – SEAWEEDS AS A SOURCE
OF FUTURE ENERGY

Production of Biofuel by Macroalgae with Preservation
  of Marine Resources and Environment [Notoya, M.] 217
Biofuel from Algae – Salvation from Peak Oil? [Rhodes, C.J.] 229

PART 6:
CULTIVATION OF SEAWEEDS IN GLOBALLY
CHANGING ENVIRONMENTS

A Review of *Kappaphycus* Farming: Prospects and Constraints
  [Hayashi, L. et al.] 251
Recycling of the Seaweed Wakame Through Degradation
  by Halotolerant Bacteria [Tang, J.-C. et al.] 285
Progressive Development of New Marine
  Environments: IMTA (Integrated Multi-Trophic
  Aquaculture) Production [Issar, A.S. and Neori, A.] 305
Reproductive Processes in Red Algal Genus *Gracilaria*
  and Impact of Climate Change [Mantri, V.A. et al.] 319
The Role of *Porphyra* in Sustainable Culture Systems:
  Physiology and Applications [Pereira, R. and Yarish, C.] 339

PART 7:
BIOTECHNOLOGICAL POTENTIAL OF SEAWEEDS

Intensive Sea Weed Aquaculture: A Potent Solution
    Against Global Warming [Turan, G. and Neori, A.] 357
The Future is Green: On the Biotechnological
    Potential of Green Algae [Reisser, W.] 373
The Potential of *Caulerpa* spp. for Biotechnological
    and Pharmacological Applications
    [Cavas, L. and Pohnert, G.] 385

PART 8:
OTHER VIEWS TO GLOBAL CHANGE

Ecology, Science, and Religion [Klostermaier, K.K.] 401
Nature and Resource Conservation as Value-Assessment
    Reflections on Theology and Ethics [Roth, H.J.] 423
Global Warming According to Jewish Law:
    Three Circles of Reference [Glicksberg, S.E.] 435
Guarding the Globe: A Jewish Approach
    to Global Warming [Rozenson, Y.] 449

Organism Index 461

Subject Index 467

Author Index 479

PREFACE

ALVARO ISRAEL[1] and RACHEL EINAV[2]
[1]*Israel Oceanographic and Limnological Research, Ltd. The National Institute of Oceanography, P.O. Box 8030, Tel Shikmona 31080. Haifa, Israel.* [2]*Blue-Ecosystems, 26 Hagat St., Zichron Yaakov, Israel*

Climate changes and global warming occurring on earth are now a widely recognized phenomena within the public and scientific communities. They will likely modify marine life dramatically as we now know it. One critical question regarding these changes is whether they occurred because of human intervention, or due to natural events on earth, or a combination of both. Irrespective of the source of these changes, it is our responsibility to understand and properly control these events so as to diminish potential or irreversible damage in the marine environment. The goal of this project, *Seaweeds and Their Role in Globally Changing Environments* was to emphasize the role of marine macroalgae, the so-called seaweeds, within the context of global changes occurring on planet Earth.

This book concentrates on the diverse aspects of the expected effects of global changes on seaweeds. First, a general overview of current changes in the oceans is given including the legal aspects associated with these modifications. While responses to global changes occur first on a species level, ultimately the modifications will arise on a community and global ecosystem levels. These aspects are discussed in Part 2. Then, Part 3 addresses short- and long-term seaweed ecophysiological responses to environmental abrupt changes, which forces marine plants to make sudden adjustments rather than adaptation processes that have occurred during millions of years of evolution. Specific and detailed aspects of seaweed responses to the UV rays are given in Part 4. Applied aspects of seaweeds follow in Parts 5 and 6. Here, the reader will find insights of potential uses of seaweeds in the future, and expected effects on seaweed cultivation practices worldwide as dictated by the globally occurring changes in the marine environment. Theoretical approaches of marine plants utilization in the future as related to modified environments are shown in Part 7. Global changes influence almost all aspects of human life, becoming daily worries/issues within the general public and scientific community. The need to enroll synergistic forces to address the problems derived from global changes and their environmental effects is apparent. Therefore, we have considered pertinent to also include spiritual/religious approaches to the issue, which are quite unique in the scientific literature. These aspects are analyzed towards the end of the book and may allow to those readers interested in such aspects of life as well.

The book begins with rather pessimistic overviews of anthropogenic and natural effects on the marine environment. Although predictions may be quite devastating, contributions presented by experts show much optimistic pictures of how the marine macroalgae will look like in terms of their ecological communities and adaptation strategies. Further, seaweeds will have a much more significant role in controlling environmental stresses caused by global change. Thus, the reader will be transferred through various levels of comprehension both scientifically and encouragingly as to how will seaweeds be viewed in the near future.

Biodata of **Alvaro Israel** and **Rachel Einav**, editors of *"Seaweeds and Their Role in Globally Changing Environments"*

Dr. Alvaro Israel is currently a Senior Scientist at the Israel Oceanographic and Limnological Research, Ltd., The National Institute of Oceanography, Haifa, Israel (www.ocean.org.il). He obtained his Ph.D. from Tel Aviv University in 1992 in Marine Botany in carbon fixation aspects of seaweeds. Then, he continued his studies and research in environmental biology of plants and algae at UCLA, USA. Dr. Israel's scientific interests are in the area of seaweed eco-physiology, global change and applied phycology. Recent scientific activities of his work include describing the effects of elevated CO_2 in photosynthesis of marine macroalgae and developing biological background for seaweed cultivation in land-based settings.

E-mail: **alvaro@ocean.org.il**

Dr. Rachel Einav is CEO of Blue Ecosystems (www.blue-ecosystems.com), a company providing marine environmental consulting. She obtained her Ph.D. from Bielefeld University (Germany) in eco-physiology and adaptation strategies of marine macroalgae. Results of her post-doc project at Bar Ilan University (Ramat Gan, Israel) were published as a book – *Seaweeds of the Eastern Mediterranean Coast*, (in Hebrew, and in English by A. R. G. Ganther Verlag K. G. (India) now being translated to Arabic. Dr. Einav scientific interests are in the area of marine environment and anthropogenic effects on seaweed communities.

E-mail: **einavr@blue-ecosystems.com**

Alvaro Israel   **Rachel Einav**

ACKNOWLEDGMENTS

We are grateful to all the authors of this project for their patience and understanding during the making of the book. We also wish to thank the reviewers involved in the evaluation of the chapters, particularly Dr. Amir Neori and Dr. Linda Whittaker, and to Guy Paz for preparing the illustration of the cover. This project was supported by Research Grant No. IS 3853-06 R from BARD, The USA–Israel Binational Agricultural Research and Development Fund.

INTRODUCTION TO GLOBALLY CHANGING ENVIRONMENT

JOSEPH SECKBACH
Hebrew University of Jerusalem, Israel

1. Weather Changes in the Past

Archeological evidence for weather changes in the past are seen in the discovery of traces of agriculture and tropical plants (bones and fossils) in desert areas. Other evidence demonstrates the inundation of settlements that today are beneath the sea. Finding marine traces and fossils at high altitudes, such as hill tops, and in other dry zones climate changes supports this assumption. Some scholars have assumed that global warming during the biblical Noah's generation may have caused the flood. Others view the Egyptian atmospheric plagues toward the exodus of the children of Israel and the splitting of the Red Sea as related to climate change by nature.

The release of current industrial pollution and other sources into the atmosphere and hydrosphere has increased at a far greater rate than any historic natural process. One serious regional environmental problem is acid rain. As long as we have been burning fossil fuels, this acidic liquid has been falling from the sky and causing damages.

Even with the biological removal of pollution caused by CO_2 which also causes atmospheric warming, the pre-existing state cannot be regained precisely. So, global warming is accelerating faster than the ability for natural repair. Lately, it has been determined that there is no link between global warming and cosmic rays or other solar activities.

2. Current Human Activities and Their Influence on the Climate Changes

A new NASA-led study shows that human-caused climate change has made an impact on a wide range of Earth's natural systems, resulting in permafrost thawing, acid rain, plants blooming earlier across Europe, and lakes declining in productivity. Researchers have linked varying forces since 1970 with rises in temperatures. Humans are influencing climate through increasing greenhouse gases emissions, among them are CO_2, N_2O, CH_4, CF_3, and CFC. Global warming is influencing physical and biological systems all over our planet but most specifically in North America, Europe, Asia, and Antarctica.

Climate change is one of the greatest challenges the world is now facing. Leaders should now deal with this disaster by calling for long-term international development programs. The ecology factors driven by man include industrial fuel burning that yields by-products and pollution spills, vehicle emissions, smoke from power stations' chimneys, huge fires, and other harmful environmental activities. All those occurrences may intensify the greenhouse effect, cause changes to climate, and directly harm global biology. Further damage caused by man includes deforestation in some developing areas. The felling of millions of trees in areas like the Amazon rain forest in addition to forest fires will reduce the green lungs of the globe. Likewise, damage could result from excess grazing by herds, excess pumping of water, and desertification. Some gases (such as CFC) released by industrial activities damage the protective screen of the stratospheric ozone layer (as, e.g., the recently discovered ozone holes over Antarctica) and cause intensive UV radiation to penetrate to earth in higher doses.

Some predictions claim that with global warming we shall have less rain and less precipitation. Or, rain will arrive in short, strong storms, so that the precipitation will not penetrate into the subsurface accumulation spaces. Such an effect would reduce and damage the subsurface water reservoirs. Others see no rain, drought, the drying out of large water supplies, dust storms on a great scale, and general damage to agriculture. All these will damage more and more genera of living creatures. There are also contrasting harmful effects, such as heavy rains and floods, or hurricanes in certain zones around the globe; intensified and ruinous damage from storms; thawing of glaciers; and the rise of sea levels, with the danger of over flooding to low lands.

The rise in temperature will influence the evaporation rates in lakes (see the current case of the drying Dead Sea [Israel], Chad lake [Africa], or Aral lake [in Asia]), which might shrink and almost vanish without sufficient income of water from their sources. Global warming also poses a severe danger for some animals, such as, the polar bear that is in danger of extinction, or the harm caused by the expansion of fire ants to areas once too cold for them. Warmer and more acidic oceanic water (due to the increase of CO_2 in the atmosphere and oceans) spells trouble for jumbo squids and other marine animals. Global warming might cause a reduction in the amount of dissolved oxygen in the oceans and lead to suffocation of marine biota. Results, for example, would be that the tuna and sword fish would turn into extinct species and corals would be harmed, resulting in the disappearance of several species of fish and reefs.

Trees in western North America are dying more quickly than they used to, but there is no corresponding increase in the number of new seedling trees. Mortality rates, which are currently of the order of 1% a year, have in many cases doubled in just a few decades. The increased mortality correlates with climate change in the region, which has warmed by an average of between 0.3°C and 0.4°C per decade since the 1970s.

Some experts claim that even if carbon emissions were stopped, temperatures around the globe would remain high until at least the year 3000. And if we continue with our current carbon dioxide discharge for just a few more decades, we could see permanent "dust bowl" conditions.

3. What Should Be Done to Curtail Human Actions That Promote Global Warming?

The responsibility of the community is to avoid desertification and repair it. There should be a shift to growing crops which need less irrigation, and the burning of large amounts of fossil crude oils and coal should be avoided by using alternative energy sources. Nations should carefully manage the consumption of fuel by the various vehicles (cars, boats, air planes, and power stations). They should maintain restrictions on deforestation and seek alternative sources of usable water (for human needs and agricultures).

4. Algae, Seaweeds, and Global Warming

Microalgae and seaweeds (see further) have enormous potential and are actively involved in lowering global warming and climate change. Algae and seaweeds (like the entire green world) absorb carbon dioxide from the atmosphere (or directly from their solution media) by the process of photosynthesis, release oxygen, and produce solar biofuel. During photosynthesis algae (and higher plants) grow; they actually drain CO_2 from the atmosphere. This gas is released again when their biomass burns. This CO_2-capturing system within the green world keeps this gas from re-entering the air (except for minor amounts released during the plant–animal respiration process). In fact, even the plant residue (e.g., the ashes) could be put to good use as mineral-rich fertilizer after being pressed into biofuel.

Marine macroalgae (seaweeds) play significant roles in the normal functioning of atmospheric environments. Even though seaweeds are restricted to the tide zones and benthic photic zones, they contribute to about 10% of the total world marine productivity. Ecologically they account for food and shelter for marine life. Seaweeds are also used as sea-vegetables for food consumption (for fish and man). In the Far Eastern countries they use *Porphyra* blades (Nori) in cuisine. In addition, elsewhere other edible seaweeds are in use, such as *Rodymenia* (Dulse), *Laminaria saccharina*, *Chodrus*, and *Ulva* (sea lettuce). There are other uses for seaweed since it is rich in vitamins, minerals, and proteins. Various marine macroalgae are potential sources of bioactive compounds, and they act as antibacterial and antiviral agents. Among them are those that may also be utilized for the treatment of human diseases such as cancer.

Globally changing environments on earth is more likely to severely modify the current equilibrated terrestrial and marine ecosystems. Specifically for the marine environment, global changes will include increased carbon dioxide which will acidify the aqueous media. It has been estimated that for CO_2, the change might be from the current 350 ppm to approximately 750 ppm within 50 years, or so. Such a difference will cause higher average seawater temperatures (within 1–3°C) and higher UV radiation on the water surface. These changes will affect marine macroalgae at different levels, namely molecular, biochemical, and population levels. While predictions of altered environments have been studied extensively for terrestrial ecosystems, comparatively much less effort has been devoted to marine habitat. Seaweeds may contribute significantly to reduce pollutants (such as CO_2, heavy metals, and excessive nutrients disposed of in the marine environments).

5. What Should Be Done to Reduce Greenhouse Gas Emission?

5.1. NO NEED FOR REDUCING THE GREENHOUSE EFFECT – FALSE PANIC ALARM

Historic records extracted from deep ice cores (taken in Antarctica drilling) show that quantities of CO_2 have varied widely in the last hundreds or thousands of years. This evidence appears to contradict the current critical view of global warming. Some voices claim that the present observation of the human-induced greenhouse effect is actually a natural occurrence. They say that the effect of carbon on the climate is overestimated and the climate crisis might be hyped. However, a new study shows that although carbon dioxide levels may have been larger in the past, the natural processes had time to react and counteract global warming.

5.2. DO IT NOW

Only good education and international enforcement applied to governments will reduce the pollution in our planet. The less greenhouse gases released to the atmosphere and hydrosphere (by various human sources), the greater and faster will be the salvation to the problem of global warming.

6. Conclusion and Summary

There are pro and con arguments about global warming and its damage to the earth's atmosphere. As a result of continuing pollution, we might witness the warming of the atmosphere, changes in the precipitation, and an increase of the CO_2 level in the atmosphere which might also cause acidification of the oceans. The elevated temperature causes the thawing and melting of the glaciers, which will raise the sea level and cause overflooding overfloating and drown the nether areas (under the sea level). Other hazards are the lowering of ground water and the salting of the aquifers near the sea shores, reducing drinking water and agricultural irrigation, desertification of large green areas, and increasing doses of UV harmful irradiation.

Activities should be designed to prevent most of the dangerous phenomena noted above; one such possible endeavor would be the search for an alternative energy source (rather than black gold or coal). A main target is utilization of alternative sources of energy, such as solar and wind energy, hydroelectric power, and "ocean energy" (using underwater vibration) for various human applications. These powers should reduce and avoid the spread of harmful gases. In some cases, the use of uranium could also be implemented as an energy source, but this means must be instituted very cautiously. Another aspect is a stricter watch over fires, and the maintenance and increase of the areas of rain forests. But above all is the education of the present and future generations to keep our Mother Earth as pure as possible.

Biodata of **Joseph Seckbach**, co-editor of this volume and author of *"Introduction to Globally Changing Environment"*

Professor Joseph Seckbach is the Founder and Chief Editor of *Cellular Origins, Life in Extreme Habitats and Astrobiology* ("COLE") book series and is the author of several chapters in this series. See www.springer.com/sereis/5775. Dr. Seckbach earned his Ph.D. from the University of Chicago, Chicago, IL (1965) in Biological sciences. Among his publications are books, scientific articles concerning plant ferritin (phytoferritin), cellular evolution, acidothermophilic algae, and life in extreme environments. He also edited and translated several popular books. Dr. Seckbach is the co-author (with R. Ikan) of the *Chemistry Lexicon* (1991, 1999, Hebrew edition) and other volumes, such as the *Proceeding of Endocytobiology VII Conference* (Freiburg, Germany, 1998) and the *Proceedings of Algae and Extreme Environments Meeting* (Trebon, Czech Republic, 2000); see:http://www.schweizerbart.de/pubs/books/bo/novahedwig-051012300-desc.ht). His new volume entitled *Divine Action and Natural Selection: Science, Faith, and Evolution*, has been edited with Professor Richard Gordon and published by World Scientific Publishing Company.

E-mail: **seckbach@huji.ac.il**

LIST OF AUTHORS FOR *"SEAWEEDS AND THEIR ROLE IN GLOBALLY CHANGING ENVIRONMENTS"*

BLEICHER-CHONNEUR GENEVIEVE
RAW MATERIALS PROCUREMENT, CARGILL TEXTURIZING SOLUTIONS, BAUPTE 50500, FRANCE.

BOUDOURESQUE CHARLES F.
CENTER OF OCEANOLOGY OF MARSEILLES CAMPUS OF LUMINY, UNIVERSITY OF THE MEDITERRANEAN, 13288 MARSEILLES CEDEX 9, FRANCE.

CAVAS LEVENT
DIVISION OF BIOCHEMISTRY, DEPARTMENT OF CHEMISTRY, FACULTY OF ARTS AND SCIENCES, DOKUZ EYLÜL UNIVERSITY, İZMIR 35160, TURKEY.

CRAIGIE JAMES S.
ACADIAN SEAPLANTS LIMITED, 30 BROWN AVENUE, DARTMOUTH, B3B1X8, NOVA SCOTIA, CANADA AND NATIONAL RESEARCH COUNCIL OF CANADA, INSTITUTE FOR MARINE BIOSCIENCES, 1411 OXFORD STREET, HALIFAX, B3H 3Z1, NS, CANADA.

CRITCHLEY ALAN T.
ACADIAN SEAPLANTS LIMITED, 30 BROWN AVENUE, DARTMOUTH, B3B1X8, NS, CANADA.

DE CLERCK OLIVIER
PHYCOLOGY RESEARCH GROUP, BIOLOGY DEPARTMENT, GHENT UNIVERSITY, GHENT 9000, BELGIUM.

EGGERT ANJA
PHYSICAL OCEANOGRAPHY AND INSTRUMENTATION, LEIBNIZ INSTITUTE FOR BALTIC SEA RESEARCH WARNEMÜNDE, ROSTOCK 18119, GERMANY.

EINAV RACHEL
BLUE ECOSYSTEMS, 26 HAGAT ST., ZICHRON YAAKOV 30900, ISRAEL.

FIGUEROA FÉLIX L.
DEPARTMENT OF ECOLOGY, FACULTY OF SCIENCE, UNIVERSITY OF MÁLAGA, 29071 MÁLAGA, SPAIN.

GAO KUNSHAN
STATE KEY LABORATORY OF MARINE ENVIRONMENTAL SCIENCE, XIAMEN UNIVERSITY, XIAMEN, 361005, CHINA.

GLICKSBERG SHLOMO E.
EFRATA COLLEGE OF EDUCATION, BEN YIFUNEH 17, BAKA'A, JERUSALEM 91102, ISRAEL.

HÄDER DONAT-P.
FRIEDRICH-ALEXANDER UNIVERSITÄT ERLANGEN/NÜRNBERG, DEPARTMENT OF BIOLOGY, STAUDTSTR. 5, ERLANGEN 91058, GERMANY.

HAYASHI LEILA
DEPTO. BEG, CENTRO DE CIÊNCIAS BIOLÓGICAS, UNIVERSIDADE FEDRAL DE SANTA CATARINA, TRINDADE, FLORIANÓPOLIS, SANTA CATARINA 88040-900, BRAZIL.

HELBLING E. WALTER
ESTACIÓN DE FOTOBIOLOGÍA PLAYA UNIÓN, CASILLA DE CORREOS Nº 15, (9103) RAWSON, CHUBUT AND CONSEJO NACIONAL DE INVESTIGACIONES CIENTÍFICAS Y TÉCNICAS, CONICET, ARGENTINA.

HURTADO ANICIA Q.
SOUTHEAST ASIAN FISHERIES DEVELOPMENT CENTER, TIGBAUAN, ILOILO 5021, PHILIPPINES.

ISRAEL ALVARO
ISRAEL OCEANOGRAPHIC & LIMNOLOGICAL RESEARCH, LTD., THE NATIONAL INSTITUTE OF OCEANOGRAPHY, P.O. BOX 8030, HAIFA 31080, ISRAEL.

ISSAR ARIE S.
BEN GURION UNIVERSITY OF THE NEGEV, J. BLAUSTEIN INSTITUTE FOR DESERT RESEARCH, ZUCKERMAN INSTITUTE FOR WATER RESOURCES, SEDE BOKER CAMPUS 84990, ISRAEL.

JHA BHAVANATH
DISCIPLINE OF MARINE BIOTECHNOLOGY AND ECOLOGY,
CENTRAL SALT AND MARINE CHEMICALS RESEARCH INSTITUTE,
COUNCIL OF SCIENTIFIC AND INDUSTRIAL RESEARCH (CSIR),
BHAVNAGAR 364002, INDIA.

KLEIN MICHA
DEPARTMENT OF GEOGRAPHY AND ENVIRONMENTAL STUDIES,
UNIVERSITY OF HAIFA, HAIFA 31905, ISRAEL.

KLOSTERMAIER KLAUS K.
DEPARTMENT OF RELIGION, UNIVERSITY OF MANITOBA,
WINNIPEG MB, R3T 2N2 CANADA.

KORBEE NATHALIE
DEPARTMENT OF ECOLOGY, FACULTY OF SCIENCE,
UNIVERSITY OF MÁLAGA, 29071 MÁLAGA, SPAIN.

KÜPPER FRITHJOF C.
SCOTTISH ASSOCIATION OF MARINE SCIENCE, OBAN,
ARGYLL PA37 1QA, SCOTLAND, UK.

LICHTER MICHAL
"COASTS AT RISK AND SEA-LEVEL RISE" RESEARCH GROUP,
THE FUTURE OCEAN EXCELLENCE CLUSTER, INSTITUTE OF
GEOGRAPHY, CHRISTIAN ALBRECHTS UNIVERSITY, KIEL 24098,
GERMANY.

MANTRI VAIBHAV A.
DISCIPLINE OF MARINE BIOTECHNOLOGY AND ECOLOGY,
CENTRAL SALT AND MARINE CHEMICALS RESEARCH INSTITUTE,
COUNCIL OF SCIENTIFIC AND INDUSTRIAL RESEARCH (CSIR),
BHAVNAGAR 364002, INDIA.

MEDINA-LÓPEZ MARCO A.
PROGRAMA DE INVESTIGACIÓN EN BOTÁNICA MARINA,
DEPARTMENTO DE BIOLOGIA MARINA, UNIVERSIDAD
AUTONOMA DE BAJA CALIFORNIA SUR APPARTADO POSTAL 19-B,
LAPAZ B.C.S 23080, MEXICO.

MASUYA FLOWER E.
INSTITUTE OF MARINE SCIENCE, UNIVERSITY OF DAR ES SALAAM, P.O. BOX 668, ZANZIBAR, TANZANIA.

NAGATA SHINICHI
ENVIRONMENTAL BIOCHEMISTRY DIVISION, RESEARCH CENTER FOR INLAND SEAS, ORGANIZATION OF ADVANCED SCIENCE AND TECHNOLOGY, KOBE UNIVERSITY, KOBE 658-0022, JAPAN.

NEORI AMIR
ISRAEL OCEANOGRAPHIC & LIMNOLOGICAL RESEARCH LTD, NATIONAL CENTER FOR MARICULTURE, P.O. BOX 1212, EILAT 88112, ISRAEL.

NOTOYA MASAHIRO
NOTOYA RESEARCH INSTITUTE OF APPLIED PHYCOLOGY, MUKOJIMA-4, SUMIDA-KU, TOKYO 131-8505, JAPAN.

OLSVIG-WHITTAKER LINDA
SCIENCE AND CONSERVATION DIVISION, ISRAEL NATURE AND NATIONAL PARKS PROTECTION AUTHORITY, 3 AM VE OLAMO ST, GIVAT SHAUL, JERUSALEM 95463, ISRAEL.

PAULY KLAAS
PHYCOLOGY RESEARCH GROUP, BIOLOGY DEPARTMENT, GHENT UNIVERSITY, GHENT 9000, BELGIUM.

PEREIRA RUI
CIIMAR/CIMAR, CENTRE FOR MARINE AND ENVIRONMENTAL RESEARCH, RUA DOS BRAGAS, 289, PORTO, PORTUGAL.

PETERS AKIRA F.
BEZHIN ROSKO, 29680 ROSCOFF, FRANCE.

POHNERT GEORG
INSTITUTE FOR INORGANIC AND ANALYTICAL CHEMISTRY, FRIEDRICH SCHILLER UNIVERSITY JENA, D-07743 JENA, GERMANY.

REDDY C.R.K.
DISCIPLINE OF MARINE BIOTECHNOLOGY AND ECOLOGY, CENTRAL SALT AND MARINE CHEMICALS RESEARCH INSTITUTE, COUNCIL OF SCIENTIFIC AND INDUSTRIAL RESEARCH (CSIR), BHAVNAGAR 364002, INDIA.

REISSER WERNER
INSTITUTE OF BIOLOGY I, GENERAL AND APPLIED BOTANY, UNIVERSITY OF LEIPZIG, D – 04103 LEIPZIG, GERMANY.

RHODES CHRIS J.
FRESH-LANDS, P.O. BOX 2074, READING, BERKSHIRE, RG4 5ZQ, UNITED KINGDOM.

RILOV GIL
ISRAEL OCEANOGRAPHIC AND LIMNOLOGICAL RESEARCH, THE NATIONAL INSTITUTE OF OCEANOGRAPHY, TEL- SHIKMONA, P.O. BOX 8030, HAIFA 31080, ISRAEL.

RIOSMENA-RODRIGUEZ RAFAEL
PROGRAMA DE INVESTIGACIÓN EN BOTÁNICA MARINA, DEPTARTAMENTO DE BIOLOGIA MARINA, UNIVERSIDAD AUTÓNOMA DE BAJA CALIFORNIA SUR APARTADO POSTAL 19-B, LA PAZ B.C.S. 23080, MÉXICO.

ROTH HERMANN JOSEF
NATURHISTORISCHER VEREIN, UNIVERSITY AT BONN, GERMANY AND EUROPAINSTITUT FÜR CISTERCIENSISCHE GESCHICHTE, PÄPSTLICHE HOCHSCHULE HEILIGENKREUZ NEAR VIENNA, AUSTRIA.

ROZENSON YISRAEL
EFRATA TEACHERS COLLEGE, 17, BEN-YEFUNEH STREET, P.O. BOX 10263, JERUSALEM 91102, ISRAEL.

SECKBACH JOSEPH
P.O. BOX 1132, EFRAT 90435, ISRAEL.

SIVAN DORIT
LEON RECANATI INSTITUTE FOR MARITIME STUDIES (RIMS), DEPARTMENT OF MARITIME CIVILIZATIONS, UNIVERSITY OF HAIFA, HAIFA 31905, ISRAEL.

TANG JING-C.
KEY LABORATORY OF POLLUTION PROCESSES AND ENVIRONMENTAL CRITERIA, MINISTRY OF EDUCATION, COLLEGE OF ENVIRONMENTAL SCIENCE AND ENGINEERING, NANKAI UNIVERSITY, TIANJIN 300071, CHINA.

TANIGUCHI HIDEJI
ENVIRONMENTAL BIOCHEMISTRY DIVISION, RESEARCH CENTER FOR INLAND SEAS, ORGANIZATION OF ADVANCED SCIENCE AND TECHNOLOGY, KOBE UNIVERSITY, KOBE 658-0022, JAPAN.

TREVES HAIM
RUPPIN ACADEMIC CENTER, SCHOOL OF MARINE SCIENCES, MIKHMORET, ISRAEL.

TURAN GAMZE
FISHERIES FACULTY, AQUACULTURE DEPARTMENT, EGE UNIVERSITY, BORNOVA, IZMIR 35100, TURKEY.

UGARTE RAUL A.
ACADIAN SEAPLANTS LIMITED, 30 BROWN AVENUE, DARTMOUTH, B3B1X8, NOVA SCOTIA, CANADA.

VERLAQUE MARC
CENTER OF OCEANOLOGY OF MARSEILLES, CAMPUS OF LUMINY, UNIVERSITY OF THE MEDITERRANEAN, 13288, MARSEILLES, CEDEX 9, FRANCE.

VILLAFAÑE VIRGINIA E.
ESTACIÓN DE FOTOBIOLOGÍA PLAYA UNIÓN, CASILLA DE CORREOS N° 15, (9103) RAWSON CHUBUT, ARGENTINA & CONSEJO NACIONAL DE INVESTIGACIONES CIENTÍFICAS Y TÉCNICAS, CONICET, ARGENTINA.

XU JUNTIAN
KEY LAB OF MARINE BIOTECHNOLOGY OF JIANGSU PROVINCE, HUAIHAI INSTITUTE OF TECHNOLOGY, LIANYUNGANG 222005, CHINA.

YARISH CHARLES
DEPARTMENTS OF ECOLOGY AND EVOLUTIONARY BIOLOGY AND MARINE SCIENCES, UNIVERSITY OF CONNECTICUT, ONE UNIVERSITY PLACE, STAMFORD, CT 06901-2315, USA.

ZHOU QIXING
KEY LABORATORY OF POLLUTION PROCESSES AND ENVIRONMENTAL CRITERIA, MINISTRY OF EDUCATION, COLLEGE OF ENVIRONMENTAL SCIENCE AND ENGINEERING, NANKAI UNIVERSITY, TIANJIN 300071, CHINA.

ZOU DINGHUI
COLLEGE OF ENVIRONMENTAL SCIENCE AND ENGINEERING,
SOUTH CHINA UNIVERSITY OF TECHNOLOGY, GUANGZHOU
510640, CHINA.

ZVIELY DOV
LEON RECANATI INSTITUTE FOR MARITIME STUDIES (RIMS),
UNIVERSITY OF HAIFA, HAIFA 31905, ISRAEL.

PART 1:
CHANGES IN THE MARINE ENVIRONMENT

Lichter
Olsvig-Whittaker

Biodata of **Michal Lichter, Dov Zviely, Micha Klein,** and **Dorit Sivan**, authors of *"Sea-Level Changes in the Mediterranean: Past, Present and Future – A Review"*

Dr. Michal Lichter is currently a Postdoctoral Researcher in the Coasts at Risk and Sea-Level Rise (CRSLR) Research Group of the Future Ocean Excellence Cluster, based at the Institute of Geography, Christian Albrechts University, Kiel, Germany. She obtained her Ph.D. in Geography and Environmental Studies from the University of Haifa, Israel, in 2009. Dr. Michal Lichter's primary research interests are Coastal geomorphology, Geomorphological mapping, spatial analysis and GIS, shoreline changes, landform processes and environmental change, climate change, and sea-level rise.

E-mail: **mlichter@gmail.com**

Dr. Dov Zviely is Researcher in the Leon Recanati Institute for Maritime Studies (RIMS), in the University of Haifa, Israel. His scientific areas are: (1) Coastal geomorphology and sedimentology; (2) continental shelf morphodynamics; (3) Quaternary coastal paleo-geography and geo-archeology; (4) Charts, maps, hydrography, and maritime history of the eastern Mediterranean Sea.

E-mail: **zviely@netvision.net.il**

Michal Lichter   Dov Zviely

Professor Micha Klein is currently Professor Assistant in the Department of Geography and Environmental Studies, in the University of Haifa, Israel. He obtained his Ph.D. in geomorphology at the school of geography, University of Leeds, UK. His scientific areas are: (a) Geomorphology with special interest in drainage basin dynamics and in coastal geomorphology and (b) Geomorphology of Israel.

E-mail: **mklein@geo.haifa.ac.il**

Dr. Dorit Sivan, Ph.D. in Geology, The Hebrew University of Jerusalem, the Faculty of Science, Institute of Earth Sciences, Department of Geology. At present, she is a Senior Lecturer and Head of the Department of Maritime Civilizations, University of Haifa, and a researcher in the Leon Recanati Institute for Maritime Studies (RIMS).
   Her research interests are mainly:

(a) Reconstruction of the coastal environment, mainly during the Holocene, and its connection to human settlement.
(b) Indications for past sea levels over the last 20,000, and mainly the last 10,000 years, during which there was intensive occupation along the Israeli coast. These two subjects, the paleogeography and sea-level curves, are essential for the understanding of mankind's historical processes, and complement historical and archeological research. This field of research is interdisciplinary and linked to the application of diverse research methods.

E-mail: **dsivan@research.haifa.ac.il**

Micha Klein

Dorit Sivan

SEA-LEVEL CHANGES IN THE MEDITERRANEAN: PAST, PRESENT, AND FUTURE – A REVIEW

MICHAL LICHTER[1], DOV ZVIELY[2], MICHA KLEIN[3], AND DORIT SIVAN[4]

[1]*Coasts at Risk and Sea-Level Rise, Research Group, The Future Ocean Excellence Cluster, Institute of Geography, Christian Albrechts University, Kiel, 24098 Germany*
[2]*Leon Recanati Institute for Maritime Studies (RIMS), University of Haifa, Haifa, 31905, Israel*
[3]*Department of Geography and Environmental Studies, University of Haifa, Haifa, 31905, Israel*
[4]*Department of Maritime Civilizations, University of Haifa, Haifa, 31905, Israel*

1. Introduction

The study of geological and historical sea-level changes constitutes an important aspect of climate change and global warming research. In addition to the imminent hazards resulting from the inundation of low-lying areas along coastal regions, the rise in sea level can also cause erosion of beaches, salt intrusion into freshwater aquifers, and other damage to the coastal environment. The utmost importance of current changes in sea level is attributed to its impact on diverse ecological systems in coastal regions (Klein et al., 2004).

On time-scales of millions of years, geological processes, such as changes in ocean basin geometry caused by plate tectonics, are dominant in affecting sea-level change, whereas on shorter time-scales of years and decades, oceanographic and climatic factors are more dominant (Lambeck and Purcell, 2005).

On time-scales of centuries and millennia, sea-level change is affected mainly by eustatic (all types of water volume variations), glacio-hydro-isostatic, and tectonic factors. Eustatic changes are global and are defined as ice volume equivalent. Isostatic sea-level changes are regional, and result from changes of ice mass balance over the crust and water and sediment over the continental shelf and ocean floor. Vertical tectonic movements are local and are caused by geological uplift or subsidence. Glacio-hydro-isostatic change has a predictable pattern, whereas tectonics is less predictable (Lambeck et al., 2004). The best way for differentiating the global, regional, and tectonic processes in long-term records is by comparing observations and glacio-hydro-isostatic models that predict the combined global and regional components (e.g., Lambeck and Purcell, 2005, and references therein). Discrepancies between the observed and the model-predicted changes are attributed to local movements, whether induced by tectonic movements, sediment compaction, or other reasons.

For short-term records of decadal scale, distinguishing between the "eustatic" component and regional–local crustal movements can be conducted only with present-day measurements. This involves simultaneous measurements of relative sea-level changes by tide-gauge and land vertical movement by GPS or other geodetic techniques. Daily and seasonal changes are caused mainly by astronomical tides and other atmospheric and oceanic forcing mechanisms.

"Eustatic" sea-level changes do not actually exist because sea-level changes are spatially heterogeneous, at least over decadal time scales (Mitrovica et al., 2001). Isostatic and local factors affecting land levels may cause relative sea-level changes that vary from place to place throughout the world (Pirazzoli, 1996).

Over the past century, sea level rose by 1–2 mm/year, with nonlinear changes ("accelerations") in different places (Woodworth and Player, 2003; Church and White, 2006; Jevrejeva et al., 2006; Woodworth, 2008), inundating flat coastal areas, and disrupting natural freshwater environments as well as human habitat in many coastal and inland communities.

2. Mediterranean Sea-Level Change Since the Middle Pleistocene

"Global" sea-level curves indicate that during the last 600,000 years (ka), the sea reached a maximum elevation of 5–10 m above present sea level (asl) (Fig. 1) at least three times and dropped to more than 100 m below present sea level (bsl) at least five times (Waelbroeck et al., 2002; Schellmann and Radtke, 2004; Rabineau et al., 2006;

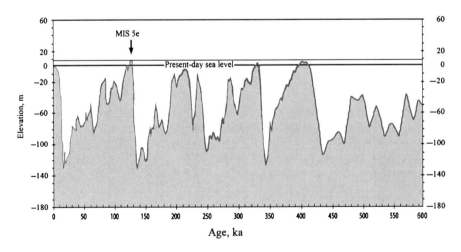

Figure 1. Global sea-level changes in the last 600 ka. (Modified after Waelbroeck et al., 2002; Rabineau et al., 2006; and Siddall et al., 2006.)

Siddall et al., 2006). The cyclic transition between glacial and interglacial cycles was about 100 ka (Shackleton, 2000), and the mean time between one glacial trough and the succeeding interglacial peak was about 20 ka.

Sea level during Marine isotope stage (MIS) 7.1, dated to between 202 and 190 ka, remained 18–9 m bsl (Bard et al., 2002), based on precise Uranium–Thorium (U-Th) dating of stalagmites from a currently submerged cave in Italy.

Sea level rose above its present level only at the peak of the last interglacial, some 125 ka ago (MIS 5e). In the Mediterranean, dating of the MIS 5e terraces is not certain, and at present there are indications of different levels, as summarized in Mauz and Antonioli (2009) and Stewart and Morhange (2009). The present elevations of the different sea-level indicators range between +175 and −125 m only in Italy because of tectonic reasons (Ferranti et al., 2006), or because of hydro-glacial adjustment of the crust (Antonioli et al., 2006). Studies in regions of the Mediterranean that are relatively tectonically stable, such as the coast of Israel (Sivan et al., 1999; Galili et al., 2007), western Sicily, and southern Sardinia, indicate that the Last Interglacial sea reached approximately 6 ± 3 m asl (Lambeck et al., 2004; Ferranti et al., 2006).

"Global" sea level later dropped to about 120 ± 5 m bsl, reaching its lowest levels during the Last Glacial Maximum (LGM), about 18 ka ago. In the Mediterranean, the longest record is found in Cosquer Cave, southern France, where Paleolithic wall paintings of horses dated to about 22 ka ago have been partially eroded by the recent rising sea water level (Lambeck and Bard, 2000 and references therein; Morhange et al., 2001).

Numerical models (Lambeck and Bard, 2000; Lambeck and Purcell, 2005) predicting sea-level changes during the last 18 ka allow estimation of the vertical movements by comparing the observations to the predictions summarized for all the Mediterranean by Stewart and Morhange (2009).

Since then, "global" sea level has been rising as a result of deglaciation and global warming (Fairbanks, 1989; Bard et al., 1990, 1996; Pirazzolli, 1991; Fleming et al., 1998; Rohling et al., 1998; Lambeck and Bard, 2000; Lambeck et al., 2002, 2004). At around 12.5 ka ago, "global" sea level rose rapidly to 70 m bsl. It continued to rise and reached 40 m bsl at the beginning of the Holocene. Levels lower than 20 bsl at the beginning of the Holocene have been observed in Israel, based on the submerged Pre-Pottery Neolithic site of Atlit Yam, situated at present 10–12 m bsl, with the bottom of one of the water wells at present 15.5 m bsl (Galili et al., 1988, 2005). Sea level continued to rise rapidly until the Mid-Holocene, when the rate slowed considerably (Lambeck and Bard, 2000; Bard et al., 1996; Lambeck et al., 2004; Poulos et al., 2009). Based on biological, sedimentological, and archeological indicators, sea-level studies around the Mediterranean indicate ±1 m bsl about 4 ka ago (Morhange et al. (2001) in the west Mediterranean, Lambeck et al. (2004) in Italy, Marriner et al. (2005) in Lebanon, and Sneh and Klein (1984), Galili et al. (1988, 2005), Nir (1997), Sivan et al. (2001, 2004), Galili and Sharvit (1998, 2000), and Porat et al. (2008) all from Israel, east Mediterranean). From about 4,000 until 2,000 years ago, there is ample archeological and biological (mainly biostructural) evidence available for sea-level reconstructions from all around the Mediterranean with better vertical accuracy of up to ±10 cm. During the Early Roman period, 2,000 years ago, sea level in the Mediterranean was 10–15 cm bsl. In Israel, Sivan et al. (2004) examined 64 coastal water wells in ancient Caesarea, and concluded that sea level was close to the present level during the Roman period. This conclusion agrees with the results found in Italy (Anzidei et al., 2008). For the Crusader period (eleventh to thirteenth centuries AD), lower levels of about 30 ± 15 cm bsl were estimated, based on the coastal water wells of Caesarea (Sivan et al., 2004; Sivan et al., 2008). These low levels are confirmed (with even lower estimated levels) by ongoing data from a few sites along the coast of Israel, based mainly on archeological evidence (Sivan et al., 2008).

3. "Global" Sea-Level Observations During the Twentieth Century

There is a consensus among sea-level researchers that the "global" sea-level rise in the past 100 years has been considerably faster than in the previous two millennia (Douglas, 2001).

"Global" sea-level rise during the twentieth century (Table 1) is estimated by most researchers to be 1–2 mm/year (Peltier, 2001; Miller and Douglas, 2004; Church and White, 2006). Church and White (2006) calculated a significant acceleration of sea-level rise of 0.013 ± 0.006 mm/year during the twentieth century.

The employment of tide-gauging facilities began in the second half of the nineteenth century, markedly improving the accuracy of sea-level measurement. Tide-gauge stations were rare prior to 1870, while spatially widespread tide-gauge records are available only for the twentieth century. Tide-gauge measurements are considered the most accurate sea-level records available for the twentieth century (Miller and Douglas, 2004), but they suffer from problems of spatial and temporal discontinuity that make calculating "global" mean trends a difficult task. Since the 1990s, sea-level measurements have also been obtained by satellite altimetry, which are often used as complementary data for tide-gauge data (Cabanes et al., 2001; Church et al., 2004). At present, more information about sea-level change

Table 1. Estimates of the "global" mean sea-level contributions from 1961 to 2003 and 1993 to 2003, compared with the observed rate of rise. (Modified after Bindoff et al., 2007.)

Source	1961 to 2003	1993 to 2003
Thermal expansion	0.42 ± 0.12	1.60 ± 0.50
Glaciers and ice caps	0.50 ± 0.18	0.77 ± 0.22
Greenland ice sheet	0.05 ± 0.12	0.21 ± 0.07
Antarctic ice sheet	0.14 ± 0.41	0.21 ± 0.35
Sum	1.10 ± 0.50	2.80 ± 0.70
Observed	1.80 ± 0.50	3.10 ± 0.70
Difference (observed-sum)	0.70 ± 0.70	0.30 ± 1.00

is available than ever before: historical data sets from tide-gauges; new understanding of postglacial rebound; precise geodetic techniques for the estimation of vertical crustal motion, and finally, more a decade of satellite altimetry, providing more precise records of recent changes in mean sea level (Cazenave and Nerem, 2004). New altimeter measurements from the TOPEX/Poseidon and Jason-1 satellites since the beginning of the 1990s have revealed a much faster rise in sea level during 1993–2003 than the average twentieth century rate. Cazenave and Nerem (2004) calculated a 2.8 ± 0.4 mm/year rise for this period (3.1 mm/year if the effects of postglacial rebound are removed).

As mentioned earlier, sea-level changes are not uniform around the Globe, e.g., Church et al. (2004) recognized a maximum sea-level rise in the eastern Pacific off-equatorial area, and minima along the equator, in the western Pacific, and in the eastern Indian Ocean.

Table 1 summarizes the contributions of thermal expansion, glaciers and ice caps, and the two ice sheets to "global" sea-level rise since 1961, according to the Fourth Assessment Report (AR4) of the Intergovernmental Panel on Climate Change (IPCC). It is, however, noted in the report that the total "global" sea-level change budget has not yet been satisfactorily closed.

From 1961 to 2003, thermal expansion accounted for only 23 ± 9% of the observed rate of sea-level rise (Bindoff et al., 2007). Since 1993, the contribution of thermal expansion of the oceans to the total rise of sea levels has been about 57%. The contribution of glaciers and ice caps decreased to about 28%, and losses from the polar ice sheets contributed the remainder (IPCC, 2007).

4. Mediterranean Sea Levels During the Twentieth Century

Tsimplis et al. (2008) examined sea-level trends and interannual variability in the Mediterranean (Genova and Trieste) during the years 1960–2000. Although the observed values did not show a rise in sea level, when removing the atmospheric and the steric contributions, the residual trends revealed a significant rise of 0.7–1.8 mm/year. This rise was not uniform, as two different trends were distinguished. Between 1960 and 1975, there was no significant change in sea level, but from 1975 to 2000, sea level rose at a rate of 1.1–1.8 mm/year. They attributed part of the residual trend to local land movements (0.3 mm/year), and its major part to a global signal, probably mass addition, after 1975.

Klein and Lichter (2009) compared observed Mediterranean rates with the "global" rate, and found the sea-level rise in the Mediterranean over the twentieth century to be in agreement with the mean "global" sea-level rise during the twentieth century (1.1–2.4 mm/year). They also found that this trend has not been consistent throughout the century (Table 2; Fig. 2). Three distinctly different sea-level trends were recognized. The first lasted from the end of the nineteenth century to 1960, when relative sea level in the Mediterranean rose by rates slightly higher than the overall "global" trend (1.3–2.8 mm/year). In the second, from 1961 to 1989,

Table 2. Twentieth century sea-level trends in the Mediterranean. Linear trends from the four tide-gauging stations with the longest record in the Mediterranean. The trends are presented for the full record and for the three different trends during the twentieth century: until 1960, from 1961 to 1989, and from 1990 to 2000 (not enough data).

PSMSL station No.	PSMSL station name	Start of record	Sea-level change until 1960 (mm/year)	Sea-level change 1961–1989 (mm/year)	Sea-level change 1990–2000 (mm/year)	Sea-level change full record (mm/year)
230051	Marseille (FR)	1885	1.72	−0.78	–	1.24
250011	Genova (FR)	1884	1.28	−0.03	–	1.22
270054	Venice (Ponte della Salute) (IT)	1909	2.77	0.44	10.11	2.40
270061	Trieste (IT)	1905	1.35	0.37	9.11	1.14

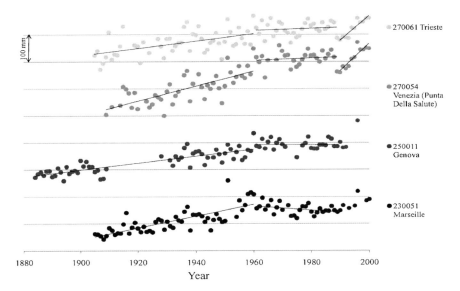

Figure 2. Linear sea-level trends from the beginning of the measurement until 1960, from 1961 to 1989, and from 1990 to 2000 in Marseilles (230051), Genoa (250011), Venice (270054), and Trieste (270061).

the observed measurements did not indicate significant changes in Mediterranean sea level. Since the beginning of the 1990s, a third, short-term trend of extremely rapid sea-level rise has been measured (4–17 mm/year). Table 2 presents sea-level trends of four Revised Local Reference (RLR) tide-gauging records in the Mediterranean, with a record of close to 100 years, available in the Permanent Service for Mean Sea Level (PSMSL) database. The stations are Marseille, Genova, Venice (with higher rates of relative sea-level rise due to subsidence in the first half of the century) and Trieste. Sea-level trends are shown for the period from the beginning of the measurement to 1960, from 1961 to 1989, from 1990 to 2000 (only Venice and Trieste had sufficient data), and for the entire record.

Klein and Lichter (2009) also found that the stability in sea level during 1961 and 1989 was the result of a rise in surface atmospheric pressure from 1961 to 1989, and that eustatic sea level has in fact been rising, but had been depressed by the rising air pressure. From 1990 onward, most gauging stations have showed an extremely high sea-level rise, 5–10 times the average twentieth century rise, and notably higher than the "global" average measured by TOPEX/Poseidon for the same years. This is in agreement with sea-level rates found in the eastern Mediterranean by Rosen (2002), who calculated a sea-level rise of 10 mm/year at the Hadera gauging station between 1992 and 2002, and Shirman (2004) who showed a 10 cm rise in sea level from 1990 to 2001 at the Ashdod and Tel Aviv tide-gauges.

New tide-gauge measurements (for location, see Fig. 3), presented in Table 3, show a slight decrease in the rate of Mediterranean sea-level rise in the first few years of the twenty-first century. The rates of sea-level rise were calculated from 27 Mediterranean PSMSL RLR tide-gauge records between 1990 and 2000, and between 1990 and 2006 (the full data sets currently available on the PSMSL website). In most stations, there has been a decrease in the rate of sea-level rise between 1990 and 2006 when compared with the trend between 1990 and 2000, but the rates remain considerably higher than the "global" and Mediterranean twentieth century rates. It is important to note that the periods considered here are short, and the

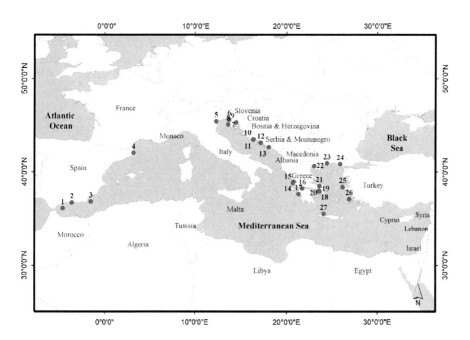

Figure 3. Location map of the tide-gauging stations presented in Table 3.

Table 3. Mediterranean sea-level trends from 1990s and onward. Linear trends are presented for 1990–2000, and where data were available trends extending to the mid-2000s were calculated.

	PSMSL station No.	PSMSL station name	Sea-level change during period 1990–2000 (mm/year)	No. of years	Sea-level change during period 1990–2006 (mm/year)	No. of years
1	220011	Algeciras (ES)	3.32	10		
2	220031	Malaga (ES)	8.62	10		
3	220041	Almeria (ES)	9.58	8		
4	220081	L'Estartit (ES)	4.50	11		
5	270054	Venezia – Ponte della Salute (IT)	10.11	11		
6	270061	Trieste (IT)	9.11	11	4.99	17
7	279003	Luka Koper (SL)	−1.20	9		
8	280006	Rovinj (HR)	10.05	11	6.24	15
9	280011	Bakar (HR)	13.52	11	8.98	15
10	280021	Split Rt Marjana (HR)	9.90	11	7.33	15
11	280031	Split Harbor (HR)	9.45	11	7.96	15
12	280046	Sucuraj (HR)	11.02	10	7.40	14
13	280081	Dubrovnik (HR)	9.51	9	7.35	13
14	290001	Preveza (GR)	11.32	11	4.92	14
15	290004	Levkas (GR)	4.21	8	2.96	12
16	290014	Patrai (GR)	14.92	8	12.88	14
17	290017	Katakolon (GR)	17.26	8	6.21	12
18	290030	North Salaminos (GR)	4.88	8		
19	290031	Piraeus (GR)	−16.15	8		
20	290033	Khalkis South (GR)	7.54	8		
21	290034	Khalkis North (GR)	1.68	10	5.71	15
22	290051	Thessaloniki (GR)	9.04	9		
23	290061	Kavalla (GR)	−0.06	9	1.70	13
24	290065	Alexandroupolis (GR)	6.46	10	5.31	15
25	290071	Khios (GR)	17.98	9	5.17	12
26	290091	Leros (GR)	−0.08	8	−1.84	11
27	290097	Soudhas (GR)	7.36	10		

trend they indicate might be merely an expression of a rising phase of an oscillating pattern. However, during no other short period in the twentieth century have tide-gauge records in the Mediterranean shown such an extreme trend.

5. Future Sea-Level Predictions

The AR4 of the IPCC (2007) predicted "global" sea-level rise of 0.18–0.59 m in 2090 and 2099, relative to 1990 and 1999 (about 2–6 mm/year) using several different future scenarios. These predictions, however, do not include uncertainties resulting from climate–carbon cycle feedbacks, or the full effects of changes in ice sheet flow. These factors are currently unknown, and therefore the upper values of these predictions are not considered upper bounds for sea-level rise.

The predictions take into consideration a contribution to sea-level rise due to increased ice flow from Greenland and Antarctica at the rates observed from 1993 to 2003. A linear increase in ice flow from the ice sheets with global average temperature change would increase the upper range of sea-level rise for these future scenarios by only 0.1–0.2 m (IPCC, 2007).

The IPCC AR4 predicts thermal expansion that contributes more than half of the average sea-level rise estimated for the twenty-first century, and land ice that loses mass increasingly rapidly. An important uncertainty relates to the question of whether discharge of ice from ice sheets will continue to increase as a consequence of accelerated ice flow, as has been observed in recent years. This would add to the sea-level rise, but quantitative predictions cannot be made with a high degree of confidence, owing to the limited understanding of the relevant processes (Bindoff et al., 2007).

The ranges of sea-level rise predictions of the AR4 are lower than those projected in the Third Assessment Report (TAR) of the IPCC (2001), because of improved information about some of the uncertainties of some contributions.

Recent attempts to predict future "global" sea-level rise confirm the ranges predicted by the IPCC reports, while others predict higher rates. Church and White (2006) estimate that if the twentieth century acceleration in sea-level rise (0.013 ± 0.006 mm/year) remains constant during the twenty-first century, sea-level would rise by 0.28–0.34 m from 1990 to 2100, a rise consistent with the middle range of the TAR and AR4 predictions.

However, Rahmstorf et al. (2007) compared sea-level predictions of the TAR with sea-level observations from the 1990s and 2000s, and found that the observations followed the upper limit of the predictions, including land-ice uncertainties. They calculated the rate of rise in the past 20 years to be 25% faster than in any other 20-year period in the last 115 years. Although they are aware of the short time interval, they conclude that these predictions may have underestimated sea-level change. Rahmstorf (2007) applied a semi-empirical methodology to project future sea-level rise by using the relations between "global" sea-level rise and global mean surface temperature. He suggests that the rate of sea-level rise is roughly proportional to the magnitude of warming above the temperatures of the pre-industrial age. This relation produced a constant of 3.4 mm/year/°C. Applying this to future IPCC scenarios, a sea-level rise of 0.5–1.4 m above the 1990 level is projected for 2100. Hence, he concludes that if the linear relations between sea-level rise and temperature that existed in the twentieth century persist through the twenty-first century, a rise of over 1 m for strong warming scenarios is not unlikely.

6. Summary

The past 600,000 years are characterized by glacial and interglacial cycles. During the glacial maxima, sea level dropped more than 100 m below its present level. Sea level in interglacial periods exceeded the present sea level by a few meters three times during that time.

During the LGM, about 18 ka ago, "global" sea level dropped by about 120 m below its present level. Since then, the transition of the global climate into an interglacial period was followed by a rapid sea-level rise until around 6,000 years ago, when there was a decrease in the rate of sea-level rise, and a relative stabilization at about the present level about 4,000 years ago.

In the Mediterranean, there are radiometric ages derived from different sea-level indicators that go back to the MIS 7.1, dated to between 202 and 190 ka ago. The last time that sea level rose above its present level was some 125 ka ago during the MIS 5e. There are ample well-dated indications for sea level during MIS 5e, located at present at different elevations due to vertical movements. There are biological, sedimentological, and mainly archeological data from the LGM, about 18 ka ago from all around the Mediterranean; the oldest being from Cosquer Cave, southern France, dated to about 22 ka ago. Sea level stabilized at almost the present level around 4,000 years ago with vertical accuracy of ±1 m. Later, 2,000 years ago, the rate of accuracy from different indicators (both biological and archeological) reaches ±10–15 cm, and fluctuations of tens of cm are recorded.

"Global" twentieth century sea-level rise is agreed by researchers to be considerably faster than in the previous two millennia. Most researchers estimate a "global" twentieth century rate of 1.0–2.5 mm/year. During the 1990s, the mean rate of "global" sea-level rise was significantly higher than the twentieth century mean rate (1.3–2.8 mm/year). Mediterranean twentieth century sea-level rise was close to the "global" rise, with the exception of the 1990s, when sea level in the Mediterranean rose at rates higher even than the unusually high "global" ones (up to three and four times the "global" 1990s rate). The rate of sea-level rise since the beginning of the 1990s has decreased in the first half of the current decade; however, it is still considerably higher than the twentieth century mean rate.

Future predictions of sea-level rise over the twenty-first century range from moderate amounts of less than 20 cm to much higher values of over 1 m. The uncertainties are attributed to the future contribution of Greenland and the Antarctic ice sheets, and uncertainties resulting from climate–carbon cycle feedback, as well as from other unpredicted proxies.

7. References

Antonioli, F., Kershaw, S., Renda, P., Rust, D., Belluomini, G., Cerasoli, M., Radtke, U. and Silenzi, S. (2006) Elevation of the last interglacial highstand in Sicily (Italy): a benchmark of coastal tectonics. Quatern. Int. **145–146**: 3–18.

Anzidei, M., Lambeck, K., Antonioli, F., Sivan, D., Stocchi, P., Spada, G., Gasperini, G., Soussi, M., Benini, M., Serpelloni, E., Baldi, P., Pondrelli, S. and Vannucci, G. (2008) *Sea Level Changes and Vertical Land Movements in the Mediterranean from Historical and Geophysical Data and Modelling. Observations and Causes of Sea-Level Changes on Millennial to Decadal Timescales.* The Geological Society, London.

Bard, E., Hamelin, B., Arnold, M., Montaggioni, L., Cabioch, G., Faure, G. and Rougerie, F. (1996) Deglacial sea-level record from Tahiti corals and the timing of global melt water discharge. Nature **382**: 241–244.

Bard, E., Hamelin, B., Fairbanks, R.G. and Zindler, A. (1990) Calibration of the ^{14}C timescale over the past 30,000 years using mass spectrometric U-Th ages from Barbados corals. Nature **345**: 405–409.

Bard, E., Antonioli, F. and Silenzi, S. (2002) Sea level during the penultimate interglacial period based on submerged stalagmite from Argentarola Cave (Italy). Earth Planet. Sci. Lett. **196**: 135–146.

Bindoff, N.L., Willebrand, J., Artale, V., Cazenave, A., Gregory, J., Gulev, S., Hanawa, K., Le Quéré, C., Levitus, S., Nojiri, Y., Shum, C.K., Talley L.D. and Unnikrishnan, A. (2007) In: S. Solomon, D. Qin, M. Manning, Z. Chen, M. Marquis, K.B. Averyt, M. Tignor and H.L. Miller (eds.) *Observations: Oceanic Climate Change and Sea Level. Climate Change 2007: The Physical Science Basis. Contribution of Working Group I to the Fourth Assessment Report of the Intergovernmental Panel on Climate Change.* Cambridge University Press, Cambridge, United Kingdom and New York, NY, USA.

Cabanes, C., Cazenave, A. and Le Provost, C. (2001) Sea-level rise during past 40 years determined from satellite and in situ observations. Science **294**: 840–842.

Cazenave, A. and Nerem, R.S. (2004) Present-day sea level change: observations and causes. Rev. Geophys. **42**: RG3001.

Church, J.A., White, N.J., Coleman, R., Lambeck, K. and Mitrovica, J.X. (2004) Estimates of the regional distribution of sea level rise over the 1950–2000 period. J. Climate **17**: 2609–2625.

Church, J.A. and White, N.J. (2006) A 20th century acceleration in global sea-level rise. Geophys. Res. Lett. **33**: L01602. doi: 10.1029/2005GL024826.

Douglas, B.C. (2001) An introduction to sea-level, In: B.C. Douglas, M.F. Kearney and S.P. Leatherman (eds.) *Sea-Level Rise History and Consequences*. Academic Press, USA, pp. 1–11.

Fairbanks, R.G. (1989) A 17,000-year Glacio-Eustatic sea level record: influence of glacial melting rates on the Younger Dryas event and deep-ocean circulation. Nature **342**: 637–642.

Ferranti, L., Antonioli, F., Mauz, B., Amorosi, A., Dai Pra, G., Mastronuzzi, G., Monaco, C., Orrù, P., Pappalardo, M., Radtke, U., Renda, P., Romanoa, P., Sanso, P. and Verrubbi, V. (2006) Markers of the last interglacial sea-level high stand along the coast of Italy: tectonic implications. Quatern. Int. **145–146**: 30–54.

Fleming, K., Johnston, P., Zwartz, D., Yokoyama, Y., Lambeck, K. and Chappell, J. (1998) Refining the eustatic sea-level curve since the Last Glacial Maximum using far- and intermediate-field sites. Earth Planet. Sci. Lett. **163**(1–4): 327–342.

Galili, E. and Sharvit, J. (1998) Ancient coastal installations and the tectonic stability of the Israeli coast in historical times, In: I.S. Stewart and C. Vita-Finzi (eds.) *Coastal Tectonics*. Geological Society, London, Special Publications, 146, pp. 147–163.

Galili, E. and Sharvit, J. (2000) The use of archeological features as indicators for coastal displacement in the Israeli coastal region, In: *The Mediterranean: Culture, Environment and Society*. University of Haifa, Faculty of Humanities, Faculty of Social Sciences and Mathematics.

Galili, E., Weinstein-Evron, M. and Ronen, A. (1988) Holocene sea-level changes based on submerged archaeological sites off the northern Carmel coast in Israel. Quatern. Res. **29**: 36–42.

Galili, E., Zviely, D. and Weinstein-Evron, M. (2005) Holocene sea-level changes and landscape evolution in the northern Carmel coast (Israel). Méditerranée **104**(1/2): 79–86.

Galili, E., Zviely, D., Ronen, A. and Mienis, H.K. (2007) Beach deposits of MIS 5e high sea stand as indicators for tectonic stability of the Carmel coastal plain, Israel. Quatern. Sci. Rev. **26**: 2544–2557.

IPCC (2001) In: J.T. Houghton, Y. Ding, D.J. Griggs, M. Noguer, P.J. van der Linden, X. Dai, K. Maskell and C.A. Johnson (eds.) *Climate Change 2001: The Scientific Basis. Contribution of Working Group I to the Third Assessment Report of the Intergovernmental Panel on Climate Change*. Cambridge University Press, Cambridge, United Kingdom and New York, NY, USA, 881 pp.

IPCC (2007) In: S. Solomon, D. Qin, M. Manning, Z. Chen, M. Marquis, K.B. Averyt, M. Tignor and H.L. Miller (eds.) *Climate Change 2007: The Physical Science Basis. Contribution of Working Group I to the Fourth Assessment Report of the Intergovernmental Panel on Climate Change*. Cambridge University Press, Cambridge, United Kingdom and New York, NY, USA, 996 pp.

Jevrejeva, S., Grinsted, A., Moore, J.C. and Holgate, S. (2006) Nonlinear trends and multiyear cycles in sea level records. J. Geophys. Res. **111**: C09012, doi:10.1029/2005JC003229.

Klein, M. and Lichter, M. (2009) Statistical analysis of recent Mediterranean Sea-level data. Geomorphology **107**(1–2): 3–9.

Klein, M., Lichter, M. and Zviely, D. (2004) Recent sea-level changes along Israeli and Mediterranean coast, In: J.O. Maos, M. Inbar and D. Shmueli (eds.) *Contemporary Israeli Geography, Special Issue of Horizons in Geography*, 60–61, pp. 247–254.

Lambeck, K. and Bard, E. (2000) Sea-level change along the French Mediterranean coast for the past 30,000 years. Earth Planet. Sci. Lett. **175**: 203–222.

Lambeck, K., Antonioli, F., Purcell, A. and Silenzi, S. (2004) Sea-level change along the Italian coast for the past 10,000 yr. Quatern. Sci. Rev. **23**: 1567–1598.

Lambeck, K. and Purcell, A. (2005) Sea-level change in the Mediterranean Sea since the LGM: model predictions for tectonically stable areas. Quatern. Sci. Rev. **24**: 1969–1988.

Lambeck, K., Yokoyama, Y. and Purcell, T. (2002) Into and out of the Last Glacial Maximum: sea-level change during oxygen isotope stages 3 and 2. Quatern. Sci. Rev. **21**: 343–360.

Marriner, N., Morhange, C., Boudagher-Fadel, M., Bourcier, M. and Carbonel, P. (2005) Geoarchaeology of Tyre's ancient northern harbour, Phoenicia. J. Archaeologi. Sci. **32**(9): 1302–1327.

Mauz, B. and Antonioli, F. (2009) Comment on "Sea level and climate changes during OIS 5e in the Western Mediterranean" by T. Bardají, J.L. Goy, J.L., C. Zazo, C. Hillaire-Marcel, C.J. Dabrio, A. Cabero, B. Ghaleb, P.G. Silva, J. Lario. Geomorphology **104**: 22–37.

Miller, L. and Douglas, B.C. (2004) Mass and volume contributions to twentieth-century global sea level rise. Nature **428**: 406–409. doi: 10.1038/nature02309.

Mitrovica, J.X., Tamisiea, M.E., Davis, J.L. and Milne, G.A. (2001) Recent mass balance of polar ice sheets inferred from patterns of global sea-level change. Nature **409**: 1026–1029.

Morhange, C., Laborel, J. and Hesnard, A. (2001) Changes of relative sea level during the past 5000 years in the ancient harbor of Marseilles, Southern France. Palaeogeogr. Palaeoclimatol. Palaeoecol. **166**: 319–329.

Nir, Y. (1997) Middle and late Holocene sea-levels along the Israel Mediterranean coast – evidence from ancient water wells. J. Quatern. Sci. **12**(2): 143–151.

Peltier, W. (2001) Global glacial isostatic adjustment and modern instrumental records of relative sea level history, In: B.C. Douglas, M.F. Kearney and S.P. Leatherman (eds.) *Sea-Level Rise History and Consequences*. Academic Press, USA, pp. 65–95.

Pirazzolli, P.A. (1991) *World Atlas of Holocene Sea-Level Changes*. Elsevier Oceanography Series 58. Elsevier, Amsterdam.

Pirazzoli, P.A. (1996) *Sea-Level Changes – The Last 20,000 Years*. Wiley, Chichester, England.

Poulos, S.E., Ghionis, G. and Maroukian, H. (2009) Sea-level rise trends in the Attico-Cycladic region (Aegean Sea) during the last 5000 years. Geomorphology **107**(1–2): 10–17. doi: 10.1016/j.geomorph.2007.05.02.

Porat, N., Sivan, D. and Zviely, D. (2008) Late Holocene embayment and sedimentological infill processes in Haifa Bay, SE Mediterranean. Israel J. Earth Sci. **57**: 21–31.

Rabineau, M., Bemé, S., Olivet, J.L., Aslanian, D., Guillocheau, F. and Joseph, P. (2006) Paleo sea levels reconsidered from direct observation of paleoshoreline position during Glacial Maxima (for the last 500,000 yr). Earth Planet. Sci. Lett. **252**: 119–137.

Rahmstorf, S. (2007) A semi-empirical approach to projecting future sea-level rise. Science **315**(5810): 368–370.

Rahmstorf, S., Cazenave, A., Church, J.A., Hansen, J.E., Keeling, R.F. and Parker, D.E. (2007) Recent climate observations compared to projections. Science **316**(5825): 709.

Rohling, E.J., Fenton, M., Jorissen, F.J., Bertrand, P., Ganssen, G. and Caulet, J.P. (1998) Magnitude of sea-level lowstands of the past 500,000 years. Nature **394**: 162–165.

Rosen, D.S. (2002) Long term remedial measures of sedimentological impact due to coastal developments on the south eastern Mediterranean coast. Littoral 2002 22–26 September, Porto, Portugal.

Schellmann, G. and Radtke, U. (2004) A revised morpho- and chronostratigraphy of the late and middle Pleistocene coral reef terraces on Southern Barbados (West Indies). Earth-Sci. Rev. **64**: 157–187.

Shackleton, N.J. (2000) The 100,000-year ice-age cycle found to lag temperature, carbon dioxide, and orbital eccentricity. Science **289**: 1897–1902.

Shirman, B. (2004) East Mediterranean sea level changes over the period 1958–2001. Israel J. Earth Sci. **53**: 1–12.

Siddall, M., Chappell, J. and Potter, E.K. (2006) Eustatic sea level during past interglacials, In: F. Sirocko, F. Litt, T. Claussen and M.-F. Sanchez-Goni (eds.) *The Climate of Past Interglacials.* Elsevier, Amsterdam, pp. 75–92.

Sivan, D., Gvirtzman, G. and Sass, E. (1999) Quaternary stratigraphy and paleogeography of the Galilee Coastal Plain, Israel. Quatern. Res. **51**: 280–294.

Sivan, D., Lambeck, K., Toueg, R., Raban, A., Porath, Y. and Shirman, B. (2004) Ancient coastal wells of Caesarea Maritima, Israel, an indicator for relative sea level changes during the last 2000 years. Earth Planet. Sci. Lett. **222**(1): 315–330.

Sivan, D., Wdowinski, S., Lambeck, K., Galili, E. and Raban, A. (2001) Holocene sea-level changes along the Mediterranean coast of Israel, based on archaeological observations and numerical model. Palaeogeogr. Palaeoclimatol. Palaeoecol. **167**: 101–117.

Sivan, D., Schattner, U., Lambeck, K., Shirman, B. and Stern, E. (2008) Lower sea-levels during the Crusader period (11th to 13th centuries AD) as witnessed in Israel: do they indicate a global eustatic component? *Observations and Causes of Sea-Level Changes on Millennial to Decadal Timescales*, The William Smith Meeting of the Geological Society, London.

Sneh, Y. and Klein, M. (1984) Holocene sea-level changes at the coast of Dor, southeast Mediterranean. Science **226**: 831–832.

Stewart, I. and Morhange, C. (2009) Coastal geomorphology and sea-level change, In: J.C. Woodward (ed.) *The Physical Geography of the Mediterranean Basin.* OUP, Cpt. 13, pp. 385–414.

Tsimplis, M., Marcos, M., Somot, S. and Barnier, B. (2008) Sea level forcing in the Mediterranean Sea between 1960 and 2000. Global Planet. Change **63**(4): 325–332. doi: 10.1016/j.gloplacha.2008.07.004.

Waelbroeck, C., Labeyrie, L., Michel, E., Duplessy, J.C., McManus, J.F., Lambeck, K., Balbon, E. and Labracherie, M. (2002) Sea-level and deep water temperature changes derived from benthonic foraminifera isotopic records. Quatern. Sci. Rev. **21**: 295–305.

Woodworth, P.L. and Player, R. (2003) The permanent service for mean sea-level: an update to the 21st century. J. Coastal Res. **19**(2): 287–295.

Woodworth, P.L. (2008) Changes in sea level on multi-decadal and century timescales as observed from global tide-gauges data set. *Observations and Causes of Sea-Level Changes on Millennial to Decadal Timescales*, The William Smith Meeting of the Geological Society, London.

Biodata of **Dr. Linda Olsvig-Whittaker**, author of *"Global Climate Change and Marine Conservation"*

Dr. Linda Olsvig-Whittaker is an Informatics Specialist in the Israel Nature and Parks Authority (INPA), headquartered in Jerusalem, Israel. She received her Ph.D. in Ecology and Evolutionary Biology at Cornell University (Ithaca, USA) in 1980. She was a Lady Davis Postdoctorate Fellow at the Technion (Haifa, Israel) during 1981–1982, followed by an year as a research scientist on the Brookfield Ecosystem Project (Flinders University, Australia). During 1984–1994, she was a research scientist for plant community ecology in the Mitrani Center for Desert Ecology (Ben Gurion University, Israel). She joined the Nature Reserves Authority in 1994 as Coordinator of Scientific Data, remaining in this position when the NRA became the Nature and Parks Authority. She is responsible for the management of the observational data in the INPA and currently heads the Israeli partnership in the EBONE project (European Biodiversity Observation Network, www.ebone.wur.nl). Dr. Olsvig-Whittaker is active in the Society for Conservation Biology, having served 5 years as communications officer for the Asia Section of SCB. She is also an editor for the journal *Plant Ecology*.

E-mail: **linda.whittaker@npa.org.il**

GLOBAL CLIMATE CHANGE AND MARINE CONSERVATION

LINDA OLSVIG-WHITTAKER
Science and Conservation Division, Israel Nature and National Parks Protection Authority, 3 Am Ve Olamo Street, Givat Shaul, Jerusalem 95463, Israel

1. Introduction

Global climate change is real. Compilations of instrumental global land and sea temperatures back to the mid-ninteenth century provide strong evidence of a warming world and recent unusual warmth, with 9 of the 10 warmest years since 1850 occurring between 1997 and 2006. The most recent projections of global climate change due to the enhanced greenhouse effect suggest that global average temperature could warm by 1.1°C to 6.4°C over 1980–1999 values by 2100, with best estimates ranging from 1.8°C to 4.0°C. These estimates are generally consistent (although not strictly comparable) with the earlier projections of 1.4°C to 5.8°C, and are based on more climate models of greater complexity and realism and better understanding of the climate system (Lough, 2007).

Global climate has always fluctuated, but the scale tends to be over tens of thousands of years. In the last few centuries, we have experienced an accelerated rate of climate change, largely due to the release of industrial gases, and especially of carbon dioxide (CO_2). By 2100, atmospheric CO_2 is expected to exceed 500 ppm, and global temperatures to rise at least 2°C, exceeding conditions of the past 420,000 years (Hoegh-Guldberg et al., 2007). The Earth's radiative heat balance is currently out of equilibrium, and mean global temperatures will continue to rise for several centuries even if greenhouse gas emissions are stabilized at present levels (IPCC, 2001).

In the marine environment, ongoing studies by NOAA (United States National Oceanic and Atmospheric Administration) scientists show that changes in surface temperature, rainfall, and sea level will be largely irreversible for more than 1,000 years after carbon dioxide emissions are completely stopped (Solomon et al., 2009). Global sea levels are predicted to rise for the next 1,000 years; the minimal irreversible global average sea level rise is predicted to be at least 0.4–1 m in the year 3000, and possibly double that if CO_2 peaks at 600 ppm. (Present concentrations are around 385 ppm.) The rise in sea level will be mainly due to two factors: thermal expansion of the ocean's water and input from melting ice.

Possibly as important as sea-level rise there will be changes in ocean chemistry (Diaz-Pulido et al., 2007). Continued emission of CO_2 will acidify sea waters. Oceanic pH is projected to decrease by about 0.4–0.5 units by 2100 (e.g. a change from pH 8.2 to 7.8).

The Mediterranean Sea, a somewhat special case as a smaller enclosed basin, will have additional problems of higher surface water temperatures and rising salinity. Sevault et al. (2004), using the high-resolution Ocean Regional Circulation Model OPAMED8, anticipated a surface temperature rise of 2.5°C by 2100, and a regionally variable salinity increase between 0.12 and 0.19 psu, with about 0.4 psu increase in the Aegean and Adriatic seas, in a scenario for years 2060 to 2100.

Israel fits the general pattern of rising sea levels. Monthly averaged sea-level changes at the Mediterranean coast of Israel during 1992–2008 show a rise of 8.5 cm in 16 years. (Data from the Hadera GLOSS station 80, operated by Israel Oceanographic and Limnological Research Institute.)

2. Responses to Global Climate Change

The impact of global climatic change on marine systems seems to be mainly felt in two areas. First is the impact on coastal waters, where rising sea-level shifts the distribution of species, and surface waters become warmer. Second, and more dramatic, is the impact on coral reefs.

2.1. COASTAL ZONES

Sixty percent of all human beings live on a 60-km wide strip of coastal zone in the world. Marine coastal water is the seat of 14–30% of the ocean's primary production, and 90% of the fishing catch. Sea-level rise will shift the habitats especially of coastal waters (UNEP-MAP-RAC/SPA, 2008).

Some local studies have been carried out on the effects of climate change in marine communities (Parmesan, 2006). In Monterey Bay, Sagarin et al. (1999) observed a decline in northern species and an increase in southern species. Similar patterns were seen in the English Channel (Southward et al., 1995, 2005) with a decline in cold-adapted fish and increase in warm-adapted fish. Similar patterns were observed in invertebrates (Parmesan, 2006).

Harley et al. (2006) predict changes in pH of oceans without precedent in the last 200–300 million years. Upwelling could either increase or decrease. Landward migration of intertidal habitats and biota may be impeded by anthropogenic infrastructure (sea walls, etc.). An increase in storm damage is expected. Biological interactions are likely to be affected (for example, sea star *Pisaster ochraceus* is quite likely to be more active in a warmer climate, with larger effects on mussel beds). Harley et al. (2006) expect "squeeze effects," with potential shifts in distribution limited by a physical barrier (sea bottom, etc.) leading to local extinctions.

There is difficulty in predicting the effects of global climate change on diversity of marine plant life. However, the competitive interaction of sea grasses and macroalgae may be predicted, with CO_2 levels rising, and intertidal macroalgae already at CO_2 saturation (Beardall et al., 1998). Review of literature so far (Short and Neckles, 1999) suggests shifts in the distribution of sea grasses. Driving factors include temperature stress (and its effects on reproduction), eutrophication, and the frequency of extreme weather events. Changing water depths redistribute habitats (zonation); change in salinity affects physiology and reproduction in sea grasses. Increased disease activity is anticipated, as is shifting competition between sea grass and algae, with the advantage going to the sea grasses.

Short and Neckles (1999) also anticipate synergistic effects: the outcome of the physical changes under global change will be complicated by interactions among biological and physical factors. For example, there is a strong interaction between temperature and CO_2 effects on calcification (the impact is greater at warmer temperatures and there is a threshold). Interactions with anthropogenic factors (overfishing, pollution) will be more easily managed.

Changes in the Mediterranean Sea have been studied by United Nations teams (UNEP-MAP-RAC/SPA, 2008). Globally, the anticipated extinction rate of species in the Mediterranean is about 15–37% by 2050. There are some observed species shifts: *Sardinella*, barracudas, and coryphenes are moving north in fisheries quantities; but sprat and anchovy (small pelagics) have collapsed and tuna and amberjack have changed in their distributions. Lessepsian migrants (Galil, 2007) are on the increase in the eastern Mediterranean. Heat stress is killing sponges and gorgonians, with crashes in extremely hot spells in 1999 and 2003. Heat has also been found to trigger virulence of *Vibrio* pathogens in sponges, cnidaria, and echnoderms; apparently by inhibition of defense mechanisms of individuals subjected to heat stress.

2.2. CORAL REEFS AND MACROALGAE

Changes in pH, CO_2, and calcium carbonate saturation state will have biggest impacts on corals, and crustose and upright calcareous macroalgae. This may shift the balance in favor of turf algae over corals. Increase of CO_2 may not only reduce calcification but ultimately dissolve calcified skeletons (Diaz-Pulido et al., 2007).

The second and more obvious impact on coral reefs comes when acidification is combined with higher sea surface temperatures. Elevated sea temperatures as small as 1°C above summer average can lead to bleaching (loss of coral algal symbiotic zooxanthellae following chronic photoinhibition). After bleaching of coral occurs, acidification of water slows recovery. It is recognized that skeleton producing corals grown in acidified experimental conditions can persist and reproduce in a sea anemone-like form, and then revert to skeleton building when the conditions permit (Fine and Tchernov, 2007). However, according to some projections, by 2050, oceans may become too acidic for corals to calcify (Caldeira and Wickett, 2003; Hoegh-Guldberg, 2005; Orr et al., 2005).

Corals are expected to become increasingly rare on reef systems, resulting in less diverse reef communities (Hoegh-Guldberg et al., 2007). Carbonate reef structures will fail to be maintained. Compounded by local stresses, functional collapse of reef systems is anticipated in some locations. This has consequences for other habitats (Hoegh-Guldberg, 1999). Coral reefs protect coastlines from storm damage, erosion, and flooding. The protection they afford enables the development of mangrove swamps and sea grass beds. As coral reefs fail, all these services will decline.

One of the anticipated effects of coral bleaching is increased substrate availability for algal turf, upright macroalgae, and crustose calcareous algae (Diaz-Pulido et al., 2007). This may be balanced by their vulnerability to terrestrial nutrient and sediment input, which may increase with erosion and desertification. Turf algae are expected to be the best competitors for the newly open spaces.

2.3. MARINE ALGAE

A summary of macroalgae response to anticipated climate change shows both positive and negative responses for nearly every climatic stress: change in ocean circulation, increased water temperature, increased CO_2, acidification, increased light and UV, sea-level rise, tropical storms, terrestrial inputs, and increased substrate availability (Diaz-Pulido et al., 2007). Algal turfs have predominantly positive response; upright macroalgae are balanced between positive and negative responses, and crustose calcareous algae, like coral, tended to have negative responses.

The direct impact of global climatic thermal rise is presumed minor due to wide temperature tolerance of macroalgae, but the high diversity of macroalgae species makes net response unpredictable. Higher temperatures may enhance turf algae as opposed to fleshy algae.

3. Recommended Conservation Measures

The general strategy for marine conservation under global climate change is best expressed by the United Nations study of the Mediterranean:

> At the end of this study, it is necessary to remember that climate change and its effects are irremediable processes. In the long term, the major issue will probably be no more than successfully predicting the future of Mediterranean biodiversity, the future composition of the fisheries and the underwater landscapes, and adapting our ways of using them accordingly. (UNEP-MAP-RAC/SPA, 2008)!

The essentially irreversible nature of global climate change has been suspected for a long time; only the magnitude of change has been questioned. Hence, conservation strategies have largely focused on amelioration of global climate changes and their effects, rather than efforts to reverse them.

So far, amelioration suggestions are sparse. In general, there are no suggestions that the direct effects of global climate change can be reversed. Instead, the suggestions are to increase system resilience by (1) reducing other stresses (such as overfishing) and (2) develop corridors and refuges for restocking.

3.1. COASTAL AREAS

Since global climate change is essentially irreversible in practice, mitigation strategies are necessary in coastal marine systems (Harley et al., 2006). Among the recommendations:

(a) Marine protected areas and no-take reserves, based on known spatial and temporal refuges that can act as buffers against climate-related stress
(b) Fisheries management
(c) Prioritization of key species (by functional role in marine communities)

3.2. MEDITERRANEAN

UNEP-MAP-RAC/SPA (2008): Conservation measures in the Mediterranean mostly focus on improving adaptability (resilience) following the model of Hulme (2005). Specific recommendations include:

(a) Widen the knowledge base about anticipated impact of global climate change on species and communities to rising temperature, rising sea level, changing rainfall regimes (river spates), increased solar radiation, modification of currents, and changes in biogeochemistry (e.g. pH).
(b) Epidemiological studies. Changing disease patterns is an anticipated concern (see also Harvell et al., 1999).
(c) Develop predictive modeling.
(d) Build federal programs.
(e) Develop economic indicators: what is the cost of global climate change and of conservation?
(f) Assist developing countries in order to assess their vulnerability.
(g) Good ecological engineering. Adaptations of infrastructure to global climate change tend to counter biodiversity conservation choices.
(h) Adapt and change fisheries patterns.
(i) Possibly implement transplantation if species decline locally.
(j) Eliminate other sources of disturbance and stress (pollutants, invaders).
(k) Enhance connectivity for refuges and restocking.
(l) Work on the scale of the whole Mediterranean basin.
(m) Protect relict, non-impacted systems by reserves.

3.3. CORAL REEFS

On its web site, the Nature Conservancy organization (TNC) has outlined its conservation strategies with respect to climate change and its impact on marine protected areas (see www.nature.org/initiatives/marine/strategies/art12286.html).

Much of the TNC focus is on coral reefs. Nature Conservancy strategies include locating areas where marine life resists bleaching and creating networks of protected areas to help nearby degraded areas to recover. Much of the strategy is to identify areas where marine life, including corals, seems relatively resistant to damage and focus on conserving these areas as refuges. In the case of coral reefs, connectivity is a consideration, with networks of protected areas allowing one area to provide colonizers to another if it should become degraded.

3.4. COASTAL MANGROVE WETLANDS

It is estimated that between human reclamation of coastal wetlands and rising sea levels due to global climatic change, by 2080 we will have lost about 80% of the world's coastal wetlands. TNC also has a focus on managing mangroves for resilience to climate change (McLeod and Salm, 2006). Most of the strategies are expected: protect coastal mangroves from other anthropogenic stressors to enhance their resilience, maintain buffer zones, restore areas with good prospects, maintain connectivity, develop adaptive management strategies, etc.

3.5. MACROALGAE

Management recommendations are mainly due to concern about expansion of algal turf, rather than loss of macroalgal cover or species. The first recommendation is to protect populations of algal herbivores, then to minimize terrestrial runoff and other sources of nutrient, sediment, and toxicant pollution. Protection of corals will also reduce expansion of macroalgae (Diaz-Pulido et al., 2007).

4. Summary

In general, the situation of marine environments under global climate change looks very bad. The factors anticipated to cause the most change in the marine environment (e.g. sea-temperature and sea-level rises) are also those least likely to be affected by amelioration, and should be seen as permanent, irreversible changes. This is grim but recognition of the situation will make practical conservation measures more effective. The standard practices for any kind of conservation (reduce environmental stress, protect key habitats, develop and protect corridors for dispersal) apply here as well. Beyond that, we simply do not have many good ideas.

5. References

Beardall, J., Beer, S. and Raven, J.A. (1998) Biodiversity of marine plants in an era of climate change: some predictions based on physiological performance. Bot. Mar. **41**(1): 113–123.

Caldeira, K. and Wickett, M.E. (2003) Anthropogenic carbon and ocean pH. Nature **425**: 365.

Diaz-Pulido, G., McCook, L., Larkim, A.W.D., Lotze, H.K., Raven, J.A., Schaffelke, B., Smith, J.E. and Steneck, R.S. (2007) Vulnerability of macroalgae of the Great Barrier Reef to climate change, In: J.E. Johnson and P.A. Marshall (eds.) *Climate Change and the Great Barrier Reef: A Vulnerability Assessment*, Great Barrier Reef Marine Park Authority, Townsville, pp. 153–192.

Fine, M. and Tchernov, D. (2007) Scleractinian coral species survive and recover from decalcification. Science **315**(5280): 1811.

Galil, B.S. (2007) Seeing red: alien species along the Mediterranean coast of Israel. Aquat. Invasions **2**(4): 281–312.

Harley, C.D.G., Hughes, A.R., Hultgren, K.M., Miner, B.G. Sorte, C.J.B., Thornber, C.S., Rodriguez, L.F., Tomanek, L. and Williams, S.L. (2006) The impacts of climate change in coastal marine systems. Ecol. Lett. **9**: 228–241.

Harvell, C.D., Kim, K., Burkholder, J.M., Colwell, R.R., Epstein, P.R., Grimes, D.J., Hofmann, E.E., Lipp, E.K., Osterhaus, A.D.M.E., Overstreet, R.M., Porter, J.W., Smith, G.W. and Vasta, G.R. (1999) Review: marine ecology – emerging marine diseases – climate links and anthropogenic factors. Science **285**: 1505–1510.

Hoegh-Guldberg, O. (1999) Climate change, coral bleaching and the future of the world's coral reefs. Maine Freshwater Res. **50**: 839–866.

Hoegh-Guldberg, O. (2005) Low coral cover in a high-CO_2 world. J. Geophys. Res. **110**: C09S06.

Hoegh-Guldberg, O., Mumby, P.J., Hooten, A.J., Steneck, R.S., Greenfield, P., Gomez, E., Harvell, C.D., Sale, P.F., Edwards, A.J., Caldeira, K., Knowlton, N., Eakin, C.M., Iglesias-Prieto, R., Muthiga, N., Bradbury, R.H., Dubi, A. and Hatziolos, M.E. (2007) Coral reefs under rapid climatic change and ocean acidification. Science **318**(5857): 1737–1742.

Hulme, P.E. (2005) Adapting to climate change: is there scope for ecological management in the face of a global threat? J. Appl. Ecol. **42**: 784–794.

Intergovernmental Panel on Climate Change (2001) *Climate Change 2001, Synthesis Report. A Contribution of Working Groups I, II, and III to the Third Assessment Report of the Intergovernmental Panel on Climate Change*. Cambridge University Press, Cambridge, UK.

Lough, J.M. (2007) Climate and climate change on the Great Barrier Reef, In: J.E. Johnson and P.A. Marshall (eds.) *Climate Change and the Great Barrier Reef*, Great Barrier Reef Marine Park Authority and Australian Greenhouse Office, Australia, pp. 15–50.

McLeod, E. and Salm, R.V. (2006) *Managing Mangroves for Resilience to Climate Change*, IUCN, Gland, Switzerland, 64 pp.

Orr, J.C., Fabry, V.J., Aumont, O., Bopp, L., Doney, S.C., Feely, R.A., Gnanadesikan, A., Gruber, N., Ishida, A, Joos, F., Key, R.M., Lindsay, K., Maier-Reimer, E., Matear, R., Monfray, P., Mouchet, A., Najjar, R.G., Plattner, G.-K., Rodgers, K.B., Sabine, C.L., Sarmiento, J.L., Schlitzer, R., Slater, R.D., Totterdell, I.J., Weirig, M.-F., Yamanaka, Y. and Yool, A. (2005) Anthropogenic ocean acidification over the twenty-first century and its impact on calcifying organisms. Nature **437**: 681–686.

Parmesan, C. (2006) Ecological and evolutionary responses to recent climate change. Ann. Rev. Ecol. Evol. Systemat. **37**: 637–669.

Sagarin, R.D., Bary, J.P., Gilman, S.E. and Baxter. C.H. (1999) Climate-related change in an intertidal community over short and long time scales. Ecol. Monogr. **69**: 465–490.

Sevault, F., Somot, S. and Déqué, M. (2004) Climate change scenario for the Mediterranean Sea. Geophysical Research Abstracts vol. 6: 02447. Ref ID: 1607-7692/gra/EGU04-A-02447. European Geosciences Union.

Short, F.T. and Neckles, H.A. (1999) The effects of global climate change on sea grasses. Aquat. Bot. **63**(3–4): 169–196.

Solomon, S., Plattner, G.-K., Knutti, R. and Friedlingstein, P. (2009) Irreversible climate change due to carbon dioxide emissions. Proc. Natl Acad. Sci. **106**(6): 1704–1709.

Southward, A.J., Hawkins, S.J. and Burrows, M.T. (1995) Seventy years' observations of changes in distribution and abundance of zooplankton and intertidal organisms in the western English Channel in relation to rising sea temperature. J. Therm. Biol. **20**: 127–155.

Southward, A.J., Langmead, O., Hardman-Mountford N.J., Aiken J., Boalch G.T., Dando P.R., Genner M.J., Joint, I., Kendall, M.A., Halliday, N.C., Harris, R.P., Leaper, R., Mieszkowska, N., Pingree, R.D., Richardson, A.J., Sims, D.W., Smith, T., Walne, A.W. and Hawkins, S.J. (2005) Long-term oceanographic and ecological research in the western English Channel. Adv. Mar. Biol. **47**: 1–105.

UNEP-MAP-RAC/SPA. (2008) In: T. Perez (ed.) *Impact of Climate Change on Biodiversity in the Mediterranean Sea*, RAC/SPA Edit., Tunis, pp. 1–90.

PART 2:
BIODIVERSITY IN MARINE ECOSYSTEMS IN THE GLOBALLY CHANGING ERA

Boudouresque
Verlaque
Rilov
Trebes
Ugarte
Craigie
Critchley

Biodata of **Dr. Charles F. Boudouresque** and **Dr. Marc Verlaque**, authors of *"Is Global Warming Involved in the Success of Seaweed Introductions in the Mediterranean Sea?"*

Dr. Charles F. Boudouresque is currently Professor of Marine Biology and Ecology at the Center of Oceanology of Marseilles (Southern France). He obtained his Ph.D. from the Aix-Marseilles University in 1970, with a study on benthic Mediterranean assemblages dominated by macrophytes. He described a dozen of new species and genera of red algae. His current scientific interests are in the area of the structure and functioning of seagrass and lagoon ecosystems, biological invasions, conservation of the biodiversity and Marine Protected areas (MPAs). He is co-author of several books on European marine algae and editor of the proceedings of nine international symposia.

E-mail: **charles.boudouresque@univmed.fr**

Dr. Marc Verlaque is currently a Senior Phycologist at the Center of Oceanology of Marseilles and CNRS (Centre National de la Recherche Scientifique) (Southern France). He obtained his Ph.D. from the Aix-Marseilles University in 1987 in Marine Ecology with a study on the relationships between the Mediterranean seaweed assemblages and large herbivores (fish, sea urchins and molluscs). His current scientific interests are in the area of the biogeography and taxonomy of the Mediterranean marine flora, species introductions, biological invasions and conservation of the biodiversity. He is co-author of several books on European marine algae.

E-mail: **marc.verlaque@univmed.fr**

Charles F. Boudouresque   Marc Verlaque

IS GLOBAL WARMING INVOLVED IN THE SUCCESS OF SEAWEED INTRODUCTIONS IN THE MEDITERRANEAN SEA?

CHARLES F. BOUDOURESQUE AND MARC VERLAQUE
Center of Oceanology of Marseilles, Campus of Luminy, University of the Mediterranean, 13288, Marseilles, cedex 9, France

1. Introduction

There is growing concern about the global warming of the Earth and about introduced species (biological invasions) (e.g. Stott et al., 2000; Oreskes, 2004; Schaffelke et al., 2006). The reasons are: (i) Both warming and biological invasions are not only in progress but are on the increase. (ii) They are more or less irreversible phenomena at human scale. In contrast, some other human impacts such as domestic pollution and oil spills are not only reversible, but also often on the decrease (Table 1; Boudouresque et al., 2005). (iii) The ecological and economic impact is huge (Pimentel et al., 2001; Boudouresque, 2002a; Goreau et al., 2005; Kerr, 2006; Sala and Knowlton, 2006), though often underestimated by stakeholders.

Politicians, decision-makers and civil servants at the ministries of the environment are often inclined to make a cause and effect connection between climate warming and the increasing rate of species introductions. Be the aim conscious or unconscious, it is not purely a matter of chance. As long as we are not able to control carbon dioxide and other greenhouse gas emissions, species introductions will be impossible to prevent. Therefore, the fact that they do not implement the international conventions they have ratified, aimed at preventing and combating species introduction, is of no importance. It is worth noting that most European countries and all Mediterranean ones have not yet drafted a single text of law to apply the recommendations of the international conventions dealing with species introduction (Boudouresque, 2002b; Boudouresque and Verlaque, 2005).

Some scientific papers also envisage, explicitly or not, a cause and effect link between climate warming and the success of biological invasions (e.g. Dukes and Mooney, 1999; Bianchi, 2007; Galil et al., 2007; Occhipinti-Ambrogi, 2007; Galil, 2008; Hellmann et al., 2008; Perez, 2008). However, they usually do not present accurate data supporting the assumption, or they only present partial and therefore possibly biased data.

The goal of this study is to revisit the possible link between climate warming and the growing flow of species introductions, their biogeographical origin and

Table 1. Time needed for recovery, after the end of the forcing disturbance.

Disturbance	Human origin?	Natural origin?	Recovery	Key references
Domestic pollution (soft substrates)	+	−	<1–10 a	Bellan et al. (1999)
Artisanal fishing (fish abundance)	+	−	<5–10 a	Ramos (1992); Roberts et al. (2001)
Oil spill	+	−	<10 a	Raffin et al. (1991)
Disease of marine species	±	+	>10 a	Moses and Bonem (2001)
Loss of long-lived species	+	±	10–100 a	Soltan et al. (2001)
Coastal development	+	−	Millennia	Meinesz et al. (1991)
Over-fishing (genetic change)	+	−	Millennia?	Conover (2000); Law (2000); Kenchington et al. (2003); Olsen et al. (2004); Jørgensen et al. (2007)
Climate warming	+	+	Glacial cycle?	Zwiers and Weaver (2000); Barnett et al. (2001)
Biological invasions	+	−	Irreversible	Bright (1998); Clout (1998)
Species neo-extinction	+	−	Irreversible	Carlton (1993); Powles et al. (2000)

their success. Here, we shall only consider the seaweeds, a polyphyletic set of multicellular photosynthetic organisms (MPOs) belonging to the Chlorobionta, Rhodobionta (kingdom Plantae) and Phaeophyceae (kingdom Stramenopiles) (Boudouresque et al., 2006; Lecointre and Le Guyader, 2006) and the Mediterranean Sea, a set of taxa and an area for which an exhaustive data set is available (Verlaque et al., 2007b).

2. Climate Change and Global Warming

Since the birth of the planet Earth, 4,560–4,540 Ma (million years) ago (Jacobsen, 2003), its climate has never stopped changing. Over the past 50 Ma, the Earth's climate has been steadily cooling. Large ice sheets appeared in the Northern Hemisphere 2.7 Ma ago (Billups, 2005). Since then, the climate has fluctuated between glacial and interglacial episodes (glacial cycles); about 850,000 years ago, the period of the glacial cycles changed from 41,000 to 100,000 years (de Garidel-Thoron et al., 2005). Glacial cycles break down into 5,000–10,000 years and ~1,500 years cycles (Cacho et al., 2002; Braun et al., 2005; Sachs and Anderson, 2005). As a rule, all these cycles are characterised by slow cooling and abrupt warming (Tabeaud, 2002; Leipe et al., 2008).

The last cold maximum of a glacial cycle (LGM, Last Glacial Maximum) occurred 21,000 years ago (Berger, 1996; Tzedakis et al., 1997). Within the current interglacial episode, the last cold maximum of a 1,500-year cycle is known as the

Little Ice Age (LIA). It peaked from the thirteenth to the early nineteenth century (Le Roy-Ladurie, 2004). The sea surface temperature conspicuously dropped (deMenocal et al., 2000), which probably favoured the Southward expansion of cold resistant species. The subsequent rapid warming, from the mid-nineteenth century, should have driven a reverse effect, i.e., a dramatic regression of cold-water affinity species and better conditions for warm-water species. Obviously, the present-day release of greenhouse gas due to human activity should have enhanced these natural trends from 1970 onwards (Stott et al., 2000; Oreskes, 2004).

Taking 1900 as the baseline, in the Mediterranean, there has been a sea-surface temperature (SST) increase of 0.2°C in the Eastern basin and 1°C in the Western basin (Moron, 2003). Since 1974, in Catalonia (Spain), the increase is 1.1°C for SST and 0.7°C at 80 m depth (Salat and Pascual, 2002). However, taking 1856 as the baseline, there is no clear trend of SST increase at Mediterranean scale. These apparent mismatches are due to the occurrence of multidecadal cycles. In the Mediterranean Sea, the temperature (SST) was relatively higher in 1875–1880, 1935–1945 and in the 2000s than around 1860, 1905–1910 and 1975–1980; the 1935–1945 warming (+0.2–0.7°C) was more pronounced in the Eastern than in the Western basin, whereas the opposite is the case for that of the 2000s (Moron, 2003). Locally, the peaks can shift to a greater or lesser degree; for example, at Marseilles (France), for the 1885 to 1967 period, SST peaked in the 1890s and 1930s–1940s (Romano and Lugrezi, 2007).

3. Introduction of Seaweed Species

An introduced species is defined here as a species, which fulfils the four following criteria (Boudouresque and Verlaque, 2002a). (i) It colonises a new area where it did not previously occur. (ii) There is geographical discontinuity between its native area and the new area (remote dispersal). This means that the occasional advance of a species at the frontiers of its native range (marginal dispersal) is not taken into consideration. Such fluctuations (advances or withdrawals) may be linked to climatic episodes. (iii) The extension of its range is linked, directly or indirectly, to human activity. (iv) Finally, new generations of the non-native species are born in situ without human assistance, thus constituting self-sustaining populations: the species is established, i.e., naturalised.

In the marine realm, the main vectors of introduction are fouling and clinging on ship hulls, solid ballast (up to the late-nineteenth century), ballast water, fishing bait, escape from aquariums, waterways and canals crossing watersheds, transoceanic canals such as the Suez Canal, aquaculture and even scientific research (Por, 1978; Zibrowius, 1991; Carlton and Geller, 1993; Verlaque, 1994; Ribera and Boudouresque, 1995; Boudouresque, 1999a; Boudouresque and Verlaque, 2002b; Olenin, 2002; Galil et al., 2007). As far as aquaculture is concerned, the introduction can occur through escape of reared and cultivated species from sea farms and from the transport of reared species, such as fish and molluscs, from one aquaculture basin to another distant one, with all the accompanying species (e.g. parasites and epibiota); when the recipient habitats are suitable, these species can survive and become established, resulting in unintentional introductions (Verlaque et al., 2007a). In the Eastern Mediterranean, the Suez Canal, which connects the Red Sea to the Levantine Basin, constitutes the main vector of species introduction. In contrast, in the Western Mediterranean, the main vector is aquaculture (Galil, 2008; Galil, 2009).

The Mediterranean is one of the areas worldwide most severely hit by biological invasions, with about 600 introduced species of MPOs and Metazoa (Boudouresque et al., 2005; Galil et al., 2007; Galil, 2008; Zenetos et al., 2008; Galil, 2009).

As far as seaweeds are concerned, 106 species were probably introduced into the Mediterranean (Table 2; Fig. 1). This is a conservative value: (i) Possible cryptogenic introductions (sensu Carlton, 1996) are not taken into account; these are species whose extensive range area might be the result of ancient introduction events, before the first inventories in the area, and whose native region (within the current area) remains unknown; they are therefore classified as native by default. (ii) In the same way, species considered as native could prove to be cryptic introductions; these are species closely resembling a native one; identification of their possibly exotic status would require an in-depth study; several species in Table 2 were at first considered as native until on the basis of a genetic study, they were assigned to a sibling exotic taxon. (iii) The introduction of exotic strains of species already present in the Mediterranean (gene introduction), e.g., *Cladosiphon zosterae*, *Desmarestia viridis*, *Ectocarpus siliculosus* var. *hiemalis* and *Pylaiella littoralis*, has not been taken into consideration here.

Since the beginning of the twentieth century, the number of seaweeds introduced into the Mediterranean has more or less doubled every 20 years (Ribera and Boudouresque, 1995; Boudouresque, 1999a; Verlaque and Boudouresque, 2004; Boudouresque et al., 2005). A similar steady increase over time has occurred for Mediterranean Metazoa (Boudouresque, 1999b; Galil, 2008), e.g., mollusc species (Zenetos et al., 2003) and in other areas, e.g., in the Bay of San Francisco (Cohen and Carlton, 1998). However, in the Mediterranean, the post-2000 increase does not fit the previous trend (Fig. 1); the possibility that the number of introduced seaweeds is reaching a plateau must be considered (see below).

4. The Relationship Between Seaweed Introduction and Climate Warming

4.1. MORE SPECIES?

The increase in the number of introduced species is clearly parallel to the twentieth century SST increase. However, as pointed out by Galil (2008), concurrent phenomena do not in themselves imply causation. This increase is parallel to that

Table 2. Seaweeds introduced into the Mediterranean. The date of first observation is the date of publication when no more information is available. For seaweed authorities, see Guiry and Guiry (2008).

Species	Date of first observation	Probability of introduction	Probable vector of introduction	Probable geographical origin	Native biogeographical distribution
Rhodobionta (Plantae)					
Acanthophora nayadiformis	1798–1801	H	?	RS, IP	Tr
Acrochaetium codicola	1952	V	FO, O	IP	NT
Acrochaetium robustum	1944	H	S	IP	NT, Tr
Acrochaetium spathoglossi	1944	H	S	IP	Tr
Acrochaetium subseriatum	1944	H	S	IP	Tr
Acrothamnion preissii	1969	V	FO	IP	NT, Tr, ST
Agardhiella subulata	1984	H	O	A	NT, Tr
Aglaothamnion feldmanniae	1975	M	FO	A	NT
Ahnfeltiopsis flabelliformis	1994	V	O	J	NC, NT, Tr
Anotrichium okamurae	?	M	FO	?	NT
Antithamnion amphigeneum	1989	V	FO	P	ST
Antithamnion nipponicum	1988	V	O	P	NC
Antithamnionella boergesenii	1937	M	?	?	NT, Tr
Antithamnionella elegans	1882	V	FO	J	Tr
Antithamnionella spirographidis	1911	H	FO	IP	NC, SC
Antithamnionella sublittoralis	1980	H	FO	IP	NT
Antithamnionella ternifolia	1926	V	FO	SH	NC, SC
Apoglossum gregarium	1992	M	FO	IP?	Tr
Asparagopsis armata	1880	V	FO	IP	ST, SC
Asparagopsis taxiformis sp. 1	1798–1801	H	S, FO	A	NT, Tr
Asparagopsis taxiformis sp. 2 invasive	1996	H	FO?	IP	ST
Bonnemaisonia hamifera	1909	V	FO	IP	NC, NT
Botryocladia madagascarensis	1991	H	FO?	IP	Tr
Ceramium bisporum	1980	M	FO	A	Tr
Ceramium strobiliforme	1991	H	FO?	A	Tr
Chondria coerulescens	1995	V	O	A	NT
Chondria curvilineata	1981	H	FO	A	NT, Tr
Chondria pygmaea	1974	V	S	RS	Tr
Chondrus giganteus f. *flabellatus*	1994	V	O	J	NC
Chrysymenia wrightii	1978	V	O	J	NC
Dasya sessilis	1984	V	O	J	NC
Dasysiphonia sp.	1998	V	O	P	NC
Feldmannophycus okamurae	1937	H	FO	IP	NC, NT, Tr
Galaxaura rugosa	1990	V	S	RS	Tr
Ganonema farinosa	1808	M	S	RS	NT, Tr, ST

(continued)

Table 2. (continued)

Species	Date of first observation	Probability of introduction	Probable vector of introduction	Probable geographical origin	Native biogeographical distribution
Goniotrichopsis sublittoralis	1989	H	FO	IP	NC, NT
Gracilaria arcuata	1931	H	S	RS, IP	Tr
Grateloupia asiatica	1984	V	O	IP	NC, NT
Grateloupia lanceolata	1982	V	O	J	NT
Grateloupia patens	1994	V	O	J	NC, NT
Grateloupia subpectinata	1997	H	O	IP	NC, SC
Grateloupia turuturu	1982	V	O	J	NC
Griffithsia corallinoides	1964	H	O	A	NC, NT
Herposiphonia parca	1997	V	O	IP	NT, Tr
Hypnea cornuta	1894	H	S	RS	NT, Tr
Hypnea flagelliformis	1956	H	S	IP	NT, Tr
Hypnea spinella	1928	H	FO?	PT	NT, Tr, ST
Hypnea valentiae	1996	H	S, FO	RS	NT, Tr, ST
Laurencia caduciramulosa	1991	M	FO	?	Tr
Laurencia okamurae	1984	H	O	P	NT, Tr
Lithophyllum yessoense	1994	V	O	P	NC
Lomentaria hakodatensis	1978	V	O	J	NC, NT, Tr
Lophocladia lallemandii	1908	M	S, FO	RS	NT, Tr
Nemalion vermiculare	2005	V	O	IP	NC
Nitophyllum stellato-corticatum	1984	V	O	J	NT
Pleonosporium caribaeum	1974	M	FO, O	PT	Tr
Plocamium secundatum	1976	M	?	SH	SC
Polysiphonia atlantica	1969–1971	H	O, FO	A	NT
Polysiphonia fucoides	1988	H	FB	A	NC, NT
Polysiphonia harveyi	1958	V	FO?	IP, A	NC, NT
Polysiphonia morrowii	1997	V	O	P	NC
Polysiphonia paniculata	1967	H	?	P	NC, NT, ST
Porphyra yezoensis	1975	V	O	J	NC
Pterosiphonia tanakae	1993	V	O	J	NT
Rhodophysema georgii	1978	H	O	A	NC, NT
Rhodymenia erythraea	1948	V	S, FO	RS, IP	Tr
Sarconema filiforme	1944	V	S	RS	Tr
Sarconema scinaioides	1945	V	S	IP	Tr
Solieria dura	1944	V	S	RS	Tr
Solieria filiformis	1922	M	?	A	NT, Tr
Symphyocladia marchantioides	1984	V	FO	IP	NC, NT, Tr, ST, SC
Womersleyella setacea	1986	V	FO	PT	Tr
Chlorobionta (Plantae)					
Caulerpa mexicana	1939	V	S	RS	NT, Tr
Caulerpa racemosa var. *cylindracea*	1990	V	AQ, BW	IP	ST
Caulerpa racemosa var. *lamourouxii*	1951	M	S	RS	Tr
Caulerpa racemosa var. *turbinata*	1926	M	S	RS	Tr

(continued)

Table 2. (continued)

Species	Date of first observation	Probability of introduction	Probable vector of introduction	Probable geographical origin	Native biogeographical distribution
Caulerpa scalpelliformis	1929	H	S	RS	Tr
Caulerpa taxifolia MAAS[a]	1984	V	AQ	PT	ST
Cladophora herpestica	1948	V	S	RS	NT, Tr, ST
Cladophora patentiramea	1991	H	S, FO	IP	Tr
Codium fragile subsp. *tomentosoides*	1946	V	FO, O	IP	NT
Codium taylori	1955	M	FO?	A	NT, Tr
Derbesia boergesenii	1972	H	S	RS	Tr
Derbesia rhizophora	1984	V	O	J	NT
Neomeris annulata	2003	H	S	RS	Tr
Ulva fasciata	1979–1984	H	O	J	NT, Tr, ST
Ulva pertusa	1984	V	O	IP	NT, Tr
Ulvaria obscura	1985	H	O	A	NC, NT
Phaeophyceae (Stramenopiles)					
Acrothrix fragilis	1998	H	O	A, P	NC
Botrytella parva	1996	H	?	?	NC
Chorda filum	1981	V	O	A, J	NC, NT
Colpomenia peregrina	1918	V	FO	IP	NC, NT, ST, SC
Fucus spiralis	1987	V	FB	A	NC, NT
Halothrix lumbricalis	1985	H	O	?	NC, NT
Leathesia difformis	(1905) 1979	H	O	A	NC, NT, ST, SC
Padina boergesenii	1962–1965	H	S	RS	Tr
Padina boryana	1974	V	S	IP	NT, Tr
Punctaria tenuissima	1985	H	O	A	NC, NT
Rugulopterix okamurae	2002	V	O	J	NT, Tr
Saccharina japonica	1976	V	O	J	NC, NT
Sargassum muticum	1980	V	O	J	NC
Scytosiphon dotyi	1960–1977	V	O	P	NT
Spathoglossum variabile	1944	V	S	RS	Tr
Sphaerotrichia firma	1970	H	O	J	NC
Stypopodium schimperi	1973?	V	S	RS	Tr
Undaria pinnatifida	1971	V	O	J	NT

Probability of introduction: V = very high, H = high, M = medium. Vector of introduction: AQ = aquariums, BW = ballast water, FB = fishing baits, FO = fouling on ship hulls, O = oyster culture, S = Suez Canal (Lessepsian species). Geographical origin: A = Atlantic, BS = Black Sea, IP = Indo-Pacific, J = Japan, P = Pacific, PT = pantropical, RS = Red Sea, SH = Southern hemisphere. Native biogéographical distribution: NC = North cold, NT = North temperate, Tr = tropical, ST = South temperate, SC = South cold (see caption to Table 3).
a MAAS = Mediterranean Aquarium and Australian Strain.

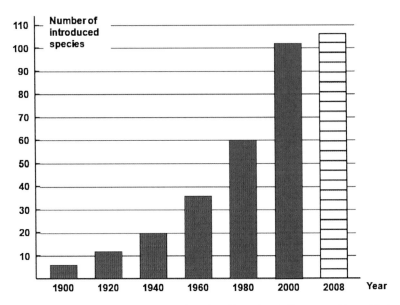

Figure 1. The cumulative number of seaweeds introduced into the Mediterranean Sea and its increase over time. Hatching for the 2008 value means that it does not correspond to the same 20-year time interval as the other values.

of the demographic pressure (Benoit and Comeau, 2005), that of the forest surface area in Western Europe and to the surge in highway traffic as well.

In fact, the increase in the number of introduced species is more probably related to the strengthening of the vectors: more aquaculture, more pleasure boats, more trade, more ships, more voyages, more speed, etc. (see Dobler, 2002; Benoit and Comeau, 2005; Briand, 2007).

4.2. MORE TROPICAL SPECIES?

Unexpectedly, the importance of tropical regions as donor areas for introductions of seaweeds to the Mediterranean was conspicuously higher in the 1800–1940 and 1941–1980 periods than later on, whereas the importance of both Southern and Northern cold regions increased from the 1980s (Table 3). Two factors, which are not mutually exclusive, may account for this. (i) Up to the 1950s, Lessepsian species, i.e., Red Sea species entering the Mediterranean via the Suez Canal, constituted the bulk of the seaweeds introduced into the Mediterranean. The Red Sea is a tropical realm. The number of new Lessepsian species peaked in the 1941–1950 period (Fig. 2), perhaps in relation with the gradual disappearance of the high-salinity barrier constituted by the Bitter Lakes up to the 1950s (see Por, 1978, 1989;

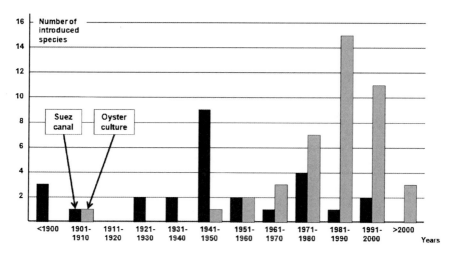

Figure 2. The number of seaweeds introduced into the Mediterranean per 10-year period (with the exception of <1900 and >2000). Two vectors are taken into account: the Suez Canal and oyster culture.

Boudouresque, 1999b). Subsequently, oyster culture took over from the Suez Canal as the main vector (Fig. 2). Massive importations of *Crassostrea gigas* oyster spat and adults from the Northern Pacific (mainly Japan), without either decontamination or quarantine, occurred in the 1970s; illegal importations (from Korea), in lesser amounts, continued up to the early 1990s (Grizel and Héral, 1991; Verlaque, 2001; Boudouresque and Verlaque, 2002b; Verlaque et al., 2007a). Donor regions were located in a cold biogeographical province. The shift from mainly tropical towards mainly cold-affinity introduced species can therefore be related to a change in the prevailing vector and the donor region. (ii) It might have been reasonable to suspect that the location of the Mediterranean phycologists changed over time, leading to phycologists working now in the Western basin rather than in the Eastern, which may have resulted in a distortion; oyster importations from cold waters of the Northwestern Pacific actually occurred mainly in Western Europe. In fact, this is the exact opposite of what actually occurred, the number of phycologists rather increasing in Eastern Mediterranean countries whereas declining in the Western ones.

If we remove the vector effect, which obviously accounts for the current biogeographical origin of most introduced species, a warmer Mediterranean should be more welcoming for a tropical than for a cold-water candidate species, and make easier its establishment. However, at the same time, cold-water candidates might be disadvantaged, so that the overall amount of new introduced species would be unchanged.

The assumption that most of the introduced species in the Mediterranean are thermophilic, originating in tropical seas (Galil, 2008; Galil, 2009), may prove to be true for Metazoa (kingdom Opisthokonts), but absolutely not for the MPOs belonging to the kingdoms Plantae and Stramenopiles. The media coverage of some introduced species believed to be of tropical origin, when they actually originate in temperate sea, has probably contributed to misleading authors. *Caulerpa taxifolia* is probably a complex of cryptic species mostly thriving in tropical seas. When it burst in on the Mediterranean, the media and scientists referred to it as 'the tropical alga' (Meinesz and Hesse, 1991; Boudouresque et al., 1995). Subsequently, molecular studies revealed the geographical origin of the strain (Mediterranean Aquarium and Australian Strain (MAAS) see Table 2): temperate Southeastern Australia (Jousson et al., 1998, 2000; Meusnier et al., 2001). Similarly, *Caulerpa racemosa* probably encompasses a complex of cryptic species. When discovered in the Mediterranean, *C. racemosa* var. *cylindracea* was at first confused with a tropical taxon already introduced into the Mediterranean, *C. racemosa* var. *turbinata* (e.g. Nizamuddin, 1991; Djellouli, 2000; Buia et al., 2001). Its true status and native area, temperate Southwestern Australia, was rapidly established (Verlaque et al., 2000, 2003). Finally, the invasive strain of *Asparagopsis taxiformis* (in fact a distinct species), which closely resembles a species common in the tropical Atlantic Ocean, actually comes from a Southern Australian temperate area (Ní Chualáin et al., 2004; Andreakis et al., 2007).

4.3. ARE THE INTRODUCED SPECIES MORE AGGRESSIVE?

As pointed out by Occhipinti-Ambrogi (2007), climate warming alters the competitive interactions between introduced and native species.

Once introduced, warm-water species (either of tropical or subtropical origin) should benefit from a warming Mediterranean (Galil, 2009). Roughly, SST is higher in the East and South than in the West and North. The current expansion of their area, Westwards and Northwards, has actually been observed (Galil, 2008). However, whatever the temperature trend, the marginal spread of an introduced species from its site of arrival constitutes a normal feature: it aims to occupy the whole of the suitable habitats and area. This spread can be very rapid, as occurred with the Chlorobionta *Caulerpa racemosa* var. *cylindracea*, which colonised the whole Mediterranean and the adjacent Atlantic coasts in less than 15 years (Verlaque et al., 2004). This spread can also take more time, as for the crustacean *Metapenaeus monoceros* (Fabricius 1798) and the fishes *Siganus luridus* (Rüppel 1829) and *S. rivulatus* (Forsskål 1775), which took 6 to 8 decades to spread from the Levant to Tunisia and Sicily (Galil, 2008). The natural marginal spread and the possible enhancement of the spread due to the SST warming are superimposed, so that unravelling their respective roles is not easy; it is therefore to be feared that premature conclusions are often drawn (i.e. 'the westwards spread of an introduced thermophilic species is due to the SST warming'). Be that as it may, it is worth noting that the current spread of native thermophilic species, such as the fishes *Sparisoma cretense* (Linnaeus 1758) and *Thalassoma pavo* (Linnaeus 1758) and the scleractinian coral *Astroides calycularis* (Pallas 1767) proves that the warming matters, whatever the degree of its contribution (Francour et al., 1994; Morri and Bianchi, 2001; Bianchi and Morri, 2004; Bianchi, 2007).

Whereas some warm-water introduced species advance, maybe partly in relation with the SST increase, such as the Rhodobionta *Womersleyella setacea* and the Phaeophyceae *Stypopodium schimperi*, the decline in abundance and the shrinking of the range of cold-water species, such as the Rhodobionta *Asparagopsis armata* (gametogenic phase) and the Chlorobionta *Codium fragile*, may be expected. Unfortunately, no data are available for the latter process: the arrival of a species at a new locality attracts more attention (and results in a scientific paper) than its absence from a previously occupied site (which may be thought to be temporary). Similarly, in the continental realm, Dukes and Mooney (1999) emphasised the components of global change (e.g. climate warming) likely to favour biological invaders, but did not consider those species, which could be disadvantaged.

Can we consider that '[algae] that are gaining ascendancy [in Mediterranean coastal ecosystems] are of tropical origin', as argued by Bianchi (2007)? Among the three seaweeds cited by the author in support of his assertion (*Stypopodium schimperi*, *Caulerpa taxifolia* and *C. racemosa* var. *cylindracea*), only the first one is actually of tropical origin.

4.4. WHAT COULD DEMONSTRATE THE IMPACT OF WARMING?

How could the impact of warming, either qualitative (which species?) or quantitative (how many species? How invasive?), on seaweed introductions, be demonstrated? (i) The increase in the number of introduced species reflects the nature and the strength of the vectors and is therefore irrelevant (see Section 4.1). (ii) The crossing of the limits of the potential area the species can occupy, a function of its physiology and competitive ability, would constitute a good criterion for a thermophilic species. However, this potential area is unknown. In addition, the spread of an introduced species is often a slow process, which takes decades, so that for many species, it may be suspected that they have not yet occupied their full potential area. Por (1989, 1990) delimited the 'Lessepsian province' corresponding to the potential expansion area of the Lessepsian species. At the moment, no strictly Lessepsian seaweeds have been reported outside this area, although a Magnoliophyta (Plantae), *Halophila stipulacea* (Forsskål) Ascherson, and several Metazoa have passed this limit. (iii) The resumption of the spread after a period of relative stasis was indicative that the potential area was reached. Four species of putatively thermophilic seaweeds might seem to meet this criterion: *Asparagopsis taxiformis*, *Caulerpa racemosa*, *Ulva fasciata* and *Lophocladia lallemandii*. The first two species proved to result in fact of new introduction events, i.e., the introduction of distinct taxa of temperate affinity, namely *Asparagopsis taxiformis* sp. 2 and *Caulerpa racemosa* var. *cylindracea* (see Table 2). The strain of *Ulva fasciata* discovered in the Northwestern Mediterranean (Thau Lagoon) is probably a new introduction

Table 3. Biogeographical affinity, in their native area, of the seaweeds introduced into the Mediterranean. Tropical = annual SST minimum >20°C. Temperate = annual SST minimum between 10°C and 20°C. Cold = annual SST minimum <10°C. See Lüning (1990) for definitions and SST maps.

| Period | Number of introduced species | Biogeographical affinity of introduced species |||
		North and South cold	North and South temperate	Tropical
1800–1940	21	4.3 (20%)	7.7 (37%)	9.0 (43%)
1941–1980	39	7.7 (20%)	14.0 (36%)	17.3 (44%)
1981–2008	46	16.2 (35%)	17.9 (39%)	11.9 (26%)

from Japan (unpublished data). Finally, the very localised new area of *Lophocladia lallemandii* (Balearic Islands, Spain), together with its proliferation, could also be indicative of a new introduction event (strain or yet unidentified taxon). (iv) The shrinking of the area of a cold-affinity species: Genetic processes such as inbreeding depression can account for this, in addition to warming. (v) The demonstration that introduced species of tropical origin are more invasive (more 'aggressive') that cold-affinity species: As pointed out by Perez (2008), this could prove to be correct for Prokaryota. As far as seaweeds are concerned, there is no indication that species of cold-water origin (such as *Sargassum muticum* and *Undaria pinnatifida*), e.g., in the Nortwestern Mediterranean Thau Lagoon, are less invasive than species of tropical origin (such as *Stypopodium schimperi*) in the warmer Eastern Mediterranean Sea.

In fact, it may simply be too early to detect a qualitative or a quantitative impact of warming on seaweed introductions, unless it is indeed an insoluble problem, or even a false problem.

5. General Discussion and Conclusion

It is difficult to give a definite answer to the question we asked ('Is global warming involved in the success of seaweed introductions?'). Several distortions may affect the data set we used. (i) Study taxa and study areas largely depend upon the phycologists and their location. (ii) Large introduced species, belonging to taxa whose delineation is not controversial, are easier to detect than tiny species whose taxonomy is confused and accessible to very few specialists. (iii) Cryptogenic introductions are by definition unknown. Taking them into consideration, where it is possible, might conspicuously modify the baseline of our data set, i.e., the panel of anciently introduced species. (iv) Cryptic introductions are not taken into account, though progress in taxonomy will progressively make this possible. (v) The native area (and biogeographical province) of a species is not always accurately known. Either it is naturally present in unknown regions and the native area is underestimated, or it constitutes a cryptogenic introduction in part of its current area, and the native area is therefore overestimated.

Even taking into account these caveats, our data do not support the assumption that climate warming enhances biological invasions in the Mediterranean, at least in the case of the seaweeds. (i) The increase over time in the number of introduced species simply reflects the development of the vectors. In the early and mid-twentieth century, the Red Sea was the main donor region (Fig. 2). Subsequently, the relative strength of this vector declined. It can be hypothesised that most of the species from the Northern Red Sea, suited to survival in Mediterranean habitats and under their present conditions, have already taken the Suez Canal. In the 1970s, oyster culture took over from the Suez Canal as the main vector (Fig. 2). Since the turn of the century, oyster culture seems to be losing ground: either because oyster importation from Northwestern Pacific is officially banned or because most of the Japanese species that were able to thrive in the Mediterranean have been already introduced. In the absence of a new leading vector, the rate of introductions seems to be slowing down (Fig. 1; see also Galil et al., 2007, for Metazoa). Is this a durable trend or just a provisional one, i.e., waiting for the occurrence of the next prevailing vector? (ii) Since the 1980s, i.e., since the undisputable warming of Mediterranean surface water, not only has the relative percentage of new introduced species of tropical origin not increased, but also it has conspicuously declined (Table 3). The reason is that what matters first is the vector (see above). (iii) The alleged 'aggressiveness' of tropical introduced species, such as *Caulerpa taxifolia* and *C. racemosa* var. *cylindracea*, is due to the fact that they are seen as of tropical origin, when they are actually native to temperate seas. Their success in the Mediterranean, a temperate sea, is therefore in no way unexpected. (iv) The warming can advantage thermophilic introduced species. However, at the same time, it can disadvantage cold water species. The overall numbers of new introduced species and the overall dominance of introduced species might therefore be unchanged.

It is interesting to note that the simulation of the effects of climate warming and biological invasions (from 1900 to 2050) on the Mediterranean continental vegetation led to the conclusion that the driving force was the introduced species, whereas warming alone or in combination with introduced species was likely to be negligible in many of the simulated ecosystems (Gritti et al., 2006).

The link between climate warming and biological invasions is therefore poorly supported by the Mediterranean seaweeds. From a quantitative point of view, there are no grounds to believe that warming is responsible for the increase in the number of introduced species, or that species of tropical origin are more 'aggressive' than those of cold-water region origin. From a qualitative point of view (i.e., which species?) together with the spread and dominance of these species, the authors who claim that warming enhances the introduction, spreading and dominance of tropical species, are simply putting Descartes before the horse: if warming becomes more pronounced, which is unfortunately highly probable, there is no doubt that they will end up being proved right.

As far as the politicians, decision-makers and civil servants are concerned, their belief that the current increase in the number of introduced species results from global warming is not supported by the available data. There is no reason for this to change in the near future, and there is therefore no excuse for not implementing the international agreements for limiting and controlling biological invasions.

6. References

Andreakis, N., Procaccini, G., Maggs, C. and Kooistra, W.H.C.F. (2007) Phylogeography of the invasive seaweed *Asparagopsis* (Bonnemaisoniales, Rhodophyta) reveals cryptic diversity. Mol. Ecol. **16**: 2285–2299.

Barnett, T.P., Pierce, D.W. and Schnur, R. (2001) Detection of anthropogenic climate change in the world's ocean. Science **292**: 270–274.

Bellan, G., Bourcier, M., Salen-Picard, C., Arnoux, A. and Casserley, S. (1999) Benthic ecosystem changes associated with wastewater treatment at Marseille: implications for the protection and restoration of the Mediterranean shelf ecosystem. Water Environ. Res. **71**(4): 483–493.

Benoit, G. and Comeau, A. (2005) *A Sustainable Future for the Mediterranean: the Blue Plan's Environment and Development Outlook*. Earthscan publ., London. pp. 464.

Berger, A. (1996) Modeling the last and next glacial–interglacial cycles. *Tendances nouvelles pour l'environnement. Journées du Programme Environnement, Vie Sociétés*, Paris, 15–17 Janvier 1996: 1–13.

Bianchi, C.N. (2007) Biodiversity issues for the forthcoming tropical Mediterranean Sea. Hydrobiologia **580**: 7–21.

Bianchi, C.N. and Morri, C. (2004) Climate change and biological response in Mediterranean Sea ecosystems. Ocean Challenge **13**(2): 32–36.

Billups, K. (2005) Snow maker for the ice ages. Nature **433**: 809–810.

Boudouresque, C.F. (1999a) Introduced species in the Mediterranean: routes, kinetics and consequences, In: *Proceedings of the Workshop on Invasive Caulerpa Species in the Mediterranean*. MAP Technical Reports Ser., UNEP, Athens, pp. 51–72.

Boudouresque, C.F. (1999b) The Red Sea – Mediterranean link: unwanted effects of canals, In: O.T. Sandlund, P.J. Schei and A. Viken (eds.) *Invasive Species and Biodiversity Management*. Kluwer, Dordrecht, pp. 213–228.

Boudouresque, C.F. (2002a) The spread of a non native species, *Caulerpa taxifolia*. Impact on the Mediterranean biodiversity and possible economic consequences, In: F. Di Castri and V. Balaji (eds.) *Tourism, Biodiversity and Information*. Backhuis publ., Leiden, pp. 75–87.

Boudouresque, C.F. (2002b) Protected marine species, prevention of species introduction and the national environmental agencies of Mediterranean countries: professionalism or amateurishness? In: *Actes du congrès international "Environnement et identité en Méditerranée"*, Corte, 3–5 July 2002, Université de Corse Pascal Paoli publ., pp. 75–85.

Boudouresque, C.F. and Verlaque, M. (2002a) Biological pollution in the Mediterranean Sea: invasive versus introduced macrophytes. Mar. Poll. Bull. **44**: 32–38.

Boudouresque, C.F. and Verlaque, M. (2002b) Assessing scale and impact of ship-transported alien macrophytes in the Mediterranean Sea, In: F. Briand (ed.) *Alien Organisms Introduced by Ships in the Mediterranean and Black Seas*. CIESM Workshop Monographs **20**: 53–61.

Boudouresque, C.F. and Verlaque, M. (2005) Nature conservation, Marine Protected Areas, sustainable development and the flow of invasive species to the Mediterranean Sea. Sci. Rep. Port-Cros Nation. Park **21**: 29–54.

Boudouresque, C.F., Meinesz, A., Ribera, M.A. and Ballesteros, E. (1995) Spread of the green alga *Caulerpa taxifolia* (Caulerpales, Chlorophyta) in the Mediterranean: possible consequences of a major ecological event. Scientia Mar. **59**(supl. 1): 21–29.

Boudouresque, C.F., Ruitton, S. and Verlaque, M. (2005) Large-scale disturbances, regime shift and recovery in littoral systems subject to biological invasions, In: V. Velikova and N. Chipev (eds.)

Large-Scale Disturbances (Regime Shifts) and Recovery in Aquatic Ecosystems: Challenges for Management Towards Sustainability. Unesco publ, pp. 85–101.

Boudouresque, C.F., Ruitton, S. and Verlaque, M. (2006) Anthropogenic impacts on marine vegetation in the Mediterranean, In: *Proceedings of the Second Mediterranean Symposium on Marine Vegetation*, Athens 12–13 December 2003. Regional Activity Centre for Specially Protected Areas publ., Tunis, pp. 34–54.

Braun, H., Christl, M., Rahmstorf, S., Ganopolski, A., Mangini, A., Kubatzki, C., Roth, K. and Kromer, B. (2005) Possible solar origin of the 1,470-year glacial climate cycle demonstrated in a coupled model. Nature **438**: 208–211.

Briand, F. (2007) (ed.) Impact of mariculture on coastal ecosystems. CIESM Workshop Monographs 32. pp. 118.

Bright, C. (1998) *Life Out of Bonds. Bioinvasion in a Borderless World*. Norton W.W. & Company publ., New York, London, 288 pp.

Buia, M.C., Gambi, M.C., Terlizzi, A. and Mazzella, L. (2001) Colonisation of *Caulerpa racemosa* along the southern Italian coast: I. Distribution, phenological variability and ecological role, In: V. Gravez, S. Ruitton, C.F. Boudouresque, L. Le Direac'h, A. Meinesz, G. Scabbia and M. Verlaque (eds.) *Fourth International Workshop on Caulerpa taxifolia.* GIS Posidonie publ., Marseilles. pp. 352–360.

Cacho, I., Grimalt, J.O. and Canals, M. (2002) Response of the Western Mediterranean Sea to rapid climatic variability during the last 50,000 years: a molecular biomarker approach. J. Mar. Syst. **33–34**: 253–272.

Carlton, J.T. (1993) Neoextinctions of marine invertebrates. Amer. Zool. **33**: 499–509.

Carlton, J.T. (1996) Biological invasions and cryptogenic species. Ecology **77**(6): 1653–1655.

Carlton, J.T. and Geller, J.B. (1993) Ecological roulette: the global transport of nonindigenous marine organisms. Science **261**: 78–82.

Clout, M. (1998) And now, the Homogocene. World Conserv. **97**(4)–**98**(1): 3.

Cohen, A.N. and Carlton, J.T. (1998) Accelerating invasion rate in a highly invaded estuary. Science **279**: 555–558.

Conover, D.O. (2000) Darwinian fishery science. Mar. Ecol. Progr. Ser. **208**: 303–307.

Garidel-Thoron, T. de, Rosenthal, Y., Bassinot, F. and Beaufort, L. (2005) Stable sea surface temperatures in the western Pacific warm pool over the past 1.75 million years. Nature **433**: 294–298.

de Menocal, P., Ortis, J., Guilderson, T. and Sarnthein, M. (2000) Coherent high- and low-latitude climate variability during the Holocene warm period. Science **288**: 2198–1202.

Djellouli, A. (2000) *Caulerpa racemosa* (Forskaal) J. Agardh en Tunisie, In: *Actes du 1ᵉʳ Symposium Méditerranéen sur la Végétation Marine*, Ajaccio, 3–4 Oct. 2000. RAC-SPA publ., Tunis, pp. 124–127.

Dobler, J.P. (2002) Analysis of shipping patterns in the Mediterranean and Black seas, In: F. Briand (ed.) *Alien Organisms Introduced by Ships in the Mediterranean and Black Seas*. CIESM Workshop Monographs 20, pp. 19–28.

Dukes, J.S. and Mooney, H.A. (1999) Does global change increase the success of biological invaders? Trends Ecol. Evol. **14**(4): 135–139.

Francour, P., Boudouresque, C.F., Harmelin, J.G., Harmelin-Vivien, M.L. and Quignard, J.P. (1994) Are the Mediterranean waters becoming warmer? Information from biological indicators. Mar. Poll. Bull. **28**(9): 523–526.

Galil, B.S. (2008) Alien species in the Mediterranean Sea – Which, when, where, why? Hydrobiologia **606**: 105–116.

Galil, B.S. (2009). Tacking stock: inventory of alien species in the Mediterranean Sea. Biol. Inv. **11**: 359–372.

Galil, B.S., Nehring, S. and Panov, V. (2007) Waterways as invasion highways – impact of climate change and globalization, In: W. Nentwig (ed.) *Biological Invasions*. Springer, Berlin, Heidelberg, pp. 59–74.

Goreau, T.J., Hayes, R.L. and McAllister, D. (2005) Regional patterns of sea surface temperature rise: implications for global ocean circulation change and the future of coral reefs and fisheries. World Resour. Rev. **17**(3): 350–370.

Gritti, E.S., Smith, B. and Sykes, M.T. (2006) Vulnerability of Mediterranean Basin ecosystems to climate change and invasion by exotic plant species. J. Biogeogr. **33**: 145–157.

Grizel, H. and Héral, M. (1991) Introduction into France of the Japanese oyster (*Crassostrea gigas*). J. Cons. Int. Explor. Mer. **47**: 399–403.

Guiry, M.D. and Guiry, G.M. (2008) *AlgaeBase*. World-wide electronic publication, National University of Ireland, Galway. http://www.algaebase.org; searched on 29 October 2008.

Hellmann, J.J., Byers, J.E., Birwagen, B.G. and Dukes, J.S. (2008) Five potential consequences of climate change for invasive species. Conserv. Biol. **22**(3): 534–543.

Jacobsen, S.B. (2003) How old is the planet Earth? Science **300**: 1513–1514.

Jørgensen, C., Enberg, K., Dunlop, E.S., Arlinghaus, R., Boukal, D.S., Brander, K., Ernande, B., Gårdmark, A., Johnston, F., Matsumura, S., Pardoe, H., Raab, K., Silva, A., Vainikka, A., Dieckmann, U., Heino, M. and Rijnsdorp, A.D. (2007) Managing evolving fish stocks. Science **318**: 1247–1248.

Jousson, O., Pawlowski, J., Zaninetti, L., Meinesz, A. and Boudouresque, C.F. (1998) Molecular evidence for the aquarium origin of the green alga *Caulerpa taxifolia* introduced to the Mediterranean Sea. Mar. Ecol. Progr. Ser. **172**: 275–280.

Jousson, O., Pawlowski, J., Zaninetti, L., Zechman, E.W., Dini, F., Di Guiseppe, G., Woodfield, R., Millar, A. and Meinesz, A. (2000) Invasive alga reaches California. Nature **408**: 157–158.

Kenchington, E., Heino, M. and Nielsen, E.E. (2003) Managing marine genetic diversity: time for action? ICES J. Mar. Sci. **60**: 1172–1176.

Kerr, R.A. (2006) A worrying trend of less ice, higher seas. Science **311**: 1698–1701.

Law, R. (2000) Fishing, selection, and phenotypic evolution. ICES J. Mar. Sci. **57**: 659–668.

Le Roy-Ladurie, E. (2004) *Histoire Humaine et Comparée du Climat. Canicules et Glaciers, XIII°–XVIII° Siècles*. Fayard publ., Paris, 740 pp.

Lecointre, G. and Le Guyader, H. (2006) *Classification Phylogénétique du Vivant*. Belin publ., Paris, pp. 559 + plates.

Leipe, T., Dippner, J.W., Hille, S., Voss, M., Christiansen, C. and Bartholdy, J. (2008) Environmental changes in the central Baltic Sea during the past 1000 years: inferences from sedimentary records, hydrography and climate. Oceanologia **50**(1): 23–41.

Lûning, K. (1990) *Seaweeds. Their Environment, Biogeography and Ecophysiology*. Wiley, New York. pp. xiii + 527.

Meinesz, A. and Hesse, B. (1991) Introduction et invasion de l'algue tropicale *Caulerpa taxifolia* en Méditerranée nord-occidentale. Oceanol. Acta **14**(4): 415–426.

Meinesz, A., Lefèvre, J.R. and Astier, J.M. (1991) Impact of coastal development on the infralittoral zone along the southern Mediterranean shore of continental France. Mar. Poll. Bull. **23**: 343–347.

Meusnier, I., Olsen, J.L., Stam, W.T., Destombe, C. and Valero, M. (2001) Phylogenetic analyses of *Caulerpa taxifolia* (Chlorophyta) and of its associated bacterial microflora provide clues to the origin of the Mediterranean introduction. Mol. Ecol. **10**(4): 931–946.

Moron, V. (2003) L'évolution séculaire des températures de surface de la Mer Méditerranée (1856–2000). C.R. Géoscience **335**: 721–727.

Morri, C. and Bianchi, C.N. (2001) Recent changes in biodiversity in the Ligurian Sea (NW Mediterranean): is there a climatic forcing? In: F.M. Faranda, L. Guglielmo and G. Spezie (eds.) *Mediterranean Ecosystems: Structures and Processes*. Springer, Milan, pp. 375–384.

Moses, C.S. and Bonem, R.M. (2001) Recent population dynamics of *Diadema antillarum* and *Tripneustes ventricosus* along the north coast of Jamaica, W.I. Bull. Mar. Sci. **68**(2): 327–336.

Ní Chualáin, F., Maggs, C.A., Saunders, G.W. and Guiry, M.D. (2004) The invasive genus *Asparagopsis* (Bonnemaisoniaceae, Rhodophyta): molecular systematic, morphology and ecophysiology of *Falkenbergia* isolates. J. Phycol. **40**: 1112–1126.

Nizamuddin, M. (1991) *The Green Marine Algae of Lybia*. Elga publ., Bern, pp. 227.

Occhipinti-Ambrogi, A. (2007) Global change and marine communities: alien species and climate change. Mar. Poll. Bull. **55**: 342–352.

Olenin, S. (2002) Black Sea – Baltic Sea invasion corridors, In: F. Briand (ed.) *Alien Organisms Introduced by Ships in the Mediterranean and Black Seas.* CIESM Workshop Monographs 20, pp. 29–33.
Olsen, E.M., Heino, M., Lilly, G.R., Morgan, M.J., Brattey, J., Ernande, B. and Dieckmann, U. (2004) Maturation trends indicative of rapid evolution preceded the collapse of northern cod. Nature **428**: 932–935.
Oreskes, N. (2004) The scientific consensus on climate change. Science **306**: 1686.
Perez, T. (2008) *Impact des Changements Climatiques sur la Biodiversité en Mer Méditerranée.* CAR/ASP publ., Tunis, pp. 62.
Pimentel, D., McNair, S., Janecka, J., Wightman, J., Simmonds, C., O'Connell, C., Wong, E., Russel, L., Zern, J., Aquino, T. and Tsomondo, T. (2001) Economic and environmental threats of alien plants, animal and microbe invasions. Agr. Ecosyst. Env. **84**: 1–20.
Por, F.D. (1978) *Lessepsian Migrations. The Influx of Red Sea Biota into the Mediterranean by Way of the Suez Canal.* Springer, Berlin, pp. viii + 228.
Por, F.D. (1989) *The legacy of the Tethys. An Aquatic Biogeography of the Levant.* Kluwer, Dordrecht, pp. viii + 214.
Por, F.D. (1990) Lessepsian migrations. An appraisal and new data. Bull. Inst. Océanogr. NS 7: 1–10.
Powles, H., Bradford, M.J., Bradford, R.G., Doubleday, W.G., Innes, S. and Levings, C.D. (2000) Assessing and protecting endangered marine species. ICES J. Mar. Sci. **57**: 669–676.
Raffin, J.P., Platel, R., Meunier, F.J., Francillon-Vieillot, H., Godineau, J.C. and Ribier, J. (1991) Etude sur dix ans (1978–1988) de populations de Mollusques (*Patella vulgata* L. et *Tellina tenuis* Da Costa), après pollution pétrolière (Amoco Cadiz). Bull. Ecol. **22**(3–4): 375–388.
Ramos, A.A. (1992) Impact biologique et économique de la Réserve marine de Tabarca (Alicante, Sud-Est de l'Espagne), In: *Economic Impact of the Mediterranean Coastal Protected Areas,* Ajaccio, 26–28 Septembre 1991, Medpan News **3**: 59–66.
Ribera, M.A. and Boudouresque, C.F. (1995) Introduced marine plants, with special reference to macroalgae: mechanisms and impact, In: F.E. Round and D.J. Chapman (eds.) *Progress in Phycological Research.* Biopress, Bristol, 11, pp. 187–268.
Roberts, C.M., Bohnsack, J.A., Gell, F., Hawkins, J.P. and Goodridge, R. (2001) Effects of marine reserve on adjacent fisheries. Science **294**: 1920–1923.
Romano, J.C. and Lugrezi, M.C. (2007) Série du marégraphe de Marseille: mesures de temperatures de surface de la mer de 1885 à 1967. C.R. Geosciences **339**: 57–64.
Sachs, J.P. and Anderson, R.F. (2005) Increased productivity in the subantarctic ocean during Heinrich events. Nature **434**: 1118–1121.
Sala, E. and Knowlton, N. (2006) Global marine biodiversity trends. Annu. Rev. Environ. Resour. **31**: 93–122.
Salat, J. and Pascual, J. (2002) The oceanographic and meteorological station at L'Estartit (NW Mediterranean). *Tracking Long-Term Hydrological Change in the Mediterranean Sea,* CIESM Workshop Series, 16, pp. 29–32.
Schaffelke, B., Smith, J.E. and Hewitt, C.L. (2006) Introduced macroalgae – a growing concern. J. Appl. Phycol. **18**: 529–541.
Soltan, D., Verlaque, M., Boudouresque, C.F. and Francour, P. (2001) Changes in macroalgal communities in the vicinity of a Mediterranean sewage outfall after the setting up of a treatment plant. Mar. Poll. Bull. **42**(1): 59–70.
Stott, P.A., Tett, S.F.B., Jones, G.S., Allen, M.R., Mitchell, J.F.B. and Jenkins, G.J. (2000) External control of 20th century temperature by natural and anthropogenic forcings. Science **290**: 2133–2137.
Tabeaud, M. (2002) Les variabilités historiques du climat en Europe. Biogeographica **78**(4): 149–157.
Tzedakis, P.C., Andrieu, V., Beaulieu, J.L. de, Crowhurst, S., Follieri, M., Hooghiemstra, H., Magri, D., Reille, M., Sadori, L., Shackleton, N.J. and Wijmstra, T.A. (1997) Comparison of terrestrial and marine records of changing climate of the last 500,000 years. Earth Plan. Sci. Lett. **150**: 171–176.

Verlaque, M. (1994) Inventaire des plantes introduites en Méditerranée: origine et répercussions sur l'environnement et les activités humaines. Oceanol. Acta **17**(1): 1–23.

Verlaque, M. (2001) Checklist of the macroalgae of Thau Lagoon (Hérault, France), a hot spot of marine species introduction in Europe. Oceanologica Acta **24**(1): 29–49.

Verlaque, M. and Boudouresque, C.F. (2004) Invasions biologiques marines et changement global, In: *Actes des 2° Journées de l'Institut Français de la Biodiversité "Biodiversité et changement global, dynamique des Interactions"*, Marseille, 25–28 Mai 2004, pp. 74–75.

Verlaque, M., Boudouresque, C.F., Meinesz, A. and Gravez, M. (2000) The *Caulerpa racemosa* complex (Caulerpales, Ulvophyceae) in the Mediterranean Sea. Bot. Mar. **43**: 49–68.

Verlaque, M., Durand, C., Huisman, J.M., Boudouresque, C.F. and Le Parco, Y. (2003) On the identity and origin of the Mediterranean invasive *Caulerpa racemosa* (Caulerpales, Chlorophyta). Eur. J. Phycol. **38**: 325–339.

Verlaque, M., Afonso-Carrillo, J., Gil-Rodriguez, M.C., Durand, C., Boudouresque, C.F. and Le Parco, Y. (2004) Blitzkrieg in a marine invasion: *Caulerpa racemosa* var. *cylindracea* (Bryopsidales, Chlorophyta) reaches the Canary Islands (NE Atlantic). Biol. Inv. **6**: 269–281.

Verlaque, M., Boudouresque, C.F. and Mineur, F. (2007a) Oyster transfers as a vector for marine species introductions: a realistic approach based on the macrophytes, In: F. Briand (ed.) *Impact of Mariculture on Coastal Ecosystems*, CIESM Workshop Monographs 32, pp. 39–47.

Verlaque, M., Ruitton, S., Mineur, F. and Boudouresque, C.F. (2007b) CIESM Atlas of exotic macrophytes in the Mediterranean Sea. Rapp. Comm. int. Mer Médit. **38**: 14.

Zenetos, A., Gofas, S., Russo, G. and Templado, J. (2003) In: F. Briand (ed.) *CIESM Atlas of Exotic Species in the Mediterranean*. Vol. 3. *Molluscs*. CIESM publ., Monaco, pp. 375.

Zenetos, A., Meriç, E., Verlaque, M., Galli, P., Boudouresque, C.F., Giangrande, A., Çinar, M.E. and Bilecenoğlu, M. (2008) Additions to the annotated list of marine alien biota in the Mediterranean with special emphasis on Foraminifera and parasites. Medit. Mar. Sci. **9**(1): 119–165.

Zibrowius, H. (1991) Ongoing modification of the Mediterranean marine fauna and flora by the establishment of exotic species. Mésogée **51**: 83–107.

Zwiers, F.W. and Weaver, A.J. (2000) The causes of 20th century warming. Science **290**: 2081–2083.

Biodata of **Dr. Gil Rilov** and **Haim Trebes**, authors of *"Climate Change Effects on Marine Ecological Communities"*

Dr. Gil Rilov is a Senior Scientist at the National Institute of Oceanography, Israel Oceanographic and Limnological Research, Haifa, Israel. He obtained his Ph.D. from Tel Aviv University in 2000 in Marine Ecology, was a post-doc Fulbright Scholar at Duke University (USA), and did a second post-doc at the Canterbury University (USA). He was an Assistant Professor – Senior Research at Oregon State University (USA) between 2005 and 2008 before returning to Israel. Dr. Rilov's scientific interests are in the areas of marine ecology and conservations and he focuses his research on benthic communities, biodiversity, species interactions, benthic–pelagic coupling, bioinvasions, and climate change.

E-mail: **rilovg@ocean.org.il**

Haim Treves is a *summa cum laude* Graduate of the Marine Sciences School of the Ruppin Academic Center, Israel. He is currently involved in research projects on the ecology of rocky shores along the Israeli shore and hopes to pursue a career in the field.

E-mail: **htreves@gmail.com**

Gil Rilov

Haim Treves

CLIMATE CHANGE EFFECTS ON MARINE ECOLOGICAL COMMUNITIES

GIL RILOV[1] AND HAIM TREVES[2]
[1] National Institute of Oceanography, Israel Oceanographic and Limnological Research, Tel-Shikmona, P.O. Box 8030, Haifa 31080, Israel
[2] Ruppin Academic Center, School of Marine Sciences, Mikhmoret, Israel

1. Introduction

It is no secret that our climate is changing – rapidly – and together with it, oceans change as well. The Intergovernmental Panel on Climate Change (IPCC), consisting of hundreds of scientists worldwide, have shown that changes in global climate have accelerated since the 1750s, causing an overall increase in temperature both on land and in the sea. The IPCC also suggests that research indicates that there is >90% chance that the change is human-mediated (IPCC, 2007). Modifications to ocean temperature, biogeochemistry, salinity, sea level, UV radiation, and current circulation patterns have all been detected within the last few decades and are expected to continue (IPCC, 2007). Increase in extreme weather is also expected, including intensification and rise in the frequency of severe storms. Less than 2 decades ago, marine ecologists could mostly speculate about the possible ecological responses of marine systems to global climate change (Lubchenco et al., 1993). Today, however, the ecological "footprint" of climate change has been observed in both terrestrial and marine ecosystems worldwide (Walther et al., 2002, 2005).

Documented ecological changes that are related, for example, to temperature alteration in the oceans include modifications to the phenology of pelagic organisms resulting in trophic "mismatches" between predators and preys (e.g., Edwards and Richardson, 2004), severe events of coral bleaching that negatively influence the structure of coral reef communities (e.g., Hughes et al., 2003), a mostly poleward shift in fish distributions in the North Sea (Perry et al., 2005), and shifts in the distributional limits of benthic organisms in temperate coastal environments (Helmuth et al., 2006b). Harley et al. (2006) provide a comprehensive review of the known and potential effects of climate change on coastal marine ecosystems. The authors demonstrate that the study of this topic is quickly accelerating, which is no surprise, given the increased rate of change in physical phenomena related to climate change in the ocean and the mounting evidence of their biological and ecological impacts. Since the publication of Harley et al.'s research, dozen more papers have been published on this issue, some with remarkable albeit worrisome findings.

In this chapter, we will illustrate some of the evidences and projections for change in the marine environment (coastal and pelagic) attributed to climate change, focusing mainly on the two most studied and experimented changes: temperature and pH. Other topics that will also be explored briefly are the predicted changes due to sea-level rise, increase of storms, and change in circulation patterns. The different aspects of climate change are expected to affect marine communities at different spatiotemporal scales and also the number of habitats impacted. For example, temperature and CO_2 will most likely have basin-scale or even global effects and can potentially affect ecosystems at all depths, while sea-level rise and increased storm frequency/intensity will probably affect mostly shallow coastal environments. Climate changes are predicted to affect ocean life from the tiniest of organisms – plankton, to the largest ones – whales (Gambaiani et al., 2009).

Several lines of research are being used by investigators to identify the links between the current (and predicted) physical changes related to global climate change and their direct and indirect effects on biological and ecological patterns and processes. On large biogeographic scales, correlative studies are mostly used to find links, for example between temperature and species distribution shifts. To predict ecological impacts under different future scenarios of ocean temperature and pH, researchers use controlled laboratory or mesocosm experiments (mostly to look at one or several species). In few cases, they could also use existing manmade (e.g., outflow areas of power plants where temperatures are increased) and natural (e.g., CO_2 vents) environments that today mimic predicted levels of these variables (mostly to look at the total ecosystem effects). Biophysical models and ecological food-web models can also be used to examine ecosystem-level effects of climate-induced increase or decrease of key species at the bottom or top of the food-web. Models and experiments also seek to find stabilizing forces that might modulate climate change effects.

2. Effects of Temperature Increases

Temperature is probably the most dominant rate-determining factor in biology; ranging from subcellular to community-level processes, with direct and indirect effects on organisms' physiology, ontogeny, trophic interactions, biodiversity, phenology, and biogeography. Increases in temperature due to climate change have the potential to impact most marine ecosystems, directly through the impact on species physiology (growth, reproduction, etc.) or indirectly through impacts on ocean dynamics (currents) or species interactions. The magnitude of ecological effects of rising temperatures would inherently vary among and even within species, as different species and even different ontogenetic stages may be unequally susceptible to thermal stress or steep fluctuations in temperature.

The most obvious and direct biological effect of global warming is attributed to the fundamental relationship between temperature and physiology. A wide range of physiological processes are influenced by temperature, among them are protein structure and function, membrane fluidity, organ function (Hochachka and Somero, 2002; Harley et al., 2006), heart function, and mitochondrial respiration (Somero, 2002). For some of these thermally sensitive traits, the acclimation of marine species to a given environment has resulted in the creation of narrow thermal optima and limits. In addition, many marine species live near their thermal tolerance limits, and small temperature increases could negatively impact their performance and survival. This was demonstrated in heat-shock response patterns of different coastal *Tegula* snails that were shown to have limited thermal tolerance, which depended on the region and habitat of the species studied (Tomanek and Somero, 1999), and again in the thermal tolerance of rocky intertidal porcelain crab species (Stillman, 2002). In the Caribbean, McWilliams et al. (2005) demonstrated that a shift of only +0.1°C resulted in 35% and 42% increases in geographic extent and intensity of coral bleaching, respectively.

Indeed, coral bleaching is one of the most well-known and studied phenomenon related to temperature stress in the marine environment. During thermal stress, corals expel most of their pigmented microalgal endosymbionts, called zooxanthellae, to become pale or white (i.e., bleached). The link between climate change and bleaching of corals is now indisputable, as episodes of coral bleaching have already increased greatly in frequency and magnitude over the past 25 years (Glynn, 1993; Hughes et al., 2003; Hoegh-Guldberg et al., 2007), strongly associated in many cases with recurrent ENSO (El Niño – Southern Oscillation) events (Baker et al., 2008). Bleaching episodes have occurred almost annually in one or more of the world's tropical or subtropical seas, resulting in catastrophic loss of coral cover in some cases, and coral community structure shift in many others. Prolonged and severe events of bleaching may result in massive mortality of overheated corals (Hughes et al., 2003). Biochemical and physiological mechanisms of symbiosis breakdown was attributed to temperature or irradiance damage to the symbionts' photosynthetic machinery, resulting in the overproduction of oxygen radicals and cellular damage to hosts and/or symbionts (Lesser, 2006). Another somewhat controversial approach addresses bleaching episodes as an important ecological process that can ultimately help reef corals to survive future warming events in which corals get rid of suboptimal algae and acquire new symbionts. This point of view defines bleaching as a strategy that sacrifices short-term benefits of symbiosis for long-term advantage (Baker, 2001).

Temperature is also a key factor in ontogenetic development, and is known to affect different ontogenetic stages distinctively (Foster, 1971; Pechenik, 1989). Hence, increased temperature can affect the timing of ontogenic transitions, sometimes resulting in a temporal mismatch between larval development and key control factors like food supply or predation intensity. An example of this is the earlier spawning of the clam *Macoma balthica* in the Wadden Sea (northwestern Europe), but not to earlier spring phytoplankton blooms (Philippart et al., 2003).

Therefore, the period between spawning and maximum food supply was extended, and food availability during the pelagic phase reduced. Furthermore, predation intensity by juvenile shrimps on juvenile *Macoma* has also increased because of earlier recruitment of juvenile shrimp to the mud flats (Philippart et al., 2003). Trophic mismatch events are a potential severe consequence of temperature rise. A phenological study across three trophic levels using five functional groups in the North Sea showed different responses to temperature changes over the years 1958–2002 (Edwards and Richardson, 2004). Using this long-term data set of 66 plankton taxa, the authors demonstrated shifts in the timing and size of seasonal peaks of different populations, related to physiological (e.g., respiration, reproduction, mortality) or environmental (e.g., stratification) conditions. Such shifts can have profound consequences to community structure and stability, like in the case of the of North Sea cod stock declines implicated to worsen by key planktonic prey declines and shifts in their seasonality (Beaugrand et al., 2003, 2008), or in the case of the northern shrimp, *Pandalus Borealis*, and its temperature-dependant timing of egg-hatching, intended to match spring phytoplankton blooms (Greene et al., 2009). On rocky intertidal shores where upwelling prevails, mussel growth responds strongly to changes in water temperature associated with ENSO and PDO (Pacific Decadal Oscillation) cycles, suggesting potential community-level effects of climate change, as mussels have important ecological roles, serving as both food and habitat for a multitude of species on the shore (Menge et al., 2008).

Rising temperatures can potentially alter significant community-controlling interactors such as predators, competitors, ecosystem engineers, mutualists, or pathogens. The behavior of a keystone predator, the sea star *Pisaster ochraceus*, in the upwelling system off the US West Coast was followed by Sanford (1999) at different water temperatures and was shown to exhibit higher mid-intertidal abundance and increased consumption rates when exposed to slightly warmer waters. The author suggested that if water temperatures rise due to climate change, more intense predation might alter the vertical extent of the prey (habitat-forming mussels) and various species inhabiting its matrix and thus affect the community as a whole (Sanford, 1999). Global warming may also reduce predation, for example in the case of the Humboldt squid, *Dusidicus gigas*, a top predator in the eastern Pacific that exhibited lower metabolic rates and activity levels when exposed to high CO_2 concentrations and temperatures, thus affecting growth, reproduction, and survival of the squid and possibly impairing predator–prey interactions in the pelagic system (Rosa and Seibel, 2008).

Another important illustration of warming water effects is the change in benthic community structure near the thermal outfall of a power-generating station on the rocky coast of California. There, communities were greatly altered in apparently cascading responses to reduced abundances of habitat-forming species like subtidal kelps and intertidal red algae (Schiel et al., 2004). In contrast, grazers showed positive response to temperature, attributed by the authors to physiological tolerances, trophic responses, space availability, and recruitment dynamics (Schiel et al., 2004).

An example of what rapid ocean warming can do on regional and community scales can be seen in the mass mortality event of 25 rocky benthic macro-invertebrate species (mainly gorgonians and sponges) in the entire Northwestern Mediterranean region that followed a heat wave in Europe in 2003 (Garrabou et al., 2009). The heat wave caused an anomalous warming of seawater, which reached the highest temperatures ever recorded in the studied regions, between 1°C and 3°C above the climatic values (both mean and maximum). Such increases are certainly within the range of expected long-term global warming of the oceans, and the authors also suggest that heat waves may become more common in the future possibly driving a major biodiversity crisis in the Mediterranean Sea.

Local or regional mortality of species is but one aspect of global climate change. Water temperature rise has already shown to drive extensive *biogeographical shifts*, expressed mostly as poleward movement of species. Significant shifts were seen, for example, in marine fish populations in the North Sea, where nearly two thirds of the species shifted in latitude or depth or both over 25 years in correlation with warming waters (Perry et al., 2005). Another example is shift in the population dynamics of the sea urchin *Centrostephanus rodgersii* along the eastern Tasmanian coastline (Ling et al., 2009). Ling et al. (2009) revealed range extension through poleward larval dispersal via atmospheric-forced ocean warming and intensification and poleward advance of the East Australian Current (EAC). Shifts are also seen in the intertidal zone, which represents a unique situation as it is situated at the interface between the land and the sea and therefore species living there are expected to be influenced by changes in both water and air temperature. On the shore, species geographic distributions are expected to shrink or shift due to changes in thermal stress and ocean circulation either directly or indirectly through species interactions. Some species could be purged from the intertidal zone by alterations in water temperature, upwelling regime (Leslie et al., 2005), or oxygen levels (Grantham et al., 2003; Chan et al., 2008). Others may be squeezed out of the system when their upper limit is reduced to the upper limit of their consumers (Harley et al., 2003). Alternatively, some species may find that environmental conditions become physiologically tolerable at regions that were previously uninhabitable, or ocean circulation changes may bring distant species to new locations, resulting in range expansion. Indeed, long-term monitoring shows that the poleward-range edges of intertidal biota have shifted by as much as 50 km per decade in some regions (Helmuth et al., 2006b). Poleward range extension was documented in various intertidal species of invertebrates and algae (Herrlinger, 1981; Weslawski et al., 1997; Lohnhart and Tupen, 2001; Zacherl et al., 2003; Helmuth et al., 2006b; Mieszkowska et al., 2006). However, change in distribution due to thermal stress may not be a simple linear/longitudinal process. Helmuth et al. (2002, 2006a) have demonstrated that thermal stress on the rocky shore exhibit a mosaic of localized hotspots that do not necessarily follow latitudes. Thermal-stress hotspots are determined mainly by the timing of low-tide during summer spring tides. These low tides on the US West coast frequently occur at the hottest time of the day at the higher latitudes (Washington and Oregon) while they happen at night time further south (California). This means that increasing water temperature may facilitate the establishment of species invading from warmer waters in complex patterns along the shore, potentially affecting community structure and function in mosaic patterns.

The link between global warming and invasion of alien species is an obvious one, as warming can allow warm water species to extend to or invade previously nonhospitable regions (Occhipinti-Ambrogi, 2007). For example, the establishment of three abundant introduced ascidians on the shores of New England was explained by the strong positive correlation between their recruitment rates and rising winter sea temperatures in the region (Stachowicz et al., 2002). In the Mediterranean, one of the hottest hotspots of marine bioinvasions, warming events, and change in circulation patterns due to climate shifts (e.g., the Eastern Mediterranean Transient) in the past century have been suggested to facilitate invasions of tropical species (Rilov and Galil, 2009).

Climate change thermal effects are not just bound to coastal or seasurface environments, but they were also shown to impact deep sea ecosystems. For example, decadal nematode community surveys conducted in the Eastern Mediterranean revealed a significant increase in nematode abundance and diversity, which was related in this case to temperature *decrease* of 0.4°C (Dennavoro et al., 2004).

3. Ocean Acidification Effects

Ocean plays a substantial role in the storage of carbon dioxide emissions through the uptake of roughly half of the fraction released by human activities up to 1994 (Sabine et al., 2004), and about 30% of recent emissions (Feely et al., 2004). Nevertheless, this regulating effect does not come without a price – continuous CO_2 uptake is estimated to create pH reduction of 0.3–0.5 units over the next 100 years in the ocean surface (Caldeira and Wickett, 2003). This magnitude of acidification is higher than any other pH fluctuations inferred from the fossil record over the past 200–300 million years (Caldeira and Wickett, 2003). With a rate of change in pH that is 100 times greater than at any

Calcification and CO_2

Atmospheric CO_2 equilibrates rapidly with the surface layer of the ocean, where most additional CO_2 combines with carbonate ions (Gattuso and Buddemeier, 2000):

$$CO_2 + CO_3^{2-} + H_2O \rightarrow 2HCO_3^-$$

This leads to a decrease in the concentration of CO_3^{2-}, one of the building blocks of calcium carbonate, and in the saturation state of calcium carbonate, Ω ($\Omega = [Ca^{2+}] \times [CO_3^{2-}]/K_{sp}$, where K_{sp} is the equilibrium constant of $CaCO_3$). Ω seems to be the controlling factor of calcification (Marubini and Thake, 1999).

time over that period, marine organisms' tolerance and ability to adapt to it is challenged and considerable impacts on the ecology of marine ecosystems are bound to happen (Guinotte and Fabry, 2008). However, impacts of these chemical changes in the ocean are still poorly understood, especially at the community to ecosystem levels (Riebesell, 2008).

It has been shown that marine plants (except seagrasses) are carbon-saturated (Gattuso and Buddemeier, 2000), and hence, are not expected to increase growth rates due to elevated CO_2 concentrations. Therefore, dissolved CO_2 concentrations rise may lead, in some localities, to macroalgae replacement by seagrasses due to carbon-limitation variations stemming from different evolvement eras of these two functional groups (Harley et al., 2006).

Furthermore, pH reduction associated with increased CO_2 levels in seawater bears profound physiological consequences in subcellular processes such as protein synthesis and ion exchange, with a disproportional extent of effects among taxa (Portner et al., 2005). Ocean acidification can also have longer-term physiological, mechanical, and structural effects, especially on organisms that build carbonate structures. For example, pH reduction manipulations have demonstrated lower metabolic rates and growth in mussels (Michaelidis et al., 2005), which involved increased hemolymph bicarbonate levels (mainly from dissolution of shell $CaCO_3$) in order to limit hemolymph acidosis, a drop in oxygen consumption rate, and an increase in nitrogen excretion (indicating net protein degradation) correlated with a slowing of growth. Another study of pH manipulation demonstrated reduced growth and survivorship in gastropods and sea-urchins (Shirayama and Thornton, 2005).

Calcification rates themselves decreased in response to increased CO_2 in coccolithophorids, coralline algae, reef-building scleractinian corals, and pteropod mollusks (Kleypas et al., 1999; Riebesell et al., 2000; Feely et al., 2004). Using laboratory and mesocosm experiments on open ocean plankton, it was shown that a decrease in the carbonate saturation state represses biogenic calcification of dominant marine calcifying organisms such as foraminifera and coccolithophorids (Riebesell et al., 2000; Riebesell, 2004). On the ecosystem level, these responses influence phytoplankton species composition and succession, favoring algal species that predominantly rely on CO_2 utilization. In benthic communities, it was predicted that calcification rates in corals and coralline red algae are very likely to drop by 10–40% with a climatically realistic doubling of the pre-industrial partial pressure of CO_2 (Feely et al., 2004). Moreover, changes in ocean chemistry may cause weakening of the existing coral skeletons and reduce the accretion of reefs (Hughes et al., 2003). Recent work actually demonstrated a 14.2% decline in coral calcification of the massive reef-building coral *Porites* along the Great Barrier Reef since 1990 (De'ath et al., 2009). The authors suggest that this decline is attributed to the increase in temperature stress and decline in saturation state of seawater aragonite. Can some coral species cope to some degree with such effect or are they doomed? Recent work on the nonreef-building hard coral *Ocullina patagonica* demonstrated the existence of physiological refugia response mechanism, allowing corals to alternate between nonfossilizing soft-body ecophenotypes and fossilizing skeletal forms in response to changes in ocean chemistry (Fine and Tchernov, 2007).

Remarkably, some of the predictions regarding high latitude regions, where planktonic shelled pteropod gastropods constitute a prominent trophic component, suggest undersaturation with respect to aragonite even within the next 50 years that may cause the collapse of their populations (Orr et al., 2005). The collapse of populations of such major components in the polar food-web may alter the structure and biodiversity of polar ecosystems.

The potential ecosystem-scale effects of change in CO_2 and pH levels was recently demonstrated in Italy at shallow coastal sites near volcanic CO_2 vents (Hall-Spencer et al., 2008). Rocky shore sites near the vents with pH levels lower by 0.5 units than the mean ocean pH (ocean acidification levels predicted by 2100 by the IPCC) exhibited remarkable community-level effects. Along pH gradient ranging from 8.1–8.2 to 7.4–7.5, communities with abundant calcareous organisms shifted to communities lacking scleractinian corals and with significant reduction in abundance of sea urchin and coralline algae. The low pH communities exhibited peaking seagrass production with no indication of adaptation or replacement of sensitive species by others capable of filling the same ecological niche (Riebesell, 2008). Another study, this time from the Pacific Northwest shores of North America, suggests that reduced pH levels in nearshore seawater over the last decade was expressed in community-level effects in the rocky intertidal (Wootton et al., 2008). There, calcareous species generally preformed more poorly than noncalcareous species in years with low pH and thus have caused change in community structure.

Ocean chemistry changes and primarily ocean acidification is a poorly understood, yet potentially crucial, factor in climate change effects on marine environments at population, community, and ecosystem scales. Scientists predict that pH reduction through the twenty-first century will exceed any other documented pH fluctuations over the last 200–300 million years and thus would have profound consequences to organisms' physiology, growth, and survivorship, along with species distribution, abundance, and biogeography. Because acidification imposes a genuine threat on organisms' tolerance and ability to adapt to it, it should be recognized as an essential research target for conservation purposes in the following years.

4. Other Potential Climate Change Effects in the Oceans

Apart from temperature and pH, oceans are expected to alter in several other ways due to the current global climate change. Climate change is predicted to influence oceanographic patterns and conditions such as current direction and velocity, depth of stratification, salinity (fresher in the higher latitudes and more salty in the subtropics), and the oxygen concentration of the ventilated thermocline (IPCC, 2007). Climate change, for instance, is predicted to modify coastal upwelling either by intensifying (Bakun, 1990) or weakening it (Vecchi et al., 2006), depending on the model used. These changes are predicted to affect, for example, survivorship and delivery of propagules to the shore as well as food supply in coastal ecosystems. On rocky shores for instance, increasing upwelling intensity and duration in intermittent upwelling regions such as the Oregon coast during the summer will reduce sessile invertebrate larval recruitment (by moving the larval pool further offshore) lowering abundances of sessile invertebrates and through higher nutrient fluxes increase macrophytes, thus making rocky intertidal habitats in Oregon more similar to those in California (Menge et al., 2004). Alternatively, if upwelling is reduced, the structure of the seaweed assemblages will change, with decreases in Laminarians and likely some red algae, and enhanced abundances of sessile invertebrates (due to higher recruitment, see Connolly and Roughgarden, 1999).

Increasing sea levels will permanently submerge some intertidal areas while others might be created changing the mosaic of communities along the shore. In areas where tidal amplitudes are small, such as the Mediterranean Sea, sea-level rise can change the structure of communities because the ratio of vertical versus horizontal surfaces will probably change and communities on different rock aspects are different (Vaselli et al., 2008). In regions where most of the rocky shore is horizontal and at mean sea level, for example where vermetid platforms are found (warm temperate seas such as the eastern Mediterranean, Bermuda, Safriel, 1974), a rapid sea-level rise would cause an inundation of most of the intertidal zone by seawater, effectively turning the platforms into subtidal reefs. Based on measurements of sea-level rise for the eastern Mediterranean (~8.5 cm between 1992 and 2008) and projections for the next 100 years of up to a meter or more (Rosen, 2008), most of the Israeli rocky shore will be underwater and that unique ecosystem will be mostly lost.

Increasing storm intensity, including tropical storms (hurricanes, cyclones), will increase the frequency and severity of disturbance inflicted on coastal communities such as mangroves, coral reefs, and rocky shores. There is already evidence that a progressive decadal increase in deep-water wave heights and periods have increased breaker heights and elevated storm wave run-up levels on beaches in the US Pacific Northwest (Allan and Komar, 2006). This of course can have substantial effects on disturbance regimes on the shore that surely will affect the structure of coastal ecological communities (Dayton and Tegner, 1984; Underwood, 1998). Larger, stronger storms are also expected to increase beach erosion. The resultant increased erosion of the shore can also affect coastal geomorphology, increase sedimentation, and therefore affect the ecology of the shore. A study on the Oregon shore that looked at effects of a cliff collapse (and with it highway 101) and reconstruction showed how rocky intertidal communities have been altered due to change in small-scale geomorphology and possibly sediment accumulation on the shore (Rilov, unpublished data).

The rate at which physical changes caused by global climate change might unfold could be slow, but they could also be fast and therefore their manifestation in the structure of communities and in biodiversity could be strong and immediate. For example, the onset of hypoxia on the shelf of Oregon coast in 2002 was relatively sudden and unprecedented (Grantham et al., 2004; Chan et al., 2008). Hypoxic conditions have since re-occurred each summer, and greatly intensified such that conditions were anoxic in 2006 (Chan et al., 2008). This resulted in massive die-offs of benthic invertebrates (e.g., crabs, sea stars) and the dwindling of reef fishes on subtidal reefs (Service, 2004). The delayed upwelling observed in 2005 on the Oregon coast (Barth et al., 2007) had not been observed in at least the previous 20 years, and had immediate consequences for concentrations of phytoplankton and larvae of sessile invertebrates. These intense coastal events seem to correspond with larger-scale oceanographic and atmospheric changes in the North Pacific that are consistent with global climate-change scenarios (e.g., Hooff and Peterson, 2006; Barth et al., 2007) and may linger and therefore have profound ecological effects on regional and potentially global scales.

5. Predictions and Projections

In the past few years, there has been great effort to develop conceptual and numerical models that aim to forecast the ecological and economical impacts of climate change on marine ecosystems. There are a dozen such projections in the current literature of which we will mention only a few examples. Paleontological studies of marine ecosystems can also aid in predicting how certain changes in ocean conditions might affect species and ecological communities. For example, Yasuhara et al. (2008) show how deep-sea benthic ecosystems communities collapsed several times during the past 20,000 years in correspondence with rapid climatic changes that lasted over centuries or less, demonstrating that climate change can have profound effects in the deep ocean and should therefore be considered in the current models. Scientists now attempt to model effects on local, regional, oceanic, or even whole-planet scales, depending on the question and information at hand.

On oceanic scales, a multispecies, functional group, coupled ocean-atmosphere model that examined mostly primary producers' response to regional biogeochemical conditions suggests significant changes by the end of this century in ecosystem structure, caused mostly by shifts in the areal extent of biomes (Boyd and Doney, 2002). Whitehead et al. (2008) examined the response of the other end of the food chain, deep-sea cetaceans, by studying their current distribution patterns in relation with sea-surface temperature, and concluded that climate change will cause declines of cetacean diversity across the tropics and increases at higher latitudes. In the coastal environment, the large, brackish, semi-enclosed Baltic Sea is predicted to freshen (owing to altered precipitation patterns) and warm up resulting in a shift in biodiversity due to the contraction of more marine species out of the system and the expansion of more freshwater species (Mackenzie et al., 2007). In several major US bays, Galbraith et al. (2002) predict that even with conservative estimates of climate change, sea-level rise will cause losses of intertidal areas that range between 20% and 70% of the current intertidal habitat that support extensive populations of migrating and wintering shorebirds. Such losses could considerably reduce the ability of these bays to support their present shorebird numbers.

Recent extensive reviews also attempt to predict the consequences of global climate change to marine ecosystems at different regions. Australia's marine life is projected to change considerably due to the multitude of present and future effects of climate change with the most serious and worrisome effects inflicted on the unique system of the Great Barrier Reef (Poloczanska et al., 2007). The small but highly diverse Mediterranean Sea is projected to transform its biological diversity due (among many other things) to climate change (Gambaiani et al., 2009). On Antarctic coasts, Smale and Barnes (2008) predict that the intensity of ice scouring will increase and later sedimentation and freshening events will become important, all leading to increased disturbance and considerable changes in benthic community structure and species distributions. And the list goes on.

Climate change is of course but one process by which humans are affecting marine biodiversity. To it, we can add invasions of alien species (that can be accelerated by climate change) and of course over-harvesting, pollution, and habitat destruction. Many of these threats may act in synergy and produce changes in biodiversity that are more pervasive than those caused by single disturbances (Sala and Knowlton, 2006). What then is the future of marine biodiversity in light of all these threats? Extinctions that are already happening will probably accelerate and the homogenization of communities due to climate effects and invasions will reduce the uniqueness of ecosystems on a global scale. Even if the current trends of destruction reverse at some point in the near future, recovery of individual species that were at the brink may take longer than expected because of Allee effects, changes in trophic community structure, difficult-to-reverse habitat changes, or a combination of several factors (Sala and Knowlton, 2006). Recovery of diversity at the community level will probably take much longer. Although the future seems grim for global biodiversity, both terrestrial and marine, we wish to conclude with a positive note that suggests that perhaps not all is doomed. Ehrlich and Pringle (2008) propose several strategies that, "if implemented soundly and scaled up dramatically, would preserve a substantial portion of global biodiversity." Those strategies include stabilization of human population, reduction of material consumption, the deployment of endowment funds, and taking major steps toward conservation using large, permanent, protected areas. This of course will require tremendous vision, effort, and mostly will by our species; however, mankind faced great challenges in the past and prevailed, and so we can only hope that it will rise again to face this climate change and biodiversity challenge.

6. References

Allan, J.C. and Komar, P.D. (2006) Climate controls on US West Coast erosion processes. J. Coastal Res. **22**: 511–529.

Baker, A.C. (2001) Ecosystems – reef corals bleach to survive change. Nature **411**: 765–766.

Baker, A.C., Glynn, P.W. and Riegl, B. (2008) Climate change and coral reef bleaching: an ecological assessment of long-term impacts, recovery trends and future outlook. Estuar. Coast. Shelf Sci. **80**: 435–471.

Bakun, A. (1990) Global climate change and intensification of coastal ocean upwelling. Science **247**: 198–201.

Barth, J.A., Menge, B.A., Lubchenco, J., Chan, F., Bane, J.M., Kirincich, A.R., McManus, M.A., Nielsen, K.J., Pierce, S.D. and Washburn, L. (2007) Delayed upwelling alters nearshore coastal ocean ecosystems in the northern California current. Proc. Natl. Acad. Sci. U.S.A. **104**: 3719–3724.

Beaugrand, G., Brander, K.M., Lindley, J.A., Souissi, S. and Reid, P.C. (2003) Plankton effect on cod recruitment in the North Sea. Nature **426**: 661–664.

Beaugrand, G., Edwards, M., Brander, K., Luczak, C. and Ibanez, F. (2008) Causes and projections of abrupt climate-driven ecosystem shifts in the North Atlantic. Ecol. Lett. **11**: 1157–1168.

Boyd, P.W. and Doney, S.C. (2002) Modelling regional responses by marine pelagic ecosystems to global climate change. Geophys. Res. Lett. **29**(16): 1806.

Caldeira, K. and Wickett, M.E. (2003) Anthropogenic carbon and ocean pH. Nature **425**: 365–365.

Chan, F., Barth, J.A., Lubchenco, J., Kirincich, A., Weeks, H., Peterson, W.T. and Menge, B.A. (2008) Emergence of anoxia in the California current large marine ecosystem. Science **319**: 920.

Connolly, S.R. and Roughgarden, J. (1999) Theory of marine communities: competition, predation, and recruitment-dependent interaction strength. Ecol. Monogr. **69**: 277–296.

Dayton, P.K. and Tegner, M.J. (1984) Catastrophic storms, El-Nino, and patch tability in a Southern-California Kelp community. Science **224**: 283–285.

De'ath, G., Lough, J.M. and Fabricius, K.E. (2009) Declining coral calcification on the Great Barrier Reef. Science **323**: 116–119.

Dennavoro, R., Dell'Anno, A. and Pusceddu, A. (2004) Biodiversity response to climate change in a warm deep sea. Ecol. Lett. **7**: 821–828.

Edwards, M. and Richardson, A.J. (2004) Impact of climate change on marine pelagic phenology and trophic mismatch. Nature **430**: 881–884.

Ehrlich, P.R. and Pringle, R.M. (2008) Where does biodiversity go from here? A grim business-as-usual forecast and a hopeful portfolio of partial solutions. Proc. Natl. Acad. Sci. U.S.A. **105**: 11579–11586.

Feely, R.A., Sabine, C.L., Lee, K., Berelson, W., Kleypas, J., Fabry, V.J. and Millero, F.J. (2004) Impact of anthropogenic CO_2 on the $CaCO_3$ system in the oceans. Science **305**: 362–366.

Fine, M. and Tchernov, D. (2007) Scleractinian coral species survive and recover from decalcification. Science **315**: 1811–1811.

Foster, B.A. (1971) On the determinants of the upper limit of intertidal distribution of barnacles (Crustacea: Cirripedia). J. Anim. Ecol. **40**: 33–48.

Galbraith, H., Jones, R., Park, R., Clough, J., Herrod-Julius, S., Harrington, B. and Page, G. (2002) Global climate change and sea level rise: potential losses of intertidal habitat for shorebirds. Waterbirds **25**: 173–183.

Gambaiani, D.D., Mayol, P., Isaac, S.J. and Simmonds, M.P. (2009) Potential impacts of climate change and greenhouse gas emissions on Mediterranean marine ecosystems and cetaceans. J. Mar. Biol. Assoc. U.K. **89**: 179–201.

Garrabou, J., Coma, R., Bensoussan, N., Bally, M., Chevaldonne, P., Cigliano, M., Diaz, D., Harmelin, J.G., Gambi, M.C., Kersting, D.K., Ledoux, J.B., Lejeusne, C., Linares, C., Marschal, C., Perez, T., Ribes, M., Romano, J.C., Serrano, E., Teixido, N., Torrents, O., Zabala, M., Zuberer, F. and Cerrano, C. (2009) Mass mortality in Northwestern Mediterranean rocky benthic communities: effects of the 2003 heat wave. Glob. Change Biol. **15**: 1090–1103.

Gattuso, J.P. and Buddemeier, R.W. (2000) Ocean biogeochemistry – calcification and CO_2. Nature **407**: 311–313.

Glynn, P.W. (1993) Coral reef bleaching: ecological perspectives. Coral Reefs **12**: 1–17.

Grantham, B.A., Chan, F., Nielsen, K.J., Fox, D.S., Barth, J.A., Huyer, A., Lubchenco, J. and Menge, B.A. (2004) Upwelling-driven nearshore hypoxia signals ecosystem and oceanographic changes in the northeast Pacific. Nature **429**: 749–754.

Grantham, B.A., Eckert, G.L. and Shanks, A.L. (2003) Dispersal potential of marine invertebrates in diverse habitats. Ecol. Appl. **13**: S108–S116.

Greene, C.H., Monger, B.C. and McGarry, L.P. (2009) Some like it cold. Science **324**: 733–734.

Guinotte, J.M. and Fabry, V.J. (2008) Ocean acidification and its potential effects on marine ecosystems. Ann. N.Y. Acad. Sci. **1134**: 320–342.

Hall-Spencer, J.M., Rodolfo-Metalpa, R., Martin, S., Ransome, E., Fine, M., Turner, S.M., Rowley, S.J., Tedesco, D. and Buia, M.C. (2008) Volcanic carbon dioxide vents show ecosystem effects of ocean acidification. Nature **454**: 96–99.

Harley, C.D.G., Hughes, A.R., Hultgren, K.M., Miner, B.G., Sorte, C.J.B., Thornber, C.S., Rodriguez, L.F., Tomanek, L. and Williams, S.L. (2006) The impacts of climate change in coastal marine systems. Ecol Lett. **9**: 228–241.

Harley, C.D.G., Smith, K.F. and Moore, V.L. (2003) Environmental variability and biogeography: the relationship between bathymetric distribution and geographical range size in marine algae and gastropods. Glob. Ecol. Biogeogr. **12**: 499–506.

Helmuth, B., Broitman, B.R., Blanchette, C.A., Gilman, S., Halpin, P., Harley, C.D.G., O'Donnell, M.J., Hofmann, G.E., Menge, B. and Strickland, D. (2006a) Mosaic patterns of thermal stress in the rocky intertidal zone: implications for climate change. Ecol. Monogr. **76**: 461–479.

Helmuth, B., Harley, C.D.G., Halpin, P.M., O'Donnell, M., Hofmann, G.E. and Blanchette, C.A. (2002) Climate change and latitudinal patterns of intertidal thermal stress. Science **298**: 1015–1017.

Helmuth, B., Mieszkowska, N., Moore, P. and Hawkins, S.J. (2006b) Living on the edge of two changing worlds: forecasting the responses of rocky intertidal ecosystems to climate change. Annu. Rev. Ecol. Evol. Syst. **37**: 373–404.

Herrlinger, T.J. (1981) Range extension of *Kelletia kelletii*. Veliger **24**: 78.

Hochachka, P.W. and Somero, G.N. (2002) Biochemical adaptation: mechanism and process in physiological evolution. Biochem. Adapt. Mec. Proc. Physiol. Evol. **i–xi**: 1–466.

Hoegh-Guldberg, O., Mumby, P.J., Hooten, A.J., Steneck, R.S., Greenfield, P., Gomez, E., Harvell, C.D., Sale, P.F., Edwards, A.J., Caldeira, K., Knowlton, N., Eakin, C.M., Iglesias-Prieto, R., Muthiga, N., Bradbury, R.H., Dubi, A. and Hatziolos, M.E. (2007) Coral reefs under rapid climate change and ocean acidification. Science **318**: 1737–1742.

Hooff, R.C. and Peterson, W.T. (2006) Copepod biodiversity as an indicator of changes in ocean and climate conditions of the northern California current ecosystem. Limnol. Oceanogr. **51**: 2607–2620.

Hughes, T.P., Baird, A.H., Bellwood, D.R., Card, M., Connolly, S.R., Folke, C., Grosberg, R., Hoegh-Guldberg, O., Jackson, J.B.C., Kleypas, J., Lough, J.M., Marshall, P., Nystrom, M., Palumbi, S.R., Pandolfi, J.M., Rosen, B. and Roughgarden, J. (2003) Climate change, human impacts, and the resilience of coral reefs. Science **301**: 929–933.

IPCC (2007) *Intergovernmental Panel on Climate Change Report – Technical Summary*, p. 74.

Kleypas, J.A., Buddemeier, R.W., Archer, D., Gattuso, J.P., Langdon, C. and Opdyke, B.N. (1999) Geochemical consequences of increased atmospheric carbon dioxide on coral reefs. Science **284**: 118–120.

Leslie, H.M., Breck, E.N., Chan, F., Lubchenco, J. and Menge, B.A. (2005) Barnacle reproductive hotspots linked to nearshore ocean conditions. Proc. Natl. Acad. Sci. U.S.A. **102**: 10534–10539.

Lesser, M.P. (2006) Oxidative stress in marine environments: biochemistry and physiological ecology. Annu. Rev. Physiol. **68**: 253–278.

Ling, S.D., Johnson, C.R., Ridgway, K., Hobday, A.J. and Haddon, M. (2009) Climate-driven range extension of a sea urchin: inferring future trends by analysis of recent population dynamics. Glob. Change Biol. **15**: 719–731.

Lohnhart, S.A. and Tupen, J.W. (2001) New range records of 12 marine invertebrates: the role of El Nino and other mechanisms in southern and central California. Bull. Southern Calif. Acad. Sci. **100**: 238–248.

Lubchenco, J., Navarrete, S.A., Tissot, B.N. and Castilla, J.C. (1993) Possible ecological responses to global climate change: nearshore benthic biota of northeastern Pacific coastal ecosystems, In: H.A. Mooney, E.R. Fuentes and B.I. Kronberg (eds.) *Earth System Responses to Global Change*. Academic Press, San Diego, CA, pp. 147–166.

Mackenzie, B.R., Gislason, H., Mollmann, C. and Koster, F.W. (2007) Impact of 21st century climate change on the Baltic Sea fish community and fisheries. Glob. Change Biol. **13**: 1348–1367.

Marubini, F. and Thake, B. (1999) Bicarbonate addition promotes coral growth. Limnol. Oceanogr. **44**: 716–720.

McWilliams, J.P., Cote, I.M., Gill, J.A., Sutherland, W.J. and Watkinson, A.R. (2005) Accelerating impacts of temperature-induced coral bleaching in the Caribbean. Ecology **86**: 2055–2060.

Menge, B.A., Blanchette, C., Raimondi, P., Freidenburg, T., Gaines, S., Lubchenco, J., Lohse, D., Hudson, G., Foley, M. and Pamplin, J. (2004) Species interaction strength: testing model predictions along an upwelling gradient. Ecol. Monogr. **74**: 663–684.

Menge, B.A., Chan, F. and Lubchenco, J. (2008) Response of a rocky intertidal ecosystem engineer and community dominant to climate change. Ecol. Lett. **11**: 151–162.

Michaelidis, B., Ouzounis, C., Paleras, A. and Portner, H.O. (2005) Effects of long-term moderate hypercapnia on acid–base balance and growth rate in marine mussels *Mytilus galloprovincialis*. Mar. Ecol-Prog. Ser. **293**: 109–118.

Mieszkowska, N., Kendall, M.A., Hawkins, S.J., Leaper, R., Williamson, P., Hardman-Mountford, N.J. and Southward, A.J. (2006) Changes in the range of some common rocky shore species in Britain – a response to climate change? Hydrobiologia **555**: 241–251.

Occhipinti-Ambrogi, A. (2007) Global change and marine communities: alien species and climate change. Mar. Pollut. Bull. **55**: 342–352.

Orr, J.C., Fabry, V.J., Aumont, O., Bopp, L., Doney, S.C., Feely, R.A., Gnanadesikan, A., Gruber, N., Ishida, A., Joos, F., Key, R.M., Lindsay, K., Maier-Reimer, E., Matear, R., Monfray, P., Mouchet, A., Najjar, R.G., Plattner, G.K., Rodgers, K.B., Sabine, C.L., Sarmiento, J.L., Schlitzer, R., Slater, R.D., Totterdell, I.J., Weirig, M.F., Yamanaka, Y. and Yool, A. (2005) Anthropogenic ocean acidification over the twenty-first century and its impact on calcifying organisms. Nature **437**: 681–686.

Pechenik, J.A. (1989) Environmental influences on larval survival and development, In: A.C. Giese, J.S. Pearse and V.B. Pearse (eds.) *Reproduction of Marine Invertebrates*. Blackwell Scientific Publications, Palo Alto, CA, pp. 551–608.

Perry, A.L., Low, P.J., Ellis, J.R. and Reynolds, J.D. (2005) Climate change and distribution shifts in marine fishes. Science **308**: 1912–1915.

Philippart, C.J.M., van Aken, H.M., Beukema, J.J., Bos, O.G., Cadee, G.C. and Dekker, R. (2003) Climate-related changes in recruitment of the bivalve *Macoma balthica*. Limnol. Oceanogr. **48**: 2171–2185.

Poloczanska, E.S., Babcock, R.C., Butler, A., Hobday, A., Hoegh-Guldberg, O., Kunz, T.J., Matear, R., Milton, D.A., Okey, T.A. and Richardson, A.J. (2007) Climate change and Australian marine life. Oceanogr. Mar. Biol. **45**(45): 407–478.

Portner, H.O., Langenbuch, M. and Michaelidis, B. (2005) Synergistic effects of temperature extremes, hypoxia, and increases in CO_2 on marine animals: from Earth history to global change. J. Geophys. Res.-Oceans **110**: C09–S09.

Riebesell, U. (2004) Effects of CO_2 enrichment on marine phytoplankton. J. Oceanogr. **60**: 719–729.

Riebesell, U. (2008) Climate change – acid test for marine biodiversity. Nature **454**: 46–47.

Riebesell, U., Zondervan, I., Rost, B., Tortell, P.D., Zeebe, R.E. and Morel, F.M.M. (2000) Reduced calcification of marine plankton in response to increased atmospheric CO_2. Nature **407**: 364–367.

Rilov, G. and Galil, B. (2009) Marine bioinvasions in the Mediterranean Sea – history, distribution and ecology, In: G. Rilov and J.A. Crooks (eds.) *Biological Invasions in Marine Ecosystems: Ecological, Management, and Geographic Perspectives*. Springer-Verlag, Heidelberg, Germany, pp. 3–11.

Rosa, R. and Seibel, B.A. (2008) Synergistic effects of climate-related variables suggest future physiological impairment in a top oceanic predator. Proc. Natl. Acad. Sci. **105**: 20776–20780.

Rosen, D. (2008) Monitoring boundary conditions at Mediterranean Basin – key element for reliable assessment of climate change, variability and impacts at Mediterranean basin shores, In: *Towards an Integrated System of Mediterranean Marine Observatories, CISEM Workshop*, La Spezia, pp. 107–111.

Sabine, C.L., Feely, R.A., Gruber, N., Key, R.M., Lee, K., Bullister, J.L., Wanninkhof, R., Wong, C.S., Wallace, D.W.R., Tilbrook, B., Millero, F.J., Peng, T.H., Kozyr, A., Ono, T. and Rios, A.F. (2004) The oceanic sink for anthropogenic CO_2. Science **305**: 367–371.

Safriel, U.N. (1974) Vermetid gastropods and intertidal reefs in Israel and Bermuda. Science **186**: 1113–1115.

Sala, E. and Knowlton, N. (2006) Global marine biodiversity trends. Annu. Rev. Env. Resour. **31**: 93–122.

Sanford, E. (1999) Regulation of keystone predation by small changes in ocean temperature. Science **283**: 2095–2097.

Schiel, D.R., Steinbeck, J.R. and Foster, M.S. (2004) Ten years of induced ocean warming causes comprehensive changes in marine benthic communities. Ecology **85**: 1833–1839.

Service R.F. (2004) Oceanography – new dead zone off Oregon coast hints at sea change in currents. Science **305**: 1099–1099.

Shirayama, Y. and Thornton, H. (2005) Effect of increased atmospheric CO_2 on shallow water marine benthos. J. Geophys. Res.-Oceans **110**.

Smale, D.A. and Barnes, D.K.A. (2008) Likely responses of the Antarctic benthos to climate-related changes in physical disturbance during the 21st century, based primarily on evidence from the West Antarctic Peninsula region. Ecography **31**: 289–305.

Somero, G.N. (2002) Thermal limits to life: underlying mechanisms and adaptive plasticity. Integr. Comp. Biol. **42**: 1316–1316.

Stachowicz, J.J., Terwin, J.R., Whitlatch, R.B. and Osman, R.W. (2002) Linking climate change and biological invasions: ocean warming facilitates nonindigenous species invasions. Proc. Natl. Acad. Sci. U.S.A. **99**: 15497–15500.

Stillman, J.H. (2002) Causes and consequences of thermal tolerance limits in rocky intertidal porcelain crabs, genus Petrolisthes. Integr. Comp. Biol. **42**: 790–796.

Tomanek, L. and Somero, G.N. (1999) Evolutionary and acclimation-induced variation in the heat-shock responses of congeneric marine snails (genus Tegula) from different thermal habitats: implications for limits of thermotolerance and biogeography. J. Exp. Biol. **202**: 2925–2936.

Underwood, A.J. (1998) Grazing and disturbance: an experimental analysis of patchiness in recovery from a severe storm by the intertidal alga *Hormosira banksii* on rocky shores in New South Wales. J. Exp. Mar. Biol. Ecol. **231**: 291–306.

Vaselli, S., Bertocci, I., Maggi, E. and Benedetti-Cecchi, L. (2008) Assessing the consequences of sea level rise: effects of changes in the slope of the substratum on sessile assemblages of rocky seashores. Mar. Ecol-Prog. Ser. **368**: 9–22.

Vecchi, G.A., Soden, B.J., Wittenberg, A.T., Held, I.M., Leetmaa, A. and Harrison, M.J. (2006) Weakening of tropical Pacific atmospheric circulation due to anthropogenic forcing. Nature **441**: 73–76.

Walther, G.R., Berger, S. and Sykes, M.T. (2005) An ecological 'footprint' of climate change. Proc. R. Soc. B – Biol. Sci. **272**: 1427–1432.

Walther, G.R., Post, E., Convey, P., Menzel, A., Parmesan, C., Beebee, T.J.C., Fromentin, J.M., Hoegh-Guldberg, O. and Bairlein, F. (2002) Ecological responses to recent climate change. Nature **416**: 389–395.

Weslawski, J.W., Zajaczkowski, M., Wiktor, J. and Szymelfenig, M. (1997) Intertidal zone of Svalbard. 3. Littoral of a subarctic, oceanic island: Bjornoeya. Polar Biol. **18**: 45–52.

Whitehead, H., McGill, B. and Worm, B. (2008) Diversity of deep-water cetaceans in relation to temperature: implications for ocean warming. Ecol. Lett. **11**: 1198–1207.

Wootton, J.T., Pfister, C.A. and Forester, J.D. (2008) Dynamic patterns and ecological impacts of declining ocean pH in a high-resolution multi-year dataset. Proc. Natl. Acad. Sci. U.S.A. **105**: 18848–18853.

Yasuhara, M., Cronin, T.M., deMenocal, P.B., Okahashi, H. and Linsley, B.K. (2008) Abrupt climate change and collapse of deep-sea ecosystems. Proc. Natl. Acad. Sci. U.S.A. **105**: 1556–1560.

Zacherl, D., Gaines, S.D. and Lonhart, S.I. (2003) The limits to biogeographical distributions: insights from the northward range extension of the marine snail, Kelletia kelletii (Forbes, 1852). J. Biogeogr. **30**: 913–924.

Biodata of **Ugarte, R.A., Craigie, J.S.,** and **Critchley, A.T.,** authors of *"Fucoid Flora of the Rocky Intertidal of the Canadian Maritimes: Implications for the Future with Rapid Climate Change"*

Dr. Raul A. Ugarte graduated as Marine Biologist from the Universidad de Concepción in Chile in 1982. He worked on the establishment of Gracilaria farms in southern Chile until 1984 and later joined the laboratory of Dr. Bernabé Santelices at the Universidad Católica de Chile in Santiago, where he worked on applied phycology until 1988. Later that year, he traveled to Canada for a training program on resource management with the Department of Fisheries and Oceans (DFO), under the direction of Dr. John Pringle and Glyn Sharp.

In 1990, he enrolled in a graduate program at Dalhousie University in Halifax, Canada, where he obtained his Ph.D. in 1994. In 1995, Raul joined Acadian Seaplants Limited, the largest independent seaweed processing company in Canada, where he remains as a research scientist until today. Dr. Ugarte is responsible for the extensive annual stock assessment program to evaluate biomass of the resource Ascophyllum nodosum along an extension of more than 2,500 km of shoreline under the company's responsibility in the Maritime Region of eastern Canada. His responsibilities with this resource also include research on habitat impact of commercial harvesting, population dynamics, and ecological research as well as the biomass assessment of other economically important seaweed species (e.g., Chondrus, Alaria, Laminaria, etc.). His work, spanning more than 20 years in the rocky intertidal of the Maritimes, has given him a unique insight into the distribution and abundance of the seaweed flora of the region.

Dr. Ugarte currently lives in Rothesay, New Brunswick, Canada.

E-mail: **rugarte@acadian.ca**

Dr. James S. Craigie currently is Researcher Emeritus, National Research Council of Canada, and Science Advisor for Acadian Seaplants Limited. He obtained his Ph.D. in 1959 from Queen's University, Kingston, ON, Canada. Additional research and studies were continued at CNRA, Versailles, and at the University College of Wales, Swansea. Dr. Craigie returned to Canada in 1960 to accept a phycology position at the Atlantic Regional Laboratory (Institute for Marine Biosciences), National Research Council of Canada, Halifax, NS. He was a Visiting Scholar 1967–1968 at the Scripps Institution of Oceanography, UC San Diego.

Dr. Craigie is a chartered member and past president of the Canadian Society of Plant Physiologists, and has served as Editor of the *Journal of Phycology* and as Associate Editor of the *Journal of the World Aquaculture Society*. His interests encompass algal aquaculture, production, primary and secondary metabolites including polysaccharides and polyphenols. He developed and taught graduate level courses in marine plant biochemistry and physiology (Biology and Oceanography) at Dalhousie University from 1964 to 2000. His research contributions have been recognized through numerous publications and awards including the Darbaker Award and Prize, National Research Council of Canada Industrial Partnership Award, the Marinalg International Honorary Certificate, Federal Partners in Technology Transfer Innovator of the Year Award, the Queen Elizabeth II Golden Jubilee Award, the Bionova Nova Scotia Award of Excellence, and the Phycological Society of America Award of Excellence. He continues to conduct research and mentor staff at Acadian Seaplants Limited and the Institute for Marine Biosciences in Halifax.

E-mail: **James.Craigie@nrc-cnrc.gc.ca**

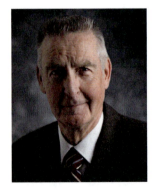

Alan T. Critchley is a reformed Academic. He graduated from Portsmouth Polytechnic, UK, and had a university career in southern Africa teaching phycology, marine ecology, and botany (KwaZulu Natal, Wits, and Namibia). He moved to the "dark side" in 2001 and took up a position in a multinational industry with Degussa Texturant Systems (now Cargill TS), where he was responsible for new raw materials for the extraction of the commercial colloid carrageenan. Since 2005, he has worked as vice president, Research for Acadian Seaplants Limited, working on value addition to seaweed extracts and on-land cultivation of seaweed for food and bioactive compounds. Not able to turn his back on the academic world entirely, he is presently adjunct professor at the Nova Scotia Agricultural College.

E-mail: **Alan.Critchley@acadian.ca**

FUCOID FLORA OF THE ROCKY INTERTIDAL OF THE CANADIAN MARITIMES: IMPLICATIONS FOR THE FUTURE WITH RAPID CLIMATE CHANGE

RAUL A. UGARTE[1], JAMES S. CRAIGIE[2], AND ALAN T. CRITCHLEY[1]

[1]*Acadian Seaplants Limited, 30 Brown Avenue, Dartmouth B3B1X8, Nova Scotia, Canada*
[2]*National Research Council of Canada, Institute for Marine Biosciences, 1411 Oxford Street, Halifax B3H 3Z1, NS, Canada*

Climate is recognized as a key driver in determining the distribution and ultimate geographical boundaries for both terrestrial and marine plant species. The western north Atlantic environment encompasses the Arctic, cold temperate, and the warm temperate Carolina regions, the first two of which influence and control the species distribution of marine flora in the Canadian Maritime Provinces. The southern limit for the Arctic and cold temperate seaweeds is along the coastline from Cape Cod, MA, to Long Island Sound. Lüning (1990) presented an extensive analysis and discussion of the limits of both intertidal and sublittoral seaweed speciation and distribution in the north Atlantic in relation to ecophysiological factors such as temperature, degree of exposure, salinity, and available light.

Fluctuations in the surface temperature of the Earth over the last 250,000–350,000 years have been reflected in cyclic changes in the greenhouse gases such as methane and carbon dioxide trapped in glacial ice (Ruddiman, 2005). Approximately 5,000 years ago, the cyclic decline in methane concentration expected for the present interglacial period was interrupted leading Ruddiman to suggest that human activity may have begun to influence the climate much earlier than had been generally considered. With the improved records over the past 150 years, trends showing increases in both atmospheric and water temperatures that correlate to human activities during the industrial age are developing (Collins et al., 2007). According to the International Panel on Climate Change – IPCC (2007), the temperature of the planet has increased $0.65 \pm 0.15°C$ between 1956 and 2005, and the ecological impact of such a climate change is already being documented worldwide in every ocean and in most major terrestrial and aquatic taxonomic groups (Parmesan, 2006). Although early in the projected trends of global warming, ecological responses to recent climate change are already clearly visible (Gian-Reto et al., 2002), meta-analyses of studies done for more than 1,400 marine and terrestrial species have demonstrated that the current increased temperature is already affecting 40% of them and that 82.3% of these species are shifting in the direction expected according to their physiological constraints (Root et al., 2003). An increase in the species diversity of the fish fauna has been already detected in the North Sea and has been related to climate change (Hiddink and Hofstede, 2008). These kinds of changes would not only impact the marine biodiversity but are expected to produce local extinctions in the subpolar regions, the tropics, and semi-enclosed seas (Cheung et al., 2008a, b).

Few reports on the effects of current climate change on macroalgae have been found in the literature, but these describe significant changes in the distribution and abundance of algae. For example, Pedersen et al. (2008) documented a significant change in seaweed community structure and related this to an increase in seawater temperature during a 28-month study in the littoral zone in Long Island Sound. Sagarin et al. (1999) reported a massive decline in *Pelvetia compressa* cover by revisiting a location in Baja California that was well monitored 60 years earlier. Simkanin et al. (2005), who revisited 63 locations along the coast of Ireland after a lapse of 45 years, also reported significant changes in the abundance of several seaweed species in the sublittoral zone. Although they were cautious about associating these changes with climate change, long-term trends seen in these kinds of surveys can be obscured by short-term fluctuations in species composition. However, Berecibar et al. (2004), by revisiting several locations after 45 years, clearly demonstrated that the phytogeographic regions of the intertidal seaweed community along the Portuguese coast have shifted northward in concert with an observed increase in surface seawater temperature (SST).

If such changes are now being detected when the global climate has warmed by an estimated average of 0.65°C in the last half-century, the effects on species and ecosystems will be obviously more drastic in response to a change in temperature as high as 6°C by 2100 as predicted (IPCC1, 2007).

Fucoid species, particularly the brown seaweed *Ascophyllum nodosum* (rockweed), are key habitat formers and energy producers, and their responses to climate change can have significant population, community, and even ecosystem consequences in the Canadian Maritime Provinces. From the economic point of view, these changes could seriously affect a 50-year-old seaweed industry that currently provides hundreds of jobs and injects millions of dollars into the local economy.

The objective of this chapter is to outline the potential changes in the fucoid flora of the Maritime Provinces that could result from the predicted increases in air and SST due to future climate change. The predictions take into consideration the current biological and ecological information of these species in the region.

1. The Canadian Maritimes

The Maritime Provinces of Canada include New Brunswick (NB), Nova Scotia (NS), and Prince Edward Island (PEI) and are zoned as a cold temperate Atlantic-boreal region (van den Hoek, 1975). These Provinces front the Atlantic Ocean and its various sub-basins such as the Gulf of Maine, the Bay of Fundy, and the Gulf of St. Lawrence (Fig. 1). Each basin presents different oceanographic conditions with the consequent temperature regimes. The SST of the Nova Scotian shore open to the Atlantic experiences average temperatures of 0.4°C during the winter and 14.5°C during the summer (range −1.4°C to 21.5°C). The Gulf of Maine directly influences the shoreline of southwestern NS to provide a summer average SST of 13.9°C and a winter average of 2.0°C (overall range 22.6–1.1°C). However, shallow embayments along the southwestern and southern shores of NS normally freeze over during the winter and experience higher temperatures than those of the open coast during the summer. The Bay of Fundy, due to its depth and exceptional tidal range with the resultant deep mixing, is the coldest basin in the region. Its maximum SST never rises above 15°C during the summer, but it also rarely experiences freezing during the winter. Finally, the Gulf of Saint Lawrence is a relatively shallow basin where SST has a much higher variation. Average winter SST in the Gulf is −1.7°C with a solid layer of ice covering much of the Gulf from January to mid-April, whereas, in August it reaches an average of 23.3°C, with a maximum of 25.6°C. Thus, the seaweeds in all these systems have to survive drastic changes of temperature, especially in the case of the intertidal seaweeds that can experience changes of more than 20°C within minutes during a tidal cycle in winter when water temperatures may be −1.5°C and air temperatures −25°C.

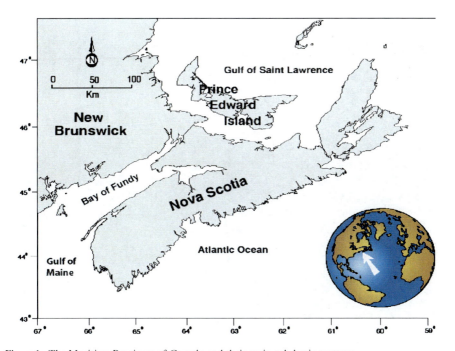

Figure 1. The Maritime Provinces of Canada and their main sub-basin systems.

1.1. THE ROCKY INTERTIDAL ZONE

The shores of PEI and NS and NB within the Gulf of Saint Lawrence (Fig. 1) present a small tidal range and a limited amount of rocky substrata. The few available locations for intertidal seaweeds suffer intensive ice abrasion during late winter and early spring (Scrosati and Heaven, 2006). Thus, most of the intertidal flora biomass of the region is located on the rocky substrata of eastern and southern NS and the Bay of Fundy shores of both NS and NB. These rocky shores extend more than 4,000 km and are dominated by the fucoid *Ascophyllum nodosum*, which forms extensive beds especially in moderately wave sheltered to sheltered areas (Fig. 2) (Sharp, 1986). *Fucus vesiculosus* is also found in the rocky intertidal zone, but in lesser quantities. It appears especially in more wave exposed or ice-scoured areas and in the upper and lower portions of the *Ascophyllum* beds in the sheltered intertidal zone (Fensom and Taylor (1974); Sharp, op. cit.; Ugarte et al., 2008). The geographical distribution of these two fucoids in the north Atlantic largely overlaps with the exception that *A. nodosum* does not occur along the Baltic Coasts, whereas, *F. vesiculosus* is well established there (Lüning, 1990). This difference is attributed to the more euryhaline characteristics of *F. vesiculosus* (Gylle et al., 2009). The red and

Figure 2. The rocky intertidal zone of New Brunswick with the dominant fucoid *Ascophyllum nodosum* (location: *Red Point* in Seal Cove, Grand Manan Island).

brown seaweeds, *Polysiphonia lanosa* and *Pilayella littoralis*, are common epiphytes of *Ascophyllum* in our region (Rawlence and Taylor, 1972; Chopin et al., 1996). Two other fucoids, *Fucus evanescens* and *F. serratus*, are also found in the lower intertidal regions in some areas. The red seaweed *Chondrus crispus* is also an important component in the lower intertidal zone in semi-exposed to exposed areas. Other red seaweeds such as *Palmaria palmata* (dulse) and *Mastocarpus stellatus* are also found in patches and sparsely intermixed with *Chondrus* in the Canadian Maritimes (Stephenson and Stephenson, 1972; Chapman and Johnson, 1990).

1.2. GEOGRAPHIC DISTRIBUTION AND CRITICAL THRESHOLDS FOR THE DOMINANT SPECIES

Ascophyllum nodosum is distributed in the western Atlantic Ocean from the Arctic Circle (66° 33′N) to Long Island (40° 42′N) in the south (Taylor, 1957; Gosner, 1978; Keser et al., 2005). The SST in this current distribution ranges from −2.1°C to 23°C (Keser et al., 2005). *Fucus vesiculosus* extends from Ellesmere Island (80° 53′N) to North Carolina (34° 48′N) (Taylor, 1957). The SST in this range of distribution varies from −2.1°C to 27.8°C.

Ascophyllum nodosum is probably one of the most studied macroalgae in the north western Atlantic, mostly because of its ecological importance as a primary producer (Baardseth, 1970; Josselyn and Mathieson, 1978, 1980; Carlson and Carlson, 1984; Cousens, 1984, 1986; Vadas et al., 2004), as a habitat for invertebrates and vertebrates (Johnson and Scheibling, 1987; Black and Miller, 1991; Rangeley and Kramer, 1998), and for its economic importance as a raw material for alginate, agricultural products, and livestock feeds (Sharp, 1986; Ugarte and Sharp, 2001). As a result, its physiological and ecological responses to environmental variables in all life stages are reasonably well documented. Its lower intertidal limits appear to be controlled by grazing pressure in sheltered environments and by competition with *Chondrus crispus* in exposed locations (Lubchenco, 1980). The upper limits of *A. nodosum* distribution on the shore are regulated by its tolerance to desiccation (Schonbeck and Norton, 1978). Exposure to long periods of drying results in visible tissue damage of *A. nodosum* after 21–28 days (op. cit.), and cell death occurs at 70% water loss (MacDonald et al., 1974). *Ascophyllum nodosum* loses 70% of its fresh weight excluding vesicles after 7.5 h at 22°C and relative humidity of 40–45% (Dorgelo, 1976). Severity of winter conditions and amount of seasonal rainfall do not seem to be critical factors for this species. A 24-year study in the environs of a thermal power plant in eastern Long Island Sound demonstrated that the growth rate of *Ascophyllum* is very sensitive to seawater temperature (Keser et al., 1998, 2005). The highest growth rate was observed during spring and mid-summer during the study period. A rapid decrease in growth occurred above 25°C, with total mortality occurring above 27°C (Keser et al., 2005).

However, the germlings and macrorecruits are rockweed's most vulnerable stages. Among the factors shown to reduce recruitment are wave action

(Vadas et al., 1990), canopy of adult plants (Vadas and Elner, 1992), and grazing pressure (Lazo et al., 1994).

2. Biomass, Harvest, Production, and Productivity of *A. nodosum* in the Maritime Provinces

2.1. BIOMASS

The *A. nodosum* resource has been divided into six geographic areas along the Nova Scotia and southern New Brunswick shores (Fig. 3). Acadian Seaplants Limited (ASL), a Canadian seaweed processing company, has been granted 11 leases in NS in the most productive rockweed areas, and the whole of the southern NB coastline. The area corresponds to 76% of the total resource for the region. For management purposes, the company has divided its leases into 340 harvesting sectors (Ugarte and Sharp, 2001) and has been assessing its resource annually since 1995 using a combination of aerial photography and extensive ground truthing. Thus, the standing stock and general condition of the resource, including its

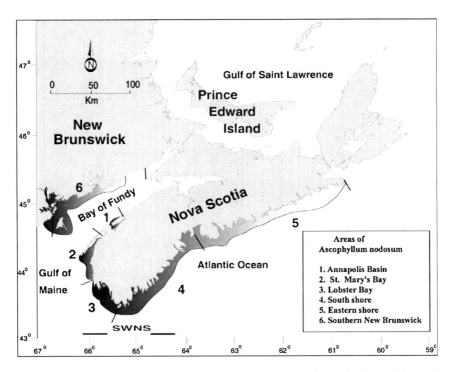

Figure 3. *Ascophyllum nodosum* distribution in Nova Scotia and southern New Brunswick. *Darkest areas* along the coastlines indicate zones having higher concentrations of biomass. (SWNS: Southwestern Nova Scotia).

population structure are well documented over most of the commercially valuable sites. The uncertainty in these values exists in area 5 of Nova Scotia, a relatively unproductive area where biomass surveys have been less intense.

The standing crop of summer (July–August) biomass for this region has been calculated at 353,395 wet tons, covering an area of close to 4,960 ha (G. Sharp, 2009, personal communication; R. Ugarte, unpublished data), to give an average biomass density of 71.3 t ha^{-1}. The highest *A. nodosum* abundance is concentrated in Lobster Bay (area 3) in southwestern Nova Scotia (SWNS) and southern New Brunswick (area 6) with a total of 91,758 and 158,811 t, respectively, corresponding to 71% of the total resource biomass (Fig. 3; Table 1). Biomass densities range between 86 and 87 t ha^{-1} in these areas.

2.2. HARVEST

Commercial exploitation of rockweed along the coastal areas of Nova Scotia began in the late 1950s when it was used as a raw material for manufacturing sodium alginate and "kelp" meal. Today, this seaweed is used as a source of a fertilizer extract and as an animal feed supplement. It is the main economic resource of the seaweed industry in the Maritime Provinces and of the country. Rockweed harvest in the region reached peak landings in 2007 and 2008 with just over 36,500 t, with 86% of the total landings corresponding to areas 3 and 6 (20,061 and 11,303 t, respectively) (Table 1). Although the total landings represent only 10.3% of the standing biomass (Table 1), the harvest in the region, with the exception of NB, is considered to be in a fully exploited condition because only 2,118 ha or 42.7% of the resource is actually accessible (Table 1). The remaining areas are either too exposed to the weather or waves, or the biomass density is too low to be profitable for the harvesters and the industry. Thus, the harvest yield varies between 13.6 and 20.5 t ha^{-2} a^{-1} in the accessible portion of the resource (Table 1). Area 3 is the most productive and has maintained this yield level since 2001. It appears that all the currently harvested areas in NS have reached or are

Table 1. Total rockweed biomass and landings in the rockweed areas of Nova Scotia and New Brunswick.

Rockweed area	Area covered (ha)	Standing stock (wet tons)	Accessible area (ha)	Landings 2008 (wet tons)	Yield (t ha^{-1})
1	35	998	35	500	14.3
2	93	9,031	47	844	18.0
3	1,073	91,758	977	20,061	20.5
4	677	42,125	229	3,839	16.8
5	1,250	50,000	0	0	0.0
6	1,832	158,811	830	11,303	13.6
Total	4,960	352,723	2,118	36,547	

very close to their maximum annual sustainable yield. Area 6 in NB could potentially increase its yield as it is currently under a fixed exploitation rate of 17% (Ugarte and Sharp, 2001).

2.2.1. Production, Productivity, and Carbon Fixation

Annual production values for *A. nodosum* for areas with different exposures to wave action in the Maritimes were estimated by Cousens (1981, 1984). His P/B (Productivity/Biomass) values varied from 0.22 to 0.79. Considering these estimates and our own observations of the distribution of the resource in the region, a P/B average of 0.54 seems reasonable. Thus, based on summer estimates of biomass, the total annual rockweed production for the region is 54,055 dry tons, equivalent to 20,000 t of carbon annually. This production would require the net absorption of 73,284 t of CO_2 from the environment each year (Table 2). Areas 6 and 3 obviously are contributing the bulk of the rockweed production and CO_2 absorption (Table 2). Our analysis shows that the highest production is in area 2 in Saint Mary's Bay with 561.5 g C m^{-2} a^{-1} and the lowest production is in area 5 (Eastern Shore) with 231.8 g C m^{-2} a^{-1} (Table 2). The annual estimates of *Ascophyllum* productivity (232–562 g C m^{-2} a^{-1}) observed in the Maritime Provinces are slightly below the range (300–894 g C m^{-2} a^{-1} 1) estimated for this species for the Northwest Atlantic (Mann, 1973; Brinkhuis, 1977; Cousens, 1981, 1984; Roman et al., 1990; Vadas et al., 2004). These earlier studies considered the spring biomass, a time when the deciduous receptacles of *A. nodosum* are most numerous and largest in size. Our results represent larger-scale measurements and should serve as a base for future comparisons during summer.

Further, estimates of carbon sequestered in the standing biomass depend on the composition of the seaweed at the time of harvest. First, the ratio of dry mass to fresh biomass must be considered. *Ascophyllum nodosum* growing in regions of significant currents and upwelling water are subjected to elevated nutrient levels, which favor a low ratio of dry to live biomass. We encounter this routinely for *A. nodosum* harvested from the lower Fundy region (area 6) when compared with biomass from the open NS coastline (areas 3–5). However, embayments with low flushing rates can suffer from reduced nutrient availability in the summer seawater and this leads to elevated dry to live biomass ratios. We have used a 29% conversion for areas 1–5 and 27.5% for area 6 (Table 2).

The composition of the *A. nodosum* must be considered, as it is known to vary significantly throughout the annual growth cycle. Analyses for *A. nodosum* provide a range of values for the major components of the seaweed (Indergaard and Minsaas, 1991). We have used an average of the values reported for the seaweed, and computed the carbon content for each major component to arrive at the 37% average overall carbon content for dry *A. nodosum*. Because the seaweed harvests occurred in summer, a period of major carbohydrate and dry matter accumulation, the use of an average carbon content may slightly underestimate the quantity of net carbon sequestered. In this regard, Vinogradov (1953; Table 10, p. 27) gives 37.99% as the carbon content for *A. nodosum*.

Table 2. Standing stock, production, and productivity values of *Ascophyllum nodosum* in the Canadian Maritime Provinces.

Rockweed area	Area covered (ha)	Standing stock (dry tons)	Production (dry tons a^{-1})	Productivity (t C a^{-1})	Productivity (g C m^{-2} a^{-1})	CO$_2$ conversion (t CO$_2$ a^{-1})
1	35	484	262	97	276.5	355
2	93	2,619	1,414	523	561.5	1,917
3	1,073	26,610	14,369	5,317	495.5	19,481
4	677	12,216	6,597	2,441	360.5	8,944
5	1,250	14,500	7,830	2,897	231.8	10,615
6	1,832	43,673	23,583	8,726	476.3	31,973
Totals	4,960	100,102	54,055	20,000	2,402	73,284

Production values assume an average P/B = 0.54 for the Maritime Provinces (Cousens, 1981). A carbon content of 37% of the seaweed dry mass was used in the calculations, and the total C × 3.664 gave the net amount of CO$_2$ assimilated.

2.3. CURRENT CONDITION OF THE RESOURCE

The rockweed resource of the Maritime Provinces has been observed and studied for almost 80 years (MacFarlane, 1931–1932, 1952; Sharp, 1981; Cousens, 1981, 1982, 1984; Ang et al., 1993, Ugarte et al., 2008). The condition of the resource was considered stable until the early 2000s when changes started to become apparent. Specifically, these changes have been in the form of bed damage due to unusual ice patterns, increased abundance of *Fucus vesiculosus*, and massive mussel recruitment in the intertidal zone.

2.4. ICE DAMAGE

Severe ice damage was observed in 2003 and 2004 in several harvesting sectors from areas 1–4 in NS (Fig. 3). In 2000, the ice pattern in the Gulf of Saint Lawrence Region began to change with a reduced period of ice cover during that year, a drastic change after that, and a historically record low in 2006 (Fig. 4). That change in ice cover corresponded well with an increase in the SST during that period for SWNS, which, in turn, corresponded to the generally increased air temperature in SWNS for the same period (Fig. 5). Rapid increases in air and SST during the spring of 2003 and 2004 produced ice breakup into rather larger pieces that scoured the intertidal zone. This scraping effect was exacerbated by early season southwest winds that normally arrive in late spring and pushed ice against the shore. Several *A. nodosum* harvesting sectors exposed to the southwest wind were considerably damaged, with some losing up to 90% of the available biomass.

Ice damage in the upper intertidal zone is a common phenomenon along the Atlantic coast of NS that, on occasion, has resulted in considerable damage to the intertidal zones of sites such as Saint Margaret Bay, near Halifax (McCook and

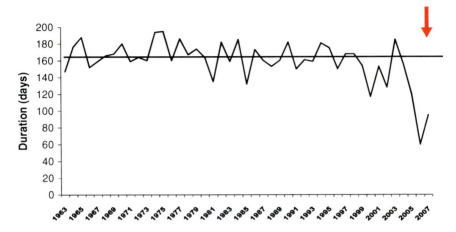

Figure 4. Ice duration in the Gulf of Saint Lawrence Region, 1963–2007. *Arrow* indicates the lowest ice duration in more than 4 decades (Data provided by Joel Chassé, DFO, Canada).

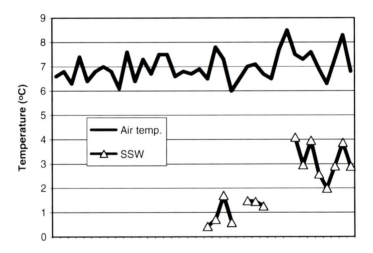

Figure 5. Annual mean air temperature (*solid line*) observed for southwestern NS (Yarmouth station) from 1970 to 2007 and winter (December–March) surface seawater temperatures (SST) (*triangles*) observed for the same area in the early 1990s and from 2000 to 2007.

Chapman, 1991). However, the magnitude of the damage observed for two consecutive years in the ASL leases has not been observed before by the local harvesters of southwestern NS, some of whom have harvested in this area for more than 35 years (J. Brennan, 2004, personal communication). In fact, during the winter of 2007–2008 and for the first time in living memory, the bays of south-western NS did not freeze over (J. Brennan, 2008, personal communication; R. Ugarte, 2008, personal observation).

2.5. INCREASE OF *FUCUS VESICULOSUS*

Historically, the rockweed beds of southwestern Nova Scotia have been composed of almost 99% *A. nodosum*, with a minor component of *F. vesiculosus* (G. Sharp, 2008, personal communication). The incidence of *F. vesiculosus* in the harvested material has been extensively monitored by ASL since 2003, after observing a slight increase from the previous two years. From an average incidence of 0.8% prior to 2003, it increased almost steadily each year to 4.6% in 2008 (approximately 500% increase). Field surveys confirmed this trend for 2005–2007 (Fig. 6). The sectors showing the highest incidence of *F. vesiculosus* were located mostly in protected areas such as Lobster Bay and the inner sectors of Saint Mary's Bay (Fig. 3).

It is tempting to attribute the increase in *F. vesiculosus* incidence to the harvest, especially since the total landings have steadily increased to a peak of over 36,500 t in 2008. However, the increased landings were the result of the opening of new areas or leases rather than increased exploitation rates within the existing harvesting areas. Some sectors in Lobster Bay (area 2) have been harvested at the same exploitation rate or higher since the early 1970s (Sharp, 1986) with no increased occurrence of *F. vesiculosus* until very recent years. The specially designed rake used by the harvesters has been proved to cut only a portion of the *A. nodosum* fronds without stripping the plants from the substratum (Ugarte and

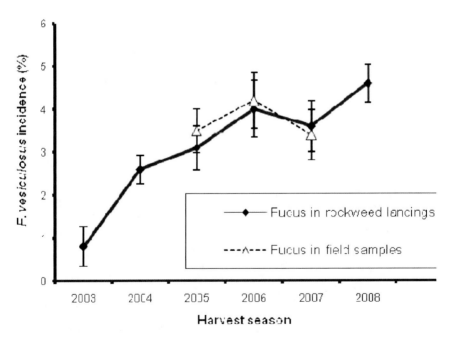

Figure 6. Incidence (percentage of biomass) of Fucus vesiculosus on rockweed (*A. nodosum*) landings from ASL lease of southwestern NS from 2003 to 2008 and from field samples taken in the same locations from 2005 to 2007 (*vertical bars* are ±2 SE).

Sharp, 2001, Ugarte et al., 2006). The deliberate targeting of *F. vesiculosus* by the harvesters also is unlikely, as it is lighter than *A. nodosum* and, therefore, it is not of economic benefit for them to harvest a *Fucus–Ascophyllum* mixture. Also, *F. vesiculosus* is a fast growing seaweed; large plants can be observed in as little as 2 years after colonization of a disturbed or available "free space" in this region (McCook and Chapman, 1991; R. Ugarte, 2005, personal observation). Therefore, the increased incidence of *F. vesiculosus* appears to be related to a recent event or physical factors. The ice damage occurring in the 2003 and 2004 season could be a possible explanation for the increased amount of *F. vesiculosus* in the landings. Though there was an increase in *F. vesiculosus* in certain denuded sectors, it was not the case in all sectors; the highest incidence was observed in protected areas with no occurrence of ice damage.

2.6. RECRUITMENT IN *ASCOPHYLLUM NODOSUM* BEDS

A heavy recruitment of the blue mussel *Mytilus edulis* on the lower part of the rockweed beds was observed during the spring of 2005 in several harvesting sectors of area 6 in southern NB (see Figs. 3 and 7). The mussels attached to the free rocky substratum and to the rockweed clumps. As the mussels gain in weight, they

Figure 7. Massive recruitment of the blue mussel, *Mytilus edulis*, in the rocky intertidal zone, and on *A. nodosum* in southern New Brunswick (location: Mc Graws Island in Letete, New Brunswick).

prevent the normal flotation of the colonized clumps and consequently interfere with the *A. nodosum* harvest. The excessive weight of the mussels also resulted in the detachment of the affected plants from the substratum during the fall storms of 2006. Losses of up to 30% of the rockweed biomass were observed in some of the worst affected sectors as recorded in the 2007 survey.

Cold surface seawater temperatures in winter and spring in the Maritime region appear to keep the spawning and recruitment of certain invertebrates under some form of control (Lemaire et al., 2006). Although temperature is only one abiotic factor controlling the recruitment of *M. edulis* (Dobretsov and Miron, 2001), the mild temperatures experienced in the springs of 2005 and 2006 apparently favored, directly or indirectly, the rapid recruitment and survival of mussels in the Passamaquoddy Bay area of NB. An inspection in June 2008 in Grand Manan Island, an important rockweed harvesting area in southern NB, detected another massive recruitment of blue mussels in the lower part of the intertidal zone. The winter of 2008 was one of the mildest on record for NB.

3. Future Potential Impact of Rapid Climate Change in the Region

Analysis of Atmosphere-Ocean General Circulation Models (AOGCMs) predicts that the average SST in the North Atlantic will increase by 2°C by 2060, and that the impact is expected to be greater during the winter period (Chmura et al., 2005). Although this increase is expected to significantly influence the distribution and abundance of marine organisms in the Northwest Atlantic, the rockweed abundance in the Maritime Provinces could in part benefit from an increase in SST, especially the stocks along the Bay of Fundy where the average SSTs would increase to around 14°C in summer. The southwestern and eastern shores of NS could reach average summer temperatures of 16–17°C, and probably up to 20°C in shallow embayments. The predicted maxima are around or below the 25°C that limits rockweed growth, and well below the lethal range of 27–28°C (Keser et al., 2005). Currently, the vegetative growth of *Ascophyllum* is almost nil from December to early March and increases by mid-March (R. Ugarte, 1999–2001, personal observation) in the Maritimes. Warmer water will promote an earlier growth season during spring and possibly slow but continuous growth during the winter period.

However, an increase in temperature will also promote the growth of other seaweeds with greater eurythermal capacity such as *F. vesiculosus* and similar opportunistic species. If an increased water temperature is the factor behind the current increase of *F. vesiculosus* in NS, then we should expect this trend to continue, with a potential *Fucus/Ascophyllum* mixture of up to 50% in some locations by 2060. Although we lack historical information for the *Fucus/Ascophyllum* proportions in NB, data collected during 2007 and 2008 by ASL showed that *F. vesiculosus* in area 6 (Fig. 3) is around 6–7% of the total fucoid biomass, and that rapid changes from *Ascophyllum* to *Fucus* have been observed in some sectors of this area (Ugarte et al., 2008).

Thus, it is probable that southern NB will also experience a significant increase in *F. vesiculosus* during the next half century. Although it is only an anecdotal observation, a shift in the fucoid composition from *Ascophyllum* to *F. vesiculosus* has also been observed in Ireland and it has been associated with a shift in prevailing wind patterns.

Invasive species are another potential problem associated with climate change and the Canadian Maritimes certainly will be affected to some degree. *Grateloupia turuturu*, a red seaweed native to Japan and Korea, has invaded the coasts of New England (Mathieson et al., 2008). The temperature tolerance of this species is 4–28°C (Simon et al., 1999, 2001) and a low SST has probably precluded its further northern invasion into Canadian waters. However, an increase in SST of 2° may well create a suitable habitat for such opportunistic species to develop along the coastline of the Maritime Provinces.

Another invasive species, *Codium fragile,* has already become established in the shallow subtidal and intertidal pools in areas 3 and 4 since the mid-1990s (Chapman et al., 2002), but so far it has been unable to colonize the waters of the Bay of Fundy due to the low summer SSTs. However, an increase in the SSTs may allow this species to survive in shallow warmer embayments of the Fundy region (Fig. 8 Photo. Raul Ugarte).

Figure 8. *Codium fragile* is commonly found in the subtidal zone and tide pools. We have also observed it as an epiphyte on *A. nodosum* in southwestern NS.

Another large change predicted by the AOGCMs models for the region by 2060 is a significant reduction of the ice season in the Gulf of Saint Lawrence during the winter (Chmura et al., 2005), a trend that we are clearly observing today (Ugarte et al., 2008). This trend will be even more dramatic in the shallow bays of NS, which will probably remain ice free during winter. Although this situation seems favorable for *Ascophyllum*, the ice also serves as a thermal insulation against sudden drops in air temperature (Scrosati and Eckersley, 2007) and as a protective barrier from winter storms. According to Environment Canada, Nova Scotia (along with Newfoundland and Labrador) has the highest storm frequency during the winter and early spring of any region in Canada owing to its proximity to the Gulf Stream. These storms can generate wave heights greater than 14 m, and storm surges in excess of 1 m. The frequency and intensity of storms have increased in the last decade in the Maritimes and this trend is expected to continue. Under this scenario, it is possible that a large percentage of rockweed biomass along the southern and eastern shores (areas 3 to 5) of NS may be lost to storm damage during the winter each year, with those in the most exposed area being unable to recoup the lost biomass during the summer months.

Information continues to be collected on the rockweed biomass and their associated flora in the Canadian Maritimes as part of the ASL harvesting responsibilities and the Company's environmental stewardship role. Such data are essential for understanding in detail the scale of changes occurring in this region, and are required when long-term retrospective analyses are carried out in the future.

4. References

Ang, P.O., Sharp, G.J. and Semple, R.E. (1993) Changes in the population structure of *Ascophyllum nodosum* (L.) Le Jolis due to mechanical harvesting. Hydrobiologia **260**: 321–326.

Baardseth, E. (1970) Synopsis of biological data on knobbed wrack, *Ascophyllum nodosum* (Linnaeus) Le Jolis. FAO Fish. Synop. **38**: 41 pp.

Berecibar, E., Ben-Hamadou, R., Tavares, M. and Santos, R. (2004) Long-term change in the phytogeography of the Portuguese continental coast. www.siam.fc.ul.pt/ECSA2008/7FEBPDF/BerecibarE.pdf. Last accessed 2009.

Black, R. and Miller, R.J. (1991) The use of the intertidal zone by fish in Nova Scotia. Environ. Biol. Fish. **31**: 109–121.

Brinkhuis, B.H. (1977) Comparison of salt-marsh fucoid production estimated from three different indices. J. Phycol. **13**: 328–335.

Carlson, D.J. and Carlson, M.L. (1984) Reassessment of exudation by fucoid macroalgae. Limnol. Oceanogr. **29**: 1077–1087.

Chapman, A.R.O. and Johnson, C.R. (1990) Distribution and organization of macroalgal assemblages in the northwest Atlantic. Hydrobiologia **192**: 77–121.

Chapman, A.S., Scheibling, R.E. and Chapman, A.R.O. (2002) Species introductions and changes in marine vegetation of Atlantic Canada, In: R. Claudi, P. Nantel and E. Muckle-Jeffs (eds.) *Alien Invaders in Canada's Waters, Wetlands and Forests*. Canadian Forest Service Science Branch, Natural Resources. Ottawa, Canada, pp. 133–148.

Cheung, W.W.L., Close, C., Lam, V., Watson, R. and Pauly, D. (2008a) Application of macroecological theory to predict effects on climate change on global fisheries potential. Mar. Ecol. Progr. Ser. **365**: 187–197.

Cheung, W.W.L., Lam, V., Sarmiento, J.L., Kearney, K., Watson, R. and Pauly, D. (2008b) Projecting global marine biodiversity impacts under climate change scenarios. FISH and FISHERIES. DOI: 10.1111/j.1467-2979.2008.00315.x.

Chmura, G.L., Vereault, S.A. and Flanary, E.A. (2005) Sea surface temperature changes in the Northwest Atlantic under a 2°C global temperature rise. Implications of a 2°C global temperature rise for Canada's natural resources. A report for the World Wildlife Fund, 30 November, 2005.

Chopin, T., Marquis, P.A. and Belyea, E.P. (1996) Seasonal dynamics of phosphorus and nitrogen contents in the brown alga *Ascophyllum nodosum* (L.) Le Jolis, and its associated species *Polysiphonia lanosa* (L.) Tandy and *Pilayella littoralis* (L.) Kjellman, from the Bay of Fundy, Canada. Bot. Mar. **39**: 543–552.

Collins, W., Colman, R., Haywood, J., Manning, M.R. and Mote, P. (2007) The physical science behind climate change. Sci. Am. **297**: 64–71.

Cousens, R. (1981) Variation in annual production by *Ascophyllum nodosum* with degree of exposure to wave action. Proc. Int. Seaweed Symp. **10**: 253–258.

Cousens, R. (1982) The effect of exposure to wave action on the morphology and pigmentation of *Ascophyllum nodosum* (L) Le Jolis in South-Eastern Canada. Bot. Mar. **25**: 191–195.

Cousens, R. (1984) Estimation of annual production by the intertidal brown alga *Ascophyllum nodosum* (L.) Le Jolis. Bot. Mar. **27**: 217–227.

Cousens, R. (1986) Quantitative reproduction and reproductive effort by stands of the brown alga *Ascophyllum nodosum* (L.) Le Jolis in South-eastern Canada. Estuar. Coast. Shelf Sci. **22**: 495–507.

Dobretsov, S.V. and Miron, G. (2001) Larval and post-larval vertical distribution of the mussel *Mytilus edulis* in the White Sea. Mar. Ecol. Prog. Ser. **218**: 179–187.

Dorgelo, J. (1976) Intertidal fucoid zonation and desiccation. Hydrobiol. Bull. **10**: 112–115.

Gian-Reto, W., Post, E., Convey, P., Menzel, A., Parmesan, C., Beebee, T.J.C., Fromentin, J.-M., Hoegh-Guldberg, O. and Bairlein, F. (2002) Ecological responses to recent climate change. Nature **416**: 389–395.

Gosner, K.L. 1978. *A Field Guide to the Atlantic Seashore: from the Bay of Fundy to Cape Hatteras. The Peterson Field Guide Series.* Houghton Mifflin Company, Boston, MA. pp. 329.

Gylle, A.M., Nygard, C.A. and Ekelund, N.G.A. (2009) Desiccation and salinity effects on marine and brackish *Fucus vesiculosus* L. (Phaeophyceae). Phycologia **48**: 156–164.

Hiddink, J.G. and Hofstede, R.T. (2008) Climate induced increases in species richness of marine fishes. Global Change Biol. **14**: 453–460.

Indergaard, M. and Minsaas, J. (1991) Animal and human nutrition, In: M.D. Guiry and G. Blunden (eds.) *Seaweed Resources in Europe: Uses and Potential.* John Wiley & Sons, Chichester, pp. 21–64.

IPCC (Intergovernmental Panel on Climate Change) (2007) Summary for policymakers, In: S. Solomon, D. Qin, M. Manning, Z. Chen, Z. and others (eds.) *Climate Change 2007: The Physical Science Basis. Contribution of Working Group I to the Fourth Assessment Report of the Intergovernmental Panel on Climate Change.* Cambridge University Press, New York, pp. 1–18.

Johnson, S.C. and Scheibling, R.E. 1987. Structure and dynamics of epifaunal assemblages on intertidal macroalgae *Ascophyllum nodosum* and *Fucus vesiculosus* in Nova Scotia, Canada. Mar. Ecol. Progr. Ser. **37**: 209–227.

Josselyn, M.N. and Mathieson, A.C. (1978) Contribution of receptacles from the fucoid *Ascophyllum nodosum* to the detrital pool of a north temperate estuary. Estuaries **1**: 258–261.

Josselyn, M.N. and Mathieson, A.C. (1980) Seasonal influx and decomposition of autochthonous macrophyte litter in a north temperate estuary. Hydrobiologia **71**:197–208.

Keser, M., Foertch, J.F. and Swenarton, J.T. (1998) A 20-year study of *Ascophyllum nodosum* population dynamics near a heated effluent in eastern Long Island Sound. J. Phycol. **34**(Suppl.: 28): Abstr. No. 66.

Keser, M., Swenarton, J.T. and Foertch, J.F. (2005) Effects of thermal input and climate change on growth of *Ascophyllum nodosum* (Fucales, Phaeophyceae) in eastern Long Island Sound (USA). J. Sea Res. **54**: 11–220.

Lazo, L., Markham, J.H. and Chapman, A.R.O. (1994) Herbivory and harvesting: effects on sexual recruitment and vegetative modules of *Ascophyllum nodosum*. Ophelia **40**: 95–113.

Lemaire, N., Pellerin, J., Fournier, M., Girault, L., Tamigneaux, E., Cartier, S. and Pelletier, E. (2006) Seasonal variations of physiological parameters in the blue mussel *Mytilus* spp. From farm sites of eastern Quebec. Aquaculture **261**: 729–751.

Lubchenco, J. (1980) Algal zonation in a New England rocky intertidal community: an experimental analysis. Ecology **61**: 333–344.

Lüning, K. (1990) *Seaweeds: Their Environment, Biogeography, and Ecophysiology*. John Wiley & Sons, New York, MacDonald, MA, pp. 527.

Fensom, D.S. and Taylor, A.R.A. (1974) Electrical impedance in *Ascophyllum nodosum* and *Fucus vesiculosus* in relation to cooling, freezing and desiccation. J. Phycol. **10**: 462–469.

MacFarlane, C. (1931–1932) Observations on the annual growth of *Ascophyllum nodosum*. Proc. Nova Scotian Inst. Sci. **28**: 27–33.

MacFarlane, C. (1952) A survey of certain seaweeds of commercial importance in southwest Nova Scotia. Can. J. Bot. **30**:78–97.

Mann, K.H. (1973) Seaweeds: Their productivity and strategy for growth. Science **182**: 975–981.

Mathieson, A.C., Dawes, C.J. Pederson, J., Gladych, R.A. and Carlton, J.T. (2008) The Asian red seaweed *Grateloupia turuturu* (Rhodophyta) invades the Gulf of Maine. Biol. Invasions **10**: 985–988.

McCook, L.J. and Chapman, A.R.O. (1991) Community succession following massive ice-scour on an exposed rocky shore: effects of *Fucus* canopy algae and of mussels during late succession. J. Exp. Mar. Biol. Ecol. **154**:137–169.

Parmesan, C. (2006) Ecological and evolutionary responses to recent climate change. Annu. Rev. Ecol. Evol. System. **37**: 637–669.

Pedersen, A., Kraemer, G. and Yarish, C. (2008) Seaweed of the littoral zone at Cove Island in Long Island Sound: annual variation and impact of environmental factors. J. Appl. Phycol. DOI 10.1007/s10811-008-9316-6.

Rawlence, D.J. and Taylor, A.R.A. (1972) A light and electron microscopic study of rhizoid development in *Polysiphonia lanosa* (L.) Tandy. J. Phycol. **8**: 15–24.

Root, T.L., Price, J.T., Hall, K.R., Schneider, S.H., Rosenzweig, C. and Pounds, J.A. (2003) Fingerprints of global warming on wild animals and plants. Nature **421**: 57–60.

Roman, C.T., Able, K.W., Lazzari, M.A. and Heck, K.L. (1990) Primary productivity of angiosperm and macroalgae dominated habitats in a New England USA salt marsh: a comparative analysis. Estuar. Coast. Shelf Sci. **30**: 35–46.

Rangeley, R.W. and Kramer, D.L. (1998) Density-dependent antipredator tactics and habitat selection in juvenile pollock. Ecology **79**: 943–952.

Ruddiman, R.F. (2005) How did humans first alter global climate? Sci. Am. **292**: 46–54.

Sagarin, R.D., Barry, J.P., Gilman, S.E. and Baxter, C.H. (1999) Climate-related change in an intertidal community over short and long time scales. Ecol. Monogr. **69**: 465–490.

Schonbeck, M. and Norton, T.A. (1978) Factors controlling the upper limits of fucoid algae on the shore. J. Exp. Mar. Biol. Ecol. **31**: 303–313.

Scrosati, R. and Heaven, C. (2006) Field technique to quantify intensity of scouring by sea ice in rocky intertidal habitats. Mar. Ecol. Progr. Ser. **320**: 293–295.

Scrosati, R. and Eckersley, L.E. (2007) Thermal insulation of the intertidal zone by the ice foot. J. Sea Res. **58**: 331–334.

Sharp, G.J. (1981) An assessment of *Ascophyllum nodosum* harvesting methods in southwestern Nova Scotia. Can. Tech. Rep. Fish. Aquat. Sci. **1012**: pp. 28.

Sharp, G.J. (1986) *Ascophyllum nodosum* and its harvesting in Eastern Canada, In: *Case Studies of Seven Commercial Seaweed Resources. FAO Technical Report* **281**: 3–46.

Simkanin, C., Power, A.M., Myers, A., McGrath, D., Southward, A.J., Mieszkowska, N., Leaper, R. and O'Riordan, R. (2005) Using historical data to detect temporal changes in the abundances of intertidal species on Irish shores. J. Mar. Biol. Assoc. U.K. **85**: 1329–1340.

Simon, C., Gall, E.A., Levasseur, G. and Deslandes, E. (1999) Effects of short-term variations of salinity and temperature on the photosynthetic response to the red alga *Grateloupia doryphora* from Brittany (France). Bot. Mar. **42**: 437–440.

Simon, C., Gall, E.A. and Deslandes, E. (2001) Expansion of the red alga *Grateloupia doryphora* along the coast of Brittany, France. Hydrobiologia **443**: 23–29.

Stephenson, T.A. and Stephenson, A. (1972) *Life Between the Tide Marks on Rocky Shores*. W.H. Freeman and Co., San Francisco, pp. 425.

Taylor, W.R. (1957) *Marine Algae of the Northeastern Coast of Northeastern Coast of North America*. University Michigan Press, Ann Arbor, MI, pp. 509.

Ugarte, R. and Sharp, G.J. (2001) A new approach to seaweed management in eastern Canada: the case of *Ascophyllum nodosum*. Cah. Biol. Mar. **42**: 63–70.

Ugarte, R., Sharp, G.J. and Moore, B.J. (2006) Changes in the brown seaweed *Ascophyllum nodosum* (L.) Le Jol., plant morphology and biomass produced by cutter rake harvests in southern New Brunswick, Canada. J. Appl. Phycol. **18**: 351–359.

Ugarte, R.A., Critchley, A.T., Serdynska, A.R. and Deveau, J.P. (2008) Changes in composition of rockweed (*Ascophyllum nodosum*) beds due to possible recent increase in sea temperature in Eastern Canada. J. Appl. Phycol. DOI 10.1007/s10811-008-9397-2.

Vadas, R.L. and Elner, R.W. (1992) Plant-animal interactions in the north-west Atlantic, In: D.M. John, S.J. Hawkins and J.H. Price (eds.) *Plant Animal Interactions in the Marine Benthos*. Systematics Association Special Volume No. 46. Clarendon Press, Oxford, UK. pp. 33–60.

Vadas, R.L., Wright, W.A. and Miller, S.L. (1990) Recruitment of *Ascophyllum nodosum*: wave action as a source of mortality. Mar. Ecol. Prog. Ser. **61**: 263–272.

Vadas, R.L., Wright, W.A. and Beal, B.F. (2004) Biomass and productivity of intertidal rockweeds (*Ascophyllum nodosum* LeJolis) in Cobscook Bay. Ecosystem modeling in Cobscook Bay, Maine: a Boreal, Macrotidal Estuary. N Naturalist **11**: 123–142.

Van den Hoek, C. (1975) Phytogeographic provinces along the coasts of the northern Atlantic Ocean. Phycologia **14**: 317–330.

Vinogradov, A.P. (1953) *The Elementary Chemical Composition of Marine Organisms*. Memoir Number II, Sears Foundation for Marine Research, Yale University, New Haven CT, pp. 647.

PART 3:
ECOPHYSIOLOGICAL RESPONSES OF SEAWEEDS

Pauly
Clerck
Zou
Gao
Rodriguez
Medina-López
Eggert
Peters
Küpper

Biodata of **Klaas Pauly** and **Olivier De Clerck**, authors of *"GIS-Based Environmental Analysis, Remote Sensing, and Niche Modeling of Seaweed Communities"*

Dr. Klaas Pauly is currently a Teaching Assistant at the Phycology Research Group, Biology Department in Ghent University, Belgium. He graduated there in 2004, presenting his dissertation on Biogeography and seasonality of macroalgal communities in the Gulf of Oman. As a Ph.D. student, he continued his research in the same group on ecology and biogeography of benthic macroalgae of the Arabian Sea and Gulf of Oman using geographical techniques including remote sensing and ecological niche modeling. His most recent work is on siphonous green algal phylogeography and the evolution of niches using the latter technique, in collaboration with Dr. Heroen Verbruggen.

E-mail: **klaas.pauly@ugent.be**

Professor Dr. Olivier De Clerck was recently appointed Director of the Phycology Research Group at the Biology Department, Ghent University, Belgium. After finishing his Ph.D. there in 1999, a revision of the brown algal genus Dictyota in the Indian Ocean, he spent 1 year at the University of Cape Town as a Smuts Memorial Post Doctoral Fellow. From 2001 to 2008, he worked in Ghent as a Fund for Scientific Research – Flanders (FWO, Belgium) post-doc fellow at the Phycology Research Group, coordinating various projects involving molecular systematics of marine red algae and sexual evolution in brown algae.

E-mail: **olivier.declerck@ugent.be**

Klaas Pauly

Olivier De Clerck

GIS-BASED ENVIRONMENTAL ANALYSIS, REMOTE SENSING, AND NICHE MODELING OF SEAWEED COMMUNITIES

KLAAS PAULY AND OLIVIER DE CLERCK
Phycology Research Group, Biology Department, Ghent University, 9000, Ghent, Belgium

1. GIS and Remote Sensing in a Nori Wrap

1.1. INTRODUCTION

In the face of global change, spatially explicit studies or meta-analyses of published species data are much needed to understand the impact of the changing environment on living organisms, for instance by modeling and mapping species' distributional shifts. A *Nature* Editorial (2008) recently discussed the need for spatially explicit biological data, stating that the absence or inaccuracy of geographical coordinates associated with every single sample prohibits, or at least jeopardizes, such studies in any research field. In this chapter, we show how geographic techniques such as remote sensing and applications based on geographic information systems (GIS) are the key to document changes in marine benthic macroalgal communities.

Our aim is to introduce the evolution and basic principles of GIS and remote sensing to the phycological community and demonstrate their application in studies of marine macroalgae. Next, we review current geographical methods and techniques showing specific advantages and difficulties in spatial seaweed analyses. We conclude by demonstrating a remarkable lack of spatial data in seaweed studies to date and hence suggesting research priorities and new applications to gain more insight into global change-related seaweed issues.

1.2. THE (R)EVOLUTION OF SPATIAL INFORMATION

The need to share spatial information in a visual framework resulted in the creation of maps as early as many thousands of years ago. For instance, an approximately 6,200-year-old fresco map covering the city and a nearby erupting volcano was found in Çatal Höyük, Anatolia (Turkey). Dating back even further, the animals, dots, and lines on the Lascaux cave walls (France) are thought to represent animal migration routes and star groups, some 15,000 years ago. Throughout written history, there has been a steady increase in both demand for and quality (i.e., the extent and amount of detail) of maps, concurrent with the ability to travel and observe one's position on earth. Like many aspects in written and graphic history, however, a revolutionary expansion took place with the introduction of (personal) computers. This new technology allowed to store maps (or any graphics) and additional information on certain map features in a digital format using an associated relational database (attribute information). It is important to note that the creation of GIS is not a goal in itself; instead, GIS are tools that facilitate spatial data management and analysis. For instance, a Nori farmer may wonder how to quantify the influence of water quality and boat traffic on the yields (the defined goals), and use GIS as tools to create and store maps and (remotely sensed) images, and perform spatial analyses to achieve these goals (Fig. 1).

At least 30,000[1] publications dating back to 1972 involve GIS (Amsterdam et al., 1972), according to ISI Web of Knowledge.[2] However, 12 years went by

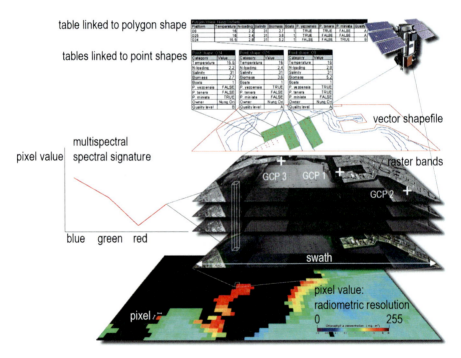

Figure 1. Schematic overview of GIS data file types and remote sensing of a Nori farm in Tokyo Bay, Japan.

---

[1]This number is based on the search term "geographic information system." The search term 'GIS' yielded 32706 records, but an unknown number of these, including the records prior to 1972, concern other meanings of the same acronym.
[2]All online database counts and records mentioned throughout this chapter, including ISI Web of Knowledge, OBIS, and Algaebase records, refer to the status on 1 July 2008.

before the first use of GIS in the coastal or marine realm was published (Ader, 1982), and since then only a meager 2,257 have followed.

Parallel to the evolution of mapping and GIS, the need to observe objects without being in physical contact with the target, remote sensing has played an important role in spatial information throughout history. In its earliest forms, it might have involved looking from a cliff to gain an overview of migration routes or cities. However, three revolutions have shaped the modern concept of remote sensing. Halfway through the nineteenth century, the development of (balloon) flight and photography allowed one to make permanent images at a higher altitude (with the scale depending on the altitude and zoom lens) and at many more times or places than were previously feasible, making remote sensing a valuable data acquisition technique in mapping. Halfway in the twentieth century, satellites were developed for Earth observation, allowing one to expand ground coverage. At the end of the twentieth century, the ability to digitally record images through the use of (multiple) CCD and CMOS sensors quickly enhanced the abilities to import and edit remote sensing data in GIS. Two kinds of remote sensing have been developed. Active remote sensing involves the emission of signals with known properties, to analyze the reflection and backscatter, with RADAR (RAdio Detecting And Ranging) as the most widespread and best-known application. Passive remote sensing means recording radiation emitted or reflected by distant objects, and most often the reflection of sunlight by objects is investigated. This chapter will only cover passive remote sensing and laser-induced active remote sensing, as sound-based active sensing (RADAR, SONAR) is limited to (3D) geomorphological and topographical studies, rather than distinguishing benthic communities and their relevant oceanographic variables.

The first remote sensing applications are almost a decade older than the first GIS publications (Bailey, 1963), and the first coastal or marine use of remote sensing appeared only few years later, starting with oceanographical applications (Polcyn and Sattinger, 1969; Stang, 1969) and followed by mapping efforts (Egan and Hair, 1971). However, out of roughly 98,500 remote sensing records in ISI Web of Knowledge, little less than 8,500 cover coastal or marine topics.

1.3. SPATIAL DATA TYPES

Analogous to manually drawn hardcopy maps, digitized maps (hardcopies transferred to computers) or computer-designed maps most often consist of three types of geometrical features expressed as vectors (Fig. 1): zero-dimensional points, one-dimensional lines, and two-dimensional polygons. For instance, a point could represent a tethered Nori platform in a bay, linked to a database containing quantitative fields (temperature, nutrients, salinity, biomass, number of active harvesting boats), Boolean fields (presence/absence of several species), and categorical fields (owner's name, quality level label). In turn, polygons encompassing several of these points may depict farms, regions, or jurisdictions. Lines could either intersect these polygons (in case of isobaths) or border them (in case of coastal structures). Vector maps and their associated databases are easy to edit, scale, reproject, and query while maintaining a limited file size.

The raster data type (Fig. 1), also called grid or image data in which all remote sensing data come, differs greatly from vector map data. Each image (whether analogously acquired and subsequently scanned, directly digitally acquired, or computer-generated) is composed of x-columns times y-rows with square pixels (or cells) as the smallest unit. Each pixel is characterized by a certain spatial resolution (the spatial extent of a pixel side), typically ranging from 1 to 1,000 m, and an intensity (z-value). The radiometric resolution refers to the number of different intensities distinguished by a sensor, typically ranging from 8 bits (256) to 32 bits (4.3×10^9). In modern remote sensing platforms, different parts (called bands) of the incident electromagnetic spectrum are often recorded by different sensors in an array. In this case, a given scene (an image with a given length and width, the latter also termed swath, determined by the focal length and flight altitude) consists of several raster layers with the same resolution and extent, each resulting from a different sensor. The amount of sensors thus determines the spectral resolution. A "vertical" profile of a pixel or group of pixels through the different bands superimposed as layers results in a spectral signature for the given pixel(s). The spectral signature can thus be visualized as a graph plotting radiometric intensity or pixel value against band number (Fig. 1). The term multispectral is used for up to ten sensors (bands), whereas hyperspectral means the presence of ten to hundreds of sensors. Some authors propose the term superspectral, referring to the presence of 10–100 sensors, and reserve hyperspectral for more than 100 sensors. Temporal resolution indicates the coverage of a given site by a satellite in time, i.e., the time between two overpasses. In the Nori farm example, one or more satellite images might be used as background layers in GIS (Fig. 1) to digitize farms and the surroundings (based on large-scale imagery in a geographic sense, i.e., with a high spatial resolution) or to detect correlations with sites and oceanographical conditions (based on small-scale imagery in a geographic sense, i.e., with a low spatial resolution).

An important aspect in GIS and remote sensing is georeferencing. By indicating a limited number of tie points or ground control points (GCPs) for which geographical coordinates have been measured in the field or for which coordinates are known by the use of maps, coordinates for any location on a computer-loaded map can be calculated in seconds and subsequently instantly displayed. Almost coincidentally with GIS evolution, portable satellite-based navigation devices (Global Positioning System, GPS) have greatly facilitated accurate measurements and storage of geographical coordinates of points of interest. In the current example, a nautical chart overlaid with the satellite images covering the Nori farms might be used as the source to select GCPs (master–slave georeferencing), or alternatively, field-measured coordinates of rocky outcrops, roads, and human constructions along the coast, serving as GCPs recognizable on the (large-scale) satellite images, might be used for direct georeferencing (Fig. 1).

2. GIS and Remote Sensing: Phycological Applications

2.1. GEOREFERENCING SPECIMENS

Acquiring GPS coordinates has become self-evident, with handheld GPS devices nowadays fitting within any budget, provided that accuracy requirements are not smaller than 10–15 m. Devices capable of handling publicly available differential correction signals like Wide Area Augmentation System (WAAS, covering North America), European Geostationary Navigation Overlay Service (EGNOS, covering Europe), and equivalent systems in Japan and India are slightly more expensive but offer accuracies between 1 and 10 m. However, accuracies are almost always found to be better in practice, especially in phycological field studies where the device would mostly be used in areas free from trees and mountains. However, field workers attempting to log shallow dives and snorkel tracks using GPS should make sure to mount the device well clear from the water, as even a single splashing wave can hamper signal reception. Accuracies within 1 m can be obtained with commercial differential GPS systems, although this increases the cost and reduces mobility of field workers as a large portable station needs to be carried along, hence restricting use on water to larger boats. However, logging GPS coordinates does not eliminate the need for textual location information, preferably using official names or transcriptions as featured on maps, and using a hierarchical format going from more to less inclusive entities (cf. GenBank locality information; NCBI, 2008). This is vital to allow for error checking (see further). Several authors have recently independently and unambiguously stated that a lack of geographic coordinates linked to each recently and future sampled specimen can no longer be excused (*Nature* Editorial, 2008; Kidd and Ritchie, 2006; Kozak et al., 2008). Moreover, recommendations were made to require a standardized and publicly available deposition of spatial meta-information on all used samples accompanying each publication, including nonspatially oriented studies. This idea is analogous to most journals requiring gene sequences to be deposited in GenBank, whenever they are mentioned in a publication (*Nature* Editorial, 2008). For instance, the Barcode of Life project, aiming at the collection and use of short, standardized gene regions in species identifications, already requires specimen coordinates to be deposited for each sequence in its online workbench (Ratnasingham and Hebert, 2007).

Adding coordinates to the existing collection databases can be a lot more challenging and time-consuming. At best, a locality description string in a certain format is already provided. In that case, gazetteers can be used to retrieve geographic coordinates. However, many coastal collections are made on remote localities without specific names, such as a series of small bays between two distant cities. Efforts have been made to develop software (e.g., GEOLocate; Rios and Bart, 1997) combining the use of gazetteers and civilian GPS databases to cope with information such as road names and distances from cities. Unfortunately, most of the existing automation efforts are specifically designed for terrestrial collection databases, lacking proper maritime names, boundaries, and functions.

For instance, the software should allow specimens to be located at a certain distance from the shoreline. For relatively small collections, coordinates can also be manually obtained by identifying landmarks described in the locality fields or known by experienced field workers using Google Earth, a free GIS visualization tool with high to very high resolution satellite coverage of the entire globe (available online at http://earth.google.com). However, manually adding specimen coordinates to database records does increase the chance of errors in the coordinates when compared with automatically retrieving and adding coordinates.

Quality control of specimen coordinates is crucial. GIS allow for overlaying collection data with administrative boundary maps such as Exclusive Economic Zone (EEZ) boundaries, and comparing respective attribute tables to check for implausible locations. A common error, for instance, involves an erroneous positive or negative sign to a coordinate pair, resulting in locations on the wrong hemisphere, on land, or in open ocean. Additionally, when used in niche modeling studies (see Section 2.3), sample localities should be overlaid with raster environmental variable maps, to check if samples are not located on masked-out land due to the often coarse raster resolution.

2.2. REMOTE SENSING

In documenting the consequences of global change, it is crucial to repeatedly and automatically obtain baseline thematic and change detection maps of (commercially or ecologically critical) seaweed beds. It has long been acknowledged that remote sensing is an ideal technique to overcome numerous problems in mapping and monitoring seaweed assemblages (Belsher et al., 1985). Accessibility of seaweed-dominated areas can be an issue if the location is remote, and the exploration of rocky intertidal shores can be hard or even hazardous. More importantly, most benthic marine macroalgal assemblages are permanently submerged, restricting their exploration to SCUBA techniques. Thus, mapping and monitoring extensive stretches on a regular basis is very time- and resource-consuming when using in situ techniques only. This section provides an overview of different remote sensing approaches, without providing procedural information. For hands-on information on image processing techniques, see Green et al. (2000).

From a technical point of view, airborne remote sensing would seem most appropriate for seaweed mapping (Theriault et al., 2006; Gagnon et al., 2008). Light fixed-wing aircrafts are relatively easy to deploy, and sensors mounted on a light aircraft flying at low to moderate altitudes (1,000–4,000 m) will typically yield data sets with a very high spatial and spectral resolution. For instance, the Compact Airborne Spectrographic Imager can resolve features measuring only 0.25×0.25 m in up to 288 bands programmable between 400 and 1,050 nm in the visible and near-infrared (VNIR) light depending on the study object characteristics. Additionally, the low acquisition altitude can result in a negligible atmospheric influence. However, light aircraft are generally not equipped with advanced autopilot capabilities and are sensitive to winds and turbulence. It takes considerable time and effort to geometrically correct images acquired from such an unstable platform. Altitude differences combined with roll and pitch (aircraft rotations around its two horizontal axes) all result in different ground pixel dimensions. Moreover, low altitude acquisitions result in a limited swath, increasing both acquisition time (and hence expense) through the use of multiple flight transects and processing time to geometrically correct and concatenate the different scenes. Alternatively, a more advanced (and hence more expensive) and stable aircraft can acquire imagery at higher altitudes covering larger areas, but this is at the cost of spatial resolution and atmospheric influence.

Overall, atmospheric and weather conditions play an important role in aerial seaweed studies, as the aircraft and the airborne and ground crew must be financed over an entire standby period in areas with unstable weather conditions (quite typical for coastal areas), as the weather conditions at the exact moment of acquisition cannot be forecasted long enough in advance during the planning stage of the campaign.

In contrast, satellites are more stable platforms that can cover much larger areas in one scene daily to biweekly, making these ideal monitoring resources (Tables 1 and 2). However, satellite-based studies of seaweed assemblages were suffering from a lack of spatial resolution until the late 1990s. Typically, seaweed assemblages are very heterogeneous due to the morphology of rocky substrates, characterized by many differences in exposure to light, temperature fluctuations, waves, grazers, and nutrients on a small area. These differences result in many microclimates and niches, creating patchy assemblages in the scale of several meters to less than a meter, while no satellite sensor resolved features less than 15 m until 2000. From that year onwards, very high resolution sensors were developed and made commercially available (Table 1), allowing for detailed subtidal seaweed mapping and quantification studies in clear coastal waters (e.g., Andréfouët et al., 2004).

With the availability of more advanced sensors in the twenty-first century, a trade-off between spatial and spectral resolution became apparent (Fig. ?) – an issue of particular relevance to seaweed studies. The trade-off situation evolved because of computer processing power and data storage capacity limitations at the time of sensor development – often 5 years prior to launch followed by another 5 years of operation. This is a long time in terms of Moore's law (Moore, 1965), describing the pace at which computer processing power doubles. These historical limitations dictated a choice between a high spatial resolution and a high spectral resolution in current sensors, but not both, whereas seaweed studies would arguably benefit from both. While the main macroalgal classes (red, green, and brown seaweeds) are theoretically spectrally separable from each other as well as from coral and seagrass in three bands, this is not the case on a generic level. Additionally, information from seaweeds at below 5–10 m depth can only be retrieved from blue and green bands owing to attenuation of red and NIR in the water column. Hence, several blue and green bands can increase thematic resolution and the resulting classification accuracies, and this is of particular value in turbid waters, characteristic of many coastal stretches. By contrast, the absence of a blue band combined with only one green band (see several sensors in Tables 1 and 2) prevents spectral

Table 1. Current and *future* space-borne remote sensors apt for seaweed mapping and monitoring: technical features.

Sensor	Platform	Scene (km)	Spatial Res.	Spectral Char.	Temp. Res.	Availability	Cost
ETM+	Landsat 7	183×170	30 m (60 m TIR, 15 m pan)	0.45–12.5 µm, 7 bands + 1 pan	16 days	1999–...	Free
ASTER	TERRA	60×60	15 m (30 m SWIR, 90 m TIR)	0.52–11.65 µm, 14 bands	16 days	2000–...	$
—	IKONOS	11.3×11.3	4 m (0.8 m pan)	0.45–0.9 µm, 4 bands + 1 pan	3–5 days off-nadir	2000–...	$$$
ALI	EO-1	37×37	30 m (10 m pan)	0.433–2.35 µm, 9 bands + 1 pan		2000–...	$$
Hyperion	EO-1	7.5×100	30 m	0.4–2.5 µm, 220 bands		2000–...	$$
—	Quickbird	16.5×16.5	2.4 m (0.6 m pan)	0.45–0.9 µm, 4 bands + 1 pan	1–3.5 days off-nadir	2001–...	$$$
CHRIS	PROBA	14×14	18 m (36 m)	0.40–1.05 µm, 18 bands (63 bands), programmable	7 days	2001–...	Free
HRG	SPOT 5	60×60	10 m (2.5 m pan)	0.5–1.75 µm, 4 bands + 1 pan	1–3 days	2002–...	$$
LISS 3-4	IRS-P6 (ResourceSat-1)	23.9×23.9	5.8 m (23.5 SWIR)	0.52–1.7, 4 bands	5 days	2003–...	$$
—	FORMOSAT-2	24×24	8 m (2 m pan)	0.45–0.9 µm, 4 bands + 1 pan	1 day	2004–...	$$$
—	KOMPSAT-2 (=Arirang-2)	15×15	4 m (1 m pan)	0.45–0.9 µm, 4 bands + 1 pan	3 days off-nadir	2006–...	$$$
AVNIR-2	ALOS	70×70	10 m (2.5 m pan)	0.42–0.89, 4 bands + 1 pan	2 days	2006–...	
—	WorldView-1	17.6×17.6	0.5 m pan	1 pan	1.7–5.4 days	2007–...	$$$
—	WorldView-2	16.4×16.4	1.84 m (0.46 m pan)	8 bands + 1 pan	1.1–3.7 days	2009–...	$$$
—	PLEIADES-HR1-2	20×20	2.8 m (0.6 m pan)	0.43–0.95 µm, 4 bands + 1 pan	1 day off-nadir using HR1-2	1: 2009–... 2: 2010–...	
OLI	LDCM	185×185?	30 m (15 m pan)	0.43–2.3 µm, 8 bands + 1 pan	16 days?	2011–2021	$?

Table 2. Current and *future* space-borne remote sensors apt for seaweed mapping and monitoring: operational and quality remarks.

Sensor	Platform	Remarks
ETM+	Landsat 7	Highest quality earth observation data: calibration within 5%; Scenes flawed with 25% gaps since 2003 failure
ASTER	TERRA	Lack of blue band limits the use to intertidal and surfacing/floating seaweeds; VNIR cross track 24° off-nadir and NIR backward looking capability for stereo 3D imaging
–	IKONOS	Cross track 60° and along-track off-nadir capability for stereo 3D imaging
ALI	EO1	ALI is a technology verification instrument. EO-1 follows same orbit as Landsat 7 by about 1 min to benefit from Landsat 7's high quality calibration. EO-1 has cross-track off-nadir capability
Hyperion	EO1	
–	Quickbird	Cross and along-track 30° off-nadir capability for stereo 3D imaging
CHRIS	PROBA	Technology verification instrument; Along track ±55° off-nadir capability for stereo 3D imaging
HRG	SPOT 5	Lack of blue band limits the use to intertidal and surfacing/floating seaweeds
LISS 3-4	IRS-P6 (ResourceSat-1)	Lack of blue band cf. SPOT 5; 26° off-nadir capability for stereo 3D imaging
–	FORMOSAT-2	Cross and along-track 45° off-nadir capability for stereo 3D imaging
–	KOMPSAT-2 (= Arirang-2)	Cross-track 30° off-nadir capability
AVNIR-2	ALOS	44° off-nadir capability; Panchromatic stereo 3D imaging
–	WorldView-1	Cross-track 45° off-nadir capability; lack of multi-spectral information limits use to texture analysis
–	WorldView-2	Successor for WV-1; Cross-track 40° off-nadir capability
–	PLEIADES-HR1-2	Planned successor in SPOT series; capable of steering 30° off-track and viewing 43° off-nadir
OLI	LDCM	Planned successor in Landsat series

discrimination of submerged seaweeds altogether and confined early remote sensing studies on seaweeds to the intertidal range (Guillaumont et al., 1993). Besides the intertidal, NIR bands are useful (in combination with red) to discriminate surfacing or floating seaweeds, and allow one to discern decomposing macroalgae, as NIR reflection decreases with decreasing chlorophyll densities (Guillaumont et al., 1997).

From Fig. 2, it should be noted that two high spatial resolution spectral imaging sensors have been developed recently, Hyperion (onboard EO-1) and CHRIS (onboard PROBA), with a spectral resolution approaching that of airborne sensors, hence forming an exception on the historical trade-off. Ongoing

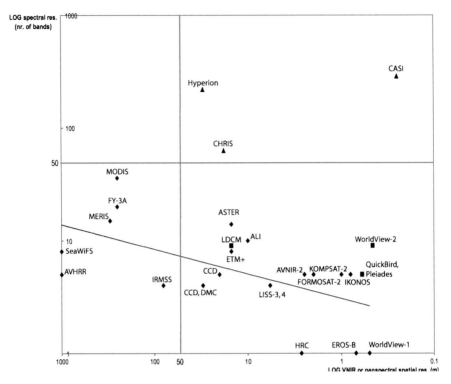

Figure 2. Trade-off between Log spectral resolution plotted against Log VNIR or pan-sharpened (where available) spatial resolution in current and future satellite sensors. All sensors are space-borne, except for the airborne CASI sensor, shown here for comparison. We consider sensors featuring a spatial resolution between 0 and 50 m and a spectral resolution above 50 bands in the visible and NIR spectrum of high value for seaweed mapping and monitoring (upper right quadrant). We therefore recommend future satellite sensor developments toward the CASI position, but note the position of the planned earth observation missions LDCM, Worldview-2, and Pleiades along the current trade-off situation (see Sections on 2.2 and 3.3). Current sensors; future sensors; current sensors forming an exception to the general trade-off situation between spectral and spatial resolution in satellite sensors (*line*).

research by the first author of this chapter suggests that CHRIS imagery can be used to map and monitor benthic communities in turbid waters at the south coast of Oman (Arabian Sea). Intertidal green, brown, and red seaweeds as well as submerged mixed seaweed beds, coral, and drifting decomposing seaweeds were discerned with reasonable accuracy during both monsoon seasons.

2.3. DISTRIBUTION AND NICHE MODELING

For centuries, biogeographical patterns have been studied in a descriptive way by delineating provinces and regions based on the presence of observed species and degrees of endemism, rather than quantifying and explaining these patterns based on environmental variables (Adey and Steneck, 2001). The question as to which environmental variables best explain seaweed species' niches and distributions is, however, one of the most important in global change research. Biogeographical models based on these variables could allow for predicting range shifts and directing field work to discover unknown seaweed species and communities.

It is widely recognized that temperature is a major forcing environmental variable for coastal macrobenthic communities, in general, and seaweeds, in particular. Temperature plays a significant role in biochemical processes, and generally species have evolved to tolerate only a (small) portion of the entire range of temperatures in coastal waters. Thus, it is evident that sea surface temperature (SST, often used as a proxy for water column temperature in shallow coastal waters) plays a prominent role in seaweed niche distribution models. Furthermore, while temperature is often measured in a time-averaged manner (daily, monthly, yearly), it is important to note that the timing of seasons differs globally (even within hemispheres due to seasonal upwelling phenomena). As some seaweed species or specific life cycles are limited by maximum and others by minimum temperatures, it is obviously essential to base models on biologically more relevant maximum, minimum, and related derived variables rather than on raw time-averaged measurements.

van den Hoek et al. (1990) gave an overview of how generalized or annual temperature isotherm maps could be used to explain the geographic distribution of seaweed species in the context of global change.

Adey and Steneck (2001) later described a quantitative model based on the maximum and minimum temperatures as the main variables, combined with area and isolation, to explain coastal benthic macroalgal species distributions. Additionally, their thermogeographic model was integrated over time as they incorporated temperatures from glacial maxima, allowing biogeographical regions to dynamically shift in response to two historical stable states of temperature regimes (glacial maxima and interglacials). In this respect, their study is of significant value in global change research, although their graphic model outputs were not based on GIS and not straightforward to interpret. Moreover, using analogous or vector isothermal SST maps, both studies suffered from a lack of resolution in SST input data, consequently compromising the resolution and accuracy of the model outputs.

Recently, two major studies demonstrated how seaweed distribution models can benefit greatly from the extensive and free availability of environmental variables on a global scale through the use of satellite data. These data are not only geographically explicit and readily usable in GIS, but also provide much more accuracy than isotherm maps due to their continuity. Schils and Wilson (2006) used Aqua/MODIS 3-monthly averaged SST data in an effort to explain an abrupt macroalgal turnover around the Arabian Peninsula. Their results pointed to a threshold of 28°C, defined by the average of the three warmest seasons, explaining diversity patterns of the seaweed floras across the entire Indian Ocean. They stressed that a single environmental factor can thus dominate the effect of other potentially interacting and complex variables. On the other hand,

Table 3. Current and *future* environmental variables retrievable from satellite data on a global scale.

Variable	Source	Resolution	Available
Sea Surface Temperature (SST)	Terra/MODIS	4 km (2 arcmin)	2000–...
Chlorophyll-a (Chl)	SeaWiFS	9 km (5 arcmin)	1997–...
	Aqua/MODIS	4 km (2 arcmin)	2002–...
Photosynthetically active Radiation (PAR)	SeaWiFS	9 km (5 arcmin)	1997–...
Euphotic Depth	Aqua/MODIS	4 km (2 arcmin)	2007–...
Surface winds	QuikSCAT/SeaWinds Scatterometer	110 km (1 arcdegree)	
Bathymetry	Various sources, assembled in ETOPO2	4 km (2 arcmin)	
Salinity	SMOS	40 km	2009–...
	Aquarius/SAC-D	100 km	2010–...

Graham et al. (2007) took several other variables in consideration to build a global model predicting the distribution of deepwater kelps. Their study was essentially a 3D mapping effort to translate the fundamental niche of kelp species, as determined by ecophysiological experiments, from environmental space into geographical space, based on global bathymetry, photosynthetically active radiation (PAR), optical depth, and thermocline depth stored in GIS. The latter was based on the interpolation of vertical profiles, whereas the former three variables were derived from satellite data sets (Table 3).

The latest development in distribution modeling approaches concerns several Species' Distribution Modeling (SDM) algorithms, also termed Ecological Niche Modeling (ENM), Bioclimatic Envelope Modeling (BEM), or Habitat Suitability Mapping (HSM). While the names are often mixed in the same context, a slight difference in meaning exists: the latter three are mostly based on presence-only data and predict the distribution of niches rather than actual species distributions, whereas the former involves presence/absence of input data and allows accurately predicting and verifying actual species distributions. Many different algorithms and software implementations exist (Maxent, GARP, ENFA, BioClim, GLM, GAD, BRT, but see Elith et al. (2006) for a review), but two fundamental properties are combined in these techniques, clearly separating them from the studies described earlier, which showed at most one of these properties. First, input data are a combination of a vector point file, representing georeferenced field observations of a species (as opposed to ecophysiological experimental data), on the one hand, and climatic variables stored in a raster GIS, on the other hand. The modeling algorithms then read the data out of GIS and use statistical functions to calculate the realized niche (as opposed to the fundamental niche; Hutchinson, 1957) in environmental space, subsequently projecting the niche back into geographical space in GIS. Second, instead of a binary identification of suitable and unsuitable areas, ENM output is a continuous probability distribution, which makes more sense from a biological point of view. Continuous probability maps may then be converted to binary maps using arbitrary thresholds. Additionally, ENM algorithms typically use several statistics to pinpoint the most important environmental variable in terms of model explanation, giving its percent contribution to the model output. Also, response curves can be calculated for the different variables, defining the niche optima.

However, care must be taken to restrict model input to uncorrelated environmental variables to obtain valid results. With a growing availability of (global, gridded) environmental data sets, which are often correlated or redundant, a data reduction strategy should be considered. One may perform a species−environment correlation analysis or ordination to make a first selection of relevant variables and perform a subsequent Pearson correlation test between environmental variables to get rid of redundant information. Alternatively, spatial principal component analysis (PCA) may be performed to obtain uncorrelated variables, using PCA components as input variables (Verbruggen et al., 2009), although the resulting variable contributions and response curves might be hard to calculate back to original variables.

Pauly et al. (2009) applied ENM using Maxent (Phillips et al., 2006) to gain insight into worldwide blooms of the siphonous green alga *Trichosolen* growing on physically damaged coral (Fig. 3). A correlation analysis was applied to

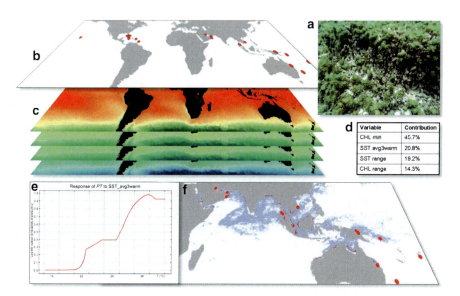

Figure 3. (a) A *Pseudobryopsis/Trichosolen* (*PT*) bloom on physically damaged coral. (b) Worldwide occurrence points of *PT* on coral. (c) Environmental grids used for model training in Maxent. (d) Relative importance of each variable in the model as identified by the algorithm. (e) Response curve of *PT* to the average of the three warmest months. (f) Binary habitat suitability map for *PT*. The gray (*blue*) shade represents suitable environment, whereas the dark (*red*) shade along the coast delineates bloom risk areas.

identify the two least correlated biologically meaningful variables derrived from SST and Chl (based on monthly data sets), adequately describing the position and extent of the distribution in environmental space. The model delineated the potential global distribution of *Trichosolen* occurring on coral based on a 95% training confidence threshold, including areas where the bloom had previously occurred. This allowed identifying areas with a high potential risk for future blooms based on environmental response curves. For instance, the response curve for the average of the three warmest months (included as a variable based on the conclusions of Schils and Wilson (2006)) shows that *Trichosolen* populations are only viable above 22°C, but only environments above 28°C are likely to sustain blooms.

3. Future Directions and Research Priorities

3.1. THE QUEST FOR SPATIAL DATA IN SEAWEED BIOLOGY

In its simplest form, "spatially explicit" seaweed data would refer to the availability of georeferenced species occurrences. While we discussed the practice of georeferencing and dissemination of spatially explicit seaweed data in depth in the second section of this chapter, we briefly show a couple of examples to demonstrate the dramatic state of the current availability of this information. For instance, looking at a random Nori species, *Porphyra yezoensis* Ueda, AlgaeBase (Guiry and Guiry, 2008) mentions 13 references to occurrence records throughout the northern hemisphere. However, the Ocean Biogeographic Information System (OBIS, an online integration of marine systematic and ecological information systems; Costello et al., 2007) contains no *P. yezoensis* records. Another random example, the siphonous (sub)tropical green alga *Codium arabicum* Kützing illustrates this further: out of 55 direct or indirect occurrence references in Algaebase, 17 are georeferenced in OBIS. However, two of the specimens wrongly have zero longitudes, hence locating the records some 400 km inland from the coast of Ghana, instead of at the Indian coast. Five out of the 17 are recorded to no better than 0.1° in both longitude and latitude, making their position uncertain within up to 120 km^2. Fifteen out of the 17 make no mention of the collector's name or publication, preventing to check the integrity of the identification. Eleven lack subcountry level locality name information, and none mention substate locality names, making it impossible to verify geographical coordinates through the use of gazetteers.

If the amount of coastal or marine publications using GIS, mapping, or remote sensing can be called minimal, averaging 8% of the total publications using these geographic techniques as previously shown, the proportion of these records mentioning seaweeds or macroalgae is statistically speaking barely existing, attaining 0.5–1% of the spatial marine studies. Studies investigating the other two best-known benthic marine communities, coral (reefs) and seagrasses, constitute up to 10%, while the remainder covers (in no particular order) mangroves and other supratidal coastal communities and structures, coastal or marine topography, and geomorphology or nautical issues. Some of the reasons accounting for this disproportion are obvious: for a start, relatively few investigate seaweeds. However, out of 12,074 studies mentioning seaweeds or macroalgae in ISI Web of Knowledge, a potential 7,279 in the fields of ecology, biogeography, phylogeography, or ecophysiology could benefit from some sort of spatial explicit information, while only 177 (2.5%) actually mention to do so in their title, abstract, or keywords. Other problems concern the nature of seaweed communities: while coral reefs and seagrass meadows usually form large and relatively homogeneous assemblages, seaweeds are spatially and spectrally very heterogeneous. This is particularly difficult to cope with in remote sensing studies, already challenged by the properties of the water column in comparison with terrestrial vegetation studies.

3.2. GEOREFERENCING SPECIMENS

Sections 1.4 and 2.1 demonstrate the need to prioritize the standardization of disseminating and linking geographical seaweed specimen information. Investigating the consequences of global change requires the availability of correct and complete global data sets. Therefore, we support the requirement of the dissemination of sample coordinates not only from geographically oriented studies, but from every study using in situ sampled seaweeds, to allow for informative and accurate meta-analyses. Coordinate pairs should be deposited in already existing global biodiversity databases such as OBIS, but minimal geographic accuracy and complete specimen information including collector's name should be required to allow vigorous quality control. The use of global biodiversity databases as a main depositing center for specimen coordinates rather than dedicated seaweed databases also opens perspectives to investigate potential correlations between seaweed and faunal distribution shifts in response to global change. However, it should also be investigated how general geographical biodiversity databases such as OBIS could be related to and synchronized with specific databases such as Algaebase and GenBank to optimize the dissemination of all kinds of specimen information.

3.3. REMOTE SENSING

No significant time gap exists between the development and deployment of airborne sensors; due to an optimal use of the most recent technologies, airborne sensors thus represent the best technical characteristics desirable for seaweed mapping to date. As time goes on, the most recent satellite sensors can benefit from the evolution in technologies to more closely resemble the properties of airborne sensors. Vahtmäe et al. (2006) used a simulation study to demonstrate that submerged seaweeds in turbid coastal waters could well be mapped using hyperspectral satellite sensors like CHRIS and Hyperion, featuring 10 nm wide bands in the visual wavelengths. However, they also postulated a signal-to-noise ratio of 1,000:1, an image quality not met by these existing sensors. It is thus vital that similar new hyperspectral, very high resolution satellite sensors should be developed for seaweed mapping and monitoring in the framework of global change research. However, Fig. 2 shows that planned sensors for the next 3 years follow the historical trade-off toward multispectral very high resolution systems. Nowadays, this seems to be motivated by two elements: the huge thrust for coral reef research, in which macroalgae are often lumped into one or few functional classes and spatial resolution is considered more important than spectral resolution, and disaster event monitoring, focusing on a near-one day site revisiting time through the use of extensive off-nadir or off-track pointing capabilities (Table 2). The latter technique also generates huge amounts of data, adding a new dimension to the historical trade-off situation: current data storage capacities allow for two image characteristics out of three (spectral, spatial, and temporal resolution) to be optimized, but not all three. Unfortunately, no significant thrust seems to exist to develop sensors ideally suitable for large-scale algal mapping and monitoring to date, explaining the characteristics of the missions in development. As a means to deal with the lack of very high resolution hyperspectral imagery, efforts have been made to combine the information from several sensors with different characteristics into one data set. This is analogous to pan-sharpening techniques, which use the high spatial detail of a panchromatic band to spatially enhance the multispectral imagery from the same sensor (see also Fig. 2). Although useful in current conditions, we suspect these techniques to become less important as more advanced sensors would be developed, since processing information from one sensor evidently is less time- and resource-consuming and more accurate than using multisensor information.

Light-based active remote sensing involves the emission of laser pulses with a known frequency and subsequently detecting fluorescence in certain wavelengths. Kieleck et al. (2001) proved this technique to be successful in discerning submerged green, brown, and red seaweeds in lab conditions. Mazel et al. (2003) used a similar prototype in-water laser multispectral fluorescence imaging system to map different coral reef bottom structures, including macroalgae, on a 1-cm resolution. Airborne laser imaging has been used extensively to provide very high resolution imagery in terrestrial applications such as forestry. In the marine realm, its applications are mostly limited to in-water (boat-mounted) transect mapping strategies, although further research to develop aerial systems could prove useful to obtain very high resolution imagery of individual seaweed patches, e.g., to map the spreading of macroalgae on coral reefs.

3.4. DISTRIBUTION AND NICHE MODELING

Presence-only data are mostly inherent to seaweed niche distribution modeling due to sampling locality bias (caused by difficulty of coastal and submerged terrain access), small sizes, or seasonally microscopic life stages of seaweed species or cryptic species. Under the title of Species' Distribution Modeling, Pearson (2007) published a general manual including (presence-only) niche modeling, mostly based on Maxent. However, the manual is based on terrestrial experiences, as niche modeling algorithms have rarely been applied to seaweed distribution to date, and some issues characteristic of marine benthic niche modeling are not elaborated. For instance, there are more global environmental GIS data available for the terrestrial realm when compared with the marine environment. Table 3 lists marine environmental variables currently available from global satellite imagery, along with data that will become available in the near future. Especially, globally gridded salinity data are lacking to date. Other variable data sets (pH, nutrients, salinity, turbidity, etc.) may also be compiled from the interpolation of in situ data, e.g., from the Worldwide Ocean Optics Database (Freeman et al., 2006) or the World Ocean Database (NOAA, 2008). These data, consisting of vertical profiles, can be advantageous for 3D modeling, but the interpolation techniques necessary to obtain gridded maps may be challenging. Furthermore, global change climate extrapolations resulted in the production of global gridded maps of environmental variables for future scenarios in the terrestrial realm, but similar data for the marine realm are not yet available. The projection of calculated niches on future distributions can greatly enhance our understanding of global change consequences, and it is therefore crucial that future research is aimed at composing similar gridded maps of future scenarios for marine environmental variables. More research should also be aimed at setting model parameters to account for spatial autocorrelation and clustering of species occurrence data. Finally, model validation and output comparison statistics are under scrutiny in recent literature (e.g., Peterson et al., 2008), and more research is needed to agree on the best statistics suitable for marine data.

Modeling on a local scale allows for including high-resolution environmental variables that are not available for the entire globe. This is particularly the case where environmental variables not available from satellite data have been measured in situ and can be interpolated locally. In other cases, one or several (very) high resolution satellite scenes can be used to provide substrate data, not relevant on a global scale with 1-km gridded environmental variables. For instance, De Oliveira et al. (2006) included substrate, flooding frequency, and wave exposure to model the distribution of several intertidal and shallow subtidal brown seaweeds along a 20 km coastal stretch in Brittany, France. Thus, it can be expected that multiscale modeling approaches will gain importance in the near future.

While human-induced effects on habitats are thought to drive short-term species dynamics, it is often stated that global climate change will influence the capacity of alien species to invade new areas on a medium to long term. Range shifts of individual species in an assemblage under climate change are based on largely the same processes driving the spread of alien species. Hence, the two can be addressed using the same approach (Thuiller et al., 2007). Although very complex processes are involved, the geographic component of species' invasions can be very well predicted using niche modeling techniques (Peterson, 2003). Once a comprehensive marine environmental data set is compiled, invaded areas and areas at risk of invasion can be successfully predicted based on the native niche of alien species (Peterson, 2005; Pauly et al., 2009), although Brönnimann et al. (2007) warn that a niche shift may occur after invasion. Nevertheless, niche modeling approaches are promising in future research of seaweeds' range shifts and invasions. Verbruggen et al. (2009) also applied niche modeling techniques to unravel the evolutionary niche dynamics in the green algal genus *Halimeda*, concluding that globally changing environments may allow certain macroalgae to invade neighboring niches and subsequently to form a divergent lineage. They also used Maxent to identify key areas to be targeted for future field work in search for new sister species – an application in biodiversity considered important in the light of global change.

To date, seaweed assemblages have often been characterized using quantitative vegetation analyses and multivariate statistics to delineate different community types and to establish the link between environmental variables and communities. With quickly developing niche modeling algorithms now regarded as the most advanced way to accomplish the latter, community niche modeling will be of particular value in global change-related seaweed research in the coming years. Ferrier and Guisan (2006) defined three ways to predict the niche of communities as a whole, rather than the niches of individual species. The assemble-first, predict-later strategy seems to be the most promising for seaweed data, since the existing floristic data have often been statistically assembled into communities. We suggest that prioritizing the development of community niche modeling algorithms can greatly speed up our insight into seaweed community response to future climate change.

4. References

Ader, R.R. (1982) A Geographic Information System for addressing issues in the coastal zone. Comput. Environ. Urban Syst. **7**: 233–243.

Adey, W.H. and Steneck, R.S. (2001) Thermogeography over time creates biogeographic regions: a temperature/space/time-integrated model and an abundance-weighted test for benthic marine algae. J. Phycol. **37**: 677–698.

Amsterdam, R., Andresen, E. and Lipton, H. (1972) Geographic Information Systems in the U.S. – an overview. Afips Conf. Proc. **40**: 511–522.

Andréfouët, S., Zubia, M. and Payri, C. (2004) Mapping and biomass estimation of the invasive brown algae *Turbinaria ornata* (Turner) J. Agardh and *Sargassum mangarevense* (Grunow) Setchell on heterogeneous Tahitian coral reefs using 4-meter resolution Ikonos satellite data. Coral Reefs **23**: 26–38.

Bailey, W.H. (1963) Remote sensing of the environment. Ann. Assoc. Am. Geogr. **53**: 577–578.

Belsher, T., Loubersac, L. and Belbeoch, G. (1985) Remote sensing and mapping, In: M.M. Littler and D.S. Littler (eds.) *Ecological Field Methods: Macroalgae. Phycological Handbook* **4**. Cambridge University Press, Cambridge, pp. 177–197.

Brönnimann, O., Treier, U.A., Müller-Schärer, H., Thuiller, W., Peterson, A.T. and Guisan, A. (2007) Evidence of climatic niche shift during biological invasion. Ecol. Lett. **10**(8): 701–709.

Costello, M.J., Stocks, K., Zhang, Y., Grassle, J.F. and Fautin, D.G. (2007) *About the Ocean Biogeographic Information System*, available online at http://www.iobis.org.

De Oliveira, E., Populus, J. and Guillaumont, B. (2006) Predictive modelling of coastal habitats using remote sensing data and fuzzy logic: a case for seaweed in Brittany (France). EARSeL eProceedings **5**(2): 208–223.

Egan, W.G. and Hair, M.E. (1971) Automated delineation of wetlands in photographic remote sensing. *Proceedings of the 7th International Symposium on Remote Sensing of Environment*, Volume III, University of Michigan, Ann Arbor, Michigan (USA) pp. 2231–2251.

Elith, J., Graham, C.H., Anderson, R.P., Dudík, M., Ferrier, S., Guisan, A., Hijmans, R.J., Huettmann, F., Leathwick, J.R., Lehmann, A., Li, J., Lohmann, L.G., Loiselle, B.A., Manion, G., Moritz, C., Nakamura, M., Nakazawa, Y., Overton, J.M., Peterson, A.T., Phillips, S.J., Richardson, K., Scachetti-Pereira, R., Schapire, R.E., Soberon, J., Williams, S., Wisz, M.S. and Zimmermann, N.E. (2006) Novel methods improve prediction of species' distributions from occurrence data. Ecography **29**: 129–151.

Ferrier, S. and Guisan, A. (2006) Spatial modelling of biodiversity at the community level. J. Appl. Ecol. **43**: 393–404.

Freeman, A.S., Chiu, C.P. and Smart, J.H. (2006) *The Office of Naval Research's Worldwide Ocean Optics Database (WOOD v4.5i), User's Guide*. The Johns Hopkins University Applied Physics Laboratory, USA, available online at http://wood.jhuapl.edu.

Gagnon, P., Scheibling, R.E., Jones, W. and Tully, D. (2008) The role of digital bathymetry in mapping shallow marine vegetation from hyperspectral image data. Int. J. Remote Sens. **29**: 879–904.

Graham, M.H., Kinlan, B.P., Druehl, L.D., Garske, L.E. and Banks, S. (2007) Deep-water kelp refugia as potential hotspots of tropical marine diversity and productivity. Proc. Natl. Acad. Sci. USA **104**: 16576–16580.

Green, E.P., Mumby, P.J., Edwards, A.J. and Clarck, C.D. (2000) Remote sensing handbook for tropical coastal management, In: A.J. Edwards (ed.) *Coastal Management Sourcebooks* 3. UNESCO, Paris, x+360 pp.

Guillaumont, B., Bajjouk, T. and Talec, P. (1997) Seaweed and remote sensing: a critical review of sensors and data processing, In: F.E. Round and D.J. Chapman (eds.) *Progress in Phycological Research* **12**. Biopress Ltd., pp. 213–282.

Guillaumont, B., Callens, L. and Dion, P. (1993) Spatial distribution and quantification of *Fucus* species and *Ascophyllum nodosum* beds in intertidal zones using SPOT imagery. Hydrobiologia **260/261**: 297–305.

Guiry, M.D. and Guiry, G.M. (2008) *AlgaeBase*. National University of Ireland, Galway, Ireland, available online at http://www.algaebase.org.

Hutchinson, G.E. (1957) Concluding remarks. Cold Spring Harb. Symp. Quant. Biol. **22**(2): 415–427.

Kidd, D.M. and Ritchie, M.G. (2006) Phylogeographic information systems: putting the geography into phylogeography. J. Biogeogr. **33**: 1851–1865.

Kieleck, C., Bousquet, B., Le Brun, G., Cariou, J. and Lotrian, J. (2001) Laser induced fluorescence imaging: application to groups of macroalgae identification. J. Phys. D Appl. Phys. **34**: 2561–2571.

Kozak, K.H., Graham, C.H. and Wiens, J.J. (2008) Integrating GIS-based environmental data into evolutionary biology. Trends Ecol. Evol. **23**: 141–148.

Mazel, C.H., Strand, M.P., Lesser, M.P., Crosby, M.P., Coles, B. and Nevis, A.J. (2003) High-resolution determination of coral reef bottom cover from multispectral fluorescence laser line scan imagery. Limnol. Oceanogr. **48**(1, part 2): 522–534.

Moore, G.E. (1965) Cramming more components onto integrated circuits. Electronics **38**(8): 4 pp.

NCBI (2008) *GenBank*. National Library of Medicine, USA, available online at http://www.ncbi.nlm.nih.gov.

NOAA (2008) *World Ocean Database*. National Oceanographic Data Center, USA, available online at http://www.nodc.noaa.gov.

Nature Editorial (2008) A place for everything. Nature **453**: 2.

Pauly, K., Verbruggen, H., Tyberghein, L., Mineur, F., Maggs, C.A., Shimada, S. and De Clerck, O. (2009) Predicting spread and bloom risk areas of introduced and invasive seaweeds. Oral presentation, 9th International Phycological Congress, Tokyo (Japan), 2–8 August 2009. Phycologia **48**(4) S, 103.

Peterson, A.T. (2003) Predicting the geography of species' invasions via ecological niche modeling. Q. Rev. Biol. **78**(4): 419–433.

Peterson, A.T. (2005) Predicting potential geographic distributions of invading species. Curr. Sci. **89**(1): 9.

Peterson, A.T., Papes, M. and Soberon, J. (2008) Rethinking receiver operating characteristic analysis applications in ecological niche modeling. Ecol. Model. **213**: 63–72.

Pearson, R.G. (2007) *Species' Distribution Modeling for Conservation Educators and Practitioners: Synthesis.* American Museum of Natural History, USA, available online at http://ncep.amnh.org.

Phillips, S.J., Anderson, R.P. and Schapire, R.E. (2006) Maximum entropy modeling of species geographic distributions. Ecol. Model. **190**: 231–259.

Polcyn, F.C. and Sattinger, I.J. (1969) Water depth determinations using remote sensing techniques. *Proceedings of the 6th International Symposium on Remote Sensing of Environment*, Volume II, University of Michigan, Ann Arbor, Michigan (USA), pp. 1017–1028.

Ratnasingham, S. and Hebert, P.D.N. (2007) BOLD: the barcode of life data system (www.barcodinglife.org). Mol. Ecol. Notes **7**(3): 355–364.

Rios, N.E. and Bart, H.L. Jr. (1997) *GEOLocate Georeferencing Software: User's Manual.* Tulane Museum of Natural History, Belle Chasse LA, USA, available online at http://www.museum.tulane.edu/geolocate.

Schils, T. and Wilson, S.C. (2006) Temperature threshold as a biogeographic barrier in northern Indian Ocean macroalgae. J. Phycol. **42**: 749–756.

Stang, F.W. (1969) Ocean and water surface temperature measurements using infrared remote sensing techniques. *Proceedings of the Spie 14 Annual Technical Symposium: Photo-Optical Instrumentation Applications and Theory*, pp. 77–83.

Theriault, C., Scheibling, R., Hatcher, B. and Jones, W. (2006) Mapping the distribution of an invasive marine alga (*Codium fragile* subsp. *tomentosoides*) in optically shallow coastal waters using the Compact Airborne Spectrographic Imager (CASI). Can. J. Remote Sens. **32**: 315–329.

Thuiller, W., Richardson, D.M. and Midgley, G.F. (2007) Will climate change promote alien plant invasions? In: W. Nentwig (ed.) *Biological Invasions. Ecological Studies* **193**. Springer Verlag, Berlin, Heidelberg, pp. 197–211.

Vahtmäe, E., Kutser, T., Martin, G. and Kotta, J. (2006) Feasibility of hyperspectral remote sensing for mapping benthic macroalgal cover in turbid coastal waters – a Baltic Sea case study. Remote Sens. Environ. **101**: 342–351.

van den Hoek, C., Breeman, A.M. and Stam, W.T. (1990) The geographic distribution of seaweed species in relation to temperature: present and past, In: J.J. Beukema, W.J. Wolff and J.J.W.M. Brouns (eds.) *Expected Effects of Climatic Change on Marine Coastal Ecosystems.* Kluwer Academic Press, Dordrecht, pp. 55–67.

Verbruggen, H., Tyberghein, L., Pauly, K., Vlaeminck, C., Van Nieuwenhuyze, K., Kooistra, W.H.C.F., Leliaert, F. and De Clerck, O. (2009) Macroecology meets macroevolution: evolutionary niche dynamics in the seaweed *Halimeda*. Global Ecol. Biogeogr. 18(4), 393–405.

Biodata of **Dinghui Zou** and **Kunshan Gao,** authors of *"Physiological Responses of Seaweeds to Elevated Atmospheric CO_2 Concentrations"*

Dr. Dinghui Zou is currently the Professor of College of Environmental Science and Engineering, South China University of Technology, China. He obtained his Ph.D. from the Institute of Hydrobiology, the Chinese Academy of Sciences, in 2001. Professor Zou's scientific interests are in the areas of: the relationship of seaweeds and global change, the mechanisms of inorganic carbon utilization in seaweeds, and the environmental regulation of growth and development of seaweeds.

E-mail: **dhzou@scut.edu.cn**

Professor Kunshan Gao is currently the distinguished Chair Professor of State Key Laboratory of Marine Environmental Science, Xiamen University, China. He obtained his Ph.D. from Kyoto University of Japan in 1989 and continued his research since then at Kansai Technical Research Institute of Kansai Electrical Co. and at University of Hawaii in USA as a postdoctoral fellow. He was appointed as Associate Professor of Shantou University in 1995, and became recognized as the outstanding young scientist in 1996 by NSFC, then as Professor for 100 talented programmes in the Institute of Hydrobiology by the Chinese Academy of Sciences in 1997. Professor Gao's scientific interests are in the areas of: ecophysiology of algae and algal photobiology, focusing on the environmental impacts of increasing atmospheric CO_2 under solar radiation.

E-mail: **ksgao@xmu.edu.cn**

Dinghui Zou   **Kunshan Gao**

PHYSIOLOGICAL RESPONSES OF SEAWEEDS TO ELEVATED ATMOSPHERIC CO_2 CONCENTRATIONS

DINGHUI ZOU[1] AND KUNSHAN GAO[2]
[1] *College of Environmental Science and Engineering, South China University of Technology, Guangzhou, 510640, China*
[2] *State Key Laboratory of Marine Environmental Science, Xiamen University, Xiamen, 361005, China*

1. Introduction

The atmospheric CO_2 concentration has been rising since the industrial revolution, and will continue to rise from the present 375 to about 1,000 ppmv by 2100 (Pearson and Palmer, 2000), increasing dissolution of CO_2 from the air and altering the carbonate system of Surface Ocean (Stumm and Morgan, 1996; Takahashi et al., 1997; Riebesell et al., 2007). For example, an increase in atmospheric CO_2 from 330 to 1,000 ppmv will lead to an increase in CO_2 concentration from 12.69 to 38.46 µM in seawater (at 15°C and total alkalinity of 2.47 eq m^{-3}) and an increase in the concentration of dissolved inorganic carbon (DIC, i.e., $CO_{2(aq)}$, HCO_3^-, and CO_3^{2-}) from 2.237 to 2.412 mM, with a concurrent decrease in the pH of the surface seawater from 8.168 to 7.735 (Raven, 1991; Stumm and Morgan, 1996). Increasing atmospheric CO_2 and its associated changes in the carbonate system can influence the physiology and ecology of seaweeds.

Seaweeds (Chlorophyta, Rhodophyta, and Phaeophyta) are usually distributed in intertidal and subtidal zones of coastal waters. They play an important role in the coastal carbon cycle (Reiskind et al., 1989) and contribute remarkably to sea-farming activities. The rate of primary production of some species is comparable with those of the most productive land plants; therefore, seaweeds have a great potential for CO_2 bioremediation (Gao and Mckinley, 1994). On the other hand, increasing pCO_2 in seawater would affect physiology of seaweeds. Therefore, a number of studies have been performed to envisage the impacts of CO_2 enrichment on photosynthesis, growth, nutrients metabolism, and cell components of seaweeds. Results showed that increased CO_2 concentration may enhance, inhibit, or not affect the growth of the species investigated. This work is intended to examine how the macroalgal species respond and acclimate to elevated CO_2 levels.

2. Inorganic Carbon Limitation

The effects of elevated CO_2 concentrations on seaweeds largely depend on the degree of carbon limitation present in natural systems. Photosynthesis of seaweeds would be severely limited under current atmospheric conditions if it were dependent only on diffusional entry of CO_2 from the medium to the site of fixation via the carbon-assimilating enzyme Rubisco. There are several aspects of CO_2 limitation of carbon assimilation in seaweeds (Beardall et al., 1998): (1) rather low dissolved CO_2 concentration; (2) low diffusion rate of CO_2 in seawater, being four orders of magnitude slower than in air; (3) the slow spontaneous formation of CO_2 from HCO_3^- dehydration; and (4) the high K_m values (40–70 µM) of Rubisco of algae. Nevertheless, photosynthesis in the investigated species can be fully or nearly saturated with the current ambient dissolved inorganic carbon (Ci) composition because of the presence of CO_2-concentrating mechanisms (CCMs) that enable the algae to efficiently utilize the bulk HCO_3^- pool in seawater (Beer, 1994; Beer and Koch, 1996; Raven, 1997; Larsson and Axelsson, 1999; Zou et al., 2004; Giordano et al., 2005), which is about 150 times more abundant than free CO_2. Some species, however, exhibit Ci-limited photosynthesis in natural seawater (e.g., Johnston et al., 1992; Andría et al., 1999a; Zou et al., 2003).

HCO_3^- is usually dehydrated extracellularly as mediated by periplasmic carbonic anhydrase (CA) to release CO_2, which is then taken up into the cell. Another important approach for Ci acquisition of algae is the active uptake of HCO_3^- through the plasma membrane facilitated by an anion exchange protein (Drechsler et al., 1993, 1994; Axelsson et al., 1995). Additionally, H^+-ATPase-driven HCO_3^- uptake has also been recognized in several marine seaweeds (Choo et al., 2002; Snoeijs et al., 2002). Seaweeds show different capacities to take advantage of the HCO_3^- pool in seawater (Axelsson and Uusitalo, 1988; Maberly, 1990; Mercado et al., 1998). Therefore, they can exhibit heterogeneous, often species-specific responses to elevated CO_2. Their physiological responses to elevated CO_2 levels can also depend on their acclimation strategies and the environmental constraints under which CO_2 enrichment is imposed.

3. Growth

When juveniles of *Porphyra yezoensis* germinated from the chonchospores were grown at enriched CO_2 levels of 1,000 or 1,600 ppmv for 20 days, their growth was significantly enhanced (Gao et al., 1991; Fig. 1). Similar findings were reported in *Gracilaria* sp., *Gracilaria chilensis*, and *Hizikia fusiforme* (Gao et al., 1993a; Zou, 2005). Although these species are capable of using bicarbonate, they still showed carbon-limited photosynthetic rates in natural seawater. Growth of a nonbicarbonate-user, the red alga *Lomentaria articulata*, was stimulated by enriched CO_2 (Kübler et al., 1999). The enhancement could be attributed to the accelerated photosynthetic carbon fixation by increasing Ci availability or the depression of photorespiration by elevating the ratio of CO_2 to O_2 in the culture medium. It was interesting that growth of a green alga, *Ulva rigida*, which showed efficient ability of HCO_3^- utilization and

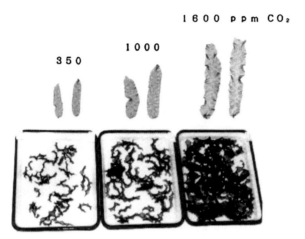

Figure 1. Enhanced growth of *Porphyra yezoensis* when 50 juveniles each (germinated from the same bunch of chonchospores released from the same chonchocelis, about 5 mm long at the beginning of the culture) were grown at different CO_2 concentrations in aeration. The photo images were taken after 20 days culture (Gao et al., 1991).

saturated photosynthesis at the current Ci concentration of seawater (Björk et al., 1993; Mercado et al., 1998), was also enhanced at high CO_2 concentrations (Björk et al., 1993; Gordillo et al., 2001). Such an enhancement of growth was suggested to be caused by the enhanced N-assimilation (Gordillo et al., 2001), but could also be attributed to downregulation of HCO_3^- uptake and consequent energy saving for its operation. On the other hand, a decrease in growth rate caused by elevated CO_2 has been reported in *G. tenuistipitata* (Garcìa-Sánchez et al., 1994), *P. leucostica* (Mercado et al., 1999), and *P. linearis* (Israel et al., 1999). Such an inhibition of growth was associated with lowered photosynthetic activity even measured at high CO_2 concentrations (Garcìa-Sánchez et al., 1994). However, such a negative effect could also be caused by acidification of the medium (Israel et al., 1999). A more recent study by Israel and Hophy (2002) reported that the growth rates of 13 species (representing Chlorophyta, Rhodophyta, and Phaeophyta) cultivated in normal seawater were comparable with their growth in CO_2-enriched seawater. The authors ascribed such nonresponsive behavior to the presence of CCMs that rely on the utilization of HCO_3^-. Obviously, researches show that enrichment of CO_2 in seawater may affect, positively, neutrally, or negatively, the growth of seaweeds in direct or indirect ways.

4. Photosynthesis

4.1. PHOTOSYNTHETIC Ci UTILIZATION

The response of macroalgal photosynthesis to elevated pCO_2 in seawater is species-specific. When cultured in high CO_2, the light-saturated photosynthetic rate was reduced in *Fucus serratus* (Johnston and Raven, 1990), *G. tenuistipitata*

(Garcìa-Sánchez et al., 1994), and *P. yezoensis* (Gao, unpublished data) when measured at normal Ci of seawater. When the photosynthetic rate was measured at elevated DIC levels, it was significantly higher in the thalli grown at enriched CO_2 levels in *P. yezoensis* (Gao et al., 1991) and *Gracilaria* sp. (Andría et al., 1999b). In *P. leucostica*, Mercado et al. (1999) found no significant difference between the maximal gross photosynthetic rates of the thalli grown at enriched and current inorganic carbon concentrations.

The photosynthetic affinity for Ci and the capacity of HCO_3^- utilization are usually lowered in seaweeds following exposures to high CO_2 (Johnston and Raven, 1990; Björk et al., 1993; Mercado et al., 1997; Andría et al., 1999a, b; Zou et al., 2003). Growing the cells at high CO_2 levels decreased activity of the external (periplasmic) or total CA activity in *Ulva* sp. (Björk et al., 1993), *G. tenuistipitata* (Garcìa-Sánchez et al., 1994), *P. leucosticta* (Mercado et al., 1997), and *H. fusimorme* (Zou et al., 2003). Such a decrease reflects a decline in the capacity of HCO_3^- utilization. Israel and Hophy (2002) showed that the enzymatic features of Rubisco did not differ in the seaweeds when compared between the CO_2-enriched and control cultures, though enrichment of CO_2 was reported to decrease the content of Rubisco in *G. tenuistipitata* (Garcìa-Sánchez et al., 1994), *Gracilaria* sp. (Andría et al., 1999a), and *P. leucosticta* (Mercado et al., 1997).

4.2. PHOTOCHEMICAL EFFICIENCY

Photosynthetic acclimation in seaweeds to high levels of Ci generally resembles their responses to high irradiances, resulting in a decrease in pigment contents. For example, the phycobiliprotein (phycoerythrin and phycocyanin) and Chl *a* contents were reduced in *Gracilaria* sp. (Andría et al., 1999b, 2001), *G. tenuistipitata* (Garcìa-Sánchez et al., 1994), and *P. leucosticta* (Mercado et al., 1999) grown at high levels of Ci than those at normal Ci level. On the other hand, both maximum quantum yield and effective quantum yield were downregulated in *P. leucostica* when grown under high Ci conditions (Mercado et al., 1999), suggesting that enriched CO_2 lowered the demand of energy for the HCO_3^- utilization mechanism.

4.3. EMERSED PHOTOSYNTHESIS OF INTERTIDAL SEAWEEDS

Intertidal seaweeds experience continual alternation of living in air and in water as the tidal level changes. Their photosynthesis undergoes dramatic environmental changes between the aquatic and terrestrial exposures. When the tide is high, they are submerged in seawater, where HCO_3^- pool is available for their photosynthesis (Beer and Koch, 1996; Beardall et al., 1998). When the tide is low, intertidal seaweeds are exposed to air, large buffering reservoir of HCO_3^- in seawater is no longer present, and atmospheric CO_2 becomes the only exogenous carbon resource for their photosynthesis. The acquisition of CO_2 is less constrained in air than in seawater, through which CO_2 diffuses about 10,000 times slower (Raven, 1999). However, this constraint can be offset by the abundance of HCO_3^-, as many intertidal algae can use HCO_3^- as the exogenous inorganic carbon source for photosynthesis (Maberly, 1990; Gao and McKinley, 1994). Thus, carbon limitation during photosynthesis in intertidal species may be potentially more important in air than in water.

It is known that intertidal seaweeds can tolerate the emersed conditions, and the photosynthesis during emersion contributes significantly to their total carbon fixation budget (e.g., Gao and Aruga, 1987; Maberly and Madsen, 1990). Our previous works (Gao et al., 1999; Zou and Gao, 2002; Zou and Gao, 2004a, b, 2005; Zou et al., 2007) showed that elevated atmospheric CO_2 might have a fertilizing effect increasing photosynthesis while exposed to air at low tide in most of the tested species, i.e. the red seaweeds *P. haitanensis, Gloiopeltis furcata*, and *Gigartina intermedia*, the brown seaweeds *Ishige okamura, H. fusiformis*, and *Sargassum hemiphyllum*, and the green seaweeds *Enteromopha linza* and *Ulva lactuca*. The relative photosynthetic enhancement by the elevated CO_2 levels increased with desiccation, although the absolute photosynthetic rate decreased with desiccation. The enhancement of daily photosynthetic production by elevated CO_2 concentration during emersion differs among species owing to their zonational depths and exposure durations and the daily timing of emersion (Gao et al., 1999; Zou and Gao, 2005; Zou et al., 2007). Additionally, the CO_2 compensation points increased with enhanced desiccation, with higher CO_2 concentrations required to maintain positive photosynthesis (Gao et al., 1999; Zou and Gao, 2002, 2005).

5. Calcification

It is estimated from more than two million surveys that the oceans have absorbed more than one third of the anthropogenic CO_2 released to the atmosphere (Sabine et al., 2004). With increasing atmospheric CO_2 concentration, CO_2 dissolves in seawater to reach new equilibrium in the carbonate system. This leads to an increase in the concentrations of HCO_3^- and H^+ and a decrease in the concentration of CO_3^{2-} and of saturation state of calcium carbonate (Gattuso et al., 1999; Gattuso and Buddemeier, 2000; Caldeira and Wickett, 2003; Orr et al., 2005). The surface water of the ocean is known to have been acidified by 0.1 pH unit (corresponding to a 30% increase of H^+) since 1800 (Orr et al., 2005), and will be further acidified by another 0.3–0.4 unit (about 100–150% increase of H^+) by 2100 (Brewer, 1997; Caldeira and Wickett, 2003). Such an ocean-acidifying process has been suggested to harm marine-calcifying organisms by reducing the rate of calcification of their skeletons or shells (e.g., Gao et al., 1993b; Gattuso et al., 1999; Riebesell et al., 2000; Orr et al., 2005).

In the coastal waters where seaweeds are distributed, pH of seawater fluctuates within a larger range than pelagic waters because of inputs from terrestrial systems and fisheries. Nevertheless, additional CO_2 input can still affect the biological activities in coastal waters, because ocean acidification will lower the pH regimes, shifting the pH range to a lower one. Therefore, increased pCO_2 and decreased pH and CO_3^{2-} will affect calcifying seaweeds. Gao et al. (1993b) showed that enrichment of CO_2 to 1,000 and 1,600 ppmv in aeration inhibited the calcification in the articulated coralline alga *Corallina pilulifera*. It has also been shown that the increase in CO_2 concentrations significantly slowed down calcification of temperate and tropical corals and coralline macroalgae (Gattuso et al., 1998; Langdon et al., 2000). For the marine-calcifying phytoplankton *Emiliania huxleyi*, calcification was reported to be reduced by the enriched CO_2 (Riebesell et al., 2000), while a recent study showed that its calcification increased with elevated CO_2 (Iglesias-Rodriguez et al., 2008). On the other hand, when pH was controlled at a constant level, elevated concentrations of DIC enhanced the calcification of *Bossiela orbigniana* (Smith and Roth, 1979) and *C. pilulifera* (Gao et al., 1993b).

6. Nitrogen Metabolism

Zou (2005) reported that both the nitrate uptake rate and the activity of nitrate reductase (NR) in the brown algae *H. fusiforme* were increased following cultures at high CO_2 levels. It was also shown that elevated CO_2 concentrations in culture stimulated the uptake of NO_3^- in *Gracilaria* sp. and *G. chilensis* (Gao et al., 1993a), *Ulva lactuca* (Zou et al., 2001), and *U. rigida* (Gordillo et al., 2001), and enhanced the activity of NR in *P. leucosticta* (Mercado et al., 1999) and *U. rigida* (Gordillo et al., 2001, 2003). This indicates that elevated CO_2 concentrations can enhance nitrogen assimilation, as more nitrogen is required to support higher growth rate. The regulation of NR activity in seaweed by CO_2 might be through a direct action on de novo synthesis of the enzyme, rather than through physiological consequences in carbon metabolism as occurring in terrestrial higher plants (Gordillo et al., 2001, 2003). Contrarily, decreased uptake rate of NO_3^- by high CO_2 in *G. tenuistipitata* (Garcìa-Sánchez et al., 1994) and *G. gaditana* (Andría et al., 1999b) was also reported. Mercado et al. (1999) stated that NO_3^- uptake and reduction might be uncoupled when algae are grown at high CO_2. Responses of macroalgal nitrogen assimilation to elevated CO_2 could be species-specific; however, the results from different studies might be also generated from different culture systems or methods.

7. C/N Ratio

Growth under enrichment of CO_2 would alter the cellular components of seaweeds. Contents of soluble proteins and phycobiliprotein were decreased in *Graciaria tenuisitipitata* (Garcìa-Sánchez et al., 1994), *Gracilaria* sp. (Andría et al., 1999b), and *P. leucosticta* (Mercado et al., 1999) when they were grown at high DIC levels. In contrast, the content of soluble carbohydrate was increased in *Gracilaria* sp. (Andría et al., 1999b). As a result of these changes, C/N ratios were increased in the seaweeds grown at elevated CO_2 levels (Garcìa-Sánchez et al., 1994; Kübler et al., 1999; Mercado et al., 1999). Although phycobiliprotein, soluble proteins, and Rubisco contents were found to decrease under DIC-enriched conditions, internal N content was not significantly affected by the DIC levels. Andría et al. (1999b) thereby suggested that the exposure and acclimation to high CO_2 would involve the reallocation of resources, such as N, away from Rubisco and other limiting components (electron transport) towards carbohydrate synthesis and nonphotosynthetic processes.

8. Summary

Atmospheric CO_2 rise leads to a proportional increase in pCO_2 of seawater and alters the carbonate chemistry, reducing the carbonate ions and pH while increasing that of bicarbonate. Physiological responses of seaweeds to elevated CO_2 concentrations are highly variable, depending on the species, growing conditions, and duration of CO_2 enrichment. In the species investigated, growth was enhanced, inhibited, or not affected by enrichment of CO_2, while photosynthetic performance varied according to Ci acquisition mechanisms or the acclimation strategies. Usually, net photosynthesis was enhanced in elevated DIC levels for the species with less efficiency in bicarbonate utilization or CCMs. Growing the seaweeds in high CO_2 downregulated their CCMs and possibly the electron transport demanded for its operation. On the other hand, calcification of calcifying seaweeds is negatively affected; nitrogen metabolism and the cellular C/N ratio would be increased in high-CO_2-grown cells. For the intertidal species, large buffering reservoir of HCO_3^- in seawater is no longer present and atmospheric CO_2 becomes the only exogenous carbon resource for their photosynthesis at low tide, elevation of atmospheric CO_2 might have a fertilizing effect, increasing their photosynthesis during emersion. More research efforts on biochemical and molecular aspects for a wider range of species grown at high CO_2/low pH conditions are needed to further evaluate the impacts of increasing atmospheric CO_2 concentrations on seaweeds. At the same time, physiological approaches are required to distinguish the effects of high CO_2 from that of lowered pH.

9. Acknowledgments

This work was supported by the Chinese 973 Project (No. 2009CB421207), the Key Project of Chinese Ministry of Education (No. 207080), and the National Natural Science Foundation (No. 40930846).

10. References

Andría, J.R., Pérez-Lloréns, J. and Vergara, J.J. (1999a) Mechanisms of inorganic carbon acquisition in *Gracilaria gaditana* nom. prov. (Rhodophyta). Planta **208**: 561–573.

Andría, J.R., Vergara, J.J. and Perez-Llorens, J.L. (1999b) Biochemical responses and photosynthetic performance of *Gracilaria* sp. (Rhodophyta) from Cadiz, Spain, cultured under different inorganic carbon and nitrogen levels. Eur. J. Phycol. **34**: 497–504.

Andría, J.R., Brun, F.G., Pérez-Lloréns, J.L. and Vergara, J.J. (2001) Acclimation responses of *Gracilaria* sp. (Rhodophyta) and *Enteromorpha intestinalis* (Chlorophyta) to changes in the external inorganic carbon concentration. Bot. Mar. **44**: 361–370.

Axelsson, L. and Uusitalo, J. (1988) Carbon acquisition strategies for marine macroalgae. I. Utilization of proton exchanges visualized during photosynthesis in a closed system. Mar. Biol. **97**: 295–300.

Axelsson, L., Ryberg, H. and Beer, S. (1995) Two modes of bicarbonate utilization in the marine green macroalga *Ulva lactuca*. Plant Cell Environ. **18**: 439–445.

Beardall, J., Beer, S. and Raven, J.A. (1998) Biodiversity of marine plants in an arc of climate change: some predictions based on physiological performance. Bot. Mar. **4**: 113–123.

Beer, S. (1994) Mechanisms of inorganic carbon acquisition in marine maroalgae (with reference to the Chlorophyta). Prog. Phycol. Res. **10**: 179–207.

Beer, S. and Koch, E. (1996) Photosynthesis of seagrasses and marine macroalgae in globally changing CO_2 environments. Mar. Ecol. Prog. Ser. **141**: 199–204.

Björk, M., Haglund, K., Ramazanov, Z. and Pedersen, M. (1993) Inducible mechanism for HCO_3^- utilization and repression of photorespiration in protoplasts and thallus of three species of *Ulva* (Chlorophyta). J. Phycol. **29**: 166–173.

Brewer, P.G. (1997) Ocean chemistry of the fossil fuel CO_2 signal: the haline signal of "business as usual". Geophys. Res. Lett. **24**: 1367–1369.

Caldeira, K. and Wickett, M.E. (2003) Anthropogenic carbon and ocean pH. Nature **425**: 365.

Choo, K.S., Snoeijs, P. and Pedersen, M. (2002) Uptake of inorganic carbon by *Cladophora glomerata* (Chlorophyta) from the Baltic Sea. J. Phycol. **38**: 493–502.

Drechsler, Z., Sharkia, R., Cabantchik, Z.I. and Beer, S. (1993) Bicarbonate uptake in the marine maxroalga *Ulva* sp. is inhibited by classical probes of anion exchange by red blood cells. Planta **191**: 34–40.

Drechsler, Z., Sharkia, R., Cabantchik, Z.I. and Beer, S. (1994) The relationship of arginine groups to photosynthetic HCO_3^- uptake in *Ulva* sp. mediated by a putative anion exchanger. Planta **194**: 250–255.

Gao, K. and Aruga, Y. (1987) Preliminary studies on the photosynthesis and respiration of *Porphyra yezoensis* under emersed condition. J. Tokyo Univ. Fish. **47**: 51–65.

Gao, K. and McKinley, K.R. (1994) Use of macroalgae for marine biomass production and CO_2 remediation: a review. J. Appl. Phycol. **6**: 45–60.

Gao, K., Aruga, Y., Asada, K., Ishihara, T., Akano, T. and Kiyohara, M. (1991) Enhanced growth of the red alga *Porphyra yezoensis* Ueda in high CO_2 concentrations. J. Appl. Phycol. **3**: 356–362.

Gao, K., Aruga, Y., Asada, K. and Kiyohara, M. (1993a) Influence of enhanced CO_2 on growth and photosynthesis of the red algae *Gracilaria* sp. and *G. chilensis*. J. Appl. Phycol. **5**: 563–71.

Gao, K., Aruga, Y., Asada, K., Ishihara, T., Akano, T. and Kiyohara, M. (1993b) Calcification in the articulated coralli alga *Corallina pilulifera*, with special reference to the effect of elevated atmospheric CO_2. Mar. Biol. **117**: 129–132.

Gao, K., Ji, Y. and Aruga, Y. (1999) Relationship of CO_2 concentrations to photosynthesis of intertidal macrioalgae during emersion. Hydrobiologia **398/399**: 355–359.

Garcìa-Sánchez, M.J., Fernández, J.A. and Niell, F.X. (1994) Effect of inorganic carbon supply on the photosynthetic physiology of *Gracilaria tenuistipitata*. Planta **194**: 55–61.

Gattuso, J.-P. and Buddemeier, R.W. (2000) Calcification and CO_2. Nature **407**: 311–312.

Gattuso, J.-P., Frankignoulle, M. and Bourge, I. (1998) Effect of calcium carbonate saturation of seawater on coral calcification. Glob. Planet Change **18**: 37–46.

Gattuso, J.-P., Allemand, D. and Frankignoulle, M. (1999) Photosynthesis and calcification at cellular, organismal and community levels in coral reefs: a review on interactions and control by carbonate chemistry. Am. Zool. **39**: 160–183.

Giordano, M., Beardall, J. and Raven, J.A. (2005) CO_2 concentrating mechanisms in algae: mechanisms, environmental modulation, and evolution. Annu. Rev. Plant Biol. **56**: 99–131.

Gordillo, F.J.L., Niell, F.X. and Figueroa, F.L. (2001) Non-photosynthetic enhancement of growth by high CO_2 level in the nitrophilic seaweed *Ulva rigida* C. Agardh (Chlorophyta). Planta **213**: 64–70.

Gordillo, F.J.L., Figueroa, F.L. and Niell, F.X. (2003) Photon- and carbon-use efficiency in *Ulva rigida* at different CO_2 and N levels. Planta **218**: 315–322.

Iglesias-Rodriguez, M.D., Halloran, P.R., Rickaby, R.E.M. et al. (2008) Phytoplankton calcification in a high-CO_2 world. Science **320**: 336–340.

Israel, A. and Hophy, M. (2002) Growth, photosynthetic properties and Rubisco activies and amounts of marine macroalgae grown under current and elevated seawater CO_2 concentrations. Glob. Change Biol. **8**: 831–840.

Israel, A., Katz, S., Dubinsky, Z., Merrill, J.E. and Friedlander, M. (1999) Photosynthetic inorganic carbon utilization and growth of *Porphyra linearis* (Rhorophyta). J. Appl. Phycol. **11**: 447–453.

Johnston, A.M. and Raven, J.A. (1990) Effects of culture in high CO_2 on the photosynthetic physiology of *Fucus serratus*. Br. Phycol. J. **25**: 75–82.

Johnston, A.M., Maberly, S.C. and Raven, J.A. (1992) The acquisition of inorganic carbon for four red macroalgae. Oecologia **92**: 317–326.

Kübler, J.E., Johnston, A.M. and Raven, J.A. (1999) The effects reduced and elevated CO_2 and O_2 on the seaweed *Lomentaria articulata*. Plant Cell Environ. **22**: 1303–1310.

Langdon, C., Takahashi, T., Sweeney, C. et al. (2000) Effect of carbonate saturation state on the calcification rate of an experimental coral reef. Glob. Biogeochem. Cycles. **14**: 639–654.

Larsson, C. and Axelsson, L. (1999) Bicarbonate uptake and utilization in marine macroalgae. Eur. J. Phycol. **34**: 79–86.

Maberly, S.C. (1990) Exogenous sources of inorganic carbon for photosynthesis by marine macroalgae. J. Phycol. **26**: 439–449.

Maberly, S.C. and Madsen, T.V. (1990) Contribution of air and water to the carbon balance of *Fucus spiralis*. Mar. Ecol. Prog. Ser. **62**: 175–183.

Mercado, J.M., Niell, F.X. and Figueroa, F.L. (1997) Regulation of the mechanism for HCO_3^- use by the inorganic carbon level in *Porphyra leucosticta* thus in Le Jolis (Rhotophyta). Planta **201**: 319–325.

Mercado, J.M., Gordillo, F.J.L., Figueroa, F.L. and Niell, F.X. (1998) External carbonic anhydrase and affinity for inorganic carbon in intertidal macroalgae. J. Exp. Mar. Biol. Ecol. **221**: 209–220.

Mercado, J.M., Javier, F., Gordillo, L., Niell, F.X. and Figueroa, F.L. (1999) Effects of different leverls of CO_2 on photosynthesis and cell components of the red alga *Porphyra leucosticta*. J. Appl. Phycol. **11**: 455–461.

Orr, J.C., Fabry, V.J., Aumont, O. et al. (2005) Anthropogenic ocean acidification over the twenty-first century and its impact on calcifying organisms. Nature **437**: 681–686.

Pearson, P.N. and Palmer, M.R. (2000) Atmospheric carbon dioxide concentrations over the past 60 million years. Nature **406**: 695–699.

Raven, J.A. (1991) Physiology of inorganic C acquisition and implications for resource use efficiency by marine phytoplankton: relation to increased CO_2 and temperature. Plant Cell Environ. **14**: 779–794.

Raven, J.A. (1997) Inorganic carbon acquisition by marine autotrophs. Adv. Bot. Res. **27**: 85–209.

Raven J.A. (1999) Photosynthesis in the intertidal zone: algae get an airing. J. Phycol. **35**: 1102–1105.

Reiskind, J.B., Beer, S. and Bowes, G. (1989) Photosynthesis, photorespiration and ecophysiological interactions in marine macroalgae. Aquat. Bot. **34**: 131–152.

Riebesell, U.L.F., Zondervan, I., Rost, B., Tortell P.D., Zeebe R.E. and Morel F.M.M. (2000) Reduced calcification of marine plankton in response to increased atmospheric CO_2. Nature **407**: 3633–3667.

Riebesell, U., Schulz, K.G., Bellerby, R.G.J., Botros, M., Fritsche, P., MeyerhÃfer, M., Neill C., Nondal, G., Oschlies, A., Wohlers, J. and ZÃllner, E. (2007) Enhanced biological carbon consumption in a high CO_2 ocean. Nature **450**: 545–548.

Sabine, L.C., Feely, R.A., Gruber, N. et al. (2004) The oceanic sink for anthropogenic CO_2. Nature **305**: 367–371.

Smith, A.D. and Roth, A.A. (1979) Effect of carbon dioxide concentration on calculation in the red coralline alga *Bossiella orbigniana*. Mar Biol. **52**: 217–225.

Snoeijs, P., Klenell, M., Choo, K.S., Comhaire, I. Ray, S. and Pedersen, M. (2002) Strategies for carbon acquisition in the red marine macroalgae *Coccotylus truncatus* from the Baltic Sea. Mar. Biol. **140**: 435–444.

Stumm, W. and Morgan, J.J. (1996) *Aquatic Chemistry*, 3rd edn. Wiley, New York.

Takahashi, T., Feely, R.A., Weiss, R.F., Wanninkhof, R.H., Chipman, D.W., Sutherland, S.C. and Timothy, T.T. (1997) Global air-sea flux of CO_2 difference. PNAS **94**: 8292–8299.

Zou, D.H. (2005) Effects of elevated atmospheric CO_2 on growth, photosynthesis and nitrogen metabolism in the economic brown seaweed, *Hizikia fusiforme* (Sargassaceae, Phaeophyta). Aquaculture **250**: 726–735.

Zou, D.H. and Gao, K.S. (2002) Effects of desiccation and CO_2 concentrations on emersed photosynthesis in *Porphyra haitanensis* (Bangiales, Rhodophyta), a species farmed in China. Eur. J. Phycol. **37**: 587–592.

Zou, D.H. and Gao, K.S. (2004a) Comparative mechanisms of photosynthetic carbon acquisition in *Hizikia fusiforme* under submersed and emersed conditions. Acta Bot. Sinica **46**: 1178–1185.

Zou, D.H. and Gao, K.S. (2004b) Exogenous carbon acquisition of photosynthesis in *Porphyra haitanensis* (Bangiales, Rhodophyta) under emersed state. Prog. Nat. Sci. **14(2)**: 34–40.

Zou, D.H. and Gao, K.S. (2005) Ecophysiological characteristics of four intertidal marine macroalgae during emersion along Shantou Coast of China, with a special reference to the relationship of photosynthesis and CO_2. Acta Oceanol. Sinica. **24(3)**: 105–113.

Zou, D.H., Gao, K.S. and Ruan, Z.X. (2001) Effects of elevated CO_2 concentration on photosynthesis and nutrients uptake of *Ulva lactuca*. J. Ocean Univ. Qingdao **31**: 877–882 (in Chinese with English abstract).

Zou, D.H., Gao, K.S. and Xia, J.R. (2003) Photosynthetic utilization of inorganic carbon in the economic brown alga, *Hizikia fusiforme* (Sargassaceae) from the South China Sea. J. Phycol. **36**: 1095–1100.

Zou, D.H., Xia, J.R. and Yang, Y.F. (2004) Photosynthetic use of exogenous inorganic carbon in the agarphyte *Gracilaria lemaneiformis* (Rhodophyta). Aquaculture **237**: 421–431.

Zou, D.H., Gao, K.S. and Run, Z.X. (2007) Daily timing of emersion and elevated atmospheric CO_2 concentration affect photosynthetic performance of the intertidal macroalga *Ulva lactuca* (Chorophyta) in sunlight. Bot. Mar. **50**: 275–279.

Biodata of **Professor Rafael Riosmena-Rodriguez** and **Professor Marco Antonio Medina-López**, authors of *"The Role of Rhodolith Beds in the Recruitment of Invertebrate Species from the Southwestern Gulf of California, México"*

Professor Rafael Riosmena-Rodriguez is currently the leader of the Marine Botany research group of Universidad Autónoma de Baja California Sur in La Paz Baja California Sur, México. He obtained his Ph.D. from the La Trobe University in 2002. Professor Riosmena-Rodríguez is deeply interested in understanding the role of marine plants (where algae are included) in coastal habitats and their evolutionary significance. His research areas include systematic, biogeography, and ecology of marine plants from subtropical habitats. He is an expert on rhodolith beds.

E-mail: **riosmena@uabcs.mx**

Professor Marco Antonio Medina-López is currently the Chairman of the Marine Biology Department of Universidad Autónoma de Baja California Sur in La Paz Baja California Sur, México. He obtained his B.Sc. from the Universidad Autónoma de Baja California Sur in La Paz Baja California Sur, México, in 1999. His scientific interests are in the areas of taxonomy and ecology of invertebrates.

E-mail: **mameditna@uabcs.mx**

Rafael Riosmena-Rodriguez　　**Marco Antonio Medina-López**

THE ROLE OF RHODOLITH BEDS IN THE RECRUITMENT OF INVERTEBRATE SPECIES FROM THE SOUTHWESTERN GULF OF CALIFORNIA, MÉXICO

RAFAEL RIOSMENA-RODRIGUEZ
AND MARCO A. MEDINA-LÓPEZ
Programa de Investigación en Botánica Marina, Dept. Biol. Mar, Universidad Autónoma de Baja California Sur, Apartado postal 19-B, La Paz, B.C.S. 23080, México

1. Introduction

Rhodoliths are free-living forms of nongeniculate coralline red algae (Corallinaceae, Rhodophyta) that form extensive beds worldwide over broad latitudinal and depth ranges (Foster, 2001). Synonymous with the maerl beds common in the northeastern Atlantic, rhodolith beds are hard benthic substrates, albeit mobile, made up of branching crustose coralline thalli. Collectively, they create a fragile biogenic matrix over carbonate sediment deposits thought to be the result of long-term accumulation of dead thalli (Bosence, 1983a). A wide morphological variation of individuals exists and appears to be a response to variation in physical factors (Bosence, 1983b; Steller and Foster, 1995). This variation in morphology and incorporation of whole rhodolith and carbonates into the fossil record has led to their use as paleoindicators of environmental conditions (Foster et al., 1997). Unconsolidated rhodolith deposits have long been harvested for human use as soil amendment in European waters (Blunden et al., 1977, 1981). However, recent studies have shown that such beds are highly susceptible to anthropogenic disturbance such as trawling harvesting and reduced water quality (review in Birkett et al., 1998). Slow rhodolith growth (Rivera et al., 2003; Steller, 2003) combined with the negative impacts of burial make recovery after disturbance predictably slow. Foster et al. (1997) found rhodolith beds to be very common in the Gulf of California and suggested that there are two main types of beds: wave beds in shallow water (0–12 m) that are influenced by wave action (Steller and Foster, 1995), and current beds in deeper water (10–>30 m) that are influenced by currents. Both types, especially current beds, are also influenced by bioturbation (Marrack, 1999). To persist, these algal beds require light, nutrients and movement from water motion (waves and currents), or bioturbation, which maintains them in an unattached and unburied state (Bosence, 1983a, b; Marrack, 1999).

The structure of individual rhodoliths influences the abundance patterns in the cryptofaunal assemblage. Intact, complex thalli, along with high rhodolith densities, are important factors driving this pattern. Complex thalli may provide more space, refuge, and resources through increased interstitial or interbranch space. As a result, rhodolith complexity appears to be a good predictor of abundance and potentially for richness. This matrix provides habitat for diverse assemblages of invertebrates and algae (Cabioch, 1968; Keegan, 1974; Bosence, 1983a, b; Steller et al., 2003). This also supports the hypothesis that the availability and shape of interstitial cavities are important for the associated crustaceans' assemblage (De Grave, 1999). Variation in physical factors, thought to influence rhodolith morphology (Bosence, 1983a; Steller and Foster, 1995), may therefore directly influence community structure. Thus, we predict that conditions that enhance structural complexity increase the available refuge among the rhodolith branches, and enhance overall species richness and abundance. Rhodolith beds support a rich community of flora and fauna found to be higher in species diversity than soft-sediment benthos alone (Steller et al., 2003). Organisms within a bed can associate with the surface of algal thalli (epi-fauna/flora), within the branches (crypto-fauna/flora) or in the underlying sediments (in-fauna/flora) (Steller et al., 2003). Factors influencing diversity patterns include increased architectural complexity and grain size, reduced sedimentation (Grall and Glemarec, 1997), and seasonal variation (Ballesteros, 1988) and reduced predation.

Bivalves have been shown to be abundant and associated with rhodolith beds and in the NE Atlantic (Hall-Spencer, 1998, 1999). Possibly, this is due to larval settlement preferences for coralline, structured, or large grain substrates, or refuge from predation. Depth stratification of bivalve species may also be related to variability in substrate type (Steller, 2003; Kamenos et al., 2004). The high density of bivalves at intermediate bed depths may reflect larval attraction to the structured settlement substrate provided by the rhodoliths or physical conditions found there. In addition, Steller and Foster (1995) found that rhodolith turnover and protection from burial was greater at shallow versus intermediate depths, suggesting that the latter affords reduced sedimentation and water flow favored by surface dwelling bivalves. Increases in summer densities may correspond to winter/spring recruitment periods of many species. It appears that rhodolith beds may positively enhance bivalve populations. However, there is a clear conservation problem between these positive attributes and the degradation resulting from commercial fishing (Hall-Spencer, 1998, 1999; Hall-Spencer and Moore, 2000).

Studies have shown that rhodolith beds support a diverse and dynamic benthic community. Community descriptions of diversity include common associated species including cryptofauna living within interstitial cavities in rhodoliths (Hinojosa-Arango and Riosmena-Rodriguez, 2004; Foster et al., 2007). The density of the associated species will vary in relation to the size of the rhodolith and the density of the bed (Steller et al., 2003). Recently, Hinojosa-Arango and Riosmena-Rodriguez (2004) have shown that criptofauna assemblages are organized independently of the main rhodolith species or growth-form that composes a bed.

Steller et al. (2003) have also shown that rhodolith beds are relevant habitats for scallop recruitment but little is known about their role in invertebrate recruitment. Because of the different physical settings, the different cryptofaunal species will differ in the proportion of juvenile and adult relative abundance. In addition to the above evaluation, seasonal trends will be evaluated as an alternative source of change, as is clear in the flora and macrofauna of the beds (Steller et al., 2003). Because of the above, the aims of this study were to determine if the rhodolith beds are a relevant habitat for cryptofaunal assemblages in the southwestern Gulf of California and if this varies with bed type or season.

2. Sites and Methods for Data Gathering

Rhodoliths were collected in four rhodolith beds (Fig. 1) in the central southwestern Gulf of California in winter 1995 and summer 1996. Sampling dates and localities were: wave bed at Isla Coronados (26°06′ N 111°17′04″ W; 7 m depth) on 17 November 1995 and 2 September 1996; wave bed at Diguet off Isla San Jose (24°53′45″ N, 110°34′45″ W; 7 m depth) on 2 February 1996 and 14 July 1996; current bed off Isla San Jose (24°52′36″ N, 110°32′07″ W; 12 m depth) on 4 February 1996 and 16 July 1996; and current bed in Canal de San Lorenzo (24°22′60″ N, 110°18′41″ W; 12 m depth) on 3 December 1995 and 14 June 1996. All localities are described in detail by Riosmena-Rodríguez et al. (in press).

A similar sampling design was used in each locality and date. Two 20-m transects at least 50 m apart were haphazardly located within each bed (origin at the boat anchor). The four largest rhodoliths were collected nearest to each 2-m

Figure 1. General view of a rhodolith bed in a wave bed.

increment along the transects (40 rhodolith/transect). Individual plants were put in separate plastic-sealed underwater at the time of sampling. After collection, all materials were fixed in 4% formalin in sea water. In the laboratory, the 40 rhodoliths from each transect were examined, and individuals were selected for standard size (3–5 cm dim.) and branch density (4 branch tips/cm^2 at the rhodolith surface). All the plants selected had the fruticose growth-form. Branch densities were determined as in Steller and Foster (1995). Twenty rhodoliths per site/date were selected and used in the analyses for a total of 160 rhodoliths for the entire study.

Each selected plant was dissected and all animals >0.3 mm extracted, segregated into groups, and placed in 70% alcohol. Taxonomic determinations to lowest taxa possible were based on Smith and Carlton (1975; Cnidarians and Amphipods), Harrison and Ellis (1991; Isopods), Sieg and Winn (1978, 1981; Tanaidaceans), Wicksten (1983; Carideans), Salazar-Vallejo et al. (Polychaetes 1989), Bastida-Zavala (1991; Polychaetes), and Brusca (1980; Echinoderms). The abundance of each species or taxon per rhodolith from each site and date was then determined. Only the data from Class Turbellaria, Class Polychaeta, Subphyllum Crustacea, Class Asteroidea, Class Ophirouridea, and Calls Equinoidea were analyzed because of their abundance and celar trend in adult/juvenile morphology. Normality (Kolmogorov $\alpha = 0.05$) and homoskedasticity (Cochran y Barlett $\alpha = 0.05$) assumptions were met for richness and for abundance data transformed log 10. Differences among life stage, location, and season (winter 1995 and summer 1996) were examined using three-way ANOVA (model 1).

3. General Findings

A total of 5,066 organisms were found in the 160 rhodoliths sampled where 85% of the fauna were possible to identify at least to genus/species level. Overall, 60% of the collected individuals were juvenile (including recently settled larvae and organisms that are starting to grow) and 40% adults. Our comparisons strongly support that most cryptofaunal species found in the samples were juveniles but have variations in relation to bed type and season (Figs. 2 and 3). Trends varied in relation to taxonomic group and probably with richness. Turbellaria and Equinodermata were found with low species number and crustacean, mollusk, and annelid were found with higher species numbers.

Three species of Turbellaria, where 100% of the specimens were new recruit in all the seasons/subhabitats, were sampled (Figs. 2 and 3). Seven species of Echinoderms were found with significant differences between seasons where only juveniles were observed in winter and a higher proportion of adults were observed in summer in both beds.

Crustacea, Mollusk, and Annelida were taxa with higher number of species where we found a tendency in the seasons and/or bed type. We found 21 species of crustaceans where juvenile individuals were more abundant than adults

THE ROLE OF RHODOLITH BEDS IN THE RECRUITMENT OF INVERTEBRATE

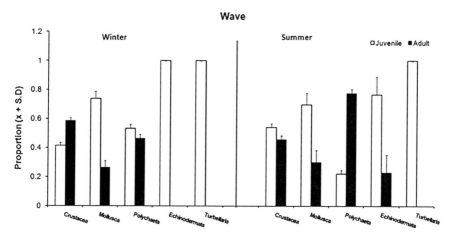

Figure 2. Comparisons in wave beds of the proportion (mean and standard) of the juvenile and adult invertebrate species from three main analyzed taxa.

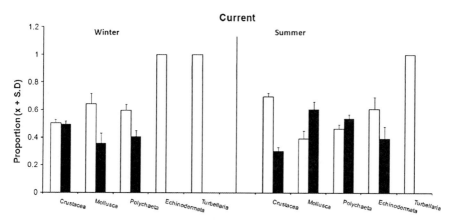

Figure 3. Comparisons in current beds of the proportion (mean and standard) of the juvenile and adult invertebrate species from three main analyzed taxa.

(Figs. 2 and 3); this trend is particularly clear in current beds where we found the higher differences among life-cycle stages (Fig. 4). In the case of wave beds, we found an inverse arrangement between seasons, with high proportion of adults in winter and high proportion of juveniles in summer. Forty mollusks present in contrasting patterns were observed. In the current beds, juvenile proportion was significantly higher in winter, and more adults were found in summer (Fig. 2), whereas in wave beds, juveniles were significantly higher in both seasons (Figs. 3 and 4). In polychaetes, seasons (summer/winter) have the most consistent differences, with higher juvenile proportion in winter and larger number of adults in summer.

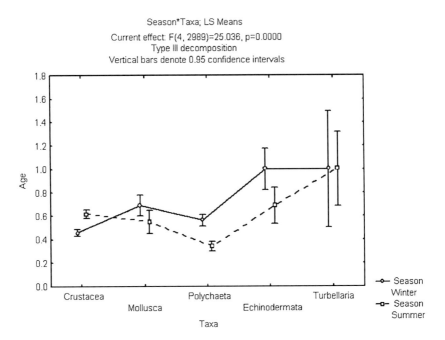

Figure 4. Statistical seasonal comparisons between the major phyla analyzed.

We also found differences between bed type and this might be an artifact of the seasonal variation. Statistically, we found significant seasonal variation in Crustacea, Polychaetes, and Equinodermata (Fig. 4), while the mollusk bed type showed the significant differences.

Rhodolith beds are one of the marine habitats that have been considered for design of marine reserves in the Gulf of California (Sala et al., 2002; Anaya-Reyna et al., 2005) and elsewhere (Birkett et al., 1998). The extensive cover of the seafloor of this habitat is the key element in their ecological value (Hetzinger et al., 2006) in relation to the associated biodiversity (Steller et al., 2003; Hinojosa-Arango and Riosmena-Rodriguez, 2004). Our study has shown that rhodolith beds are also the habitat for recruitment of many of the invertebrate associated species where some of them (scallops) are part of the fishery of the area (Steller, 2003). Also, there are species considered in some category of protection (as many species of coral and equinoderm) that are continuously present in the beds (Riosmena-Rodríguez et al., in press) and started from their recruitment as cryptofauna. A similar situation is found in several beds around the world (Hopkins et al., 1991; Birkett et al., 1998) where rhodolith beds are not only the habitat for adult species but also the recruitment habitats. This situation might enhance the number of species present and their abundance over years; we could not determine mortality among the samples, but it was clear that the scallops larvae were present in very abundant numbers inside the sea anemone, which lives commonly as part of the cryptofauna.

It is well known that the thalli surface of coralline algae is a common space for recruitment for many invertebrates (Johnson et al., 1991) used as a "signal" to find a hard structure (whether the hard structure is a free-living species or a species attached to the rocks). In some cases, rhodolith beds might be just the contact point for these larvae because in a later stage they will migrate to the surrounding areas (Steller, 2003). The continuous presence of juvenile stages within a rhodolith is more related with the trophic structure of each bed (Grall and Glemarec, 1997). This close relationship between scallops and rhodolith as a recruitment habitat explains the extensive mollusk bank that has been a formal fishery (Hall-Spencer, 1998). The high impact of several fisheries might produce the lack of larvae in the water column and thus inhibit recruitment. An example of this situation can be seen when, according to historical analysis, rhodolith beds were the recruitment habitats of the pearl oyster at the beginning of the twentieth century but not present, since the fisheries collapse. A similar situation occurred in 1991 when rhodolith beds were heavily impacted by a combination of factors (Steller et al., 2003) related with high number of fishing permits and the lack of critical analysis of the collecting procedures because most fishermen used to dive with Hokka and leather shoes, which crashed and buried rhodoliths, making their yearly recovery impossible and maintaining that condition over the last 17 years.

Rhodolith (maerl) in the Gulf of California needs urgent consideration in the management plan of benthic fisheries (as scallops, octopus, and trawling fisheries as shrimp) because it has almost no existence. In fact, a management plan for rhodolith beds might be a good strategy because of their presence in several habitats and depths (Riosmena-Rodríguez et al., in press). In addition, their value as monitoring objects more than justifies their being based on fisheries needs but there are species under protection in this habitat (where they spend most of their lives) as is currently in place for European waters (Birkett et al., 1998).

4. Summary

The role of rhodolith beds as recruitment habitats as part of their cryptofauna was evaluated based on seasonal sampling over a series of beds that represent two different microhabitats (current and wave beds). Our basic goal was to understand if the cryptofauna were composed mostly by juvenile or adult individuals of each species as part of the understanding of the rhodolith beds as a critical habitat for conservation worldwide. We have collected 160 rhodoliths from more than 5,000 individuals of 116 species were found. Turbellaria and Equinodermata are phyla with low number of species, but mostly juveniles were present with no seasonal variation for the former and a small variation for the latter. Crustacea, Mollusk, and Annelid were taxa with higher number of species wherein we found a tendency in the seasons and/or bed type. Crustaceans were mostly juvenile individuals than adults in current beds and the inverse pattern was observed in wave beds. In current beds, juveniles were significantly higher in winter, and more adults were found in summer, whereas in wave beds, juveniles were significantly higher in both seasons. In polychaetes, seasons (summer/winter) have the most consistent differences, with higher juvenile proportion in winter and higher proportion of adults in summer. Currently, rhodolith beds are considered as a relevant habitat in the near shore areas in the Gulf of California but the subdivision between current and wave beds need to be taken into consideration for protection proposes. The present results have shown the value of rhodolith beds as a recruitment habitat for many species and the urgent need to not only consider them in the management of the fisheries but also the need of a management plan for the habitat itself.

5. References

Anaya-Reyna, G., Weaver, A.H. and Palmeros-Rodriguez, M.A. (2005) *Propuesta Para la Creación del Parque Nacional Espiritu Santo*. Niparaja, México, 15 pp.

Bastida-Zavala, J.R. (1991) Poliquetos (Annelida: Polychaeta) del sureste de la Bahía de La Paz, B.C.S. México: dhesives y aspectos biogeográficos. Bachelour Thesis, Universidad Autónoma de Baja California Sur, La Paz, México.

Ballesteros, E. (1988) Composición y estructura de los fondos de maerl de Tossa de Mar (Gerona, Espana). Collect Bot (Barcelona) **17**: 161–182.

Birkett, D., Maggs, C. and Dring, M. (1998) Maerl (volume V). An overview of dynamic and sensitivity characteristics for conservation management of marine SACs. Scottish Association for Marine Science (UK Marine SACs Project).

Blunden, G., Farnham, W., Jephson, N., Barwell, C., Fenn, R. and Plunkett, B. (1981) The composition of maerl beds of economic interest in Northern Brittany, Cornwall and Ireland. Int. Sea. Symp. **10**: 651–656.

Blunden, G., Farnham, W., Jephson, N., Fenn, R. and Plunkett, B. (1977) The composition of maerl from the Glenan Islands of Southern Brittany. Bot. Mar. **20**: 121–125.

Bosence, D.W. (1983a) Ecological studies on two carbonate sediment-producing algae, In: T.M. Peryt (ed.) *Coated Grains*. Springer, Heidelberg, Germany, pp. 270–278.

Bosence, D.W.J. (1983b) The occurrence and ecology of recent rhodoliths – a review, In: T.M. Peryt (ed.) *Coated Grains*. Springer, Berlin, pp. 225–242.

Brusca, R.C. (1980) *Common Intertidal Invertebrates of the Gulf of California*. Univetsity of Arizona Press, Tucson.

Cabioch, J. (1968) Contribution á la connaissance des peuplements benthiques de La Manche occidentale. Cah. Biol. Mar. **9**: 493–711.

De Grave, S. (1999) The influence of sedimentary heterogeneity on within maerl bed differences in infaunal crustacean community. Est. Coast. Shelf Sci. **49**: 153–163.

De Grave, S., Fazakerley, H., Kelly, L., Guiry, M., Ryan, M. and Walshe, J. (2000) A study of selected maerl beds in Irish waters and their potential for sustainable extraction. Mar Res. Ser. **10**: 1–44.

Foster, M.S. (2001) Rhodoliths: between rocks and soft places. J. Phycol. **37**: 659–667.

Foster, M.S., McConnico, L.M., Lundsten, L., Wadsworth, T., Kimball, T., Brooks, L.B., Medina-Lopez, M.A., Riosmena-Rodriguez, R., Hernandez-Carmona, G., Vasquez-Slizondo, R.M., Johnson, S. and Steller, D.L. (2007) The diversity and natural history of a Lithothamnion muelleri-Sarassum horridum community in the Gulf of California. Cienc. Mar. **33**: 367–384.

Foster, M.S., Riosmena-Rodriguez, R., Steller, D.L. and Woelkerling, W.J. (1997) Living rhodolith beds in the Gulf of California and their implications for paleoenvironmental interpretation, In: M.E. Johnson and J. Ledesma-Vazquez (eds.) *Pliocene Carbonates and Related Facies Flanking*

the Gulf of California, Baja California, Mexico. Geological Society of America Special Paper no. 318. Boulder, Colorado.
Grall, J. and Glemarec, M. (1997) Biodiversite des fonds de maerl en Bretagne: approache fonctionnelle et impacts anthropogeniques. VIE MILIEU **47**: 339–349.
Hall-Spencer, J. (1998) Conservation issues relating to maerl beds as habitats for molluscs. J. Conch Spec. Publ. **2**: 271–286.
Hall-Spencer, J.M. (1999) Effects of towed demersal fishing gear on biogenic sediments: a 5-year study, In: O. Giovanardi (ed.) *Impact of Trawl Fishing on Benthic Communities.* ICRAM, Rome, pp. 9–20.
Hall-Spencer, J.M. and Moore, P.G. (2000) Impact of scallop dredging on maerl grounds, In: M.J. Kaiser and S.J. de Groot (ed.) *Effects of Fishing on Non-Target Species and Habitats: Biological, Conservation and Socio-Economic Issues.* Blackwell Science, Oxford, pp. 105–117.
Harrison, K. and Ellis, J.P. (1991) The genera of Sphaeromatidae (Crsutacea: Isopoda): a key and distribution list. Invertebr. Taxon. **5**: 915–952.
Hetzinger, S., Halfar, J., Riegl, B. and Godinez-Orta, L. (2006) Sedimentology and acoustic mapping of modern rhodolith beds on a non-tropical carbonate shelf (Gulf of California, Mexico). J. Sed. Res. **76**: 670–682.
Hinojosa-Arango, G. and Riosmena-Rodriguez, R. (2004) The influence of species composition and growth-form of rhodolith beds in cryptofaunal assemblages in the Gulf of California. PSZN Mar. Ecol. **54**: 234–244.
Hopkins, T.S., Valentine, J.F., McClintock, J.B., Marion, K.R. and Watts, S.A. (1991) Echinoderms associated with a rhodolith community on the Alabama OCS: management considerations for a unique environmental setting, In: *Proceedings of the Eleventh Annual Gulf of Mexico Information Transfer Meeting, November, 1990.* U.S. Dept. of the Interior, Mineral Management Service. New Orleans, Louisiana, pp. 443–448.
Johnson, C.R., Sutton, D.C., Olson, R.R. and Giddins, R. (1991) Settlement of crown-of-thorns starfish: role of bacteria on surfaces of coralline algae and a hypothesis for deepwater recruitment. Mar. Ecol. Progr. Ser. **71**: 143–162.
Kamenos, N.A., Moore, P.G. and Hall-Spencer, J.M. (2004) Nursery-area function of maerl grounds for juvenile queen scallops Aequipecten opercularis and other invertebrates. Mar. Ecol. Prog. Ser. **274**: 183–189.
Keen, M.A. (1971) *Sea Shell of Tropical West America. Marine Mollusks from Baja California to Peru.* Stanford University Press, California.
Keegan, B.F. (1974) The macrofauna of maerl substrates on the west coast of Ireland. Cah. Biol. Mar. **15**: 513–530.
Marrack, E. (1999) The relationship between water motion and living rhodolith beds in the southwestern Gulf of California, Mexico. Pal. **14**: 159–171.
Riosmena-Rodríguez, R., Steller, D.L., Hinojosa-Arango, G. and Foster, M.S. (in press) Reefs that rock and roll: biology and conservation of rhodolith beds in the gulf of California, In: R. Brusca (ed.) *Marine Biodiversity and Conservation in the Gulf of California.* University of Arizona Press and Sonoran Desert Museum, Tuscon, AZ.
Rivera, M.G., Riosmena-Rodriguez, R. and Foster, M.S. (2003) Edad y crecimiento de Lithothamnion muellerii (Corallinales, Rhodophyta) en el suroeste del Golfo de California, México. Cienc. Mar. **30**(1B): 235–249.
Sala, E., Aburto-Oropeza, O., Paredes, G., Parra, I., Barrera, J.C. and Dayton, P.K. (2002) A general model for designing networks of marine reserves. Science **298**: 1991–1993.
Salazar-Vallejo, S.J., de León-González, J.A. and Salices-Polanco, H. (1989) Poliquetos (Annelida: Polychaeta) de México. Libros Universitarios. U.A.B.C.S. La Paz, B.C.S., México.
Sieg, J. and Winn, R. (1978) Key to suborders and families of Tanaidacean (Crustacea). Proc. Biol. Soc. Wash. **4**: 840–846.
Sieg, J. and Winn, R. (1981) Key the Tanaidae (dhesives: Tanaidacea) of California, with a key to the world genera. Proc. Biol. Soc. Wash. **94**: 315–343.

Smith, R. and Carlton, J. (1975) *Ligth's Manual: Intertidal Invertebrates of the Central California Coast*. California Press, Berkeley.

Steller, D.L. (2003) Rhodoliths in the Gulf of California: growth, demography, disturbance and effects on population dynamics of catarina scallops. Ph.D. University of California, Santa Cruz.

Steller, D.L. and Foster, M.S. (1995) Environmental factors influencing distribution and morphology of rhodoliths in Bahia Concepcion, B.C.S., Mexico. J. Exp. Mar. Biol. Ecol. **194**: 201–212.

Steller, D.L., Riosmena-Rodriguez, R., Foster, M.S. and Roberts, C.A. (2003) Rhodolith bed diversity in the Gulf of California: the importance of rhodolith structure and consequences of disturbance. Aquatic Conser. Mar. Fresh. Ecosys. **13**: S5–S20.

Wicksten, M.K. (1983) A monograph on the shallow water caridean srimps of the Gulf of California, México. All Hac. Publ. **13**: 1–59.

Biodata of **Anja Eggert**, **Akira F. Peters,** and **Frithjof C. Küpper**, authors of *"The Potential Impact of Climate Change on Endophyte Infections in Kelp Sporophytes"*

Dr. Anja Eggert is Scientific Assistant at the Chair of Applied Ecology, Institute of Biological Sciences, University of Rostock, Germany, since 2003. She obtained her Diploma in Biology in 1997 from the University of Bremen and her Ph.D. in Marine Botany in 2002 from the University of Groningen; both theses were related to the ecophysiology of polar and tropical macroalgae. Anja Eggert continued her ecophysiological research on algae in various projects at the University of Rostock, and extended her interests to the fields of biochemistry and molecular biology. Her main research focus is related to the adaptation and acclimation potential of marine algae to environmental parameters such as temperature, salinity and UV-stress. She is now working in the field of ecosystem modelling at the Leibniz Institute for Baltic Sea Research Warnemünde.

E-mail: **anja.eggert@io-warnemuende.de**

Dr. Akira F. Peters is a Gentleman Scientist in his enterprise Bezhin Rosko at Roscoff, Brittany, France. He obtained his Ph.D. in 1986 from Konstanz University where in 1981 he had begun to study reproduction, phylogeny, taxonomy and ecology of brown algae, including pathological disorders of macroalgae caused by endophytes. He worked as post-doc, assistant and associate university professor in Chile, the Netherlands, Germany and France. Following 2001, his research interest has shifted to algal genetics and development, in the context of the development of a brown algal model (*Ectocarpus*) at the Station Biologique de Roscoff where he worked until 2006. In 2008–2009, he was a Ray Lankester Fellow at the Marine Biological Association at Plymouth, UK.

E-mail: **akirapeters@gmail.com**

Anja Eggert

Akira F. Peters

Dr. Frithjof C. Küpper currently holds the position of Reader in Algal Ecology at the Scottish Association for Marine Science in Oban, Argyll, Scotland. He joined the organisation in 2003 as Lecturer and Head of Culture Collection of Algae and Protozoa. Following undergraduate and graduate studies at the universities of Konstanz and Paris-Sud XI/Orsay and at the Station Biologique de Roscoff (Ph.D. in 2001), respectively, he worked at the University of California, Santa Barbara, for 2 years as a post-doctoral research associate with Prof. Alison Butler. Frithjof's research interests are chemical ecology, inorganic biochemistry and physiology of marine plants and microbes, especially in the context of defence, stress and biogeochemical cycles. During his time in Roscoff and Konstanz, he developed an interest in algal pathologies and halogen metabolism.

E-mail: **fck@sams.ac.uk**

THE POTENTIAL IMPACT OF CLIMATE CHANGE ON ENDOPHYTE INFECTIONS IN KELP SPOROPHYTES

ANJA EGGERT[1], AKIRA F. PETERS[2], AND FRITHJOF C. KÜPPER[3]
[1] *Physical Oceanography and Instrumentation, Leibniz Institute for Baltic Sea Research Warnemünde, Rostock, 18119, Germany*
[2] *Bezhin Rosko, Roscoff, 29680, France*
[3] *Scottish Association of Marine Science, Oban, Argyll, PA37 1QA, Scotland, UK*

1. Introduction

There is a strong scientific consensus that coastal marine ecosystems are threatened by global climate change. These ecosystems are particularly vulnerable as many disturbances act at the terrestrial–marine interface and are predicted to increase, such as increased land run-off after floods or higher wave energies owing to increased storm frequency (Helmuth et al., 2006; IPCC, 2007). An alarming decrease in the density and biomass of canopy-forming kelps has been reported worldwide (Dayton et al., 1999; Steneck et al., 2002; Connell et al., 2008) and recent European monitoring programs indicate substantial losses of *Laminaria digitata* in France (Morizur, 2001) and of *Saccharina latissima* (formerly *L. saccharina*) along the Southwest coast of Norway and Sweden (survey in 1996–2006, Norwegian Institute for Water Research, 2007) and on the German island Helgoland (Pehlke and Bartsch, 2008). For instance, the losses of *S. latissima* at the Norwegian West and Skagerrak coasts are estimated to be 50% and 90%, respectively. Here, the decline in kelp abundance is most pronounced in sheltered waters, where the kelp forest in large areas has been replaced by a silty turf community dominated by filamentous algae. Anthropogenic influences, such as eutrophication and global climate change, have been postulated as possible causes for the loss of canopy-forming kelps. However, substantial scientific evidence is still lacking.

The genus *Laminaria sensu lato* is one of the most important macroalgal genera of the order Laminariales (= "kelp") in temperate to polar rocky coastal ecosystems, especially in the northern hemisphere (Bartsch et al., 2008). This is reflected by its high species numbers, its considerable overall biomass and its dominance and economic significance. Since low levels of redundancy in functional species traits exist in many coastal marine systems, including kelp forests, changes in species diversity, and the final loss of habitat-forming species would severely affect ecosystem functioning (Micheli and Halpern, 2005). Targets would include the diversity and abundance of the associated fauna that directly or indirectly use the structure provided by macroalgae as a habitat or as food source (Graham, 2004) as well as a variety of trophic levels via changes in the amount and source of detritus inputs into the food web (Duggins et al., 1989).

The linear warming trend over the last 50 years from 1956 to 2005 amounts to 0.10–0.16°C per decade. The temperature increase is widespread over the globe and is greater at higher northern latitudes (IPCC, 2007). Additionally, heat waves, such as in summer 2003 in Europe, are likely to occur more frequently (Schär and Jendritzky, 2004). Moreover, a regional climate model for the Baltic Sea area predicts an increase in the mean summer temperatures of 3–5°C within the next century, i.e., above the global warming average (BACC, 2008). Even though some negative effects of ocean warming on coastal biota are already evident (Harley et al., 2006), their consequences on marine ecosystems are far less understood than they are in terrestrial environments. It is assumed that macroalgal-based coastal ecosystems with low species redundancy are particularly vulnerable to global warming (Micheli and Halpern, 2005; Ehlers et al., 2008). To predict the fate of coastal ecosystems, an assessment of the vulnerability of key species to climate change is necessary (Harley et al., 2006).

Perhaps the most pervasive changes in terrestrial and marine biota to recent global warming are the shifts in geographical ranges of species (Parmesan and Yohe, 2003). Latitudinal range shifts of marine species have been demonstrated in several studies and forecasts suggest that even greater impacts can be expected in the future (several NE Atlantic taxa including kelps: Southward et al., 1995; Californian gastropod: Zacherl et al., 2003; Caribbean coral: Precht and Aronson, 2004; North Sea fishes: Perry et al., 2005; Chilean gastropods and chitons: Rivadeneira and Fernández, 2005, Portuguese macroalgae: Lima et al., 2007). The southern boundaries of the cold-temperate kelp species in the NE Atlantic are either set by summer lethal limits or by winter "reproduction" limits. As it has been shown for *L. hyperborea*, its southern boundary is set by a summer lethal limit at the 18°C-August isotherm and a winter reproduction limit at the 13°C-February isotherm at the Iberian Peninsula (Fig. 1). If only these temperature limitations were to be considered, the projected increase in sea surface temperatures by 2°C during winter or summer would lead to lethal summer temperature regimes, and consequently the disappearance at the coasts of the Iberian Peninsula, Brittany, S. England, S. Ireland, and Helgoland (marked in pink in Fig. 1, Breeman, 1990).

Comprehensive field surveys to study latitudinal shifts of benthic marine macroalgae are limited to one. Lima et al. (2007) examined shifts in macroalgal species inhabiting the Portuguese rocky coast. It is known that many cold-temperate species, e.g., *L. hyperborea* as described earlier, have their southern boundaries at this coast (André, 1971; Breeman, 1988). In comparison with the reported distribution boundaries of 26 cold-water species by André (1971) in this region, Lima et al. (2007) describe significant northward shifts for seven species (e.g., *Palmaria palmata*: 358 km, *Himanthalia elongata*: 219 km), significant southward shifts for

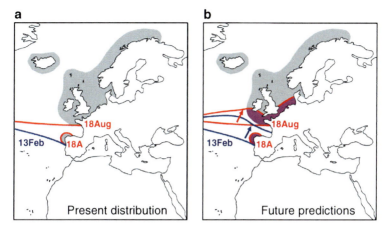

Figure 1. (a) Present southern distribution boundary of *Laminaria hyperborea*. (b) Boundary shift after the expected rise in seawater temperature of 2°C. Shown are the present 13°C February winter isotherm (*blue line*) and the 18°C August summer isotherm (*red line*, after Breeman, 1990). Marked in pink are those populations that would disappear under such a scenario.

seven species (e.g., *Fucus vesiculosus*: 157 km, *Ahnfeltia plicata*: 330 km) and non-significant changes for 12 species (e.g., *L. hyperborea*, *S. latissima*, *Ascophyllum nodosum*). Therefore, at present no generalisations about poleward range shifts of cold-temperate algae due to increasing temperature can be made. It is obvious that a range of factors not only related to changes in temperature such as eutrophication, or which are rather indirectly related (e.g., biotic interactions), could explain some of the observed distributional shifts.

2. Biotic Interactions

Environmental factors (e.g., temperature) might alter species composition, and the strength or even sign of interspecific interactions (Graham, 1992; Davis et al., 1998), predicting ecological responses to climate change, requires additional information on how abiotic changes are mediated by biotic interactions. Benthic marine macroalgae are subjected to a variety of biotic stress factors, such as intra- and interspecific competition, colonization, grazing, or pathogenic diseases. In marine algae, competition and grazing are considered as the major factors determining the structure of natural populations (e.g., Lubchenco and Gaines, 1981). Experimental proof for direct impacts of climate change on biotic interactions in coastal ecosystems are limited to the impact of thermal stress on predation (Sanford, 1999) and grazing (Leonard, 2000; Morelissen and Harley, 2007). However, interactions with pathogens can also have significant impacts on marine algae. Algae can be attacked by viruses (*Emiliania huxleyi*: Bratbak et al., 1996, marine plankton: Culley and Steward, 2007, *Ectocarpus siliculosus*: Müller et al., 1990), bacteria

(*L. japonica*: Sawabe et al., 1998, *Gracilaria conferta*: Weinberger and Friedlander, 2000, *L. religiosa*: Vairappan et al., 2001), fungi and oomycetes (*S. latissima*: Schatz, 1984, *Porphyra* sp.: Uppalapati and Fujita, 2000, *Pylaiella littoralis*: Küpper and Müller, 1999, Küpper et al., 2006) or other algae (*Rhodomela confervoides* ←→ *Harveyella mirabilis*: Kremer, 1983; *Chondrus crispus* ←→ *Acrochaete operculata*: Bouarab et al., 1999, *Mazzaella laminarioides* ←→ *Endophyton ramosum* and *Pleurocapsa* sp.: Faugeron et al., 2000, filamentous red algae ←→ kelp gametophytes: Hubbard et al., 2004, kelp species ←→ filamentous brown algae: e.g., Peters and Schaffelke, 1996; Küpper et al., 2002). Most of the research on infectious diseases in algae has focused on characterizing the parasites and describing morphological aspects of host–parasite interactions. Studies going beyond the description stage, particularly into the effect of infections on the host performance and fitness, are scarce. Infected thalli of the red alga *Mazzaella laminarioides* are more susceptible to wave action. The endophytes also negatively affect the reproductive output of this red alga (Faugeron et al., 2000). A similar effect has also been reported for virus-infected species of *Ectocarpus* (Müller et al., 1990). The field study by Schatz (1984) showed that infected *S. latissima* grow more slowly than healthy plants, and it is hypothesized by the author that the fungus limits long-distance transport of photosynthates. The photosynthetic performance of pathogen-infected organisms has been studied so far only in *Pylaiella littoralis* with the fungus *Chytridium polysiphoniae* (Gachon et al., 2006). Virtually, nothing is known about potential ecological consequences of climate change on algal–pathogen interactions. Primarily based on the knowledge from terrestrial systems, Harvell et al. (2002) predict that most host–pathogen interactions will become more frequent or disease impacts more severe with global warming. The interactions will be affected by (1) more rapid pathogen development owing to increased growth rates of the pathogen at higher temperatures (e.g., coral pathogen: Alker et al., 2001), (2) increased winter survival of the pathogen (e.g., oyster disease: Cook et al., 1998) and (3) increased host susceptibility at higher temperatures (e.g., coral pathogens: Harvell et al., 2001).

Field observations in kelp species from different parts of the world document massive prevalence of infection by endophytic algae. In the 1990s, infection rates of *L. hyperborea* and of *S. latissima* (formerly *L. saccharina*) in the NE Atlantic and the western Baltic Sea were as high as 25–100% and 70–100%, respectively (Lein et al., 1991; Peters and Schaffelke, 1996; Ellertsdóttir and Peters, 1997). Frequent infections have also been reported in other geographical regions, such as in *Laminaria* species in the NW Pacific (Yoshida, 1980) and in several members of the Laminariales in the NE and SE Pacific (Apt, 1988; Peters, 1991). According to kelp trawlers, dark spots (as shown in Fig. 2a) have been observed on the kelps earlier and infections by endophytic algae have been described as a common disease of kelp species for many decades (Lein et al., 1991). Accordingly, infection of kelps by endophytic brown, filamentous algae has been regarded as a pathogenic disease. Publications on endophytic algae in kelps focus on the mere description of the prevalence, on the taxonomy of the endophytes or defence mechanisms of the hosts (see later). However, little is known about the ecological

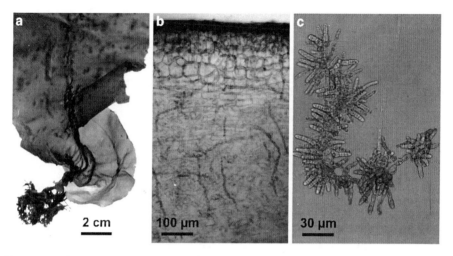

Figure 2. *Saccharina latissima* infected with *Laminariocolax aecidioides* at Kiel Bight (western Baltic Sea) in 1993. (**a**) Silhouette of the host with twisted stipe, distorted meristem and dark spots, (**b**) light microscopy of a hand section of the host, well-pigmented filaments of the endophyte growing in the intercellular space of the host medulla, (**c**) light microscopy of the isolated endophyte in unialgal culture, showing many plurilocular sporangia and a phaeophycean hair.

significance of this host–pathogen interaction. Nevertheless, it is reasonable to assume that endophytes negatively influence the fitness and productivity of *Laminaria* sporophytes.

Endophyte species of *Laminaria* are mainly microscopic, morphologically little differentiated, filamentous brown algae (e.g., *Laminariocolax aecidioides*, *L. tomentosoides*, *Laminarionema elsbetiae*), recently classified in the family Chordariaceae within the order Ectocarpales (Peters, 2003, Fig. 2b). European *S. latissima* is mainly infested by *Laminarionema elsbetiae* and *L. hyperborea* with *Laminariocolax aecidioides*, respectively (Peters and Ellertsdottir, 1996; Burkhardt and Peters, 1998; Peters and Burkhardt, 1998; Table 1). The endophytes are distributed among host plants via zoospores from plurilocular sporangia. Zoospores of *Laminariocolax aecidioides* and *Laminarionema elsbetiae* attach to and penetrate the healthy host surface; no wounds or other openings are required for successful invasion of the host and no epiphytic stage precedes infection (Heesch and Peters, 1999). Thus, these endophytes are immediately invasive. This is noteworthy since a great number of epiphytic algae occur on kelps but most of these are unable to penetrate into the host. Thus, endophytes must have developed special attributes to achieve infection. Endophyte spores settle with their anterior end on the host surface and fibrillar adhesive material is formed around the attaching end. As no inward deflection of the host surface was observed, Heesch and Peters (1999) proposed that the surface is locally dissolved by enzymes. A similar mechanism has been described for the green endophyte *Acrochaete operculata*, which infects the red alga *Chondrus crispus* (Correa and McLachlan, 1994).

Table 1. Distribution of endophytic brown algae in three most abundant European kelp species. ×: presence, ××: high prevalence.

	Laminaria hyperborea	*Laminaria digitata*	*Saccharina latissima*
Laminariocolax aecidioides (Rosenvinge)	××[a,b,c]	×[c]	×[b,d,e]
Laminariocolax tomentosoides (Farlow) Kylin (1947)	×[a]	××[a,b]	×[a]
Laminariocolax tomentosoides subsp. *deformans* (Dangeard) Peters (2003)		×[c]	
Laminarionema elsbetiae Kawai and Tokuyama (1995)		×[a]	××[a,c]

[a] Ellertsdóttir and Peters (1997).
[b] Burkhardt and Peters (1998).
[c] Peters (2003).
[d] Schaffelke et al., 1996.
[e] Peters and Schaffelke, 1996.

Figure 3. *Laminaria digitata* infected by *Laminariocolax tomentosoides* and twisted stipe (Ile de Batz off Roscoff in the English Channel, 1997).

Endophytic infections are described as a common disease of kelps. Hosts show either weak symptoms such as dark spots on the lamina or warts on the stipes or severe morphological changes like twisted stipes or crippled laminae. Endophytic infection of *S. latissima* by *Laminariocolax aecidioides* has been divided into three disease categories according to Peters and Schaffelke (1996): (1) Thalli are infected microscopically and disease symptoms are absent. (2) Moderate symptoms (i.e., dark spots, ridges, and small wart-like structures) are visible. Consequently, this endophytic infestation is called "dark spot disease." (3) Severe morphological changes, such as distorted stipes or crinkled blades may occur (Fig. 3).

Even though the presence of endophytes is not necessarily harmful to the host, blades with severe morphological changes are less flexible and hence more susceptible to wave action (Lein et al., 1991; Dr. Akira F. Peters, 2009 personal communication). Furthermore, negative effects of both endophytes and their polar and non-polar extracts on the growth rates of *Laminaria* sporophytes have been shown, indicating direct chemical interactions with the host tissue (A.F. Peters, 2009, personal communication). In addition, endophytic infestation possibly interferes with the fertility of kelp sporophytes. Sporangia can cover about 70% of the blade surface (Kain, 1975), so that 20% coverage of the blade surface by the "dark spot disease" will diminish the potential reproductive area significantly (Lein et al., 1991). On the other hand, Lein et al. (1991) suggested that endophyte infection per se might inhibit the formation of sori in *L. hyperborea*, which was also observed in *L. digitata* (Lüning et al., 2000) and in *S. latissima* (Peters and Schaffelke, 1996). However, the underlying mechanisms are completely unstudied. Furthermore, brittle thalli (especially twisted stipes, Fig. 3) are more likely to be detached during storms, again reducing the potential reproductive area or period. The importance of this effect increases under the scenario of climate change, as changes in atmospheric circulation might also change storm frequency. An increase in the frequency of winter storms has already been observed in coastal oceans (Bromirski et al., 2003) and the trend is expected to continue (IPCC, 2007). Only two field studies considered interactions between abiotic factors and endophyte infections in kelp hosts. *Laminaria/Saccharina* sporophytes in the western Baltic and at Helgoland exhibited considerably stronger disease symptoms in the shallow than in the deep water (Schaffelke et al., 1996; Ellertsdóttir and Peters, 1997). It was hypothesised that a reduced fitness of the host tissue, caused by higher UV radiation at lower water depth or higher photosynthetically active radiation resulting in higher growth rates of the photosynthetic parasite may have caused stronger disease symptoms. Accordingly, how climate change may modulate this biotic interaction remains speculative.

3. Defence Responses

Laminaria sporophytes can potentially recognize attacks by endophytes and initiate effective defence responses within minutes. On the other hand, endophytes like *Laminariocolax tomentosoides* also seem to have developed mechanisms that either eliminate the defence response of *L. digitata* or detoxify reactive oxygen species by having a high peroxide-scavenging capacity (Küpper et al., 2002). The so-called oxidative burst, the rapid release of reactive oxygen species, facilitated the resistance of *Macrocystis pyrifera* and *L. digitata* against the pathogenic brown algal endophytes *Laminariocolax tomentosoides* and *Laminariocolax macrocystis*. This response took 7 days to occur and possibly involved induction or up-regulation of other structural or chemical defences (Küpper et al., 2002). The oxidative burst in response to oligoalginates, which are released from the host cell wall after endophyte attack, is a good example of activated defence and currently the best described example of a chemical defence mechanism in kelps. In sporophytes of *L. digitata*, the presence of oligoguluronates – degradation products of alginate – resulted in a massive release of reactive oxygen species ($O_2^{-\cdot}$, H_2O_2) by epidermal cells (Küpper et al., 2001). The same response was also observed in sporophytes, but not gametophytes, of *S. latissima*, *L. hyperborea*, *L. ochroleuca*, *L. pallida*, *M. pyrifera*, *Saccorhiza polyschides*, *Chorda filum*, and *Lessonia nigrescens*, so that the response appears to be universal in kelp and kelp-like sporophytes but not in their gametophytes (Küpper et al., 2002). H_2O_2 concentrations in the range released by *L. digitata* were toxic to alginate-degrading bacteria (Küpper et al., 2002) and axenic *M. pyrifera* was rapidly infected by pathogenic bacteria when the oxidative burst response was blocked with an NAD(P)H-oxidase inhibitor (Küpper et al., 2002). Treatment of non-axenic *M. pyrifera* or *L. digitata* with the inhibitor also resulted in rapid degradation by their natural bacterial flora, which indicates that the oxidative burst must play an important role in the algal defence against harmful bacteria and the maintenance of harmless, eventually even protective biofilms. Recently, it was reported that *L. digitata* can recognize not only oligoalginates (i.e., endogenous elicitors), but also lipopolysaccharides from the outer cell envelope of a range of gram-negative bacterial taxa as exogenous elicitors for an oxidative burst and other early defence responses (Küpper et al., 2006). Similarly, free polyunsaturated fatty acids and methyl jasmonate can trigger an oxidative burst and induce resistance against pathogenic endophytes in *L. digitata* (Küpper et al., 2009).

Striking differences exist in endophyte host prevalence and disease symptoms among *Laminaria/Saccharina* sporophytes. A field survey in 1994 at Helgoland (North Sea) showed that *S. latissima* has a higher percentage of infected thalli and of severe thallus deformations than *L. hyperborea* (Schaffelke et al., 1996). It is still an open question whether these observations originate in different resistance of the hosts or in different virulence of the endophytes. First indications point in both directions. The most common endophyte of *S. latissima* at Helgoland (North Sea) is *Laminarionema elsbetiae*, whereas *Laminariocolax aecidioides* preferentially infects *L. hyperborea*. Furthermore, field-collected sporophytes of *L. hyperborea* exhibit high constitutive rates of peroxide release; whereas accumulation of H_2O_2 is very transient in *S. latissima* and occurs only after the addition of elicitor (Küpper et al., 2002).

Reactive oxygen species generated during the oxidative burst may play a role in biotic defence not only through direct cytotoxicity, but also through their peroxidase-catalyzed reactions. The release of hypohalous acids (HOX) and, subsequently, halogenated organic compounds by *Laminaria digitata* increases after oxidative burst elicitation – in particular, of iodinated compounds (Palmer et al., 2005; Chance et al., 2009). Hypohalous acids generated by *L. digitata* inactivated bacterial quorum-sensing signals and thereby caused dispersal of biofilms. In contrast, a direct role of halogenated organic compounds in the defence of kelps has not been demonstrated so far, even though it appears likely. Iodinated compounds – which are produced at increased rates following elicitation compared with the unstressed steady state (Palmer et al., 2005; Chance et al., 2009) – exhibit a higher toxicity than brominated and chlorinated compounds owing to the higher effectiveness of iodine as a leaving group in nucleophilic substitutions compared with bromine and chlorine, respectively (Küpper et al., 2008). In this context, it should also be noted that bromoform, which is the main volatile halocarbon produced by *L. digitata* (Carpenter et al., 2000) and most other seaweeds (Carpenter and Liss, 2000), contributes to the defence against bacterial and algal epiphytes in red seaweeds (Ohsawa et al., 2001; Paul et al., 2006) and a similar effect in kelps may be possible.

Additionally, Küpper et al. (2006, 2009) reported for *L. digitata* that the early events of the defence reactions included an activation of fatty acid oxidation cascades, i.e., the release of free saturated and unsaturated fatty acids (FFAs) and accumulation of hydroxyl derivatives of fatty acids (oxylipins). The biosynthetic pathway of oxylipins in *L. digitata* is not fully identified yet. These compounds are generated by lipoxygenases in mammals and some of them have been reported to be anti-inflammatory mediators (Miller et al., 1990), besides a wide range of other physiological roles. Thus, oxylipins may well be another component of the defence response in kelps.

4. Biogeochemical Feedback

It is well established that aerosol particle bursts occur in the atmosphere over kelp beds at low tide during daytime (O'Dowd et al., 2002). In coastal ecosystems, Laminariales are major contributors to the iodine and volatile halocarbon flux from the ocean to the atmosphere (Giese et al., 1999; Carpenter et al., 2000). Release of molecular iodine occurs at rates around five orders of magnitude higher than that of volatile organic iodine compounds and is thus the major source of coastal new particle production, providing precursors for cloud condensation nuclei (Palmer et al., 2005; Küpper et al., 2008). While bromine emissions are high in the unstressed steady state (Carpenter et al., 2000), iodine emissions are strongly increased during oxidative stress – both in the context of defence reactions (Palmer et al., 2005; Chance et al., 2009) and ozone exposure (Palmer et al., 2005; Küpper et al., 2008). In fact, it has recently become clear that kelps utilize iodide accumulation in a unique inorganic antioxidant system to protect their apoplast and thallus surface against oxidative stress (Küpper et al., 2008), effectively linking antioxidant protection to atmospheric processes and coastal climate.

Thus, comparable with the oceanic release of dimethyl sulfide (DMS) by planktonic algae (Larsen, 2005), coastal primary productivity, cloud cover, ultraviolet radiation are linked via biogeochemical feedback mechanisms. They make it extraordinarily difficult to predict both future climate change and its impacts at a regional and local level. It has been shown that oxidative stress caused increased halocarbon and I_2 production by *Laminaria digitata* (Palmer et al., 2005; Chance et al., 2009). While gaseous ozone caused the maximum rates of I_2 release observed so far, oligoguluronates elicited the highest release of iodine-containing halocarbons including CH_2I_2. Thus, a potential link between endophyte infection and the biogeochemical cycle of iodine in coastal systems can be postulated.

5. Conclusion

A dramatic decrease in kelp forests has been reported in parts of the world. Kelp species are subjected to a variety of biotic stress factors, one of them being the interaction with pathogens. Bacteria are the causative agents of the red spot disease in kelp species, whereas the known pathogens of the dark spot disease are endophytic algae. Field observations from different parts of the world document massive rates of infection prevalence by endophytic brown algae. In recent years, major advances in identifying the taxonomic position of the pathogens and describing the biochemical basis of the defence responses of the host have been achieved. However, many aspects of this biotic interaction are still poorly understood, including pathogenicity of the endophytes. Nothing is known about

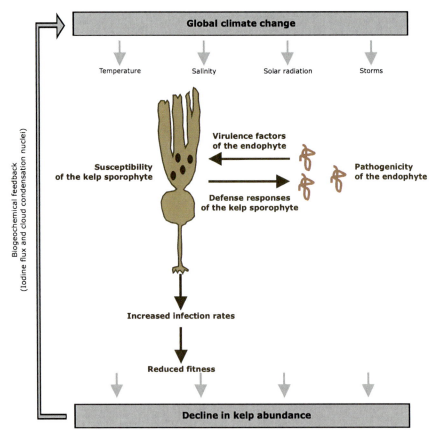

Figure 4. Global climate change (e.g., increase in temperature, freshwater input, UV radiation, storm frequency) may impact several components of the host–pathogen interaction (i.e., susceptibility of the kelp sporophyte, pathogenicity of the endophyte). Increased infection rates may cause a reduced fitness of the kelp sporophytes. This may contribute to the disappearance of kelp forests, which in turn might feedback with global climate change via the iodine metabolism of the kelps.

the ecological significance, i.e., the effect of infections on host physiological performance and fitness. Accordingly, how climate change may modulate this biotic interaction remains speculative (Fig. 4). Further research effort needs to deepen our mechanistic understanding of such host–pathogen interactions, to estimate its ecological impact and to predict and assess the effects of climate change.

6. References

Alker, A.P., Smith, G.W. and Kim, K. (2001) Characterization of *Aspergillus sydowii* (Thom et Church), a fungal pathogen of Caribbean sea fan corals. Hydrobiologia **460**: 105–111.
Apt, K.E. (1988) Etiology and development of hyperplasia induced by *Streblonema* sp. (Phaeophyta) on members of the Laminariales (Phaeophyta). J. Phycol. **24**: 28–34.
Ardré, F. (1971) Contribution a l'etude des algues marines du Portugal ii. Ecologie et chorologie. Bulletin du Centre d'Etudes et de Recherches Scientifiqes. Biarritz **8**: 359–574.
BACC (Baltex Assessment of Climate Change) (2008) *Assessment of Climate Change for the Baltic Sea Basin*. Series: Regional Climate Studies. Springer, Berlin, Germany.
Bartsch, I., Wiencke, C., Bischof, K., Buchholz, C.M., Buck, B.H., Eggert, A. et al. (2008) The genus Laminaria *sensu lato*: recent insights and developments. Eur. J. Phycol. **43**: 1–86.
Bouarab, K., Potin, P., Correa, J. and Kloareg, B. (1999) Sulfated oligosaccharides mediate the interaction between a marine red alga and its green algal pathogenic endophyte. Plant Cell **11**: 1635–1650.
Bratbak, G., Wilson, W. and Heldal, M. (1996) Viral control of *Emiliania huxleyi* blooms? J. Mar. Syst. **9**: 75–81.
Breeman, A.M. (1988) Relative importance of temperature and other factors in determining geographic boundaries of seaweeds: experimental and phonological evidence. Helgolander Meeresun. **42**: 199–241.
Breeman, A.M. (1990) Expected effects of changing seawater temperatures on the geographic distribution of seaweed species, In: J.J. Beukema et al. (eds.) *Expected Effects of Climatic Change on Marine Coastal Ecosystems*. Kluwer, The Netherlands, pp. 69–76.
Bromirski, P.D., Flick, R.E. and Cayan, D.R. (2003) Storminess variability along the California coast: 1858–2000. J. Climate **16**: 982–993.
Burkhardt, E. and Peters, A.F. (1998) Molecular evidence from nrDNA sequences that *Laminariocolax* (Phaeophyceae, Ectocarpales *sensu lato*) is a worldwide clade of closely related kelp endophytes. J. Phycol. **34**: 682–691.
Carpenter, L.J. and Liss, P.S. (2000) On temperate sources of bromoform and other reactive organic bromine gases. J. Geophys. Res. Atmos. **105**: 20539–20547.
Carpenter, L.J., Malin, G., Küpper, F.C. and Liss, P.S. (2000) Novel biogenic iodine-containing trihalomethanes and other short-lived halocarbons in the coastal east atlantic. Global Biochem. Cycles **14**: 1191–1204.
Chance, R., Baker, A.R., Küpper, F.C., Hughes, C., Kloareg, B. and Malin, G. (2009) Release and transformations of inorganic iodine by marine macroalgae. Estuar. Coast. Shelf Sci. **82**: 406–414.
Intergovernmental Panel on Climate Change, IPCC (2007) Synthesis report. Nadi, Fiji. June 2007. 280 pp.
Connell, S.D., Russell, B.D., Turner, D.J., Shepherd, S.A., Kildea, T., Miller, D., Airoldi, L. and Cheshire, A. (2008) Recovering a lost baseline: missing kelp forests from a metropolitan coast. Mar. Ecol. Progr. Ser. **360**: 63–72.
Cook, T., Folli, M., Klinck, J., Ford, S. and Miller, J. (1998) The relationship between increasing sea-surface temperature and the northward spread of *Perkinsus marinus* (Dermo) disease epizootics in oysters. Estuar. Coast. Shelf Sci. **46**: 587–597.
Correa, J.A. and McLachlan, J.L. (1994) Endophytic algae of *Chondrus crispus* (Rhodophyta). 5. Fine-structure of the infection by *Acrochaete operculata* (Chlorophyta). Eur. J. Phycol. **29**: 33–47.
Culley, A.I. and Steward, G.F. (2007) New genera of RNA viruses in subtropical seawater, inferred from polymerase gene sequences. Appl. Environ. Microbiol. **73**: 5937–5944.

Davis, A.J., Jenkinson, L.S., Lawton, J.H., Shorrocks, B. and Wood, S. (1998) Making mistakes when predicting shifts in species range in response to global warming. Nature 391: 783–786.

Dayton, P.K., Tegner, M.J., Edwards, P.B. and Riser, K.L. (1999) Temporal and spatial scales of kelp demography: the role of oceanographic climate. Ecol. Monogr. 69: 219–250.

Duggins, D.O., Simenstad, C.A. and Estes, J.A. (1989) Magnification of secondary production by kelp detritus in coastal marine ecosystems. Science 245: 170–173.

Ehlers, A., Worm, B. and Reusch, T.B.H. (2008) Importance of genetic diversity in eelgrass *Zostera marina* for its resilience to global warming. Mar. Ecol. Progr. Ser. 355: 1–7.

Ellertsdóttir, E. and Peters, A.F. (1997) High prevalence of infection by endophytic brown algae in populations of *Laminaria* spp. (Phaeophyceae). Mar. Ecol. Progr. Ser. 146: 135–143.

Faugeron, S., Martínez, E.A., Sánchez, P.A. and Correa, J.A. (2000) Infectious diseases in *Mazzaella laminarioides* (Rhodophyta): estimating the effect of infections on host reproductive potential. Dis. Aquat. Organ. 42: 143–148.

Gachon, C.M.M., Küpper, H., Küpper, F.C. and Šetlík, I. (2006) Single-cell chlorophyll florescence kinetic microscopy of *Pylaiella littoralis* (Phaeophyceae) infected by *Chytridium polysiphoniae* (Chytridiomycota). Eur. J. Phycol. 41: 395–403.

Giese, B., Laturnus, F., Adams, F.C. and Wiencke, C. (1999) Release of volatile iodinated c1–c4 hydrocarbons by marine macroalgae from various climate zones. Environ. Sci. Technol. 33: 2432–2439.

Graham, R.W. (1992) Late Pleistocene faunal changes as a guide to understanding effects of greenhouse warming on the mammalian fauna of North America, In: R.L. Peters and T.E. Lovejoy (eds.) *Global Warming and Biological Diversity*. Yale University Press, New Haven, pp. 76–87.

Graham, M.H. (2004) Effects of local deforestation on the diversity and structure of southern California giant kelp forest food webs. Ecosystems 7: 341–357.

Harley, C.D.G., Hughes, A.R., Hultgren, K.M., Miner, B.G., Sorte, C.J.B., Thorber, C.S., Rodriguez, L.F., Tomanek, L. and Williams, S.L. (2006) The impacts of climate change in coastal marine systems. Ecol. Lett. 9: 228–241.

Harvell, D., Kim, K., Quirolo, C., Weir, J. and Smith, G. (2001) Coral bleaching and disease: contributors to 1998 mass mortality in *Briareum asbestinum* (Octocorallia, Gorgonacea). Hydrobiologia 460: 97–104.

Harvell, C.D., Mitchell, C.E., Ward, J.R., Altizer, S., Dobson, A.P., Ostfeld, R.S. and Samuel, M.D. (2002) Climate warming and disease risks for terrestrial and marine biota. Science 296: 2158–2162.

Heesch, S. and Peters, A.F. (1999) Scanning electron microscopy observation of host entry by two brown algae endophytic in *Laminaria saccharina* (Laminariales, Phaeophyceae). Phycol. Res. 47: 1–5.

Helmuth, B., Mieszkowska, N., Moore, P. and Hawkins, S.J. (2006) Living on the edge of two changing worlds: forecasting the responses of rocky intertidal ecosystems to climate change. Annu. Rev. Ecol. Evol. Syst. 37: 373–404.

Hubbard, C.B., Garbary, D.J., Kim, K.Y. and Chiasson, D.M. (2004) Host specificity and growth of kelp gametophytes symbiotic with filamentous red algae (Ceramiales, Rhodophyta). Helgol. Mar. Res. 58: 18–25.

Kain, J.M. (1975) The biology of *Laminaria hyperborea*. VII. Reproduction of the sporophyte. J. Mar. Biol. Assoc. 55: 567–582.

Kawai, H. and Tokuyama, M. (1995) *Laminarionema elsbetiae* gen. et sp. nov. (Ectocarpales, Phaeophyceae), a new endophyte in *Laminaria* sporophytes. Phycological Research 43: 185–190.

Kremer, B.P. (1983) Carbon economy and nutrition of the alloparasitic red alga *Harveyella mirabilis*. Mar. Biol. 76: 231–239.

Küpper, F.C. and Müller, D.G. (1999) Massive occurrence of the heterokont and fungal parasites *Anisolpidium*, *Eurychasma* and *Chytridium* in *Pylaiella littoralis* (Ectocarpales, Phaeophyceae). Nova Hedwigia 69: 381–389.

Küpper, F.C., Kloareg, B., Guern, J. and Potin, P. (2001) Oligoguluronates elicit an oxidative burst in the brown algal kelp *Laminaria digitata*. Plant Physiol. 125: 278–291.

Küpper, F.C., Müller, D.G., Peters, A.F., Kloareg, B. and Potin, P. (2002) Oligoalginate recognition and oxidative burst play a key role in natural and induced resistance of sporophytes of Laminariales. J. Chem. Ecol. 28: 2057–2081.

Küpper, F.C., Gaquerel, E., Boneberg, E.-M., Morath, S., Salaün, J.-P. and Potin, P. (2006) Early events in the perception of lipopolysaccharides in the brown alga *Laminaria digitata* include an oxidative burst and activation of fatty acid oxidation cascades. J. Exp. Bot. **57**: 1991–1999.

Küpper, F.C., Carpenter, L.J., McFiggans, G.B., Palmer, C.J., Waite, T.J., Boneberg, E.-M., Woitsch, S., Weiller, M., Abela, R., Grolimund, D., Potin, P., Butler, A., Luther III, G.W., Kroneck, P.M.H., Meyer-Klaucke, W. and Feiters, M.C. (2008) Iodide accumulation provides kelp with an inorganic antioxidant impacting atmospheric chemistry. Proc. Natl. Acad. Sci. **105**: 6954–6958.

Küpper, F.C., Gaquerel, E., Cosse, A., Adas, F., Peters, A.F., Müller, D.G., Kloareg, B., Salaün, J.-P. and Potin, P. (2009) Free fatty acids and methyl jasmonate trigger defense reactions in *Laminaria digitata*. Plant Cell Physiol. **50**: 789–800.

Kylin, H. (1947) Die Phaeophyceen der schwedischen Westküste. Acta. Univ. Lund. **43**: 1–99.

Larsen S.H. (2005) Solar variability, dimethyl sulphide, clouds, and climate. Global Biogeochem. Cycles **19**: 1–12.

Lein, T.E., Sjotun, K. and Wakili, S. (1991) Mass-occurrence of a brown filamentous endophyte in the lamina of the kelp *Laminaria hyperborea* (Gunnerus) Foslie along the southwestern coast of Norway. Sarsia **76**: 187–193.

Leonard, G.H. (2000) Latitudinal variation in species interactions: a test in the New England rocky intertidal zone. Ecology **81**: 1015–1030.

Lima, F.P., Ribeiro, P.A., Queiroz, N., Hawkins, S.J. and Santos, A.M. (2007) Do distributional shifts of northern and southern species of algae match the warming pattern? Global Change Biol. **13**: 2592–2604.

Lubchenco, J. and Gaines, S.D. (1981) A unified approach of marine plant-herbivore interactions. I. populations and communities. Annu. Rev. Syst. Ecol. **12**: 405–437.

Lüning, K., Wagner, A. and Buchholz, C. (2000) Evidence for inhibitors of sporangium formation in *Laminaria digitata* (Phaeophyceae) during the season of rapid growth. J. Phycol. **36**: 1129–1134.

Micheli, F. and Halpern, B.S. (2005) Low functional redundancy in coastal marine assemblages. Ecol. Lett. **8**: 391–400.

Miller, C.C., Ziboh, V.A., Wong, T. and Fletcher, M.P. (1990) Dietary supplementation with oils rich in (n-3) and (n-6) fatty acids influences *in vivo* levels of epidermal lipoxygenase products in guinea pigs. J. Nutrition. **120**: 36–44.

Morelissen, B. and Harley, C.D.G. (2007) The effects of temperature on producers, consumers, and plant-herbivore interactions in an intertidal community. J. Exp. Mar. Biol. Ecol. **348**: 162–173.

Morizur, Y. (2001) Changements climatiques ou surexploitation? Gros temps sur les algues brunes. Les nouvelles de l'ifremer **25**: 1.

Müller, D.G., Stache, B. and Lanka, S. (1990) A virus infection in the marine brown alga *Ectocarpus siliculosus* (Phaeophyta). Bot. Acta. **103**: 72–82.

O'Dowd, C.D., Jimenez, J.L., Bahreini, R., Flagan, R.C., Seinfeld, J.H., Hämeri, K., Pirjola, L., Kulmala, M., Jennings, S.G. and Hoffmann, T. (2002) Marine aerosol formation from biogenic iodine emissions. Nature **417**: 632–636.

Ohsawa, N., Ogata, Y., Okada, N. and Itoh, N. (2001) Physiological function of bromoperoxidase in the red marine alga, *Corallina pilulifera*: production of bromoform as an allelochemical and the simultaneous elimination of hydrogen peroxide. Phytochemistry **58**: 683–692.

Palmer, C.J., Anders, T.L., Carpenter, L.J., Küpper, F.C. and McFiggans, G.B. (2005) Iodine and halocarbon response of *Laminaria digitata* to oxidative stress and links to atmospheric new particle production. Environ. Chem. **2**: 282–290.

Parmesan, C. and Yohe, G. (2003) A globally coherent fingerprint of climate change impacts across natural systems. Nature **421**: 337–42.

Paul, N.A., de Nys, R. and Steinberg, P.D. (2006) Chemical defence against bacteria in the red alga *Asparagopsis armata*: linking structure with function. Mar. Ecol. Progr. Ser. **306**: 87–101.

Pehlke, C. and Bartsch, I. (2008) Changes in depth distribution and biomass of sublittoral seaweeds at Helgoland (North Sea) between 1970 and 2005. Climate Res. **37**: 135–147.

Perry, A.L., Low, P.J., Ellis, J.R. and Reynolds, J.D. (2005) Climate change and distribution shifts in marine fishes. Science **308**: 1912–1915.

Peters, A.F. (1991) Field and culture studies of *Streblonema macrocystis* sp. nov (Ectocapales, Phaeophyceae) from Chile: a sexual endophyte of giant kelp. Phycologia **30**: 365–377.

Peters, A.F. (2003) Molecular identification, taxonomy and distribution of brown algal endophytes, with emphasis on species from Antarctica, In: A.R.O. Chapman et al. (eds.) *Proceedings of the 17th International Seaweed Symposium*. Oxford University Press, New York, pp. 293–302.

Peters, A.F. and Burkhardt, E. (1998) Systematic position of the kelp endophyte *Laminarionema elsbetiae* (Ectocarpales *sensu lato*, Phaeophyceae) inferred from nuclear ribosomal DNA sequences. Phycologia **37**: 114–120.

Peters, A.F. and Ellertsdottir, E. (1996) New record of the kelp endophyte *Laminarionema elsbetiae* (Phaeophyceae, Ectocarpales) at Helgoland and its life history in culture. Nova Hedwigia **62**: 341–349.

Peters, A.F. and Schaffelke, B. (1996) *Streblonema* (Ectocarpales, Phaeophyceae) infection in the kelp *Laminaria saccharina* (Laminariales, Phaeophyceae) in the western Baltic. Hydrobiologia **327**: 111–116.

Precht, W.F. and Aronson, R.B. (2004) Climate flickers and range shifts of reef corals. Front Ecol. Environ. **2**: 307–314.

Rivadeneira, M. and Fernández, M. (2005) Shifts in southern endpoints and distribution in rocky intertidal species along the south-eastern Pacific coast. J. Biogeogr. **32**: 203–209.

Sanford, E. (1999) Regulation of keystone predation by small changes in ocean temperature. Science **283**: 2095–2097.

Sawabe, T., Makino, H., Tatsumi, M., Nakano, K., Tajima, K., Iqbal, M.M., Yumoto, I., Ezura, Y. and Christen, R. (1998) *Pseudoalteromonas bacteriolytica* sp. nov., a marine bacterium that is the causative agent of red spot disease of *Laminaria japonica*. Int. J. Syst. Bacteriol. **48**: 769–774.

Schaffelke, B., Peters, A.F. and Reusch, T.B.H. (1996) Factors influencing depth distribution of soft bottom inhabiting *Laminaria saccharina* (L.) Lamour. In Kiel Bay, western Baltic. Hydrobiologia **326/327**: 117–123.

Schär, C. and Jendritzky, G. (2004) Climate change: hot news from summer 2003. Nature **432**: 559–560.

Schatz, S. (1984) The *Laminaria-Phycomelaina* host–parasite association: seasonal patterns of infection, growth and carbon and nitrogen storage in the host. Helgolander Meeresun. **37**: 623–631.

Southward, A.J., Hawkins, S.J. and Burrows, M.T. (1995) Seventy years' observation of changes in distribution and abundance of zooplankton and intertidal organisms in the western English channel in relation to rising sea temperature. J. Therm. Biol. **20**: 127–155.

Steneck, R.S., Graham, M.H., Bourque, B.J., Corbett, D., Erlandson, J.M., Estes, J.A. and Tegner, M.J. (2002) Kelp forest ecosystems: biodiversity, stability, resilience and future. Environ. Conserv. **29**: 436–459.

Uppalapati, S.R. and Fujita, Y. (2000) Carbohydrate regulation of attachment, encystment, and appressorium formation by *Pythium porphyrae* (Oomycota) zoospores on *Porphyra yezoensis* (Rhodophyta). J. Phycol. **36**: 359–366.

Vairappan, C.S., Suzuki, M., Motomra, T. and Ichimura, T. (2001) Pathogenic bacteria associated with lesions and thallus bleaching symptoms in the japanese kelp *Laminaria religiosa* Mmiyabe (Laminariales, Phaeophyceae). Hydrobiologia **445**: 183–191.

Norwegian Institute for Water Research (2007) Long-range transported nutrients and the status of the sugar kelp on the south coast of Norway. Report 2358.

Weinberger, F. and Friedlander, M. (2000) Response of *Gracilaria conferta* (Rhodophyta) to oligoagars results in defense against agar-degrading epiphytes. J. Phycol. **36**: 1079–1086.

Yoshida, T. (1980) Distribution of *Streblonema aecidioides* around Japan and its host. Jpn. J. Phycol. **27**: 182.

Zacherl, D., Gaines, S.D. and Lonhart, S.I. (2003) The limits to biogeographical distributions: insights from the northward range extension of the marine snail, *Kelletia kelletii* (Forbes, 1852). J. Biogeogr. **30**: 913–924.

PART 4:
THE EFFECTS OF UV RADIATION ON SEAWEEDS

Figueroa
Korbee
Gao
Xu
Helbling
Villafañe
Häder

Biodata of **Félix L. Figueroa** and **Nathalie Korbee**, authors of *"Interactive Effects of UV Radiation and Nutrients on Ecophysiology: Vulnerability and Adaptation to Climate Change"*

Professor Félix L. Figueroa is Full Professor of the Department of Ecology, Faculty of Sciences in the University of Málaga, Spain. He obtained his Ph.D. on the photocontrol of photosynthetic pigment synthesis in macroalgae. The investigation concerns on photobiology, i.e., photoregulation of nitrogen metabolism in algae and the effects of UV radiation on ecophysiology of algae. His research also includes bio-optics of both continental and oceanic waters. From 1996, he is leading an interdisciplinary research group concerned with photobiology and biotechnology of marine organisms. This group combines basic research on photoprotection with applied research, i.e., use of algal substances as photoprotectors, antioxidants, and immunostimulants.

E-mail: **felix_lopez@uma.es**

Dr. Nathalie Korbee is currently Assistant Professor in the Department of Ecology, Faculty of Sciences in the University of Málaga, Spain. She obtained her Ph.D. from the University of Málaga in 2003 and is continuing her studies and research in the same university. Her scientific areas of interest are photobiology, ecophysiology, and photoprotective mechanisms against UV radiation in marine macroalgae: Accumulation of UV-absorbing compounds (mycosporine-like amino acids), antioxidant, and carotenoids.

E-mail: **nkorbee@uma.es**

Félix L. Figueroa   Nathalie Korbee

INTERACTIVE EFFECTS OF UV RADIATION AND NUTRIENTS ON ECOPHYSIOLOGY: VULNERABILITY AND ADAPTATION TO CLIMATE CHANGE

FÉLIX L. FIGUEROA AND NATHALIE KORBEE
Department of Ecology, Faculty of Sciences, University of Málaga, 29071, Málaga, Spain

1. Ozone Depletion and Climate Change in Aquatic Ecosystems: Interaction Among Variables at Different Scales and Ecological Status of Coastal Waters

The IV Report of the Intergovernmental Panel of Climate Change (IPCC, 2007a, b) has concluded that the warming of the Earth is unequivocal as is now evident from observations of the increasing global air and ocean temperatures, the reduction of ice and snow in polar region and high mountains, and rising global average sea level. It is crucial to know the vulnerability and the ecosystem capacity of adaptation to climate change. Most of the studies are being conducted in land ecosystems and oceanic waters; meanwhile, it is necessary for a greater research effort in coastal waters, lakes, and lagoons. The IV-IPCC report defines the adaptation capacity as the capability of a system to adapt or adjust to climate change (including the climate variability and the climate extremes) to take the advantages of the opportunities or to carry the consequences. On the other hand, the vulnerability of the systems is the grade in which the systems are not capable of carrying the adverse effects of climate change. The vulnerability is a function of character, magnitude, and change rate of climate change, and also of the submitted environmental variations, the sensitivity, and capacity of adaptation of the ecosystems.

The IV IPCC (2007a, b) reported changes in the abundance of macroalgae and phytoplankton, a higher acidity of the ocean waters and an increase in the surface water temperature. The availability of drinking water will decrease (10–30%) in the Mediterranean area of Europe in which at present the water stress already exists. The vulnerability of the coastal areas in Southern Europe will be increased by other human factors such as marine contamination, urban and agricultural effluents, and the urban development.

No implemented regulations are presently addressing the protection of marine environment as a whole against multiple stresses, including climate change. However, a number of water- and marine-related directives have been established with specific issues like water quality and sustainable management of marine resources in response to political concern in restricting domains (e.g., bathing, drinking, water, fisheries).

On the other hand, the Water Framework Directive (2000/60/EC) of the European Parliament and Council establishing a framework for community action in the field of water policy (WFD) provides a good example of an integrated management and allows great flexibility in meeting good ecological and chemical status not only in continental waters (river and lakes) but also transitional (lagoons and estuarine) and coastal waters. The ecological status is an expression of the quality of the structure and functioning of aquatic ecosystems associated with surface waters. The ecological status is directly related to human activities: urban, industrial, and agricultural effluents, urban pressure on the line coast among others, but recently it is also related to climate change impacts. Its successful implementation would increase the ecosystem capacity for resilience and reduce the vulnerability of these waters to climate change stresses (Hoepffner, 2006).

The ecological status of coastal waters must be evaluated by

1. Biological elements: Composition, abundance, and biomass of phytoplankton, other aquatic flora, and benthic invertebrate fauna
2. Hydromorphologic elements supporting the biological elements as (a) morphological conditions (depth variation structure and substrate of coastal bed, structure of the intertidal zone), (b) tidal regime: direction of dominant currents, water exposure, and chemical and physicochemical elements supporting the biological elements as general (transparency, thermal, oxygenation, and nutrient conditions and salinity), and specific pollutants (pollution by all priority substances identified as being discharged into the body of the water and pollution by other substances identified in significant quantities into the body water).

The coastal waters in Europe as in other parts of the world, i.e., USA and Eastern Asia, continuously exposed to increasing human pressure through activities such as fisheries, energy production, trade, commercial, and tourism. Thus, the effect of climatic change is difficult to untangle from direct anthropogenic activities. The latter often reduces the resilience property of the marine and coastal ecosystems, which then become more vulnerable to stresses due to climate forcing. Macroalgae have been used as a good indicator of the water quality because their sedentary condition integrates the effects of long-term exposure of nutrient or other pollutants resulting in a decrease or even disappearance of the most sensitive species and its replacement by highly resistant, nitrophilic, or opportunistic species (Murray and Littler, 1978). Macrophytes have been used as biological indicators in different European geographical areas such as region 1 (Atlantic Ocean) by research groups of United Kingdom and Ireland (Wells and Wilkinson, 2002; Wells et al., 2007) and region 6 (Mediterranean) by groups of Greece (Orfanidis et al., 2001), France (Thibaut et al., 2005), or Spain, mainly in the Catalonia coastal waters (Ballesteros et al., 2007; Arévalo et al., 2007) among others. The investigations reported the importance to define the ecological status of the water, not only based on the composition, abundance, and biomass of phytoplankton or macrophytes but also by using new indicators. There is still no agreement at the European level on the evaluation design and the specific indicators by using macrophytes; in contrast, the indicators for macroinvertebrate evaluations are already decided, i.e., seven different ISO regulations.

The intertidal macroalgae communities respond to changes in nutrient status when they are exposed to eutrophication, toxic substances, and other habitat modification known as general ambient stresses. Specifically, the WFD outlines the criteria that need to be related to type-specific reference (undisturbed area) conditions for macroalgae: (1) taxonomic composition corresponds totally or nearly totally to undisturbed conditions, (2) there are no detectable changes in macroalgae abundance owing to anthropogenic activities. Regarding the composition of macrophytes, the WFD states that for high quality, all sensitive taxa must be present. The requirements stipulated for reference and high quality conditions by WFD create two main problems: (1) It is not well known which species are the sensitive ones in any particular situation, as sensitivity species tend to be less abundant members of the community, or such that they will not constantly present even under good water-quality conditions and (2) species composition can be naturally highly variable.

At present, there is a controversy on the indicators used to evaluate the ecological status of aquatic ecosystems, as new indicators are being proposed in addition to species composition such as

1. Specific richness. Wilkinson and Tittley (1979) reported that the richness remains broadly constant in the absence of environmental alterations, over days, months, seasons, and years. Although different ecological communities do not contain the same number of species (Krebs, 1978), there is a particular range of species richness which can be expected within intertidal communities (Wells et al., 2007). Using data for 100 rocky shores, Wilkinson et al. (1988) founded a link between species richness and localized intertidal variables.
2. Proportion of Chlorophyta and Rhodophyta taxa. The Chlorophyta species constitute a high proportion of small filamentous and delicate species and show an increase in species numbers with decreasing environmental quality. Generally, the Chlorophyta species although small and often filamentous are able to adapt more rapidly to changes in the environments, whereby proportions increase with decreasing quality status. Consequently, the changes in the proportion of Rhodophyta and Chlorophyta species have been considered to be indicative of human influences and shift in quality status, i.e., in high ecological status, the proportion of Chlorophyta is 20–25%, whereas that of Rhodophyta is 45–55%. There are exceptions to this pattern, for example, the increase in the red algae Ceramiales under stress conditions; this is a group of red algae with filamentous and simple morphology. Thus, Gorostiaga et al. (2008) proposed the proportion of species of simple morphology/complex morphology as an indicator independent of the taxonomic identity.

3. Ratio of ecological–functional status group. Wells (2002) proposed the functional groups according to the classification of Littler et al. (1983). ESG1: late successional or perennial and ESG2: opportunistic and annuals. In high ecological status, the proportion of ESG1/ESG2 is about 0.5–0.9, whereas the values are 0.1–0.4 in low ecological status (Wells, 2002). Arévalo et al. (2007) applied methods based on functional-form group of macroalgae. They reported that changes in the species composition and structure of Mediterranean macroalgal-dominated communities form upper sublittoral zone described along a gradient of nutrient enrichment form urban sewage outfall. *Ulva*-dominated communities only appear close to sewage outfall, *Corallina*-dominated communities replace ulvacean at intermediate levels of nutrient enrichment, and *Cystoseira*-dominated communities thrive in the reference site. The functional group approach is adequate, since it is linked to the concept of bioindicator species and to the progressive increase in the structural complexity of aquatic ecosystems (Gorostiaga et al., 2008).
4. CARLIT index. It combines community cartography and available information about the value of the community as indicators of water quality, using GIS technology, to provide an index that fulfills the requirements of the WFD, i.e., it takes into account sites in reference conditions and it is expressed as numerical values ranging between zero and one. Ballesteros et al. (2007) used this approach to express the ecological status of the coastal waters of Catalonia in the 37 areas in which the coast was parceled.

In the context of climate change and its influence in the marine benthos, the effort of research conducted has been scarce, owing to the newness of the field, the smaller number of studies with large scale as references (30–50 years), and the intrinsic high environmental temporal–spatial variation in aquatic ecosystems. In spite of this, scientific attention is being devoted to the prediction of biological changes in benthic marine communities (Bhaud et al., 1995; Alcock, 2003) as well as to the evaluation of effects already attributed to climate change with new advances in the statistical approaches (Hiscock et al., 2004; Helmuth et al., 2006).

Improvements in water quality in Spain has recently been followed by noticeable changes in species composition and vegetation structure (Ballesteros et al., 2007; Arévalo et al., 2007; Pérez-Ruzafa et al., 2007; Gorostiaga et al., 2008) – species richness significantly increased throughout the study area, whereas algal cover only increased at the most degraded sites. Pollution removal promoted the development of morphologically more complex species. Intertidal vegetation at the degraded sites became progressively more similar to that at the reference site. In the Basque coast, five recovery stages discriminated by different species (SIMPER routine) were characterized from ordination (MDS) analyses (Gorostiaga et al., 2008): (1) extremely degraded: *Gelidium pusillum* is the most abundant species, which is accompanied by *Bachelotia antillarum* at the low intertidal level (0.75 m); (2) heavily degraded: *Gelidium pusillum* remains dominant and accompanied by *Caulacanthus ustulatus* at the high intertidal level (1.4 m); (3) moderately degraded: *Corallina elongata* becomes dominant, *C. ustulatus* remains abundant at the high level; (4) slightly degraded: *C. elongata* remains dominant in both tidal levels, *Chondracanthus acicularis* and *Lithophyllum incrustans* are abundant at the high level. *Pterosiphonia complanata* and *Stypocaulon scoparium* become abundant at the low level; (5) reference stage: *Lithophyllum incrustans* and *Laurencia obtusa* are abundant together with *C. elongata* at the high level, whereas *Stypocaulon scoparium* dominates the low level, with *Bifurcaria bifurcata*, *Jania rubens*, and *Cystoseira tamariscifolia* as abundant species. Thus, this study reveals that phytobenthic communities are useful indicators of water quality and provide real data that contribute to the assessment of the ecological status of rocky open shores on the Basque coast species richness, algal cover, and proportion of species with complex morphology have been used as good indicators of the ecological status of coastal waters. The bioindicator capacity of certain species of subtidal systems has been tested in the frame of climate change, i.e., retraction due to temperature increase and high irradiance of marine macroalgae such as *G. sesquipedale*, *Pterosiphonia complanata*, or the increase of *Cystoseira baccata*, *Codium decorticatum* y *Peyssonelia* sp. The biomass is the other key parameter but with high sampling cost, i.e., the biomass and production of *Gelidium* meadows have decreased in the last 10 years. It is evident that the Production/Biomass (P/B) ratio in a stressed community will be lower than that in communities under optimal conditions.

The evaluation of the vulnerability requires the knowledge of the structure–function of the aquatic ecosystems (Tilman et al., 2002). One of the best indicators of the human activity, i.e., urban sewage, which leads to the decline of biodiversity (Wilson, 2003), is not only related to the species extinction but also the loss of genetic and functional diversity at different levels of organization (Naeem et al., 1999).

In terrestrial systems, habitat distribution models have been applied with success to define protected areas relating the direct field observations on the species distribution with the predictor variables of the ecosystems according to theoretical models and statistical tools (Guisan and Zimmermann, 2000; Seoane et al., 2006) and to detect key environmental variables on the species abundance (Luoto et al., 2001). These approaches are scarce in marine communities (Kaschner et al., 2006) and especially in coastal ecosystems (Robertson et al., 2003; Calvo Aranda, 2007).

The macroalgae of Southern Iberian Peninsula, both Mediterranean and Atlantic species, present higher mechanism for photoprotection compared with algae of Northern latitudes, i.e., dynamic photoinhibition (Figueroa et al., 1997a, 2002; Flores-Moya et al., 1998; Jiménez et al., 1998) and accumulation of photoprotectors (Karsten et al., 1998; Pérez-Rodríguez et al., 1998, 2001; Korbee et al., 2005; Abdala-Díaz et al., 2006).

There is very scarce information about adaptation and ecophysiological responses. It is urgent to investigate ecophysiological responses at the molecular, metabolic, and individual levels to global climate change. Functional indicators can be not only a good basis for ecophysiological studies but also in the management of the aquatic environment, i.e., ecological status according to WFD.

The use of new indicators to both evaluate the ecological status and the vulnerability and adaptation capacity of the macrophyte community is proposed as follows:

1. Structural, biological, and ecological indicators: species richness, biodiversity, and other structural parameters
2. Functional indicators of photosynthesis: optimal quantum yield and maximal, electron transport rate as in vivo chlorophyll fluorescence is associated to Photosystem II (PAM fluorometry)
3. Functional indicator of nutrient status: stoichiometry, i.e., C:N:P ratios
4. Stress indicators: heat shock proteins (HSP), proteases, and reactive oxygen species (ROS)

The integration of ecological and ecophysiological approaches will give the basis for the evaluation of ecological status and the prediction of variations of the structure–function of aquatic ecosystems owing to climate change.

2. Nutrient Status and the Capacity of Acclimation to Increased UV Radiation: UV Screen Substances (Mycosporine-Like Amino Acids and Phenolic Compounds)

2.1. INTERACTION OF FACTORS OF CLIMATE CHANGE: TEMPERATURE, PHOTOSYNTHETIC IRRADIANCE, AND UV RADIATION AT DIFFERENT SCALES (ORGANISMS VERSUS ECOSYSTEMS)

There is an accumulating body of evidence to suggest that many marine ecosystems, both physically and biologically, are responding to changes in regional climate caused by the warming of air and sea surface temperatures and to a lesser extent by the modification of precipitation regime and wind patterns (Hoepffner, 2006).

Recent evidences indicate that the increase in temperature over the last decade has had a primary role in influencing the ecology of European seas in intertidal rocky shore populations (Hawkinks and Jones, 1992). Ecological changes in the Southern North Sea (Perry et al., 2005) and English channel also appear to be closely related to climate-driven sea temperature fluctuations (Southward et al., 2005). Other documented range shifts and recent appearance of warm-water species new to marine environment include tropical macroalgae in the Mediterranean (Walther et al., 2002). Successive heat waves over the Mediterranean Sea and subsequent peaks in the water temperature field have been lethal to some invertebrates like sponges and gorgonias, and it could be also affecting supralittoral macroalgae (Laubier et al., 2003). Seagrasses beds, which have an important role in the marine store carbon and stabilizing the bottom sediment against erosion, may suffer considerably through intensification of extreme weather patterns through storms, wave action, resuspension of sediment in the water column, as well as sudden pulses of freshwater runoff (Pergent et al., 1994). After such events, the recolonization of the benthos can take several years. In spite of the importance of biodiversity for ecological functioning, we have still scarce knowledge on the effects of climate change on this aspect of our seas. Even as our oceans cover more than 70% of the surface of our planet, less than 10% of published research on biodiversity dealt with marine systems (Parry et al., 2007).

Changes in the abundance of certain species in the Spanish coast have been related to thermal shocks (B. Martínez, 2009, personal communication). The results of a demographic study, which is currently being undertaken in the Cantabric coast, show clear differences in the structure and demography of the brown algae *Fucus serratus* in the southern distributional of Asturias compared with historical data (Anadón and Niell, 1981). Transplant experiments done in 1990 showed that *F. serratus* can grow out from the limit of distribution (Arrontes, 1993). However, the last studies suggested that marginal population of *F. serratus* are above their limit of environmental tolerance and a retraction in the distribution is taking place in this geographical area. Other vulnerable communities are the aquatic angiosperms both from marine and continental waters. The meadows of seagrasses *Posidonia oceanica* are essential in the protection of the marine environments on the Spanish Mediterranean coast (Medina et al., 2001); thus, the retraction of these communities due to anthropogenic factors including climate change would have catastrophic consequences.

The concentration of ozone depleting substances in the atmosphere are now decreasing but the recovery of the ozone layer as 1980 is still far to produce. The area affected each year by ozone hole seems to reach a constant maximal level, but there is greater uncertainty about future UV-B radiation than future ozone, since UV-B radiation will be additionally influenced by climate change (Mckenzie et al., 2007). At some sites of northern hemisphere, UV-B irradiance may continue to increase because of continuing reduction in aerosol extinction since 1990. The recovery of the ozone layer is expected to delay to 2070 due to the decrease in the temperature in the stratosphere as an influence of the climate change. Calculations based on the absorption characteristics of O_3 suggest that a 10% decrease in the ozone layer produces an increase of 5% of irradiance at 320 nm but an increase of 100% at 300 nm (Frederick et al., 1989). The decrease in ozone layer in Southern Iberian Peninsula has been about 0.3% per year, i.e., 0.5–0.75% increase in biological weighted irradiance related to DNA damage and algal photoinhibition (Häder et al., 2007). The ozone depletion is affecting the UVB/UVA ratios since only UV-B is increased. The increasing in this ratio can have an important effect on repair capacity and biochemical cycles (Jeffrey et al., 1996; Zepp et al., 2007). The turbidity and beam attenuation coefficient (c) are good indicators of the penetration of UVR. The decrease in transparency of the water by an increase in the concentration of dissolved and particulate material due to different activities (urban sewage outflow, aquacultural effluents) together with the increasing of temperature can be the probable reason for the significantly negative effect of macroalgae marine angiosperm communities. In aquatic ecosystems, the increase in UVB (280–315 nm) by ozone depletion together the high exposure to UVA (315–400 nm) according to the latitude and altitudes have been related to the damage of DNA, RNA, proteins, and lipids in aquatic organisms (Buma et al., 1995, 1997; Bischof et al., 1998; Helbling et al., 2001) as the increase of oxygen radicals (ROS) and subsequent stimulation of antioxidant systems (Bischof et al., 2006), decrease of enzyme activities as Rubisco (Bischof et al., 2002), nitrate reductase, and carbonic anhydrase (Gomez et al., 1998; Flores-Moya et al., 1998; Viñegla, 2000; Figueroa and Viñegla, 2001), pigment photodamage (Figueroa et al., 1997a; Häder and Figueroa, 1997; Aguilera et al., 2002), photoinhibition of photosynthesis (Figueroa et al., 1997a; Figueroa et al., 2002), inhibition of growth (Altamirano et al., 2000a, b), and inhibition of different reproductive stages (Altamirano et al., 2003a, b; Wiencke et al., 2000, 2006). The algae present different sensitivity to UVB according to species, morphology, and life cycle (Dring et al., 1996; Altamirano et al., 2003a, b). This pattern is related to the action of photoprotection mechanisms as the photorepair of DNA is mediated by PAR and UVR (Mitchell and Karentz, 1993), accumulation of lipidic and water-soluble antioxidants, and the activation of antioxidant enzymes (Cockell and Knowland, 1999), and finally the accumulation of UV-screen photoprotectors as mycosporine-like amino acids (MAAs) in red macroalgae (Karsten et al., 1998; Korbee Peinado et al., 2004, 2005), phenolic compounds in brown algae (Pavia et al., 1997; Connan et al., 2004; Abdala-Díaz et al., 2006), and trihydroxicoumarins in the green algae *Dasycladus vermicularis* (Pérez-Rodríguez et al., 1998, 2001) or other phenolic compounds in *Ulva pertusa* (Han and Han, 2005). Temperature can affect the repair process and it has an antagonist effect with UV radiation on growth pattern in macroalgae (Altamirano et al., 2003a, b).

The factors can change the physiological and ecological responses with antagonist or synergic effects. The analysis of multiple factors influencing at different rates and scales is a hot research point in the research on the effects of Global Climate Change (Breitburg et al., 1999; Xenopoulos et al., 2002; Doyle et al., 2005). One of the most important challenge of the ecological research is to understand and predict the effect of multiple factors of abiotic stress (increase in CO_2, photosynthetic irradiance, UV and temperature, input of nutrient by pulses, decrease in pH, eutrophication) on species, communities, and ecosystems (Breitburg et al., 1999).

Most of the studies on the effect of global climate change on aquatic organisms has been conducted with one or two variables and the number of studies on interaction of more number of factors are very scarce (Bischof et al., 2006; Häder et al., 2007). In the last 15 years, our research group has investigated on the effect of the increase in ultraviolet radiation (UVR), temperature, and CO_2 on the photosynthetic metabolism, nutrient incorporation, and growth in microalgae, macroalgae, and aquatic macrophytes of both coastal and continental waters. The investigation has been conducted in the field (in situ) monitoring the main environmental factors (solar radiation, nutrient, and temperature), in experimental controlled conditions under solar radiation (outdoor) and under artificial conditions (indoor). The mechanisms for acclimation to global climate change have been evaluated including as follows: photoinhibition, photoprotection, nutrient uptake systems, and patterns of growth, reproduction, and morphogenesis (Häder and Figueroa, 1997; Gómez and Figueroa, 1998; Conde-Álvarez, 2001; Figueroa et al., 2002; Villafañe et al., 2003; Gómez et al., 2004).

The studies on the adaptation capacity and vulnerability of the coastal ecosystems of Southern Europe to climate change are still scarce. One of the best monitored ecosystems is the high altitude lake ecosystems (Sierra Nevada, Southern Spain). The structure–function and complex interaction has been studied in microbial food webs of these lakes submitted to stress, i.e., high solar UV radiation and nutrient availability (Carrillo et al., 2002, 2006, 2008). However, it is necessary to study other lakes submitted to other interactive stress conditions, i.e., saline lagoons. In coastal waters, Mercado et al. (1998) is a pioneer study on the use of carbon by the macroalgae of Southern Iberian Peninsula; this study reported that in green algae the assimilation of Ci is saturated at the present CO_2 concentration, whereas 70% of the red and brown algae analyzed are not saturated, consequently an increase in CO_2 favored the primary production.

The main investigation has been conducted at species level and there is still scarce information on the interactive effect of climate change variables on the structure, diversity, and primary production of the algal and aquatic macrophyte communities (Lüning, 1990; Häder and Figueroa, 1997; Bischof et al., 2006). The increase in UVB due to ozone depletion has been related to the alterations at primary production and community structure (Roleda et al., 2006), reproductive stages (Wiencke et al., 2000), and succession (Santas et al., 1998; Wahl et al., 2004; Dobrestov et al., 2005). The effects of UVR are only important in the early phases of succession as spore germination or the growth of juvenile algae, i.e., spores of supralittoral dominant species are more resistant to UVR than that of subtidal grown algae (Wiencke et al., 2006). The same results were observed on the resistance of the pollen to UVR of macrophytes of the Laguna of Fuente de Piedra, i.e., the species with pollinization in surface waters presented higher survival rate to UVR than plants with underwater pollinization (Conde-Álvarez et al., 2008).

The depth distribution (zonation) and consequently the structure of the coastal system is correlated to the sensitivity of the algae to UV radiation, i.e., the supralittoral present less DNA damage and higher repair rate than the algae grown in the subtidal area (van de Poll et al., 2001; Bischof et al., 1998; Gómez et al., 2004; Wiencke et al., 2006).

Studying the interaction among factors of climate change is very important to know the bio-optical properties of both macrophytes and aquatic ecosystem. Our research group has conducted studies on bio-optical properties of diverse aquatic ecosystem (Mediterranean, Atlantic, and Antarctic). The evaluation of absorptances of macroalgae is very important to estimate the photosynthetic capacity of macrophytes. In a laboratory study, it has been reported that specific attenuation coefficient (K_c) was a good indicator of the acclimation capacity of algae in terms of photosynthesis to increase photosynthetic and UV radiation (Figueroa et al., 1997a; Figueroa et al., 2002). In macrophytes, the absorptance has been related to morpho-functional aspects related to light and nutrients (López-Figueroa, 1992;

Enríquez et al., 1994; Salles et al., 1996). The capacity of acclimation to UV radiation has been related to morpho-functional characteristics of macrophytes in tropical ecosystems (Larkum and Wood, 1993; Hanelt et al., 1994), temperate latitudes as Mediterranean (Santas et al., 1998; Gómez and Figueroa, 1998), North Sea (Dring et al., 1996), Arctic (Hanelt, 1998), Antarctic (Bischof et al., 1998), and Patagonia of Chile (Gómez et al., 2004).

The preliminary studies suggest that the influence of UVB at ecosystem level could be more pronounced on the structure of the community and a trophic level with the subsequent effects on biogeochemical cycles than that on the levels of biomass per se (Häder et al., 2007). Thus, in the next 50 years, the levels of UVR can still affect the structure and function of macrophyte communities and consequently to a higher scale the biogeochemical cycles.

2.2. INTERACTION OF FACTORS OF CLIMATE CHANGE: NUTRIENT AVAILABILITY, STOICHIOMETRY, AND STRUCTURE–FUNCTION OF AQUATIC ECOSYSTEMS

Although one of the most evident effects of the human activities on aquatic ecosystems is the decrease in biodiversity, this decline also includes the loss of genetic and functional diversity at different organization levels. It is a consensus among ecologists to increase the focus to other field of diversity of ecosystems (Schulze and Mooney, 1993; Naeem and Wright, 2003). Most of the studies on the relation between biodiversity and the functioning of ecosystem have been conducted in terrestrial ecosystems and with communities of higher plants (Naeem et al., 1999; Naeem and Wright, 2003). In addition, the number of studies on the interaction between factors of global climate change (increased CO_2, temperature, and UVR) at regional scale and the relation of the pattern of biodiversity and function of ecosystems are still more scarce (Naeem et al., 1999).

There are a great number of reports on the individual effects of UVR and nutrients on organisms (Helbling and Zagarese, 2003; Häder et al., 2007), but the number of studies on the interaction between UV radiation and nutrient availability (Villafañe et al., 2003) is scarce. Sterner et al. (1997) proposed the hypothesis light: nutrients (LNH) with predictive models. However, the investigation has treated partial aspects such as structure and succession of nannoplankton (Bergeron and Vincent, 1997; Xenopoulos and Frost, 2003) or functional aspects of algal community (Xenopoulos et al., 2002; Litchman et al., 2002).

Most of the adaptive mechanisms and the sensitivity of algae to UVR depend on the nutrient availability or cell nutrient status, nitrogen (N) in marine (Shelly et al., 2002; Korbee et al., 2005; Villafañe, 2004; Medina-Sánchez et al., 2006) and phosphorus (P) in freshwater (Hiriart et al., 2002; Xenopoulos and Frost, 2003; Medina-Sánchez et al., 2006) ecosystems. The nitrogen limitation affects many processes in macrophytes, among which are not only the photosynthetic capacity (Pérez-LLoréns et al., 1996), but also the content of proteins

(Henley et al., 2001; Vergara et al., 1995) and photoprotection mechanisms (Korbee Peinado et al., 2004; Korbee et al., 2006; Huovinen et al., 2006). The limitation of N reduces the cell size (Doucette and Harrison, 1990; García-Pichel, 1994). The increase in the relation cell surface/volume not only favored nutrient incorporation but also decreased the attenuation of UVR (antagonist effect). Increasing sensitivity to UVR has been reported for N-limited algae in short- and mid-term studies (Litchman et al., 2002; Korbee Peinado et al., 2004) and P-starved algae in shorter-term (Hiriart et al., 2002) and longer-term (Shelly et al., 2005) studies.

The accumulation of MAAs (photoprotectors) with the increase in both nitrate and ammonium availability (Table 1) suggest that these substances can have other functions beyond UV-screen capacity as also reservoir of N (Korbee et al., 2005, 2006; De la Coba, 2007). Thus, in addition to UV radiation, the nitrogen availability is crucial as trigger and substrate for the enhancement of photoprotective system. In scenarios of increased UV radiation, combined eutrophication is expected in the investment of energy to accumulate MAAs. The nitrogen of MAAs can be used when the N sources is reduced (anticipating strategy), as it has been suggested for phycobiliproteins (Algarra and Rüdiger, 1993; Tandeau de Marsac and Houmardd, 1993; Sinha and Häder, 1998; Talarico and Maranzana, 2000).

The content of MAAs is much higher in algae grown about 1 week under high nitrogen conditions (300–500 µM inorganic nitrogen) than that grown under low nitrogen concentration (50 µM). The increase is higher in the presence of UV radiation indicating an interaction between nutrient supply and UV radiation (Korbee Peinado et al., 2004). Thus, these data show that the algae grown under eutrophication conditions could be more photoprotected against increased UV radiation.

Table 1. Concentration of mycosporine-like amino acids (expressed as mg gDW^{-1}) in red macroalgae grown at different nitrogen conditions (high and low N), and the percentage of the increment under high nitrogen availability.

Species	High N	Low N	Increase (%)	Reference
Porphyra columbina				
Indoor (PAR + UV)	8.9 ± 0.9	3.7 ± 0.76	140.0	Korbee Peinado et al. (2004)
Porphyra leucosticta				
Indoor (PAR + UV)	9.7 ± 0.45	4.3 ± 0.63	126.2	Korbee et al. (2005)
Porphyra umbilicalis				
Indoor (PAR + UV)	8.2 ± 0.48	5.3 ± 1.66	55	Korbee et al. (2005)
Grateloupia lanceola				
Indoor (PAR + UV)	3.4 ± 0.25	1.9 ± 0.12	78.9	Huovinen et al. (2006)
Grateloupia dichotoma				
Indoor (PAR)	3.1 ± 0.30	0.9 ± 0.14	244.4	Figueroa et al. (inedit)
Gracilaria cornea				
Indoor (PAR)	0.6 ± 0.01	0.4 ± 0.05	50.0	Figueroa et al. (inedit)
Outdoor (PAR + UV)	1.8 ± 0.18	0.8 ± 0.09	125.0	
Gracilaria tenuistipitata				
Indoor (PAR)	0.2 ± 0.02	0.15 ± 0.01	33.3	Bonomi et al. (inedit)
Indoor (PAR + UV)	4.2 ± 0.3	1.75 ± 0.09	142.3	

In addition, MAAs can be true "multipurpose" secondary metabolites, which have many additional functions in the cell beyond their well-known UV screen (see review by Oren and Gunde-Cimerman, 2007) and nitrogen source (Korbee et al., 2006). MAAs can be functional under osmotic stress in cyanobacteria (Oren, 1997), in desiccation resistance (Tirkey and Adhikary, 2005; Wright et al., 2005), and against free oxygen radicals due to antioxidant activity as scavengers of free radicals (Dunlap and Yamamoto, 1995; Suh et al., 2003). Recently, de la Coba et al. (2008) reported that several standard in vitro assays were performed to determine the potential antioxidant capabilities of purified aqueous extracts of the mycosporine-like amino acids (MAAs) porphyra 334 plus shinorine (P-334 + SH), isolated from the red alga *Porphyra rosengurttii*, asterina 330 plus palythine (AS + PNE), from the red alga *Gelidium corneum*, shinorine (SH), from the red alga *Ahnfeltiopsis devoniensis*, and mycosporine–glycine (M-Gly), isolated from the marine lichen *Lichina pygmaea*. The scavenging potential of hydrosoluble radicals (ABTS⁺ decolorization method), inhibition of lipid peroxidation (β-carotene linoleate model system), and the scavenging capacity of superoxide radicals (pyrogallol autooxidation assay) were evaluated. In terms of hydrosoluble radicals scavenging, the antioxidant activity of all MAAs studied was dose-dependent and it increased with the alkalinity of the medium (pH 8.5). M-Gly presented the highest activity in all pH tested; at pH 8.5 its IC_{50} was eightfold L-ascorbic acid (L-ASC). AS (+PNE) showed higher activity than L-ASC at pH 8.5, as well. P-334 (+SH) and SH showed scarce activity of scavenging of hydrosoluble free radicals. AS (+PNE) showed high activity for the inhibition of lipid peroxidation, relative to vitamin E, and superoxide radical scavenging, whereas the activities of P-334 (+SH) and SH were moderate (de la Coba et al., 2008). According to these results, the potential of MAAs in photoprotection is high due to a double function: (1) UV chemical sunscreens with high efficiency for UVB and UVA bands (2) their antioxidant capacity. MAAs as secondary metabolites can be key substances against stress conditions provoked by different phenomena, i.e., climate change, pollution, etc.

The limitation of P also reduced the photosynthetic rate in macroalgae (Flores-Moya et al., 1997). The interaction UVR-P is very complex with both synergic and antagonist effect depending on the temporal scale (Medina-Sánchez et al., 2006). In phytoplankton communities, there are some published results about the interaction of UV and P inputs. These papers pointed out that P-enrichment unmasked the inhibitory UVR effect on primary production of algae in a similar way to other studies (Xenopoulos et al., 2002; Doyle et al., 2005; Carrillo et al., 2008; Korbee et al., in preparation). Xenopoulos and Frost, (2003) found stronger negative UVR effects on total phytoplankton growth under P enrichment. Doyle et al. (2005) did not observe a negative effect of UVR on Alpine lakes phytoplankton growth in the absence of nutrient additions; only with the addition of nutrients did UVR exposure depress the growth of some species.

According to the previous arguments, it is relevant to conduct synthetic studies analyzing the interactive effects of climate change factors such as temperature, photosynthetic irradiance, UV radiation, and nutrient availability (input pulses), and the structure (biodiversity) and function (physiological process and capacity of adaptation) of ecosystems.

3. Ecophysiology Approach: Experimental Design of the Studies of Interaction Between Ultraviolet (UV) Radiation and Nutrients

Design and application of novel indicators for an integrated evaluation of the ecological status and vulnerability of coastal waters, lagoons, and hypersaline lakes to climate change according to enviromental gradients (temperature, photosynthetic irradiance, UV radiation, nutrient availability, and confinement).
For this purpose, it is neccesary to study the abundance and species composition as well as functional elements.

Abundance and species composition: Comparison with previous historical data. Systemic or ecological indicators: species richness and Shannon diversity using multiple regressions, permutational multivariate analysis of variance (Permanova), and considering the natural scales variation through an appropriate multiscale sampling design. Correlation between biological changes and climatic variables (period 1970–2008). Habitat Distribution models.

Functional elements: Bio-optical indicators, maximal quantum yield by in vivo Chl *a* fluorescence as an indicator of physiological status, stoichiometry (C:N:P) as an indicator of nutrient status, and stress indicators: heat shock proteins, proteases, and reactive oxygen species (ROS).

For the evaluation of the effect of stress factors on the capacity of acclimation of aquatic macrophytes, it is neccesary to determine the interaction of physico-chemical and biological variables. For this purpose, we suggest the development of several studies:

1. Ecophysiological–multifactorial studies focusing on factors of climate change (temperature, photosynthetic irradiance, UV radiation, and nutrient availability): dynamic photoinhibition, accumulation of UV screen photoprotectors, antioxidant capacity.
2. Field measurements of environmental factors (temperature, PAR, and turbidity) and its relation with stress symptoms in subtidal macrophytes: the case of the decline of the stands of *Gelidium corneum* or *Fucus serratus*. Complementary field experiments to elucidate the main sources of stress (irradiance and nutrients).
3. Effect of UVR, temperature, and salinity on seed germination, seed production, and growth in macrophytes. Relations between renovation rate of seeds and climate conditions of each hydrological year.

Among functional indicators, in vivo Chlorophyll fluorescence is not only used to estimate photosynthesis but also the general physiological status of marine macrophytes and phytoplanktons. This nonintrusive technique is being applied since 15 years in macrophytes (Hanelt et al., 1994; Häder and Figueroa, 1997; Figueroa et al., 2003). Although PAM fluorescence has been used in field studies of both intertidal and subtidal algae, it is still necessary to investigate the use of this technique as an indicator of ecological process as primary production.

In addition, we proposed the following stress indicators:

Heat shock proteins (HSP). They are molecular chaperones for protein molecules. They are usually cytoplasmic proteins and they perform functions in various intracellular processes. They play an important role in protein–protein interactions such as folding and assisting in the establishment of proper protein conformation (shape) and prevention of unwanted protein aggregation. By helping to stabilize partially unfolded proteins, HSPs aid in transporting proteins across membranes within the cell. Some members of the HSP family are expressed at low to moderate levels in all organisms because of their essential role in protein maintenance (Lam et al., 2000). However, the peculiarity of HSPs is that they are a group of proteins whose expression is increased when the cells are exposed to elevated temperatures or other stress. The upregulation of the HSPs induced mostly by heat shock factor (HSF) is a key part of the heat shock response. The HSPs are named according to their molecular weights. Production of high levels of HSPs can also be triggered by exposure to different kinds of environmental stress conditions, such as ultraviolet light, nitrogen starvation, or desiccation. Consequently, the HSPs are also referred to as stress proteins and their upregulation is sometimes described more generally as part of the stress response (Burdon, 1986). Scientists have not discovered exactly how heat-shock (or other environmental stressors) activates the heat-shock factor. However, some studies suggest that an increase in damaged or abnormal proteins brings HSPs into action. The use of the expression of HSPs, HSP70 and HSP90, both related to thermotolerance to cell on exposure to heat and UV stress, as stress indicator are of great interest in the scenario of climate change.

Cellular oxidation. Reactive oxygen species (ROS). Oxidative stress is caused by an imbalance between the production of reactive oxygen and a biological system's ability to readily detoxify the reactive intermediates or easily repair the resulting damage. All forms of life maintain a reducing environment within their cells. This reducing environment is preserved by enzymes that maintain the reduced state through a constant input of metabolic energy. Oxidative burst seems to be an important regulator of the stress response, as the production of antioxidants inhibits the expression of HSPs. The oxidative burst is a rapid transient production of reactive oxygen species (ROS) such as superoxide (O_2^-), hydrogen peroxide (H_2O_2), or hydroxyl radicals (OH·). The generation of ROS around the photosystem II (PS II) diminishes the efficiency of the photosynthetic apparatus: the D1 protein from the PS II is damaged by ROS and the Calvin cycle is disrupted, hence CO_2 fixation is arrested (Lesser, 2006).

Proteases. These are also known as proteinases or proteolytic enzymes and are a large group of enzymes. Proteases belong to the class of enzymes known as hydrolases, which catalyze the reaction of hydrolysis of various bonds with the participation of a water molecule. These enzymes determine the lifetime of other proteins playing important physiological roles among which stress response appears as a key point (Van der Hoorn, 2008). This is one of the fastest "switching on" and "switching off" regulatory mechanisms in the physiology of an organism. By complex cooperative action, the proteases may proceed as cascade reactions, which result in rapid and efficient amplification of an organism's response to a physiological signal. It has been shown that proteases play an important role in photosynthetic organisms subjected to stress. Proteases have been reported to act on nitrogen starvation, irradiance stress, and light deprivation (Berges and Falkowski, 1998), and are essential components of cells and participate in processes ranging from photoacclimation and nutrient acquisition to development and stress responses. It has been reported that applying the methods to a wide range of macroalgae from intertidal zones Southwestern Spain and Northern Ireland (Pérez-Lloréns et al., 2003), macroalgal proteases are easily measurable but highly variable. A major source of variability that has not been assessed is differing environmental conditions. Therefore, measurements of proteolytic enzymes may provide a valuable tool for examining biologically relevant changes in environments.

4. Future Research to Evaluate the Vulnerability of the Macrophytes to Climate Change

The studies on the ecological status and vulnerability to climate change of aquatic ecosystems need integration and coordination. At present, most of the studies of aquatic enviroments are using macrophytes as biological indicators focusing at the community level, i.e., monitoring abundance, percentage covered, chlorophyta/rhodophyta ratios, and in a few cases, species richness and biodiversity (Gorostiaga and Díez, 1996; Wells et al., 2007; Ballesteros et al., 2007; Arévalo et al., 2007; Pérez-Ruzafa et al., 2007). In contrast to pioneer papers, other papers also included energetic dimension as production in the succession studies (Niell, 1979; Lapointe et al., 1981). However, plankton studies combine trophic and biotic approaches in the studies of the vulnerability of aquatic ecosystems to climate change (Carrillo et al., 2006) with a focus on stoichiometric and functional approaches.

On the other hand, there are a good number of reports on ecophysiology of macrophytes with focus on functional approach such as photosynthesis or nutrient assimilation (Häder and Figueroa, 1997; Figueroa et al., 2002; Bischof et al., 2006) including also morphofunctional aspects and growth-reproductive studies (Altamirano et al., 2003a, b; Gómez et al., 2004). However, the research on the effect factors of climate change and the adaptation strategies on the structure–function of macrophyte communities and ecosystems are still very scarce (Bischof et al., 2006). Previous investigations have demonstrated that temperature and UVBR increases, applied separately, can alter aquatic food-web structure and function (Mostajir et al., 1999).

Only experimental studies under controlled conditions in large mesocosms can (1) control forcing factors to mimic model predictions, (2) evaluate the response of quasi entire food webs to climate change (e.g., microbial food-webs), and (3) acquire pertinent data toward the development of predictive global change models integrating both physical and biological components (Nouguier et al., 2007).

Several investigations using mesocosms have demostrated the effects of UVR on the structure–function of aquatic food-web (Wängberg et al., 2001; Hernando et al., 2006) (Fig. 1).

The response of the phytoplankton and bacterial spring succession to the predicted warming of sea surface temperature in temperate climate zones during winter was studied using an indoor-mesocosm approach (Hoppe et al., 2008). Also, research groups are studying the interactive effect of UV and P pulses using mesocosms with natural phytoplankton communities in high mountain lakes (Carrillo et al., 2008).

In addition, the experimental multifactorial approach with potent statistical basis can give information on the vulnerability and the capacity of adpatation to environmental changes including climate change. The approach also gives information on the use of novel experimental designs based on mesocosms to evaluate the interactive effects of enviromental variables related with climate change (temperature, photosynthetic irradiance, UV radiation, and nutrient availability) in macrophytes. Liboriussen et al. (2005) designed a flow-through shallow lake mesocosm to evaluate climate change; this system could be adjusted to be applied in other aquatic ecosystems. Recently, a new automatically operated system that can provide accurate simulations of the increases in UV-B radiation and temperature predicted by climate change scenarios was reported; the system follows the natural

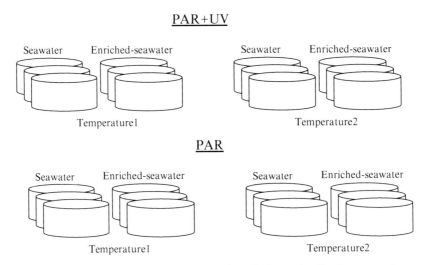

Figure 1. Experimental design using mesocosms to evaluate the interactive effect between environmental variables: UV radiation, temperature, and nutrient availability in macrophytes.

fluctuations of these variables and so reproduces the high degree of environmental variability that is inherent to aquatic systems (Nouguier et al., 2007).

It is necessary to adapt or create mesocosms for their application in the research on marine macrophytes. It is crucial to link ecological and ecophysiological studies to evaluate the ecological status of coastal waters, lagoon, and hypersaline lakes by using macrophyte as bioindicators not only at structural level (community ecology) but also at funtional one (energetic ecology).

5. References

Abdala-Díaz, R.T., Cabello-Pasini, A., Pérez-Rodríguez, E., Conde Álvarez, R.M. and Figueroa, F.L. (2006) Daily and seasonal variations of optimum quantum yield and phenolic compounds in *Cystoseira tamariscifolia* (Phaeophyta). Mar. Biol. **148**: 459–465.

Aguilera, J., Bischof, K., Karsten, U., Hanelt, D. and Wiencke, C. (2002) Seasonal variation in ecophysiological patterns in macroalgae from an Arctic fjord. II. Pigment accumulation and biochemical defence systems against high light stress. Mar. Biol. **140**:1087–1095.

Alcock, R. (2003) The effects of climate change on rocky shore communities in the Bay of Biscay, 1895–2050. Ph.D. thesis, University of Southampton: 296 p.

Algarra, P. and Rüdiger, W. (1993) Acclimation processes in the light harvesting complex red alga *Pophyridium purpureum* (Bory) Drew *et* Ross according to irradiance and nutrient availability. Plant Cell Environ. **16**:149–159.

Altamirano, M., Flores-Moya, A. and Figueroa, F.L. (2003a) Effects of UV radiation and temperature on growth germlings of three species of Fucus (Phaeophyceae). Aquat. Bot. **75**: 9–20.

Altamirano, M., Flores-Moya, A., Külenkamp, R. and Figueroa, F.L. (2003b) Stage dependent sensitivity to ultraviolet radiation on zygotes of the brown alga *Fucus serratus*. Zygote **11**:101–106.

Altamirano, M., Flores-Moya, A., Conde, F. and Figueroa, F.L. (2000a) Growth seasonality, photosynthetic pigments and C and N content in relation to environmental factors: a field study on *Ulva olivascens* (Ulvales, Chlorophyta). Phycologia **39**: 50–58.

Altamirano, M., Flores-Moya, A. and Figueroa, F.L. (2000b) Long-term effect of natural sunlight under various ultraviolet radiation conditions on growth and photosynthesis of intertidal *Ulva rigida* (Chlorophyceae) cultivated *in situ*. Bot. Mar. **43**: 119–126.

Anadón, R. and Niell, F.X. (1981) Distribución longitudinal de macrófitos en la costa asturiana (N de España). Investig Pesq (Barc) **45**:143–156.

Arévalo, R., Pinedo, S. and Ballesteros, E. (2007) Changes in the composition and structure of Mediterranean rocky-shore communities following a gradient of nutrient enrichment: descriptive study and test of proposed methods to assess water quality regarding macroalgae. Mar. Pollut. Bull. **55**: 104–113.

Arrontes, J. (1993) Nature of the distributional boundary of *Fucus serratus* on the north shore of Spain. Mar. Ecol. Prog. Ser. **93**:183–193.

Ballesteros, E., Torras, X., Pinedo, S., García, M., Mangialajo, L. and de Torres, M. (2007) A new methodology base don litoral community cartography by macroalgae for the implementation of the European Water Framework Directive. Mar. Pollut. Bull. **55**: 172–180.

Bergeron, M. and Vincent, W.E. (1997) Microbial food web responses to phosphorus supply and solar UV radiation in a subarctic lake. Aquat. Microb. Ecol. **12**: 239–249.

Berges, J.A. and Falkowski, P.G. (1998) Physiological stress and cell death in marine phytoplankton: induction of proteases in response to nitrogen or light limitation. Limnol. Oceanogr. **43**: 129–135.

Bhaud, M., Cha, J.H., Duchene, J.C. and Nozais, C. (1995) Influence of temperature on the marine fauna – what can be expected from a climatic-change. J. Therm. Biol. **20**: 91–104.

Bischof, K., Gómez, I., Molis, M., Hanelt, D., Karsten, U., Lüder, U., Roleda, M.Y., Zacher, K. and Wiencke, C. (2006) Ultraviolet radiation shapes seaweed communities. Rev. Environ. Sci. Biotechnol. **51**: 141–166.

Bischof, K., Hanelt, D. and Wiencke, C. (1998) UV-radiation can affect depth-zonation of Antarctic macroalgae. Mar. Biol. **131**:597–605.

Bischof, K., Kräbs, G., Wiencke, C. and Hanelt, D. (2002) Solar ultraviolet radiation affects activity of ribuloses-1,5 bisphosphate carboxilase–oxygenase and the composition of the xantophyll cycle pigments in the intertidal green alga *Ulva lactuca* L. Planta **215**: 502–509.

Bischof, K., Rautenberg, R., Brey, L. and Pérez-Lloréns, J.L. (2006) Physiological acclimation along gradients of solar irradiance within mats of the filamentous green macroalga *Chaetomorpha linum* form southern Spain. Mar. Ecol. Prog. Ser. **306**: 165–175.

Breitburg, D., Rose, K. and Cowan, J. (1999) Linking water quality to larval survival: predation mortality of fish larvae in an oxygen-stratified water column. Mar. Ecol. Prog. Ser. **178**: 39–54.

Buma, A.G.J., Engelen, A.H. and Gieskes, W.W.C. (1997) Wavelength-dependent induction of thymine dimers and growth rate reduction in the marine diatom *Cyclotella* sp. exposed to ultraviolet radiation. Mar. Ecol. Prog. Ser. **153**: 91–97.

Buma, A.G.J., Haneneb, E.J., Van Roza, L., Veldhuis, M.J.W. and Gieskes, W.W.C. (1995) Monitoring Ultraviolet B-induced DNA damage in individual diatom cells by immunofluorescence thymine dimer detection. J. Phycol. **31**:314–321.

Burdon, R.H. (1986) Heat shock and the heat shock proteins. Biochem. J. **240**: 313–324.

Calvo Aranda, S. (2007) Modelo de distribución de hçabitat de Himanthala elongata (Phaeophyta, Fucales) en el Norte de la Península Ibérica. Proyecto Fin de Carrera. Universidad Autónoma de Madrid.

Carrillo, P., Medina-Sánchez, J.M. and Villar-Argaiz, M. (2002) The interaction of phytoplankton and bacteria in a high mountain lake: importance of the spectral composition of solar radiation. Limnol. Oceanogr. **47**: 1294–1306.

Carrillo, P., Medina-Sánchez, J.M., Villar-Argaiz, M., Delgado-Molina, J.A. and Bullejos, F.J. (2006) Complex interactions in microbial food webs: stoichiometric and functional approaches. Limnetica **25**(1–2): 189–204.

Carrillo, P., Delgado-Molina, J.A., Medina-Sanchez, J.M., Bullejos, F.J. and Villar-Argaiz, M. (2008) Phosphorus inputs unmask negative effects of ultraviolet radiation on algae in a high mountain lake. Global Change Biol. **14**: 423–439.

Cockell, C.S. and Knowland, J. (1999) Ultraviolet radiation screening compounds. Biol. Rev. **74**: 311–345.

Conde-Álvarez, R.M. (2001) Variaciones espacio-temporales y ecofisiología de los Macrófitos acuáticos de la Laguna atalasohalina de Fuente de Piedra (Sur de la Península Ibérica). Tesis Doctoral. Universidad de Málaga, Málaga.

Conde-Álvarez, R.M., Figueroa, F.L., Bañares-España, E. and Nieto-Caldera, J.M. (2008) Photoprotective role of inflorescence and UV-effects on pollen viabillity of different freshwater plants. Aquat. Sci. **70**: 57–64.

Connan, S., Goulard, F., Stiger, V., Deslandes, E. and Ar Gall, E. (2004) Interspecific and temporal variation in phlorotanin levels in an assemblage of brown algae. Bot. Mar. **47**: 410–416.

De la Coba Luque, F. (2007) Evaluación de la capacidad fotoprotectora y antioxidante de aminoácidos tipo micosporina: aplicaciones biotecnológicas. Tesis Doctoral. Universidad de Málaga, Málaga.

De la Coba, F., Aguilera, J., Figueroa, F.L., de Gálvez, M.V. and Herrera, E. (2009) Antioxidant activity of mycosporine-like amino acids isolated from three red macroalgae and one marine lichen. J. Appl. Phycol. **21**: 161–169.

Directive 2000/60/EC of the European Parliament and of the Council establishing a framework for the Community action in the field of water policy Official Journal (OJ L 327) on 22 December 2000: 1–72.

Dobrestov, S.V., Qian, P.Y. and Wahl, M. (2005) Effects of solar radiation on the formation of shallow. Early successional biofouling communities in Hong-Kong. Mar. Ecol. Prog. Ser. **290**: 55–65.

Doucette, G.J. and Harrison, P.J. (1990) Some effects of iron and nitrogen stress on the red tide dinoflagellate *Gymnodinium sanguineum*. Mar. Ecol. Prog. Ser. **62**: 293–306.

Doyle, S.A., Saros, J.E. and Williamson, C. (2005) Interactive effects of temperature and nutrient limitation on the response of alpine phytoplankton growth to ultraviolet radiation. Limnol. Oceanogr. **50**: 1362–1367.

Dring, M.J., Wagner, A., Boesko, J. and Lüning, K. (1996) Sensitivity of intertidal and subtidal red algae to UVA and UVB radiation, as monitored by chlorophyll fluorescence measurements: influence of collection depth, season and lenght of radiation. Eur. J. Phycol. **31**: 293–302.

Dunlap, W.C. and Yamamoto, Y. (1995) Small-molecule antioxidants in marine organisms: antioxidants in marine organisms: antioxidant activity of mycosporine-glicine. Comp. Biochem. Physiol. B Biochem. Mol. Biol. **112**: 105–114.

Enríquez, S., Agustí, S. and Duarte, C.M. (1994) Light absorption by marine macrophytes. Oecologia **98**: 121–129.

Figueroa, F.L., Mercado, J., Jiménez, C., Salles, S., Aguilera, J., Sánchez-Saavedra, M.P., Lebert, M., Häder, D.-P., Montero, O. and Lubián, L. (1997a) Relationship between bio-optical characteristics and photoinhibition of phytoplankton. Aquat. Bot. **59**: 137–251.

Figueroa, F.L., Salles, S., Aguilera, J., Jiménez, C., Mercado, J., Viñegla, B., Flores-Moya, A. and Altamirano, M. (1997b) Effects of solar radiation on photoinhibition and pigmentation in the red alga *Porphyra leucosticta*. Mar. Ecol. Prog. Ser. **151**: 81–90.

Figueroa, F.L. and Viñegla, B. (2001) Effects of solar UV radiation on photosynthesis and enzyme activities (carbonic anhydrase and nitrate reductase) in marine macroalgae from Southern Spain. Rev. Chil. Hist. Nat. **74**: 237–249.

Figueroa, F.L., Jiménez, C., Viñegla, B., Pérez-Rodríguez, E., Aguilera, J., Flores-Moya, A., Altamirano, M., Lebert, M. and Häder, D.-P. (2002) Effects of solar UV radiation on photosynthesis of the marine angiosperm *Posidonia oceanica* from southern Spain. Mar. Ecol. Prog. Ser. **230**: 59–70.

Figueroa, F.L., Conde-Álvarez, R. and Gómez, I. (2003) Relations between electron transport rates determined by pulse amplitude modulated fluorescence and oxygen evolution in macroalgae under different light conditions. Photosynth. Res. **75**: 259–275.

Flores-Moya, A., Altamirano, M., Cordero, M., González, M.E. and Perez, M.G. (1997) Phosphorous-limited growth in the seasonal winter red algae *Porphyra leucosticta*. Bot. Mar. **40**: 187–191.

Flores-Moya, A., Gómez, I., Viñegla, B., Altamirano, M., Pérez-Rodríguez, E., Maestre, C., Caballero, R.M. and Figueroa, F.L. (1998) Effects of solar radiation on photosynthetic performance, pigment content and enzymatic activities related to nutrient uptake, of the endemic Mediterranean red alga *Rissoella verruculosa*. New Phytol. **139**: 673–683.

Frederick, J.E., Snell, H.E. and Haywood, E.K. (1989) Solar ultraviolet radiation at the Earth's surface. Photochem. Photobiol. **51**: 443–450.

García-Pichel, F. (1994) A model for internal self-shading in planktonic organism and its implications for the usefulness of ultraviolet sunscreens. Limnol. Oceanogr. **39**:1704–1717.

Gómez, I. and Figueroa, F.L. (1998) Effects of solar UV stress on chlorophyll fluorescente kinetics of intertidal macroalgae from southern Spain: a case study in *Gelidium* species. J. Appl. Phycol. **10**: 285–294.

Gomez, I., Perez-Rodriguez, E., Vinegla, B., Figueroa, F.L. and Karsten, U. (1998) Effects of solar radiation on photosynthesis, UV-absorbing compounds and enzyme activities of the green alga *Dasycladus vermicularis* from southern Spain. J. Photobiol. B: Biol. **47**: 46–57.

Gómez, I., Figueroa, F.L., Ulloa, N., Morales, V., Lovengreen, C., Huovinen, C. and Hess, S. (2004) Patterns of photosynthetic performance in intertidal macroalgae from southern Chile in relation to solar radiation. Mar. Ecol. Prog. Ser. **270**: 103–116.

Gorostiaga, J.M. and Díez, I. (1996) Changes in the sublittoral benthic marine macroalgae in the poluted area of Abra de Bilbao and proximal coast. Mar. Ecol. Prog. Ser. **130**: 157–167.

Gorostiaga, J.M., Muguerza, N., Novoa, S., Santolaria, A., Secilla, A. and Díez, I. (2008) Changes in the marine sublittoral vegetation at the western Basque Coast between 1982 and 2007: a

consequence of the climate change? Abstracts XI International Symposium on Oceanography of the Bay of Biscay, San Sebastián (España).
Guisan, A. and Zimmermann, N.E. (2000) Predictive habitat distribution models in Ecology. Ecol. Model. **135**: 147–186.
Häder, D.-P. and Figueroa, F.L. (1997) Annual review: photoecophysiology of marine macroalgae. Photochem. Photobiol. **66**: 1–14.
Häder, D.P., Kumar, H.D., Smith, R.C. and Worest, R.C. (2007) Effect of solar UV radiation on aquatic ecosystems and interactions with climate change. Photochem. Photobiol. Sci. **6**: 267–285.
Han, Y.S. and Han, T. (2005) UV-B induction of UV-B protection in U*lva pertusa* (Chlorophyta). J. Phycol. **41**: 523–530.
Hanelt, D., Jaramillo, I.M., Nultsch, W., Senger, S. and Westermeier, R. (1994) Photoinhibition as regulative mechanism of photosynthesis in marine algae of Antarctica. Ser Cient INACH **44**: 57–77.
Hanelt, D. (1998) Capability of dynamic photoinhibition in Arctic macroalgae is related to their depth distribution. Mar. Biol. **131**: 361–369.
Hanelt, D., Li, J. and Nultsch, W. (1994) Tidal dependence of photoinhibition of photosynthesis in marine macrophyte of south China Sea. Bot. Acta **106**: 66–72.
Hawkinks, S.J. and Jones, H.D. (1992) *Rocky Shores: Marine Field Study Course. Guide I*. Immel Publishing, London.
Helbling, E.W., Villafañe, V.E., Buma, A.G.J., Andrade, M. and Zaratti, F. (2001) DNA damage and photosynthetic inhibition induced by solar ultraviolet radiation in tropical phytoplankton (Lake Titicaca, Bolivia). Eur. J. Phycol. **36**: 157–166.
Helbling, E.W. and Zagarese, H. (2003) *UV Effects in Aquatic and Ecosystems*. Comprehensive Series in Photosciences. The Royal Society of Chemistry, Cambridge.
Helmuth, B., Broitman, B.R., Blanchette, C.A., Gilman, S., Halpin, P., Harley, C.D.G., O'Donnell, M.J., Hoffmann, G.E., Menge, B.A. and Strickland, D. (2006) Mosaic pattern of thermal stress in the rocky intertidal zone: implicatiuons for climate change. Ecol. Monogr. **76**(4): 461–479.
Henley, W.J., Levavasseur, G., Franklin, L.A., Osmond, C.B. and Ramus, J. (2001) Photoacclimation and photoinhibition in *Ulva rotundata* as influenced by nitrogen availability. Planta **184**: 235–43.
Hernando, M., Schloss, I., Roy, S. and Ferreyra, G. (2006) Photoacclimation to long-term ultraviolet radiation exposure of natural Sub-Antarctic Phytoplankton communities: fixed surface incubations versus mixed mesocosms. Photochem. Photobiol. **82**: 923–935.
Hiriart, V.P., Greenberg, B.M., Guildford, S.J. and Smith, R.E.H. (2002) Effects of ultraviolet radiation on rates and size distribution of primary production by Lake Erie phytoplankton. Can. J. Fish. Aquat. Sci. **59**: 317–328.
Hiscock, K., Southward, A., Tittley, I. and Hawkins, S. (2004) Effects of changing temperature on benthic marine life in Britain and Ireland. Aquat. Conserv. **14**: 333–362.
Hoepffner, N. (ed.) (2006) *Marine and Coastal Dimension of Climate Change in Europe*. A report of the European Water Directors. European Communities. Institute for Environmental and Sustainability. Office for Official Publications of the European Communities, Luxembourg, 87 pp.
Hoppe, H.G., Breithaupt, P., Walther, K., Koppe, R., Bleck, S., Sommer, U. and Jurgens, K. (2008) Climate warning in winter affects the coupling between phytoplankton and bacteria during the spring bloom: a mesocosm study. Aquat. Microb. Ecol. **51**: 105–115.
Huovinen, P., Matos, J., Sousa-Pinto, I. and Figueroa, F.L. (2006) The role of nitrogen in photoprotection against high irradiance in the Mediterranean red alga *Grateloupia lanceola*. Aquat. Bot. **84**: 208–316.
IPCC (2007a) "Contribución del grupo de trabajo I al IV Informe de Evaluación del grupo intergubernamental de expertos sobre cambio climático": Bases Científicas y Físicas. Valencia (2007) documento publicado en www.mma.es
IPCC (2007b) "Contribución del grupo de trabajo II al IV Informe de Evaluación del grupo intergubernamental de expertos sobre cambio climático" Cambio Climático 2007: impactos, adaptación y vulnerabilidad al cambio climático. Valencia (2007) documento publicado en www.mma.es

Jeffrey, W.H., Aas, P., Lyons, M.M., Coffin, R.B., PledgerJand, R.J. and Mitchell, D.L. (1996) Ambient solar radiation-induced photodamage in marine bacterioplankton. Photochem. Photobiol. **64**: 419–427.

Jiménez, C., Figueroa, F.L., Salles, S., Aguilera, J., Mercado, J., Viñegla, B., Flores-Moya, A., Lebert, M. and Häder, D.-P. (1998) Effects of solar radiation on photosynthesis and photoinhibition in red macrophytes from an intertidal system of southern Spain. Bot. Mar. **41**: 329–338.

Karsten, U., Sawall, T., Hanelt, D., Bischof, K., Figueroa, F.L., Flores-Moya, A. and Wiencke, C. (1998) An inventory of UV-absorbing mycosporine-like amino acids in macroalgae from polar to warm-temperate regions. Bot. Mar. **41**: 443–453.

Kaschner, K., Watson, R., Trites, A.W. and Pauly, D. (2006) Mapping world-wide distributions of marine mammal species using relative environmental suitability (RES) model. Mar. Ecol. Prog. Ser. **316**: 285–310.

Korbee Peinado, N., Abdala-Díaz, R., Figueroa, F.L. and Helbling, E.W. (2004) Ammonium and UVR stimulate the accumulation of mycosporine-like aminoacids (MAAs) in *Porphyra columbina* (Rhodophyta) from Patagonia, Argentina. J. Phycol. **40**: 248–259.

Korbee, N., Huovinen, P., Figueroa, F.L., Aguilera, J. and Karsten, U. (2005) Availability of ammonium influences photosynthesis and the accumulation of mycosporine-like amino acids in two Porphyra species (Bangiales, Rhodophyta). Mar. Biol. **146**: 645–654.

Korbee, N., Figueroa, F.L. and Aguilera, J. (2006) Accumulation of mycosporine-like amino acids (MAAs): biosynthesis, photocontrol and ecophysiological functions. Rev. Chil. Hist. Nat. **79**: 119–132.

Krebs, C.J. (1978) *Ecology: The Experimental Analysis of Distribution and Abundance*, 2nd edn. Harper & Row, New York.

Lam, E., Fukuda, H. and Greenberg, J. (eds.) (2000) *Programmed Cell Death in Higher Plants*. Kluwer, Dordrecht, The Netherlands.

Lapointe, B.E., Niell, F.X. and Fuentes, J.M. (1981) Community structure, succession, and production of seaweeds associated with mussel –rafts in the Ría de Arosa. Mar. Ecol. Prog. Ser. **5**: 243–253.

Larkum, A.W.D. and Wood, W.F. (1993) The effect of UV-B radiation on photosynhesis and respiration of phytoplankton, benthic macroalgae and seagrasses. Photosynth. Res. **36**: 17–23.

Laubier, L., Pérez, T., Lejeusne, C., Garrabou, J., Chevaldonné, P., Vacelet, J., Boury-Esnault, N. and Harmelin, J.G. (2003) La Méditerranée se réchauffe-t-elle? Is the Mediterranean warming up? Marine Life **13**(1–2): 71–81.

Lesser, M.P. (2006) Oxidative stress in marine environments: biochemistry and physiological ecology. Annu. Rev. Physiol. **68**: 253–278.

Liboriussen, L., Landkildehus, F., Meerhoff, M., Bramm, M.E., Sondergaard, M., Christoffersen, K., Richardson, K., Sondergaard, M., Lauridsen, T.L. and Jeppesen, E. (2005) Global warning: design of a flow-through shallow lake mesocosm climate experiment. Limnol. Oceanogr. Methods **3**: 1–9.

Litchman, E., Neale, P.J. and Banaszak, A.T. (2002) Increased sensitivity to ultraviolet radiation in nitrogen-limited dinoflagellates: photoprotection and repair. Limnol. Oceanogr. **47**: 86–94.

Littler, M.M., Littler, D.S. and Taylor, P.R. (1983) Evolutionary strategiews in a tropical barrier reef system: functional form groups of marine macroalgae. J. Phycol. **19**: 229–237.

López-Figueroa, F. (1992) Diurnal variations in pigment contents in *Porphyra laciniata* and in *Chondrus crispus* and its relation to the diurnal changes of underwater light quality and quantity. Mar. Ecol. **13**: 285–305.

Lüning, K. (1990) *Seaweeds. Their Environment, Biogeography and Ecophysiology*. Wiley, New York.

Luoto, M., Kuusaari, M. and Rita, H. (2001) Detrminants of distrbution ad abundante in the clouded apollo butterfly: a landscape ecological approach. Ecography **24**: 601–617.

McKenzie, R.L., Aucamp, P.J., Bais, A.F., Björn, L.O. and Ilyas, M. (2007) Changes in biologically active ultraviolet radiation reaching the Earth's surface. Photochem. Photobiol. Sci. **6**(3): 218–231.

Medina, J.R., Tintoré, J. and Duarte, C.M. (2001) Las Praderas de *Posidonia oceanica* y regeneración de playas. Rev. de Obras Públicas **3409**: 31–43.

Medina-Sánchez, J.M, Villar-Argaiz, M. and Carrillo, P. (2006) Solar radiation-nutrient interaction enhances the resource and predation algal control on bacterioplankton: a short-term experimental study. Limnol. Oceanogr. **51**: 913–924.

Mercado, J.M., Gordillo, F.J., Figueroa, F.L. and Niell, F.X. (1998) External carbonic anhydrase and affinity to inorganic carbon in intertidal macroalgae. J. Exp. Mar. Biol. Ecol. **221/2**: 209–220.

Mitchell, D.L. and Karentz, D. (1993) The induction and repair of DNA photodamage in the environment, In: A.R. Young, L.O. Björn, J. Moan and W. Nultsch (eds.) *Environmental UV Photobiology*. Plenum, New York, pp. 345–377.

Mostajir, B., Demers, S., De Mora, S., Belzile, C., Chanut, J.P., Gosselin, M., Roy, S., Villegas, P.Z., Fauchot, J., Bouchard, J., Bird, D., Monfort, P. and Levasseur, M. (1999) Experimental test of the effect of ultraviolet-B radiation in a planktonic community. Limnol. Oceanogr. **44**: 586–596.

Murray, S.N. and Littler, M.M. (1978) Patterns of algal succession in a perturbated marine intertidal community. J. Phycol. **14**: 506–512.

Naeem, S. and Wright, J.P. (2003) Disentangling biodiversity effects on ecosystem functioning: deriving solutions to a seemingly insurmountable problem. Ecol. Lett. **6**: 567–579.

Naeem, S., Byers, D., Tjossem, S.F., Bristow, C. and Li, S. (1999) Plant neighborhood diversity and production. Ecoscience **6**: 355–365.

Niell, F.X. (1979) Structure and succession in rocky algal communities of a temperate intertidal ecosystem. J. Exp.Mar. Biol. Ecol. **36**: 185–200.

Nouguier, J., Mostajir, B., Le Floc'h, E. and Vidussi, F. (2007) An automatically operated system for simulating global change temperature and ultraviolet B radiation increases: application to the study of aquatic ecosystem responses in mesocosm experiments. Limnol. Oceanogr. Methods **5**: 269–279.

Orfanidis, S., Panayotidis, P. and Stamatis, N. (2001) Ecological evaluation of transitional and coastal waters: a marine benthic macrophytes-based model. Mar. Sci. **2**: 45–65.

Oren, A. (1997) Mycosporine-like amino acids as osmotic solutes of halophilic cyanobacteria. Geomicrobiol. J. **14**: 231–240.

Oren, A. and Gunde-Cimerman, N. (2007) Mycosporines and mycosporine-like amino acids: UV protectants multipurpose secondary metabolites? FEMS Microbiol. Lett. **269**: 1–10.

Parry, M.L., Canziani, O.F., Palutikof, J.P., van der Linden, P.J. and Hanson, C.E. (eds.) *Climate Change 2007: Impacts, Adaptation and Vulnerability*. Contribution of working group II to the Fourth Assessment report of the Intergovernmental panel on climate change.

Pavia, H., Cevin, G., Lindgren, A. and Aberg, P. (1997) Effects of UV-B radiation and stimulated herbivory on phlorotannins in the brown alga *Ascophyllum nodosum*. Mar. Ecol. Prog. Ser. **157**: 139–146.

Pérez-Lloréns, J.L., Vergara, J.J., Pino, R.R., Hernández, I., Peralta, G. and Niell, F.X. (1996) The effect of photoacclimation on the photosynthetic physiology of *Ulva curvata* and *Ulva rotundata* (Ulvales, Chlorophyta). Eur. J. Phycol. **331**:349–59.

Pérez-Lloréns, J.L., Benítez, E., Vergara, J.J. and Berges, J.A. (2003) Characterisation of proteolytic enzyme activities in macroalgae. Eur. J. Phycol. **38**: 55–64.

Pérez-Rodríguez, E., Gómez, I. and Figueroa, F.L. (1998) Effects of UV radiation on photosynthesis and excretion of UV-absorbing pigments of *Dasycladus vermicularis* (Chlorophyta, Dasycladales) from Southern Spain. Phycologia **37**: 379–387.

Pérez-Rodríguez, E., Aguilera, J., Gómez, I. and Figueroa, F.L. (2001) Excretion of coumarins by the Mediterranean green alga *Dasycladus vermicularis* in response to environmental stress. Mar. Biol. **139**: 633–639.

Pérez-Ruzafa, A., Marcos, C., Pérez-Ruzafa, I.M., Barcala, E., Hegazi, M.I. and Qispe, J. (2007) Detecting changes resulting form human pressure in a naturally quick-changing and heterogeneous environment: spatial and temporal scales of variability in coastal lagoons. Estuar. Coast. Shelf Sci. **75**: 178–188.

Pergent, G., Romero, J.M., Pergent-Martini, C., Mateo, M.A. and Boudeouresque, C.F. (1994) Primary production, stocks and fluxes in the Mediterranean seagrass *Posidonia oceanica*. Mar. Ecol. Prog. Ser. **106**: 139–146.

Perry, A.L., Low, P.J., Ellis, J.R. and Reynolds, J.D. (2005) Climate change and distribution shifts in marine fishes. Science **308**: 1912–1915.

Robertson, M.P., Peter, C.I., Villeet, M.H. and Ripley, B.S. (2003) Comparing models for predicting species' potential distribution: a case study using correlative and mechanistic predictive modeling techniques. Ecol. Model. **164**: 153–157.

Roleda, M.Y., Hanelt, D. and Wiencke, C. (2006) Growth and DNA damage in young *Laminaria* sporophytes exponed to ultraviolet radiation: implications for depth zonation of kelps in Helgoland (Norh Sea). Mar. Biol. **148**: 1201–1211.

Salles, S., Aguilera, J. and Figueroa, F.L. (1996) Light field in algal canopies: changes in spectral light ratios and growth of *Porphyra leucosticta* Thur. In Le Jol. Sci. Mar. **60**(Suppl. 1): 29–38.

Santas, R., Korda, A., Lianou, C. and Santas, P. (1998) Community response to UV radiation. I Enhanced UVB effects on biomass and community structure of filamentous algal assemblages growing in a coral reef mesocom. Mar. Biol. **131**: 153–162.

Schulze, E.-D. and Mooney, H.A. (1993) Ecosystem function of biodiversity: a summary, In: E.-D. Shulze and H.A. Mooney (eds.) *Biodiversity and Ecosystem Function*. Springer, Berlin, Germany, pp. 497–510.

Seoane, J., Justribo, J.H., García, F., Retamar, J., Rabadán, C. and Atienza, J.C. (2006) Habitat-suitability modelling to assess the effects of land-use changes on Dupont's lark *Chersophilus duponti*: a case study in the Layna Important Bird Area. Biol. Conserv. **128**: 241–252.

Shelly, K., Heraud, P. and Beardall, J. (2002) Nitrogen limitation in Dunaliella tertiolecta (Chlorophyceae) leads to increased susceptibility to damage by Ultraviolet-B radiation but also increased repair capacity. J. Phycol. **38**: 713–720.

Shelly, K., Roberts, S., Heraud, P. and Beardall, J. (2005) Interactions between UV-exposure and phosphorus nutrition. I. effects on growth, phosphate uptake, and chlorophyll fluorescence. J. Phycol. **4**: 1204–1211.

Sinha, R.P. and Häder, D.-P. (1998) Phycobilisomes and environmental stress, In: B.N. Verma, A.N. Kargupta and S.K. Goyal (eds.) *Advances in Phycology*. APC Publications, New Delhi, pp. 71–80.

Southward, A.J., Langmead, O., Hardman-Mountford, N.J., Aiken, J., Boalch, G.T., Dando, P.R., Genner, M.J., Joint, I., Kendall, M.A., Halliday, N.C., Harris, R.P., Leaper, R., Mieszkowska, N., Pingree, R.D., Richardson, A.J., Sims, D.W., Smith, T., Walne, A.W. and Hawkins, S.J. (2005) Long-term oceanographic and ecological research in the western English Channel. Adv. Mar. Biol. **47**: 1–105.

Sterner, R.W., Elser, J.J., Fee, E.J., Guildford, S.J. and Chrzanowski, T.H. (1997) The light: nutrient ratio in lakes: the balance of energy and materials affects ecosystem functioning. Am. Nat. **150**: 663–684.

Suh, H.-J., Lee, H.-W. and Jung, J. (2003) Mycosporine–glycine protects biological systems against photodynamic damage by quenching singlet oxygen with high efficiency. Photochem. Photobiol. **78**: 109–113.

Talarico, L. and Maranzana, G. (2000) Light and adaptative responses in red macroalgae: an overview. J. Photochem. Photobiol. **56**: 1–11.

Tandeau de Marsac, N. and Houmardd, J. (1993) Adaptation of cyanobacteria to environmental stimuli: new steps towards molecular mechanisms. FEMS Microbiol. Rev. **104**: 119–190.

Thibaut, T., Pinedo, S., Torras, X. and Ballesteros, E. (2005) Long-term decline of the populations of Fucales (*Cystoseira* spp. and *Sargassum* spp.) in the Albéres coast (France), north-western Mediterranean). Mar. Pollut. Bull. **50**: 1472–1489.

Tilman, D., Knops, J., Wedin, D. and Reich, P. (2002) Plant diversity and composition: effects on productivity and nutrient dynamics of experimental grasslands, In: M. Loreau, S. Naeem and P. Inchausti (eds.) *Biodiversity and Ecosystem Functioning. Synthesis and Perspectives*. Oxford University Press, Oxford, pp. 21–35.

Tirkey, J. and Adhikary, S.P. (2005) Cyanobacteria in biological soil crust of India. Curr. Sci. **89**: 515–521.

Van de Poll, W.H., Eggert, A., Buma, A.G.J. and Breeman, M. (2001) Effects of UV-B induced DNA damage and photoinhibition f growth of temperate marine red macrophytes: habitat-related differences in UV-B tolerance. J. Phycol. **37**: 30–37.

Van der Hoorn, R.A.L. (2008) Plant proteases: from phenotypes to molecular mechanisms. Annu. Rev. Plant Biol. **59**: 191–223.

Vergara, J.J., Bird, K.T. and Niell, F.X. (1995) Nitrogen assimilation following NH_4^+ pulses in the red alga *Gracilariopsis lemaneiformis*: effect on C metabolism. Mar. Ecol. Prog. Ser. **122**: 253–263.

Villafañe, V.E. (2004) Ultraviolet radiation and primary productivity in temperate aquatic environments of Patagonia (Argentina). Ph.D. thesis, University of Groningen, Groningen, 133 pp.

Villafañe, V.E., Sundbäck, K., Figueroa, F.L. and Helbling, E.W. (2003) Photosynthesis in the Aquatic Environment as Affected by UVR, In: E.W. Helbling and H.E. Zagarese (eds.) *UV Effects in Aquatic Organisms and Ecosystems*. The Royal Society of Chemistry, Cambridge, pp. 357–397.

Viñegla, B. (2000) Efecto de la radiación ultravioleta sobre actividades enzimáticas relacionadas con el metabolismo del carbono y nitrógeno en macroalgas y fanerógamas marinas. Tesis Doctoral. Universidad de Málaga, Málaga.

Wahl, M., Molis, M., Davis, A., Drobestov, S., Durr, S.T., Johansson, J., Kinley, J., Kirugara, D., Langer, M., Lotze, H.K., Thiel, M., Thomanson, J.C., Word, B. and Ben-Yosef, D.Z. (2004) UV effects that come and go: a global comparison of marine benthic community level impacts. Global Change Biol. **10**: 1962–1972.

Wängberg, S.A., Wulff, A., Nilson, C. and Stagell, U. (2001) Impact of UVBR on microalgae and bacteria – a mesocosm study with computer modulated UVB Radiation. Aquat. Microbiol. Ecol. **25**: 75–86.

Wells, E. (2002) Seaweeds species biodiversity on intertidal rocky seashores in the British Isles. Ph.D. Thesis, Heriot-Watt University, Edinburgh.

Wells, E. and Wilkinson, M. (2002) Intertidal seaweed biodiversity in relation to environmental factors – a case study form Northern Ireland. Mar. Biodiversity in Ireland and adjacent waters, Ullster Museum. Belfast.

Wells, E., Wilkinson, M., Word, P. and Scanlan, C. (2007) The use of macroalgal richness and composition on intertidal rocky shores in the assessment of ecological quality under the European water framework directive. Mar. Pollut. Bull. **55**: 151–156.

Walther, G.R., Post, E., Convey, P., Menzel, A., Parmesan, C., Beebee, T.J.C. et al. (2002) Ecological responses to recent climate change. Nature **416**: 389–395.

Wiencke, C., Gómez, I., Pakker, H., Flores-Moya, A., Altamirano, M., Hanelt, D., Bischof, K. and Figueroa, F.L. (2000) Impact of UV radiation on viability, photosynthetic characteristics and DNA on algal zoospores: implications for depth zonation. Mar. Ecol. Prog. Ser. **197**: 217–219.

Wiencke, C., Roleda, M.Y., Gruber, A., Clayton, M. and Bischof, K. (2006) Susceptibility of zoospores to UV radiation determines upper distribution limit of Arctic kelp: evidence through field experiments. J. Ecol. **94**: 455–463.

Wilkinson, M. and Tittley, I. (1979) The marine algae of Elie: a reassessment. Bot. Mar. **22**: 249–256.

Wilkinson, M., Fuller, I.A., Telfer, T.C., Moore, C.G. and Kingston, P.F. (1988) *A Conservation Oriented Survey of the Intertidal Seashore of Northern Ireland*. Institute of Offshore Engineering, Heriot-Watt University, Edinburgh.

Wilson, J.G. (2003) Evaluation of estuarine quality status at system level using the biological quality index and the pollution load index. Biol. Environ. **103B**: 49–57.

Wright, D.J., Smith, S.C., Joardar, V., Scherer, S., Jervis, J., Warren, A., Helm, R.F. and Potts, M. (2005) UV radiationand desiccation modulate the three-dimensional extracellular matrix of *Nostoc commune* (cyanobacteria). J. Biol. Chem. **280**: 40271–40281.

Xenopoulos, M.A. and Frost, P.C. (2003) UV radiation, phosphorus, and their combined effects on the taxonomic composition of phytoplankton in a boreal lake. J. Phycol. **39**: 291–302.

Xenopoulos, M.A., Frost, P.C. and Elser, J.J. (2002) Joint effects of UV radiation and phosphorus supply on algal growth rate and elemental composition. Ecology **83**: 423–435.

Zepp, R.G., Erickson, D.J., Paul, N.D. and Sulzberger, B. (2007) Interacrive effects of solar UV radiation and climate change on biogeochemical cycling. Photochem. Photobiol. Sci. **6**: 286–300.

Biodata of **Kunshan Gao** and **Juntian Xu**, authors of *"Ecological and Physiological Responses of Macroalgae to Solar and UV Radiation"*

Professor Kunshan Gao is currently the distinguished Chair Professor of State Key Laboratory of Marine Environmental Science, Xiamen University, China. He obtained his Ph.D. from Kyoto University of Japan in 1989 and continued his research since then at Kansai Technical Research Institute of Kansai Electrical Co. and at University of Hawaii in USA as a postdoctoral fellow. He was appointed as associate professor of Shantou University in 1995, and became recognized as the outstanding young scientist in 1996 by NSFC, then as professor for 100 talented programs in the Institute of Hydrobiology by the Chinese Academy of Sciences in 1997. Professor Gao's scientific interests are in the areas of ecophysiology of algae and algal photobiology, focusing on the environmental impacts of increasing atmospheric CO_2 under solar radiation.

E-mail: **ksgao@xmu.edu.cn,**

Dr. Juntian Xu is currently the Lecturer of the Key Lab of Marine Biotechnology of Jiangsu Province, Huaihai Institute of Technology, China. He obtained his Ph.D. from Shantou University in 2008. Dr. Xu's scientific interests are in the areas of physiology and photobiology of macroalgae, focusing on the environmental impacts of increasing atmospheric CO_2 under solar radiation.

E-mail: **juntianxu@163.com**

Kunshan Gao **Juntian Xu**

ECOLOGICAL AND PHYSIOLOGICAL RESPONSES OF MACROALGAE TO SOLAR AND UV RADIATION

KUNSHAN GAO[1] AND JUNTIAN XU[2]
[1]*State Key Laboratory of Marine Environmental Science, Xiamen University, Xiamen, 361005, China*
[2]*Key Lab of Marine Biotechnology of Jiangsu Province, Huaihai Institute of Technology, Lianyungang, 222005, China*

1. Introduction

Solar visible radiation (400–700 nm, photosynthetically active radiation, PAR) drives photosynthesis and, therefore, is indispensable for all forms of life. In aquatic ecosystems, photosynthetic carbon fixation contributes to nearly 50% of the global primary production (Behrenfeld et al., 2006). Within the euphotic zone (down to 1% of surface PAR), cells receive not only PAR but also ultraviolet radiation (UVR, 280–400 nm), which can penetrate to considerable depths (Hargreaves, 2003). UVR is usually considered harmful at either organism or community levels (Häder et al., 2007). However, UV-A can be utilized for photosynthetic carbon fixation by phytoplankton, playing negative (at high levels) and positive roles like a double-edged sword (Gao et al., 2007).

Solar UVR can reduce photosynthetic rate (Helbling et al., 2003), damage cellular components such as D1/D2 protein (Sass et al., 1997) and DNA molecule (Buma et al., 2003), alter the rate of nutrient uptake (Fauchot et al., 2000), and affect growth (Villafañe et al., 2003) and fatty acids composition (Goes et al., 1994) of algae. Recently, it has been found that natural levels of UVR can even alter the morphology of the cyanobacterium *Arthrospira* (Spirulina) *platensis* (Wu et al., 2005; Gao and Ma, 2008). On the other hand, positive effects of UV-A (315–400 nm) have also been reported. UV-A enhances carbon fixation under reduced (Barbieri et al., 2002) or fast-fluctuating (Helbling et al., 2003) solar radiation and allows photorepair of UV-B-induced DNA damage (Buma et al., 2003); it can even drive photosynthetic carbon fixation in the absence of PAR (Gao et al., 2007). However, to date, estimations of aquatic biological production have been carried out in incubations considering only PAR (i.e., using UV-opaque vials made of glass or polycarbonate) (van Donk et al., 2001) without UVR being considered (Hein and Sand-Jensen, 1997; Schippers et al., 2004; Zou and Gao, 2002). Seaweeds, as the major group of primary producers in coastal waters, experience dramatic changes of solar radiation during a day or through their different life stages. It is of general concern to know how macroalgal species respond to fluctuation or changes of the solar radiation from a physiological or photobiological point of view. In addition, UV-B (280–315 nm) irradiance at the surface of the earth has been raised due to the reduction of stratospheric ozone layer associated with the CFC pollutants (Kerr and McElroy, 1993; Aucamp, 2007). Therefore, impacts of solar UVR as well as enhanced levels of UV-B on macroalgae have also been widely investigated.

2. Physiological Responses to Short- and Long-Term Exposures

Growth is an important parameter that incorporates stress effects in biochemical and physiological processes within the cell. Growth measurements are useful in estimating possible change in productivity due to enhanced UV-B irradiance associated with ozone depletion. UVR is generally known to inhibit macroalgal growth (Dring et al., 1996; Altamirano et al., 2000; Henry and Van Alstyne, 2004; Han and Han, 2005; Davison et al., 2007). Enhanced levels of UV-B can further inhibit macroalgal growth as found in the brown algae *Ectocarpus rhodochondroides* (Santas et al., 1998) and *Dictyota dichotoma* (Kuhlenkamp et al., 2001). Michler et al. (2002) reported that most of the investigated 13 macroalgal species showed reduced growth rates in the presence of UVR. On the other hand, UVR resulted in a higher growth rate in *Gracilaria lemaneiformis* under reduced levels (25%) of solar radiation (Xu, J and Gao, K, unpublished data). UVR was found to result insignificant impact on the growth of *Ulva lactuca* during winter period (Xu and Gao, 2007), whereas growth of *Ulva expansa* was largely inhibited by UV-B during the summer period (Grobe and Murphy, 1998). It is possible that UVR-related impacts on macroalgal growth depend on seasonal environmental changes, such as temperature, which is much lower during winter season. UVR-induced inhibition of photosynthetic O_2 evolution in a *Gracilaria* plant increased with increased seawater temperature (Gao and Xu, 2008).

UVR affects macroalgal growth to different degrees. UV-B radiation, though with the strength of less than 1% of the total solar radiation, can significantly reduce the growth rates of most of the species investigated so far (Grobe and Murphy, 1994; Pang et al., 2001; Michler et al., 2002; Jiang et al., 2007; Gao and Xu, 2008). In contrast to the impacts of UVR in short-term experiments, hardly any lasting difference was found in growth between samples exposed to solar radiations with or screened off UV-B in long-term experiments in *Ulva rigida* (Altamirano et al., 2000) and *Fucus serratus* (Michler et al., 2002). Nevertheless, UVR was found to affect the growth in long-term and field experiments in *Laminaria* spp. (Michler et al., 2002; Roleda et al., 2006a) and *Gracilaria lemaneiformis* (Gao and Xu, 2008). In the green alga, *Codium fragile*, no change in the biomass was observed during the first week; however, when exposed to UVR, it increased by about 70% in the following 16 days (Michler et al., 2002). In another green alga, *Ulva lactuca*, presence of UV-B only caused inhibition of photochemical yield in the initial 2 days, whereas UV-A showed insignificant effect; solar PAR resulted in most of the inhibition during noon time (Xu and Gao, 2007). UV-A was shown to reduce the photosynthetic rate at higher irradiance levels around noontime, but enhanced it during sunrise at low irradiance levels in *Gracilaria lemaneiformis* (Gao and Xu, 2008). Photosynthesis in macroalgae is known to be negatively affected by UV-B as well as UV-A (Cordi et al., 1997; Aguilera et al., 1999b; Han et al., 2003). However, short-term experiments about the responses to UVR do not usually provide enough information to relate to the long-term effects. For example, *Laminaria ochroleuca* showed partial acclimation to chronic UVR exposure in photosynthesis, but did not in growth (Roleda et al., 2004). On the other hand, solar UV-B radiation could play a role in the recovery process of the inhibited photosynthesis in a brown alga *Dictyota dichotoma* (Flores-Moya et al., 1999), and UV-A (315–400 nm) radiation could aid in DNA repair (Pakker et al., 2000a,b) and enhance growth (Henry and Van Alstyne, 2004) in macroalgae.

Species of macroalgae are distributed to different depths in the intertidal zone and exposed to different levels of PAR and UVR because of the attenuation of seawater. Their vertical distribution is closely related to their sensitivity to UV-B (Hanelt et al., 1997; Bischof et al., 1998) as well as to their recovery capacity after being damaged by UVR (Gómez and Figueroa, 1998). Upper species are more resistant to solar UVR. UV-B had little effect on eulittoral species but significantly inhibited the growth of sublittoral red macrophytes (van de Poll et al., 2001).

3. UV-Regulated Photosynthetic Performance Under the Sun

Solar UV radiation is a permanently existing environmental factor that macroalgae are usually exposed to. However, photosynthetic performance under the sun has been mainly investigated in the absence of UVR owing to the ignorance of the transmitting characteristics of vessels used (glass, polyethylene, and polycarbonate materials do not allow UV-B and part of UV-A to penetrate, Van Donk et al., 2001).

In the studies without considering the effects of UVR, diurnal photosynthesis of macroalgae was depressed in the afternoon on sunny days in *Macrocystis pyrifera* surface canopy (Gerard, 1986), *Sargassum* spp. (Gao and Umezaki, 1989; Gao, 1990), *Ulva curvata*, *Codium decorticatum*, *Dictyota dichotoma*, *Petalonia fascia*, and *Gracilaria foliifera* (Ramus and Rosenberg, 1980). The photosynthetic efficiency of O_2 evolution was found to be higher in the morning than in the afternoon in *Ulva rotundata* (Henley et al., 1991) and *Sargassum horneri* (Gao, 1990) under solar PAR. Such an afternoon photosynthetic depression was not found on rainy or highly cloudy days (Gao and Umezaki, 1989) and may be largely removed by superimposing a light fluctuation on the diurnal regime as demonstrated in phytoplanktons (Marra, 1978). Contrarily, the red alga *Gelidiella acerosa* was found to photosynthesize inefficiently in the morning compared with that of midday and afternoon (Ganzon-Fortes, 1997). Owing to light-transmission characteristics of the incubation vessels used, these previous findings demonstrated the asymmetrical diurnal photosynthesis under PAR only, without UV-B or UVR being considered. Although both UV-A and UV-B might cause less damages than PAR under natural solar radiation (Dring et al., 2001), the highest photoinhibition was found at noon in macroalgae under full spectrum of solar radiation (Huppertz et al., 1990; Hanelt, 1992). However, the effects caused by UVR have only infrequently been differentiated from that of PAR (Hanelt et al., 1997; Flores-Moya et al., 1999).

Recently, it has been found that involvement of UV-B depressed the apparent photosynthetic efficiency on sunny days, whereas UV-A enhanced the apparent photosynthetic efficiency during sunrise period in *Gracilaria lemaneiformis* (Gao and Xu, 2008). Daytime photosynthetic performance usually depends on the extent and pattern of fluctuating solar irradiance, especially that of UV-A, which can refurbish damaged photosynthetic apparatus and ameliorate the afternoon P_{max} depression in view of the balance between damage and repair. High levels of UV-A radiation at midday caused photosynthetic inhibition of some macroalgae (Häder et al., 2001), but low levels of UV-A radiation have been found to enhance the growth of the brown alga *Fucus gardneri* embryos (Henry and Van Alstyne, 2004) as well as photosynthetic CO_2 fixation by phytoplanktons (Gao et al., 2007). Recently, absorption of UV-A energy has been found to be transferred to Chl. *a* in a diatom (Orellana et al., 2004) and *Porphyra* spp. (Zheng, Y and Gao, K, unpublished data). It has been recently proved that both *Gracilaria* and *Porphyra* plants can utilize UV-A for photosynthesis (Xu, J and Gao, K, unpublished data). The double-edged (positive at low and negative at high levels) effects of UV-A could magnify the discrepancy between the estimations of photosynthetic production and growth according to weather conditions.

4. Impacts of UVR on Different Life Stages

Different life stages of macroalgae showed different sensitivity to irradiation stress (Dring et al., 1996; Hanelt et al., 1997; Altamirano et al., 2003; Roleda et al., 2004; Véliz et al., 2006). Most studies conducted to evaluate the impact of UV-B on seaweeds used macro-thallus stages; however, early developmental life stages of intertidal algae seemed to be more sensitive to UVR than adult stages (Major and Davison, 1998; Coelho et al., 2000; Hoffman et al., 2003; Véliz et al., 2006). Studies to establish the sensitivity of early developmental stages are critical, since the survival and growth of these stages will determine the recruitment of a species and thus productivity.

Different life stages of *Porphyra* plants have been found to exhibit different photosynthetic characteristics.

Light utilization efficiency in thallus was much higher than that in conchocelis stage of *P. yezoensis* (Zhang et al., 1997), and maximal net photosynthetic rate of the conchocelis was lower than that of the thalli (Tanaka, 1985; Gao and Aruga, 1987). Electron transfer inhibitor DCMU blocked the energy transfer from PS II to PS I in the thalli, but not in the conchocelis of *P. yezoensis* (Pan et al., 2001). UVR was found to degrade photosynthetic pigments in both *P. leucosticta* (Figueroa et al., 1997) and *P. umbilicalis* (Aguilera et al., 1999a) and to reduce the effective quantum yield of *P. leucosticta* (Figueroa et al., 1997), but it resulted in insignificant photoinhibition in *P. umbilicalis* (Gröniger et al., 1999). These studies have focused on the thallus stage of *Porphyra* spp. In nature, *Porphyra*-conchocelis lives in shells, which are often found in shallow coastal waters (Jao, 1936) and must be exposed to certain extent of UVR. The conchocelis stage of *Porphyra haitanensis* contained much less UV-screening compounds and its PSII activity was more damaged even under reduced levels of solar radiation compared with the thallus stage (Jiang and Gao, 2008).

UVR is known to affect early development (Huovinen et al., 2000; Henry and Van Alstyne, 2004; Roleda et al., 2005, 2007; Wiencke et al., 2006, 2007) and spore germination (Wiencke et al., 2000; Altamirano et al., 2003; Han et al., 2004; Steinhoff et al., 2008) of macroalgae. It can effectively delay photosynthetic recovery in arctic kelp zoospores following photo-exposures (Roleda et al., 2006b). UV-B rather than UV-A negatively affected the germination of the zygotes of *Fucus serratus* (Altamirano et al., 2003) and the zoospores of *Laminaria hyperborean* (Steinhoff et al., 2008). Different life stages of some macroalgal species showed different degrees of enduring UVR, with increased tolerance as individuals differentiate. UV-A was found to play an important role in the morphogenesis of sporelings in *Porphyra haitanensis*, enhancing transverse cell division from conchospores (Jiang et al., 2007). Morphological differences among life stages can affect the energy transfer of PAR and UVR in the tissue, resulting in different responses to UVR in *Laminaria* spp. (Dring et al., 1996; Roleda et al., 2006c) as well as *Porphyra haitanensis* (Jiang and Gao, 2008). Longer path-length for the absorbed UVR energy in tissues can reduce its damaging effects.

5. Accumulation of UV-Absorbing Compounds as a Strategy Against UVR

Adaptation to UVR has equipped macroalgae with defensive mechanisms to minimize UV-induced damages. Macroalgae can protect themselves via avoidance, repair, and screening mechanisms (Karentz, 1994; Franklin and Foster, 1997; Karentz, 2001). In addition to photoreactivation and nucleotide excision repair of UV-induced DNA damage (Buma et al., 1995; Pakker et al., 2000a; Lud et al., 2001), an important mechanism to reduce the damaging impact of UVR in marine macroalgae is the synthesis and accumulation of UV-absorbing compounds (UVAC) (Karsten et al., 1998). These compounds, such as mycosporine-like amino acids (MAAs) (Karentz et al., 1991; Dunlap and Shick, 1998), scytonemin (Garcia-Pichel and Castenholz, 1991; Dillon et al., 2002), and phlorotannins (Pavia et al., 1997; Pavia and Brock, 2000), have been found in many photosynthetic organisms.

They increased in cellular content with increased UV exposure (Brenowitz and Castenholz, 1997; Pavia et al., 1997; Han and Han, 2005; Zheng and Gao, 2009) to reduce UV-related photoinhibition and damage, playing a protective role against solar UVR (Oren and Gunde-Cimerman, 2007).

Seaweeds often exhibit high levels of UVAC, such as MAAs in the red alga *Porphyra columbina* (Korbee-Peinado et al., 2004), an unknown UV-B absorbing substance in the green alga *Ulva pertusa* (Han and Han, 2005), and phlorotannin in the brown algae *Ascophyllum nodosum* and *Fucus gardneri* (Pavia et al., 1997; Henry and Van Alstyne, 2004). Higher levels of UVAC have been found in the red alga *Gracilaria lemaneiformis* under full spectrum of solar radiation than UVR-free treatments, reflecting a responsive induction (Gao and Xu, 2008). Synthesis of UVAC has been found to be induced by UV-B in *Chondrus crispus* (Karsten et al., 1998), *Porphyra columbina* (Korbee-Peinado et al., 2004), and *Ulva pertusa* (Han and Han, 2005). Such stimulation is dependent on both dose and wavelength, with higher accumulation of UVAC under high daily doses (Karsten et al., 1998; Franklin et al., 2001). UVR was suggested to trigger some photoreceptors (active wavelengths between 280 and 320 nm) in the algae to sense the need for UVAC synthesis (Han and Han, 2005; Oren and Gunde-Cimerman, 2007). Accumulation of UVAC is often associated with decreased Chl a, resulting in an increased ratio of UVAC to Chl a (Gao and Xu, 2008).

MAAs, the most common UV-screening compounds, are water-soluble substances with absorption maxima ranging from 310 to 360 nm (Nakamura et al., 1982). Although their UVR-protective function is not yet completely clear, the most acceptable interpretation is that they play a role as a screen against UVR (Conde et al., 2000; Karsten et al., 2005). Some of these compounds may also function as antioxidants (Dunlap and Yamamoto, 1995; Suh et al., 2003), osmosis-regulating substances (Oren, 1997), antenna pigments channeling the energy to the photosynthetic apparatus (Sivalingam et al., 1976; Gao et al., 2007), or an intracellular nitrogen storage (Korbee-Peinado et al., 2004; Korbee et al., 2006). Accumulation of MAAs could be induced by different radiation treatments (Karsten et al., 1999; Korbee-Peinado et al., 2004; Karsten et al., 2005) or affected by osmotic stress (Oren, 1997; Klisch et al., 2002) and nutrient availability (Korbee-Peinado et al., 2004; Korbee et al., 2005; Zheng and Gao, 2009). The accumulation of MAAs was found to be dependent on both dose and wavelength of incident solar radiation, with higher accumulation of MAAs associated with high daily doses in *Chondrus crispus* (Karsten et al., 1998; Franklin et al., 2001). Nutrient availability was also found to affect the accumulation of MAAs (Karsten and Wiencke, 1999; Korbee-Peinado et al., 2004); enrichment of nitrate enhanced the content of MAAs in *Gracilaria tenuistipitata* (Zheng and Gao, 2009). *Porphyra* plants contain high levels of MAAs (up to 1% of the dry weight), mainly porphyra-334, which accumulates to the highest concentrations among the species of red algae studied so far (Gröniger et al., 1999; Hoyer et al., 2001). However, some studies showed that contents of MAAs did not increase in response to UVR or PAR and could not completely protect *Porphyra umbilicalis* and *Gracilaria cornea* against UVR (Gröniger et al., 1999; Sinha et al., 2000).

Distribution of seaweed at different zonational depths affects the accumulation of MAAs. Intertidal species are usually more resistant to UV stress (i.e., inhibition of photosynthesis) than subtidal species that have less or no MAAs (Maegawa et al., 1993). It was found that deep-water polar macroalgal species did not have MAAs, whereas supra- and eulittoral species contained MAAs to high concentrations (Hoyer et al., 2002). Total MAAs content in *Mastocarpus stellatus* was sixfold higher than in *Chondrus crispus* that was generally found at a greater depth; quantum yield and maximal electron transport rate were more reduced in *C. crispus* than *M. stellatus* by UV-B radiation (Bischof et al., 2000). MAAs content in *Devaleraea ramentacea* increased with decreased depth, being correlated with a higher photosynthetic capacity under UVR treatment (Karsten et al., 1999). The macroalgal zonation patterns relate to their ability to resist high radiation stress (Hanelt, 1998).

Macroalgal species distributed at the upper part of intertidal zone may be exposed to much higher solar radiation during emersion if the low tide coincides with local noon. Recently, it was shown that desiccation or dehydration of *Porphyra haitanensis* thalli led to higher absorptivity of the UVAC (Jiang et al., 2008). The ability for *Porphyra haitanensis* thalli to increase its cellular content of UVAC during such emersion period allows them to cope with UVR stress. The possible strategy for macroalgal species to survive at the upper levels of intertidal zone is to increase its content of UVAC, which play roles in both sunscreening and osmosis regulation.

6. Summary

PAR drives photosynthesis, whereas UVR is usually known to harm physiological processes in macroalgae as well as phytoplankton. UV-A, however, at reduced levels, has been shown to enhance photosynthesis and repairing processes of photodamaged molecules, whereas UV-B mostly results in harmful effects. During their long history of evolution, seaweeds have developed protective strategies against harmful UV irradiances, such as synthesizing and accumulating UVAC and the repair of DNA damage. Different life stages of seaweeds show different sensitivity to solar UVR, with less-differentiated forms being more sensitive to UVR. Species distributed at different depths in the intertidal zone also show different responses to solar UVR; upper species, that are usually exposed to higher levels of solar radiation and accumulate higher contents of UVAC (such as MAAs), are more tolerant of UVR. On the other hand, diurnal photosynthesis can be underestimated during twilight period or cloudy days and overestimated during noontime if the effects of UVR are ignored owing to positive and negative effects caused by UV-A, respectively, at low and high irradiance levels.

7. References

Aguilera, J., Jiménez, C., Figueroa, F.L., Lebert, M. and Häder, D.P. (1999a) Effect of ultraviolet radiation on thallus absorption and photosynthetic pigments in the red alga *Porphyra umbilicalis*. J. Photochem. Photobiol. B Biol. **48**: 75–82.

Aguilera, J., Karsten, U., Lippert, H., Vögele, B., Philipp, E., Hanelt, D. and Wiencke, C. (1999b) Effects of solar radiation on growth, photosynthesis and respiration of marine macroalgae from the Arctic. Mar. Ecol. Prog. Ser. **191**: 109–119.

Altamirano, M., Flores-Moya, A. and Figueroa, F.L. (2000) Long-term effects of natural sunlight under various ultraviolet radiation conditions on growth and photosynthesis of intertidal *Ulva rigida* (Chlorophyceae) cultivated in situ. Bot. Mar. **43**: 119–126.

Altamirano, M., Flores-Moya, A., Kuhlenkamp, R. and Figueroa, F.L. (2003) Stage-dependent sensitivity to ultraviolet radiation in zygotes of the brown alga *Fucus serratus*. Zygote **11**: 101–106.

Aucamp, P.J. (2007) Questions and answers about the effects of the depletion of the ozone layer on humans and the environment. Photochem. Photobiol. Sci. **6**: 319–330.

Barbieri, E.S., Villafañe, V.E. and Helbling, E.W. (2002) Experimental assessment of UV effects on temperate marine phytoplankton when exposed to variable radiation regimes. Limnol. Oceanogr. **47**: 1648–1655.

Behrenfeld, M.J., O'Malley, R.T., Siegel, D.A., McClain, C.R., Sarmiento, J.L., Feldman, G.C., Milligan, A.J., Falkowski, P.G., Letelier, R.M. and Boss, E.S. (2006) Climate-driven trends in contemporary ocean productivity. Nature **444**: 752–755.

Bischof, K., Hanelt, D. and Wiencke, C. (1998) UV-radiation can affect depth zonation of Antarctic macroalgae. Mar. Biol. **131**: 597–605.

Bischof, K., Kräbs, G., Hanelt, D. and Wiencke, C. (2000) Photosynthetic characteristics and mycosporine-like amino acids under UV radiation: a competitive advantage of *Mastocarpus stellatus* over *Chondrus crispus* at the Helgoland shoreline? Helgoland Mar. Res. **54**: 47–52.

Bischof, K., Kräbs, G., Wiencke, C. and Hanelt, D. (2002) Solar ultraviolet radiation affects the activity of ribulose-1,5-bisphosphate carboxylase-oxygenase and the composition of photosynthetic and xanthophyll cycle pigments in the intertidal green alga *Ulva lactuca* L. Planta **215**: 502–509.

Bischof, K., Gómez, I., Molis, M., Hanelt, D., Karsten, U., Lüder, U., Roleda, M.Y., Zacher, K. and Wiencke, C. (2006) Ultraviolet radiation shapes seaweed communities. Rev. Environ. Sci. Biotechnol. **5**: 141–166.

Brenowitz, S. and Castenholz, R.W. (1997) Long-term effects of UV and visible irradiance on natural populations of a scytonemin-containing cyanobacterium (Calothrix sp.). FEMS Microbiol. Ecol. **24**: 343–352.

Brouwer, P.E.M., Bischof, K., Hanelt, D. and Kromkamp, J. (2000) Photosynthesis of two Arctic macroalgae under different ambient radiation levels and their sensitivity to enhanced UV radiation. Polar Biol. **23**: 257–264.

Buma, A.G.J., van Hannen, E.J., Roza, L., Veldhuis, M.J.W. and Gieskes, W.W.C. (1995) Monitoring ultraviolet-B-induced DNA damage in individual diatom cells by immunofluorescent thymine dimer detection. **31**: 314–321.

Buma A.G.J., De Boer, M.K. and Boelen, P. (2001) Depth distributions of DNA damage in Antarctic marine phyto- and bacterioplankton exposed to summertime UV radiation. J. Phycol. **37**: 200–208.

Buma, A.G.J., Boelen, P. and Jeffrey, W.H. (2003) UVR-induced DNA damage in aquatic organisms, In: E.W. Helbling and H.E. Zagarese (eds.) *UV Effects in Aquatic Organisms and Ecosystems*. The Royal Society of Chemistry, Cambridge, pp. 291–327.

Coelho, S.M., Rijstenbil, J.W. and Brown, M.T. (2000) Impacts of anthropogenic stresses on the early development stages of seaweeds. J. Aquat. Ecosyst. Stress Recov. **7**: 317–333.

Conde, D., Aubriot, L. and Sommaruga, R. (2000) Changes in UV penetration associated with marine intrusions and freshwater discharge in a shallow coastal lagoon of the Southern Atlantic Ocean. Mar. Ecol. Prog. Ser. **207**: 19–31.

Cordi, B., Depledge, M.H., Price, D.N., Salter, L.F. and Donkin, M.E. (1997) Evaluation of chlorophyll fluorescence, in vivo spectrophotometric pigment absorption and ion leakage as biomarkers of UV-B exposure in marine macroalgae. Mar. Biol. **130**: 41–49.

Davison, I.R., Jordan, T.L., Fegley, J.C. and. Grobe, C.W. (2007) Response of *Laminaria saccharina* (Phaeophyta) growth and photosynthesis to simultaneous ultraviolet radiation and nitrogen limitation. J. Phycol. **43**: 636–646.

Dillon, J.I., Tatsumi, C.I., Tandingan, P.I. and Castenholz, R.I. (2002) Effect of environmental factors on the synthesis of scytonemin, a UV-screening pigment, in a cyanobacterium (*Chroococcidiopsis* sp.). Arch. Microbiol. **177**: 322–331.

Dring, M.J., Makarov, V., Schoschina, E., Lorenz, M. and Lüning, K. (1996) Influence of ultraviolet radiation on chlorophyll fluorescence and growth in different life-history stages of three species of *Laminaria* (Phaeophyta). Mar. Biol. **126**: 183–191.

Dring, M.J., Wagner, A. and Lüning, K. (2001) Contribution of the UV component of natural sunlight to photoinhibition of photosynthesis in six species of subtidal brown and red seaweeds. Plant Cell Environ. **24**: 1153–1164.

Dunlap, W.C. and Shick, J.M. (1998) UV radiation absorbing mycosporine-like amino acids in coral reef organisms: a biochemical and environmental perspective. J. Phycol. **34**: 418–430.

Dunlap, W.C. and Yamamoto, Y. (1995) Small-molecule antioxidants in marine organisms: antioxidant activity of mycosporine-glycine. Comp. Biochem. Phys. B **112**: 105–114.

Fauchot, J., Gosselin, M., Levasseur, M., Mostajir, B., Belzile, C., Demers, S., Roy, S. and Villegas, P.Z. (2000) Influence of UV-B radiation on nitrogen utilization by a natural assemblage of phytoplankton. J. Phycol. **36**: 484–496.

Figueroa, F.L., Salles, S., Aguilera, J., Jiménez, C., Mercado, J., Viñegla, B., Flores-Moya, A. and Altamirano, M. (1997) Effects of solar radiation on photoinhibition and pigmentation in the red alga *Porphyra leucosticta*. Mar. Ecol. Prog. Ser. **151**: 81–90.

Flores-Moya, A., Hanelt, D., Figueroa, F.L., Altamirano, M., Viñegla, B. and Salles, S. (1999) Involvement of solar UV-B radiation in recovery of inhibited photosynthesis in the brown alga *Dictyota dichotoma* (Hudson) Lamouroux. J. Photochem. Photobiol. B: Biol. **49**: 129–135.

Foyer, C.H., Lelandais, M. and Kuner, K.J. (1994) Photooxidative stress in plants. Physiol. Plant **92**: 696–717.

Franklin, L.A. and Foster, R.M. (1997) The changing irradiance environment: consequences for marine macrophyte physiology. productivity and ecology. Eur. J. Phycol. **32**: 207–232.

Franklin, L.A., Kräbs, G. and Kuhlenkamp, R. (2001) Blue light and UV-A radiation control the synthesis of mycosporine like amino acids in *Chondrus crispus* (Florideophyceae). J. Phycol. **37**: 257–270.

Ganzon-Fortes, E.T. (1997) Diurnal and diel patterns in the photosynthetic performance of the Agarophyte *Gelidiella acerosa*. Bot. Mar. **40**: 93–100.

Gao, K. (1990) Diurnal photosynthetic performance of *Sargassum horneri*. Jpn. J. Phycol. **38**: 163–165 (in Japanese with English summary).

Gao, K. and Aruga, Y. (1987) Preliminary studies on the photosynthesis and respiration of *Porphyra yezoensis* under emersed condition. J Tokyo Univ Fish **47**: 51–65.

Gao, K. and Ma, Z. (2008) Photosynthesis and growth of *Arthrospira (Spirulina) platensis* (Cyanophyta) in response to solar UV radiation, with special reference to its minor variant. Environ. Exp. Bot. **63**: 123–129.

Gao, K. and Umezaki, I. (1989) Studies on diurnal photosynthetic performance of *Sargassum thunbergii*. Changes in photosynthesis under natural sunlight. Jpn. J. Phycol. **37**: 89–98.

Gao, K. and Xu, J. (2008) Effects of solar UV radiation on diurnal photosynthetic performance and growth of *Gracilaria lemaneiformis* (Rhodophyta). Eur. J. Phycol. **43**: 297–307.

Gao, K., Aruga, Y., Asada, K., Ishihara, T., Akano, T. and Kiyohara, M. (1991) Enhanced growth of the red alga *Porphyra yezoensis* Ueda in high CO_2 concentrations. J. Appl. Phycol. **3**: 356–362.

Gao, K., Aruga, Y., Asada, K. and Kiyohara, M. (1993) Influence of enhanced CO_2 on growth and photosynthesis of the red algae *Gracilaria* sp. and G. Cilensis. J. Appl. Phycol. **5**: 563–571.

Gao, K., Wu, Y., Li, G., Wu, H., Villafañe, V.E. and Helbling, E.W. (2007) Solar UV radiation drives CO_2 fixation in marine phytoplankton: a double-edged sword. Plant Physiol. **144**: 54–59.

Garcia-Pichel, F. and Castenholz, R.W. (1991) Characterization and biological implications of scytonemin, a cyanobacterial sheath pigment. J. Phycol. **27**: 395–409.

Gerard, V.A. (1986) Photosynthetic characteristics of giant kelp (*Macrocystis pyrifera*) determined *in situ*. Mar. Biol. **90**: 473–482.

Gledhill, M., Nimmo, M., Hill, S.J. and Brown, M. (1997) The toxicity of copper (II) species to marine algae, with particular reference to macroalgae. J. Phycol. **33**: 2–11.

Goes, M., Martins, C.B., Teles, F.F.F., Matos, F.J.A., Guedes, Z.B.L. and Oria, H.F. (1994). Moisture content and fatty acid composition of five tropical fruits. Revista Ceres **41**: 234–243.

Gómez, I. and Figueroa, F.L. (1998) Effects of solar UV stress on chlorophyll fluorescence kinetics of intertidal macroalgae from southern Spain: a case study in *Gelidium* species. J. Appl. Phycol. **10**: 285–294.

Grobe, C.W. and Murphy, T.M. (1994) Inhibition of growth of *Ulva expansa* (Chlorophyta) by ultraviolet-B radiation. J. Phycol. **30**: 783–790.

Grobe, C.W. and Murphy, T.M. (1998) Solar ultraviolet-B radiation effects on growth and pigment composition of the intertidal alga *Ulva expansa* (Setch.) S. & G. (Chlorophyta). J. Photochem. Photobiol. B Biol. **225**: 39–51.

Gröniger, A. and Häder, D.P. (1999) Stability of mycosporine-like amino acids. Recent Res. Dev. Photochem. Photobiol. **2000**: 247–252.

Gröniger, A., Hallier, C. and Häder, D.P. (1999) Influence of W radiation and visible light on *Porphym umbilicalis*: photoinhibition and MAA concentration. J. Appl. Phycol. **11**: 437–445.

Häder, D.P., Lebert, M. and Helbling, E.W. (2001) Effects of solar radiation on the Patagonian macroalga *Enteromorpha linza* (L.) J. Agardh – Chlorophyceae. J. Photochem. Photobiol. B: Biol. **62**: 43–54.

Häder, D.P., Kumar, H.D., Smith, R.C. and Worrest, R.C. (2007) Effects of solar UV radiation on aquatic ecosystems and interactions with climate change. Photochem. Photobiol. Sci. **6**: 267–285.

Han, Y.S. and Han, T. (2005) UV-B induction of UV-B protection in *Ulva pertusa* (Chlorophyta). J. Phycol. **41**: 523–530.

Han, T., Han, Y.S., Kain, J.M. and Häder, D.P. (2003) Thallus differentiation of photosynthesis, growth, reproduction, and UV-B sensitivity in the green alga *Ulva pertusa* (Chlorophyceae). J. Phycol. **39**: 712–721.

Han, T., Kong, J.A., Han, Y.S., Kang, S.H. and Häder, D.P. (2004) UV-A/blue light-induced reactivation of spore germination in UV-B irradiated *Ulva pertusa* (Chlorophyta). J. Phycol. **40**: 315–322.

Hanelt, D. (1992) Photoinhibition of photosynthesis in marine macrophytes of the South Chinese Sea. Mar. Ecol. Prog. Ser. **82**: 199–206.

Hanelt, D. (1998) Capability of dynamic photoinhibition in Arctic macroalgae is related to their depth distribution. Mar. Bio. **131**: 361–369.

Hanelt, D., Wiencke, C. and Nultsch, W. (1997) Influence of UV radiation on the photosynthesis of arctic macroalgae in the field. J. Photochem. Photobiol. B Biol. **38**: 40–47.

Hargreaves, B.R. (2003) Water column optics and penetration of UVR, In: E.W. Helbling and H.E. Zagarese (eds.) *UV Effects in Aquatic Organisms and Ecosystems*. The Royal Society of Chemistry, Cambridge.

Hein, M. and Sand-Jensen, K. (1997) CO_2 increases oceanic primary production. Nature **388**: 526–527.

Helbling, E.W., Gao, K., Goncalves, R.J., Wu, H. and Villafañe, V.E. (2003) Utilization of solar ultraviolet radiation by phytoplankton assemblages from the Southern China Sea when exposed to fast mixing conditions. Mar. Ecol. Prog. Ser. **259**: 59–66.

Helbling, E.W., Barbieri, E.S., Sinha, R.P., Villafañe, V.E. and Häder, D.P. (2004) Dynamics of potentially protective compounds in Rhodophyta species from Patagonia (Argentina) exposed to solar radiation. J. Photochem. Photobiol. B Biol. **75**: 63–71.

Henley, W.J., Levavasseur, G., Franklin, L.A., Lindley, S.T., Ramus, J. and Osmond, C.B. (1991) Diurnal responses of photosynthesis and fluorescence in *Ulva rotundata* acclimated to sun and shade in outdoor culture. Mar. Ecol. Prog. Ser. **75**: 19–28.

Henry, B.E. and Van Alstyne, K.L. (2004) Effects of UV radiation on growth and phlorotannins in *Fucus gardneri* (Phaeophyceae) juveniles and embryos. J. Phycol. **40**: 527–533.
Hoffman, J.R., Hansen, L.J. and Klinger, T. (2003) Interactions between UV radiation and temperature limit inferences from single-factor experiments. J. Phycol. **39**: 268–272.
Hoyer, K., Karsten, U., Sawall, T. and Wiencke, C. (2001) Photoprotective substances in Antarctic macroalgae and their variation with respect to depth distribution, different tissues and developmental stages. Mar. Ecol. Prog. Ser. **211**: 117–129.
Hoyer, K., Karsten, U. and Wiencke, C. (2002) Induction of sunscreen compounds in Antarctic macroalgae by different radiation conditions. Mar. Biol. **141**: 619–627.
Huovinen, P.S., Oikari, A.O.J., Soimasuo, M.R. and Cherr, G.N. (2000) Impact of UV radiation on the early development of the giant kelp (*Macrocystis pyrifera*) gametophytes. Photochem. Photobiol. **72**: 308–313.
Huppertz, K., Hanelt, D. and Nultsch, W. (1990) Photoinhibition of photosynthesis in the marine brown alga *Fucus serratus* as studied in field experiments. Mar. Ecol. Prog. Ser. **66**: 175–182.
Jao, C.C. (1936) New Rhodophyceae from Woods Hole. Bull Torrey Bot. Club **63**: 237–257.
Jiang, H. and Gao, K. (2008) Effects of UV radiation on the photosynthesis of conchocelis of *Porphyra Haitanensis* (Bangiales, Rhodophyta). Phycologia **47**: 241–248.
Jiang, H., Gao, K. and Helbling, E.W. (2007) Effects of solar UV radiation on germination of conchospores and morphogenesis of sporelings in *Porphyra haitanensis* (Rhodophyta). Mar. Biol. **151**: 1751–1759.
Jiang, H., Gao, K. and Helbling, E.W. (2008) UV-absorbing compounds in *Porphyra haitanensis* (Rhodophyta) with special reference to effects of desiccation. J. Appl. Phycol. **20**: 387–395.
Kamer, K. and Fong, P. (2001) Nitrogen enrichment ameliorates the negative effects of reduced salinity on the green macroalga. J. Exp. Marine Biol. and Ecol. **218**: 87–93.
Karentz, D. (1994) Ultraviolet tolerance mechanisms in Antarctic marine organisms. Antarctic Res. Ser. **62**: 93–110.
Karentz, D. (2001) Chemical defenses of marine organisms against solar radiation exposure: UV-absorbing mycosporine-like amino acids and scyotnemin, In: J.B. McClintock and B.J. Baker (eds.) *Marine Chemical Ecology*. CRC Press, Birmingham, pp. 481–486.
Karentz, D., Mceuen, F.S., Land, M.C. and Dunlap, W.C. (1991) Survey of mycosporine-like amino acid compounds in Antarctic organisms: potential protection from ultraviolet exposure. Mar. Biol. **108**: 157–166.
Karsten, U. and West, J.A. (2000) Living in the intertidal zone – seasonal effects on heterosides and sun-screen compounds in the red alga *Bangia atropurpurea* (Bangiales). J. Exp. Mar. Biol. Ecol. **254**: 221–234.
Karsten, U. and Wiencke, C. (1999) Factors controlling the formation of UV-absorbing mycosporine-like amino acids in the marine red alga *Palmaria palmata* from Spitsbergen (Norway). J. Plant Physiol. **155**: 407–415.
Karsten, U., Franklin, L.A., Lüning, K. and Wiencke, C. (1998) Natural ultraviolet radiation and photosynthetically active radiation induce formation of mycosporine-like amino acids in the marine macroalga *Chondrus crispus* (Rhodophyta). Planta **205**: 257–262.
Karsten, U., Bischof, K., Hanelt, D., Tüg, H. and Wiencke, C. (1999) The effect of ultraviolet radiation on photosynthesis and ultraviolet-absorbing substances in the endemic Arctic macroalga *Devaleraea ramentacea* (Rhodophyta). Physiol. Plant **105**: 58–66.
Karsten, U., Friedl, T., Schumann, R., Hoyer, K. and Lembcke, S. (2005) Mycosporine-like amino acids and phylogenies in green algae: *prasiola* and its relatives from the trebouxiophyceae (Chlorophyta). J. Phycol. **41**: 557–566.
Kerr, J.B. and McElroy, C.T. (1993) Evidence for large upward trends of ultraviolet-B radiation linked to ozone depletion. Science **262**: 1032–1034.
Klisch, M., Sinha, R.P. and Häder, D.P. (2002) UV-absorbing compounds in algae. Curr. Top. Plant Biol. **3**: 113–120.

Korbee, N., Huovinen, P., Figueroa, F.L., Aguilera, J. and Karsten, U. (2005) Availability of ammonium influences photosynthesis and the accumulation of mycosporine-like amino acids in two *Porphyra* species (Bangiales, Rhodophyta). Mar. Biol. **146**: 645–654.

Korbee, N., Figueroa, F.L. and Aguilera, J. (2006) Accumulation of mycosporine-like amino acids (MAAs): biosynthesis, photocontrol and ecophysiological functions. Revista Chilena de Historia Natural **79**: 119–132.

Korbee-Peinado, N., Abdala-Díaz, R.T., Figueroa, F.L. and Helbling, E.W. (2004) Ammonium and UV radiation stimulate the accumulation of mycosporine-like amino acids in *Porphyra columbina* (Rhodophyta) from patagonia, argentina. J. Phycol. **40**: 248–259.

Kuhlenkamp, R., Franklin, L.A. and Lüning, K. (2001) Effect of solar ultraviolet radiation on growth in the marine macroalga *Dictyota dichotoma* (Phaeophyceae) at Helgoland and its ecological consequences. Helgol. Mar. Res. **55**: 77–86.

Larned, S.T. (1998) Nitrogen-versus phosphorus-limited growth and sources of nutrients for coral reef macroalgae. Mar. Biol. **132**: 409–421.

Litchman, E., Neale, P.J. and Banaszak, A.T. (2002) Increased sensitivity to ultraviolet radiation in nitrogen-limited dinoflagellates: photoprotection and repair. Limnol. Oceanogr. **47**: 86–94.

Lobban, C.S. and Harrison, P.J. (eds.) (1994) *Seaweed Ecology and Physiology*. Cambridge University Press, Cambridge.

Lud, D., Huiskes, A.H.L., Moerdijk T.C.W. and Rozema, J. (2001) The effects of altered levels of UV-B radiation on an Antarctic grass and lichen. Plant Ecol. **154**: 87–99.

Maegawa, M., Kunieda, M. and Kida, W. (1993) Difference of the amount of W absorbing substance between shallow and deep water red algae. Jpn. J. Phycol. **41**: 351–354.

Major, K.M. and Davison, I.R. (1998) Influence of temperature and light on growth and photosynthetic physiology of *Fucus evanescens* (Phaeophyta) embryos. Eur. J. Phycol. **33**: 129–138.

Marra, J. (1978) Effect of short-term variation in light intensity on photosynthesis of a marine phytoplankter: a laboratory simulation study. Mar. Biol. **46**: 191–202.

Michler, T., Aguilera, J., Hanelt, D., Bischof, K. and Wiencke, C. (2002) Long-term effects of ultraviolet radiation on growth and photosynthetic performance of polar and cold-temperate macroalgae. Mar. Biol. **140**: 1117–1127.

Nakamura, H., Kobayashi, J. and Hirata, Y. (1982) Separation of mycosporinelike amino acids in marine organisms using reverse-phase high performance liquid chromatography. J. Chromatogr. **250**: 113–118.

Nilawati, J., Greenberg, B.M. and Smith, R.E.H. (1997) Influence of ultraviolet radiation on growth and photosynthesis of two cold ocean diatoms. J. Phycol. **33**: 215–224.

O'Neal, S.W. and Prince, J.S. (1988) Seasonal effects of light, temperature, nutrient concentration and salinity on the physiology and growth of *Caulerpa paspaloides* (Chlorophyceae). Mar. Biol. **97**: 17–24.

Oliver, M.J., Tuba, Z. and Mishler, B.D. (2000) The evolution of vegetative desiccation tolerance in land plants. Plant Ecol. **151**: 85–100.

Orellana, M.V., Petersen, T.W. and Van Den Engh, G. (2004) UV-excited blue auto-fluorescence of *pseudo-nitzschia multiseries* (Bacillariophyceae). J. Phycol. **40**: 705–710.

Oren, A. (1997) Mycosporine-like amino acids as osmotic solutes in a community of halophilic cyanobacteria. Geomicrobiol. J. **14**: 231–240.

Oren, A. and Gunde-Cimerman, N. (2007) Mycosporines and mycosporine-like amino acids: UV protectants or multipurpose secondary metabolites? FEMS Microbiol. Lett. **269**: 1–10.

Pakker, H., Beekman, C.A.C. and Breeman, A.M. (2000a) Efficient photoreactivation of UVBR-induced DNA damage in the sublittoral macroalga *Rhodymenia pseudopalmata* (Rhodophyta). Eur. J. Phycol. **35**: 109–114.

Pakker, H., Martins, R., Boelen, P., Buma, A.G.J., Nikaido, O. and Breeman, A.M. (2000b) Effects of temperature on the photoreactivation of ultraviolet-B induced DNA damage in *Palmaria palmata* (Rhodophyta). J. Phycol. **36**: 334–341.

Pan, J., Shi, D., Chen, J., Peng, G., Zeng, C. and Zhang, Q. (2001) Excitation energy distribution between two photosystems in *Porphyra yezoensis* and its significance in photosynthesis evolution. Chinese Sci. Bull. **46**: 49–52.

Pang, S., Gómez, I. and Lüning, K. (2001) The red macroalga *Delesseria sanguinea* as a UVB-sensitive model organism: selective growth reduction by UVB in outdoor experiments and rapid recording of growth rate during and after UV pulses. Eur. J. Phycol. **36**: 207–216.

Pavia, H. and Brock, E. (2000) Extrinsic factors influencing phlorotannin production in the brown alga *Ascophyllum nodosum*. Mar. Ecol. Prog. Ser. **193**: 285–294.

Pavia, H., Cervin, G., Lindgren, A. and Åberg, P. (1997) Effects of UV-B radiation and simulated herbivory on phlorotannins in the brown alga *Ascophyllum nodosum*. Mar. Ecol. Prog. Ser. **157**: 139–146.

Ramus, J. and Rosenberg, G. (1980) Diurnal photosynthetic performance of seaweeds measured under natural conditions. Mar. Biol. **56**: 21–28.

Rautenberger, R. and Bischof, K. (2006) Impact of temperature on UV-susceptibility of two *Ulva* (Chlorophyta) species from Antarctic and Subantarctic regions. Polar Biol. **29**: 988–996.

Roleda, M.Y., Hanelt, D., Kraebs, G. and Wiencke, C. (2004) Morphology, growth, photosynthesis and pigments in *Laminaria ochroleuca* (Laminariales, Phaeophyta) under ultraviolet radiation. Phycologia **43**: 603–613.

Roleda, M.Y., Wiencke, C., Hanelt, D., van de Poll, W.H. and Gruber, A. (2005) Sensitivity of *Laminariales* zoospores from Helgoland (North Sea) to ultraviolet and photosynthetically active radiation: implications for depth distribution and seasonal reproduction. Plant Cell Envion. **28**: 466–479.

Roleda, M.Y., Wiencke, C. and Hanelt, D. (2006a) Thallus morphology and optical characteristics affect growth and DNA damage by UV radiation in juvenile Arctic *Laminaria* sporophytes. Planta **223**: 407–417.

Roleda, M.Y., Hanelt, D. and Wiencke, C. (2006b) Exposure to ultraviolet radiation delays photosynthetic recovery in Arctic kelp zoospores. Photosynth. Res. **88**: 311–322.

Roleda, M.Y., Hanelt, D. and Wiencke, C. (2006c) Growth and DNA damage in young *Laminaria* sporophytes exposed to ultraviolet radiation: implication for depth zonation of kelps on Helgoland (North Sea). Mar. Biol. **148**: 1201–1211.

Roleda, M.Y., Wiencke, C., Hanelt, D. and Bischof, K. (2007) Sensitivity of the early life stages of macroalgae from the northern hemisphere to ultraviolet radiation. Photochem. Photobiol. **83**: 851–862.

Santas, R., Korda, A., Lianou, C. and Santas, P. (1998) Community responses to UV radiation: I. Enhanced UVB effects on biomass and community structure of filamentous algal assemblages growing in a coral reef mesocosm. Mar. Biol. **131**: 153–162.

Sass, L., Spetea, C., Máté, Z., Nagy, F. and Vass, I. (1997) Repair of UV-B-induced damage of photosystem II via de novo synthesis of the D1 and D2 reaction centre subunits in *Synechocystis* sp. PCC 6803. Photosynth. Res. **54**: 55–62.

Schippers, P., Lürling, M. and Scheffer, M. (2004) Increase of atmospheric CO_2 promotes phytoplankton productivity. Ecol. Lett. **7**: 446–451.

Shick, J.M. and Dunlap, W.C. (2002) Mycosporine-like amino acids and related gadusols: biosynthesis, accumulation, and UV-protective functions in aquatic organisms. Annu. Rev. Physiol. **64**: 223–262.

Sinha, R.P., Klisch, M., Gröniger, A. and Häder, D.P. (2000) Mycosporine-like amino acids in the marine red alga *Gracilaria cornea* – effects of UV and heat. Environ. Exp. Bot. **43**: 33–43.

Sivalingam, P.M., Ikawa, T. and Nisizawa, K. (1976) Physiological roles of a substance 334 in algae. Bot. Mar. **19**: 9–21.

Steinhoff, F.S., Wiencke, C., Müller, R. and Bischof, K. (2008) Effects of ultraviolet radiation and temperature on the ultrastructure of zoospores of the brown macroalga *Laminaria hyperborean*. Plant Biol. **10**: 388–397.

Suh, H.J., Hyun-Woo Lee, H.W. and Jung, J. (2003) Mycosporine glycine protects biological systems against photodynamic damage by quenching singlet oxygen with a high efficiency. Photochem. Photobiol. **78**: 109–113.

Tanaka, H. (1985) *Growth and Physiology of Porphyra Yezoensis Free-Living Conchocelis*. Master's thesis, Tokyo University of Fisheries, pp. 1–28.

Todd, C.D. and Lewis, J.R. (1984) Effects of low air temperature on *Laminaria digitata* in southwestern Scotland. Mar. Ecol. Prog. Ser. **16**: 199–201.

van de Poll, W.H., Eggert, A., Buma, A.G.J. and Breeman, A.M. (2001) Effects of UV-B-induced DNA damage and photoinhibition on growth of temperate marine red macrophytes: habitat-related differences in UV-B tolerance. J. Phycol. **37**: 30–37.

van Donk, E., Faafeng, B.A., de Lange, H.J. and Hessen, D.O. (2001) Differential sensitivity to natural ultraviolet radiation among phytoplankton species in Arctic lakes (Spitsbergen, Norway). Plant Ecol. **154**: 247–259.

Vass, I. (1997) Adverse effects of UV-B light on the structure and functions of the photosynthetic apparatus, In: M. Pessarakali (ed.) *Handbook of Photosynthesis*. Marcel Dekker, New York, pp. 931–946.

Véliz, K., Edding, M., Tala, F. and Gómez, I. (2006) Effects of ultraviolet radiation on different life cycle stages of the south Pacific kelps, *Lessonia nigrescens* and *Lessonia trabeculata* (Laminariales, Phaeophyceae). Mar. Biol. **149**: 1015–1024.

Villafañe, V.E., Sunbäck, K., Figueroa, F.L. and Helbling, E.W. (2003) Photosynthesis in the aquatic environment as affected by ultraviolet radiation, In: E.W. Helbling and H.E. Zagarese (eds.) *UV Effects in Aquatic Organisms and Ecosystems*. The Royal Society of Chemistry, Cambridge, pp. 357–397.

Wiencke, C., Gómez, I., Pakker, H., Flores-Moya, A., Altamirano, M., Hanelt, D., Bischof, K. and Figueroa, F. (2000) Impact of UV radiation on viability, photosynthetic characteristics and DNA of brown algal zoospores: implications for depth zonation. Mar. Ecol. Prog. Ser. **197**: 217–229.

Wiencke, C., Roleda, M.Y., Gruber, A., Clayton, M.N. and Bischof, K. (2006) Susceptibility of zoospores to UV radiation determines upper depth distribution limit of Arctic kelps: evidence through field experiments. J. Ecol. **94**: 455–463.

Wiencke, C., Lüder, U.H. and Roleda, M.Y. (2007) Impact of ultraviolet radiation on physiology and development of zoospores of the brown alga *Alaria esculenta* from Spitsbergen. Physiol. Plantarum **130**: 601–612.

Wu, H., Gao, K., Villafañe, V.E., Watanabe, T. and Helbling, E.W. (2005) Effects of solar UV radiation on morphology and photosynthesis of filamentous cyanobacterium *Arthrospira platensis*. Appl. Environ. Microbiol. **71**: 5004–5013.

Xu, J. and Gao, K. (2007) Effects of solar UVR on the effective quantum yield of *Ulva lactuca*. Acta Oceanol. Sin. **29**: 127–132 (in Chinese with English summary).

Zacher, K., Roleda, M.Y., Hanelt, D. and Wiencke, C. (2007) UV effects on photosynthesis and DNA in propagules of three Antarctic seaweeds (*Adenocystis utricularis*, *Monostroma hariotii* and *Porphyra endiviifolium*). Planta **225**: 1505–1516.

Zhang, X., Brammer, E., Pedersén, M. and Fei, X. (1997) Effects of light photon flux density and spectral quality on photosynthesis and respiration in *Porphyra yezoensis* (Bangiales, Rhodophyta). Phycol. Res. **45**: 29–37.

Zheng, Y. and Gao, K. (2009) Impact of solar UV radiation on the photosynthesis, growth and UV-absorbing compounds in *Gracilaria lemaneiformis* (Rhodophyta) grown at different nitrate concentrations. J. Phycol. **45**: 314–323.

Zou, D. and Gao, K. (2002) Effects of desiccation and CO_2 concentrations on emersed photosynthesis in *Porphyra haitanensis* (Bangiales, Rhodophyta), a species farmed in China. Eur. J. Phycol. **37**: 587–592.

Biodata of **E. Walter Helbling**, **Virginia E. Villafañe**, and **Donat-P. Häder**, authors of *"Ultraviolet Radiation Effects on Macroalgae from Patagonia, Argentina"*

Dr. E. Walter Helbling is currently the Director of Estación de Fotobiología Playa Unión (EFPU, Argentina) and a researcher from Consejo Nacional de Investigaciones Científicas y Técnicas (CONICET, Argentina). He obtained his Ph.D. from Scripps Institution of Oceanography, University of California, San Diego (USA). His scientific interests are in ecophysiology of plankton and photobiology of aquatic systems in relation to climate change.

E-mail: **whelbling@efpu.org.ar**

Dr. Virginia E. Villafañe is currently a Researcher from Consejo Nacional de Investigaciones Científicas y Técnicas (CONICET, Argentina). She obtained her Ph.D. from University of Groningen (The Netherlands) and continued her research in Patagonia at Estación de Fotobiología Playa Unión (EFPU, Argentina). Dr. Villafañe's scientific interests are in the areas of ecophysiology of plankton and photobiology.

E-mail: **Virginia@efpu.org.ar**

E. Walter Helbling   Virginia E. Villafañe

Professor Donat-P. Häder holds the Chair of Plant Ecophysiology at the Department for Biology at the Friedrich-Alexander University in Erlangen-Nürnberg. He obtained his Ph.D. from the University of Marburg in 1973. After a Postdoc year in East Lansing Michigan state, he became Researcher in Marburg. Professor Häder's scientific interests are in the areas of the effects of stratospheric ozone depletion and resulting increasing solar UV-B radiation at the Earth's surface on the biota. He concentrates on these effects in combination with global climate change on aquatic ecosystems in many habitats over the globe.

E-mail: **dphaeder@biologie.uni-erlangen.de**

ULTRAVIOLET RADIATION EFFECTS ON MACROALGAE FROM PATAGONIA, ARGENTINA

E. WALTER HELBLING[1,2], VIRGINIA E. VILLAFAÑE[1,2], AND DONAT-P. HÄDER[3]
[1]Estación de Fotobiología Playa Unión, Casilla de Correos N° 15, (9103), Rawson, Chubut, Argentina
[2]Consejo Nacional de Investigaciones Científicas y Técnicas, CONICET, Argentina
[3]Department of Biology, Friedrich-Alexander Universität Erlangen/ Nürnberg, Staudtstr. 5, 91058, Erlangen, Germany

1. General Considerations

Aquatic ecosystems account for almost half of the primary production on our planet, matching the combined productivity of all terrestrial ecosystems (Siegenthaler and Sarmiento, 1993). Though most of the aquatic productivity is due to phytoplankton, macroalgae contribute to a significant share, especially in coastal areas. In their natural environment, macroalgae are generally exposed to excessive solar PAR (photosynthetic active radiation, 400–700 nm) as well as to ultraviolet radiation (UV-B, 280–315 nm, and UV-A, 315–400 nm), especially in the upper eulittoral and the supralittoral (Hanelt, 1998). The coincidence of low tides and high solar angles results in the highest radiation stress, generally reflected as photoinhibition, i.e., the reduction in photosynthetic rates. Photoinhibition (Dring et al., 1996; Franklin and Forster, 1997; Häder et al., 2001b), is determined not only in macroalgae from the tropics and temperate zones but also in Arctic and Antarctic environments (Hanelt et al., 1997; Hanelt, 1998). Most of the observed photoinhibition is due to PAR, as this waveband has a high proportion of solar radiation energy reaching the Earth's surface. However, in the top meters of the water column, a significant percentage of photoinhibition is caused by UV-B, and to a lesser extent by UV-A (Dring et al., 1996; Häder, 1997).

In this chapter, we will review our knowledge about the effects of solar radiation on macroalgae from Patagonia, Argentina. This area is especially important from a photobiological point of view, as it presents high heliophany and episodic ozone depletion events (Helbling et al., 2005). Although relatively few studies were conducted in this area, most of them were performed under in situ conditions, and thus the information is highly valuable as it reflects the natural situation of the area.

2. Macroalgae Diversity in Patagonia

Extensive literature on the different macroalgae groups present in the Patagonia area is available (see Boraso and Zaixso, 2008 and references therein). Particularly, there are many studies regarding macroalgal communities of the Chubut Province. In Golfo San José, which represents the limit between the oceanic biogeographic provinces of Patagonia and Argentina, ca. 30% of the surface at depths <10 m is covered by *Dictyota dichotoma*. In the intertidal zone of Golfo Nuevo, *Blidingia minima* and *Enteromorpha* spp. are characteristic. In tidal pools, *Cladophora falklandica*, *Ulva rigida*, and *Polysiphonia brodiaei* are commonly found; the Rhodophyta *Corallina officinalis* has been found very important to fix and protect the shore from the incoming waves. In the past 10 years, rocky seabeds have been dominated by the invasive Phaeophyta *Undaria pinnatifida* (Casas and Piriz, 1996; Martin and Cuevas, 2006; Casas et al., 2008) in several areas of the Patagonia coast. Moreover, in sandy beaches of Golfo Nuevo, beach-cast seaweeds were dominated by green algae from the genera *Ulva* and *Codium* but lately *U. pinnatifida* displaced *Codium* spp. (Piriz et al., 2003). Farther south, and in the intertidal zone of Golfo San Jorge, *Enteromorpha* spp. and *Porphyra columbina* are characteristic. In the lower intertidal, *C. officinallis* dominates, but other species from the genera *Cladophora*, *Ulva*, *Adenocystis*, *Bryopsis*, *Codium*, *Chondria*, *Leathesia*, *Colpomenia*, *Spongomorpha*, and *Urospora* are also found.

Macroalgae of commercial interest have also been the focus of several investigations (Boraso de Zaixso et al., 1998 and references therein). The most important commercial species found in Patagonian waters are *Gracilaria gracilis* and *Macrocystis pyrifera*, which are usually found in the center and south of the Chubut Province.

In Santa Cruz Province, and particularly in the Ría de Puerto Deseado, many studies have focused on the taxonomy of macroalgal species: So far, more than 200 species have been identified (Boraso de Zaixso, 1995). In Tierra del Fuego, and in the intertidal zone near Ushuaia, the genera *Rama*, *Rhizoclonium*, *Cladophoropsis*, *Porphyra*, *Bostrychia*, *Iridaea*, *Hildenbrandtia*, and *Caepidium* are common, as also Ulvales and Cladophorales are diverse; in the subtidal zone, diverse corallinaceae are representative.

3. Solar Radiation and Ozone Conditions Over the Area of Patagonia

Solar radiation is an environmental factor that strongly affects organisms living in aquatic ecosystems. The radiation levels at which aquatic organisms are exposed depend on several factors, such as the solar irradiance reaching the Earth's surface, type and concentration of atmospheric gases (i.e., mainly ozone), altitude, and particulate and dissolved material in the water column (Blumthaler and Webb, 2003; Hargreaves, 2003). In Patagonia, there is a clear trend of high radiation values during summer and low during winter (Orce and Helbling, 1997; Villafañe et al., 2004).

However, there is high variability in irradiance and daily doses values over Patagonia owing to the large latitudinal coverage with concomitant changes in solar zenith angle, day length, and atmospheric aerosols content, among other factors (Holm-Hansen et al., 1993) (more detailed information about latitudinal differences in radiation levels in temperate and sub-Antarctic sites of Patagonia as well as along Argentina is presented in Orce and Helbling (1997)). Representative patterns of solar radiation over Patagonia are presented in Fig. 1: daily doses of PAR (Fig. 1a) vary from ~14 MJ m^{-2} in summer to <1 MJ m^{-2} in winter; UVR daily doses follow a similar pattern, with UV-A (Fig. 1b) ranging from ~2 to 0.15 MJ m^{-2}, whereas UV-B (Fig. 1c) ranges from ~45 to <5 KJ m^{-2}.

UV-B radiation is additionally affected by ozone concentrations (Madronich, 1993; Blumthaler and Webb, 2003). During the past 2 decades, there was an increasing interest in evaluating the effects of enhanced UV-B radiation due to the thinning of the ozone layer (i.e., ozone "hole") on aquatic biota (Häder et al., 2007). Total column ozone concentration over mid-Patagonia varies throughout the year, with low values (~220–230 Dobson Units, D.U.) in April–May, and high ones (~400 D.U.) during September (Fig. 1d), which is in agreement with the reported dynamics of photochemical production of ozone over the stratosphere (Molina and Molina, 1992). During early spring, however, there are some days (ovals in Fig. 1d) characterized by relatively low ozone concentrations (220–270 D.U.) that are associated with the Antarctic polar vortex and to the ozone "hole"; in fact, the signaling of the Antarctic ozone "hole" over Patagonia has been determined as far north as 38°S (Orce and Helbling, 1997). Several studies have shown the presence of low-ozone air masses over Patagonia, either because the Antarctic polar vortex covers the tip of South America for short periods of time (Frederick et al., 1993; Díaz et al., 1996), or because ozone-depleted air masses detach from the polar vortex (i.e., end of November to early December) and circulate northward (Atkinson et al., 1989; Kirchhoff et al., 1996). However, these studies have highlighted the temporal and spatial variability of low-ozone air masses over Patagonia. Moreover, Helbling et al. (2005) used TOMS data to determine the aerial coverage of low-ozone air masses (<275 D.U.) that were related to the Antarctic ozone "hole" over Patagonia and they found that, in general, they covered ~20% of the Patagonia area during 1979 and then increased up to ~95% during 2002. However, it should be stressed that these data represent the maximum coverage of low-ozone air masses during 1 day, and the dynamics of the polar vortex is such that it influences Patagonia only during a few days per year (Orce and Helbling, 1997).

4. Impact on Macroalgae Photosynthesis

The photosynthetic performance under solar radiation of several macroalgae has been measured on site on the coast of Patagonia. On the rocky shore, the organisms cover a wide habitat from the supralittoral to the sublittoral with substantial differences in exposure to PAR and UVR. The Chlorophyte *Ulva rigida* grows

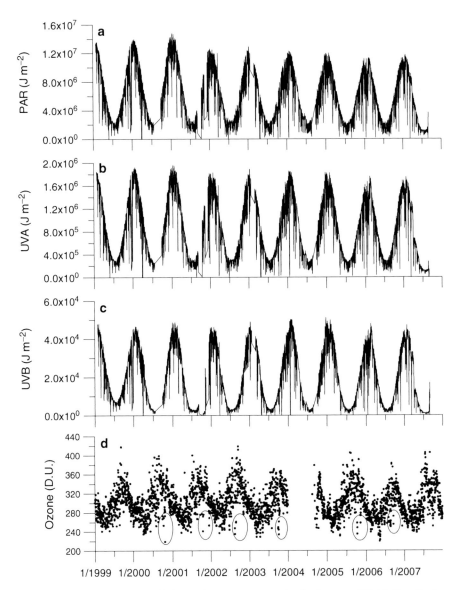

Figure 1. Daily doses of solar radiation and ozone concentration in the area of Bahía Engaño from 1999 to 2007. (a) PAR (400–700 nm); (b) UV-A (315–400 nm); (c) UV-B (280–315 nm); (d) Ozone concentrations (in D.U.). Solar radiation data were obtained with a broadband ELDONET radiometer (www.eldonet.org) permanently installed on the roof of Estación de Fotobiología Playa Unión. Ozone concentrations data were obtained from NASA (http://jwocky.gsfc.nasa.gov). Note the low ozone values during springtime (*inside ovals*).

throughout the upper eulittoral to the supralittoral zone in shallow rock pools. Using a pulse amplitude modulated fluorometer (PAM, 2000, Walz, Effeltrich, Germany), it was found that the nonphotochemical quenching (qN) rises with increasing irradiances of actinic light starting as low as 10 W m^{-2}, whereas the photochemical quenching (qP) decreases antagonistically (Häder et al., 2000). This physiological response has also been found in higher plants (Niyogy et al., 1998). When exposed to solar radiation for 15 min during low tide and at solar noon on a bright day, the photosynthetic yield (*Y*) decreased to about 50% of its value in dark-adapted plants, but recovered rapidly within 30 min in dim light. When solar UV-B radiation was excluded using filter foils (Montagefolie, Folex, Dreieich, Germany), *Y* was significantly less impaired.

The Chlorophyte *Enteromorpha linza* occurs in the same habitat, but during low tide, it is exposed on the rocky surface as it rarely grows inside the rock pools. When exposed to solar radiation for 15 min during low tide and at solar noon, the decrease in *Y* is even more pronounced than in *U. rigida* and decreases to ~0.2 from the initial dark-adapted value of 0.7 (Häder et al., 2001a). Cutting off the UV-B wavelength band resulted in a less pronounced reduction in *Y* and cutting off the total UV band (using Ultraphan UV Opak filter, Digefra, Munich, Germany) caused an even less pronounced inhibition. The effects of the UV-B or total UV components of solar radiation were still visible throughout the recovery period in dim light. In the experiments described here, the thalli were confined to a flow-through holder and constantly exposed to solar radiation. However, when exposed free floating in a mesocosm, there was a decrease in *Y* by only ~35% during clear days and less pronounced on cloudy days; but in any case, the effects of especially the UV-B band were noticeable.

The filamentous Rhodophytes *Ceramium* sp. and *Callithamnion gaudichaudii* were found in the lower eulittoral inside rock pools. Both were strongly affected by solar radiation (Häder et al., 2004). Both UV-A and UV-B had pronounced effects compared with that of PAR (Fig. 2). Recovery was much slower than in the Chlorophytes and the *Y* was back to the dark-adapted value only during the night. However, it is interesting to note that the increase in qN and the decrease in qP started at much higher irradiances than in the Chlorophytes (about 50 W m^{-2}). Similar results were found in the Rhodophyte *Porphyra columbina*, which grew on the shaded sides of the rock pools in the lower eulittoral.

Two growth forms of *Corallina officinalis* were found in the middle and low eulittoral (only accessible during very low tides), which differed in their morphology and calcification so that the skeleton of the low-eulittoral *Corallina* was less calcified than that in the mid-eulittoral algae (45% (w/w) and 49% of the total dry weight (DW), respectively (Richter et al., 2006)). Moreover, it was found that their photosynthetic parameters were different as well (Häder et al., 2003). The induction curves with quenching analysis showed a faster decrease in the current and maximal fluorescence (Ft and Fm) in the low eulittoral strain compared with the mid eulittoral growth form. Simultaneously, qN rose much faster and higher in the low eulittoral strain (Fig. 3).

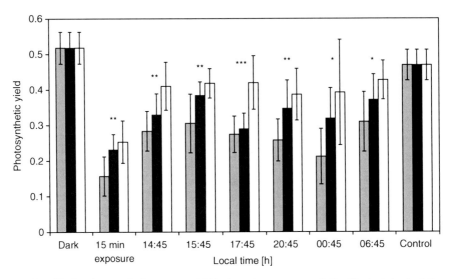

Figure 2. Effective photosynthetic quantum yield in *Ceramium* sp. measured after 30 min dark adaptation, 15 min exposure, and after increasing recovery times in the shade. *Gray bars*, specimens exposed to unfiltered solar radiation. *Black bars*, specimens exposed to UV-A and PAR. *Open bars*, specimens exposed to PAR only. (After Häder et al., 2004.)

The Phaeophyte *Dictyota dichotoma* was found in rock pools in the mid eulittoral. After exposure to 15 min of solar radiation in a fixed position, the effective Y decreased dramatically and did not fully recover until the next morning (Häder et al., 2001b). Free floating thalli were not affected as much as those in a fixed position.

Overall, there is a wide range of photosynthetic responses of Patagonian macroalgae to solar radiation. All results from Patagonia and other coasts confirm that the sensitivity to solar radiation increases with their depth of growth (Häder, 1997). It can be discussed whether this is due to the fact that more sensitive species select a habitat lower in the water column or whether resistance increased with higher exposure to solar radiation. In any case, macroalgae have developed a number of protective mechanisms against excessive solar radiation. In addition to UV-absorbing compounds (see Section 5), most macroalgae use an effective repair mechanism for damaged DNA and proteins in the photosynthetic apparatus. To prove that the D1 protein in the reaction center of photosystem II is resynthesized after photodamage and proteolysis, streptomycin or chloramphenicol were applied during recovery of several macroalgae (*Ulva*, *Porphyra*, *Dictyota*). Both delayed the recovery indicating that the D1 protein resynthesis was inhibited (Häder et al., 2002). Several macroalgae groups use the xanthophyll cycle to dispose of excessive radiation by thermal dissipation (Niyogi et al., 1998). Dithiothreitol is an inhibitor of the violaxanthin de-epoxidase, so when administered to the same algae it affected both photoinhibition and recovery (Häder et al., 2002). However, this was also found in the red algae, which are believed not

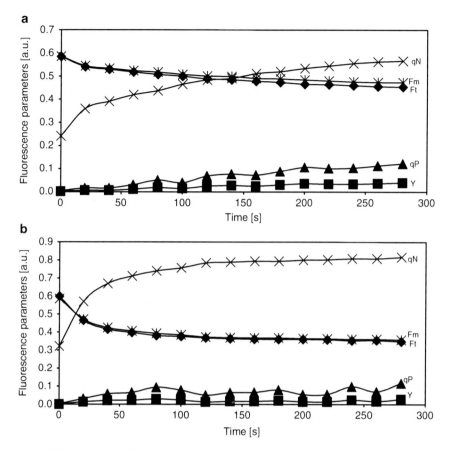

Figure 3. Induction curve with quenching analysis in the mid-eulittoral strain (a) and the low eulittoral strain (b) of *Corallina officinalis*. (After Häder et al., 2003.)

to possess the xanthophyll cycle. This indicates that the inhibitor should be used with caution since it seems to have strong side effects.

5. Presence and Dynamics of UVR-Absorbing Compounds in Patagonian Macroalgae

Macroalgae constitute a source of UV-absorbing compounds, typically mycosporine-like amino acids (MAAs). These compounds are known to protect organisms against UVR stress because of their ability to absorb short wavelengths, but other ecophysiological functions such as protectors against desiccation or as osmotic regulators, antioxidants, and even as accessory pigments have been reported (Korbee Peinado et al., 2006). Therefore, the capacity of synthesizing and accumulating these compounds would provide an adaptive advantage for organisms exposed to different ambient stressors. This is especially important for Patagonian macroalgae that are subjected to high radiation levels and suffer desiccation, especially during low tides and at summer time. Studies dealing with these compounds in Patagonian macroalgae have focused on two main aspects: (a) to determine their presence and abundance in key species of the community and (b) to assess their dynamics throughout daily cycles and considering the tide effects. These studies are particularly important in the context of climate change, as they give insight into the capacity of different algae to cope with increasing solar radiation.

Specific studies were carried out with the two strains of the Rhodophyte *Corallina officinalis* (i.e., low- and mid-eulittoral forms). High-performance liquid chromatography (HPLC) analysis indicated the presence of two MAAs, shinorine and palythine, with the absolute concentration of the latter being about tenfold higher than that of the former (Fig. 4). The amount of MAAs in low-eulittoral samples was significantly lower than that in the mid-eulittoral strain. Significant diurnal changes in the MAAs concentration and in the ratio between shinorine and palythine of the low-eulittoral *Corallina* algae were also observed (Richter et al., 2006): Both MAAs concentration increased around local noon, but the ratio between shinorine and palythine decreased during midday owing to a higher increase in palythine over shinorine (Fig. 4). In the afternoon, the MAAs concentration decreased again. In the mid-eulittoral strain, MAAs dynamics showed an opposite pattern so that around noon the palythine concentration decreased compared with shinorine, whereas in the afternoon the palythine concentration tended to increase. Although the data indicate a strong influence of solar radiation on MAAs synthesis in *C. officinalis*, it is still an open question, whether endogenous circadian or circatidal rhythms are also involved in this process.

Concentration of UV-absorbing compounds and photosynthetic pigments as a function of different radiation treatments throughout daily cycles were done with the Rhodophyte *Porphyra columbina*. Five MAAs were identified: mycosporine–glycine, shinorine, porphyra-334, palythine, and asterina. Porphyra-334 was the most abundant MAA (~80% of the total concentration) and it was always present regardless of the conditions under which the algae were exposed. Shinorine was also present in high concentrations (~20%), whereas the remaining MAAs occurred at much lower concentrations. UV-absorbing compounds in *P. columbina* generally decreased throughout the daily cycles in the two radiation treatments implemented (i.e., PAR+UVR and PAR only) but, in contrast to *Corallina officinalis*, higher values were determined at night; also, and in general, slightly lower values at the end of the experiment were determined in samples exposed only to PAR. The concentration of photosynthetic pigments, on the other hand, remained low throughout the experiment. Results from ammonium-enrichment experiments on the synthesis of MAAs and photosynthetic pigments (Korbee Peinado et al., 2004) showed no significant increase in the concentration of MAAs during a 6-day exposure at concentrations of 0 and 50 mM NH_4^+. On the other hand, samples

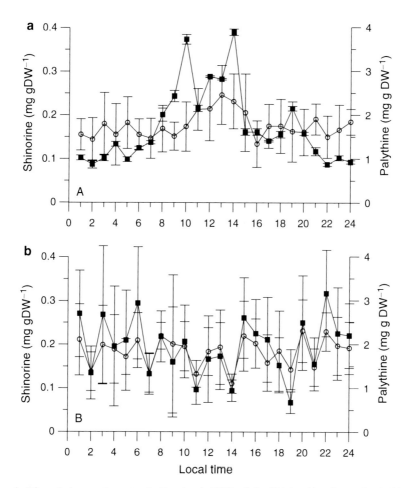

Figure 4. Diurnal changes in concentration (mg/g DW) of the MAAs shinorine and palythine in *Corallina officinalis* growing in the low eulittoral (**a**) and in the mid-eulittoral (**b**) zones. Data are the means and standard deviation of five independent measurements of different samples collected at the corresponding time. Differences between midday low eulittoral samples and morning or evening samples in shinorine are significant ($P < 0.05$) and highly significant ($P < 0.001$) in palythine. Differences between midday samples and morning or evening samples were only significant ($P < 0.05$) for palythine but not for shinorine. (After Helbling et al., 2004.)

grown at 300 mM NH_4^+ had a significant increase compared with the initial value and other treatments at day 6. In addition, and after 3 days of exposure, the content of MAAs was significantly lower in thalli exposed only to PAR compared with treatments receiving additionally UV-A and UVR, indicating a stimulation of MAA synthesis in these treatments.

The daily variations of UV-absorbing compounds in *Ceramium* sp. exposed to full solar radiation followed approximately the daily irradiance cycle, with high concentrations during the day and decreasing in the evening; during the day, their concentration in samples exposed to UVR was significantly higher than in those exposed only to PAR. *Callithamnion gaudichaudii* displayed high variability in the concentration of UV-absorbing compounds in algae exposed to full solar radiation, with significantly higher values during early morning and decreasing during the day.

A comparison of the co-variation of UV-absorbing compounds as a function of chlorophyll *a* in seven macroalgae species is shown in Fig. 5. UV-absorbing compounds had a wide range of responses according to the species: In *C. officinalis* and *P. columbina* exposed to full solar radiation, a significant positive correlation was observed. On the other hand, in *Ceramium* sp and in *C. gaudichaudii* a poor correlation between these compounds was found. Small amounts of UV-absorbing compounds were found in *Ulva rigida*, *Dictyota* sp., and *Enteromorpha linza*. Carotenoids, however, showed a significant positive correlation with chlorophyll *a* in species studied (carotenoids = 0.9 * chl *a*, $R^2 = 0.89$, $P < 0.0001$).

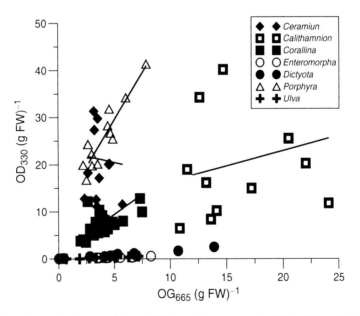

Figure 5. Mean absorption characteristics of UV-absorbing compounds (OD at 330 nm/fresh weight (FW)) of *Ceramium* sp., *Callithamnion gaudichaudii*, *Corallina officinalis*, *Porphyra columbina*, *Enteromorpha linza*, *Dictyota* sp., *and Ulva rigida* exposed to UVR as a function of chl *a* concentration (OD(665 nm)/FW). (After Helbling et al., 2002.)

6. Algae as Food for Other Organisms

Macroalgae, as well as the environment where they grow, offer shelter and food for diverse organisms, mainly invertebrates. Although many studies have been carried out in different locations to evaluate the effects of solar radiation on the interactions between macroalgae and other organisms, very few studies have addressed this topic in the Patagonia area (Menchi, 2001; Helbling et al., 2002). Particularly, these studies evaluated the impact of UVR on the survival of the amphipod *Ampithoe valida* and the isopod *Idothea baltica* feeding on different macroalgae and differentially bioaccumulating UV-absorbing compounds. The relationship between the optical density at 334 nm (i.e., an estimator of the concentration of UV-absorbing compounds) in the crustaceans and that of their diets is shown in Fig. 6. In *I. baltica* (Fig. 6a), there was

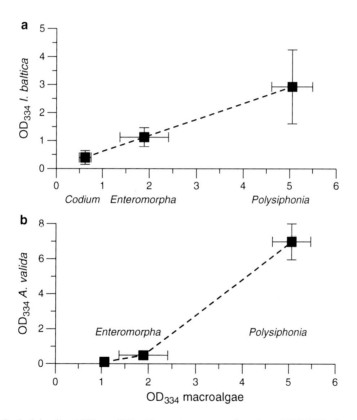

Figure 6. Optical density at 334 nm (OD_{334}) in crustaceans as a function of OD334 in the macroalgae diet. (**a**) *I. baltica*; (**b**) *A. valida*. The symbols indicate the different algae used in this study: *Codium* sp. (▲) collected during February 2001; *Enteromorpha* sp. (●) collected during February and June 2000 and February 2001, and *Polysiphonia* sp. (■) collected during February 2000. The *vertical* and *horizontal lines* are the standard deviation.

a significantly higher concentration of UV-absorbing compounds when individuals were feeding on the Rhodophyte *Polysiphonia* sp. than when they were feeding on Chlorophytes. In *A. valida* (Fig. 6b), there was also an increase in the optical density at 334 nm, being low when the organisms were feeding on *Enteromorpha* sp. and significantly higher when they were feeding on *Polysiphonia* sp. Moreover, a higher concentration of UV-absorbing compounds was found in *A. valida* compared with that in *I. baltica* when feeding on *Polysiphonia* sp. This situation, however, was reversed when the two crustacean species were collected from Chlorophyte species. Survival experiments carried out with both species of crustaceans indicated a different ecological role of these compounds. In *A. valida*, and since a significant higher survival was observed when organisms were feeding on Rhodophytes compared with Chlorophytes, MAAs seem to provide an effective protection against UV-B radiation. In *I. baltica*, however, mortality was high and not significantly different in individuals feeding on rich and poor MAA diets. However, high amounts of MAAs in eggs/embryos of *I. baltica* suggested that these compounds might provide protection to the progeny rather than to adults.

7. Conclusions

The results of the in situ experiments summarized above indicate that the studied macroalgae are shade plants adapted to low light conditions during high tide favored by strong absorption and scattering of solar radiation in the water column. However, during low tide, organisms are damaged by high solar radiation exposure. Any further increase in solar UVR – for example, due to the continue decrease of the stratospheric ozone layer or the extent of influence of the Antarctic ozone 'hole' over Patagonia – would worsen this situation, leading to more inhibition of the algae. However, and so far, the studies have shown that the thalli protect themselves by actively shutting down the photosynthetic electron transport to recover during the subsequent low light phase. It is obvious that different species are adapted to different heights on the coast, and it can be concluded that the duration and intensity of solar radiation is a decisive factor in the habitat zonation of macroalgae in the Patagonian region.

8. Acknowledgments

This work was supported by Agencia Nacional de Promoción Científica y Tecnológica – ANPCyT (Project PICT N° 2005-32034 to VEV), Proalar (Project N° 2000-104 to EWH), the United Nations Global Environmental Fund (PNUD Project N° B-C-39 to EWH), Fundación Antorchas (Project A-13955/3 to EWH), the Deutsche Akademische Austauschdienst (Project Proalar N° T332 408 138 415-RA to D.-P.H), and Fundación Playa Unión. This is contribution N° 114 of Estación de Fotobiología Playa Unión.

9. References

Atkinson, R.J., Matthews, W.A., Newman, P.A. and Plumb, R.A. (1989) Evidence of the mid-latitude impact of Antarctic ozone depletion. Nature **340**: 290–294.
Blumthaler, M. and Webb, A.R. (2003) UVR climatology, In: E.W. Helbling and H.E. Zagarese (eds.) *UV Effects in Aquatic Organisms and Ecosystems*. The Royal Society of Chemistry, Cambridge, pp. 21–58.
Boraso de Zaixso, A. (1995) Algas bentónicas de Puerto Deseado (Santa Cruz), Composición de la flora luego de la erupción del volcán Hudson. Nat Patagon Cienc Biol. **3**: 129–152.
Boraso de Zaixso, A., Cianca, M. and Cerezo, A.S. (1998) The seaweed resources of Argentina, In: A.T. Critchley and M. Ohno (eds.) *Seaweed Resources of the World*. Japan International Cooperation Agency, Tokyo, pp. 372–384.
Boraso, A. and Zaixso, J.M. (2008) Algas marinas bentónicas, In: D. Boltovskoy (ed.) *Atlas de sensibilidad ambiental de la costa y el Mar Argentino*. Secretaría de Ambiente y Desarrollo Sustentable, República Argentina.
Casas, G.N. and Piriz, M.L. (1996) Surveys of *Undaria pinnatifida* (Laminariales, Phaeophyta) in Golfo Nuevo, Argentina. Hydrobiologia **326/327**: 213–215.
Casas, G.N., Piriz, M.L. and Parodi, E.R. (2008) Population features of the invasive kelp *Undaria pinnatifida* (Phaeophyceae: Laminariales) in Nuevo Gulf (Patagonia, Argentina). J. Mar. Biol. Assoc. UK **88**: 21–28.
Díaz, S.B., Frederick, J.E., Lucas, T., Booth, C.R. and Smolskaia, I. (1996) Solar ultraviolet irradiance at Tierra del Fuego: comparison of measurements and calculations over full annual cycle. Geophys. Res. Lett. **23**: 355–358.
Dring, M.J., Wagner, A., Boeskov, J. and Lüning, K. (1996) Sensitivity of intertidal and subtidal red algae to UVA and UVB radiation, as monitored by chlorophyll fluorescence measurements: influence of collection depth and season, and length of irradiation. Eur. J. Phycol. **31**: 293–302.
Franklin, L.A. and Forster, R.M. (1997) The changing irradiance environment: consequences for marine macrophyte physiology, productivity and ecology. Eur. J. Phycol. **32**: 207–232.
Frederick, J.E., Soulen, P.F., Diaz, S.B., Smolskaia, I., Booth, C.R., Lucas, T. and Neuschuler, D. (1993) Solar ultraviolet irradiance observed from Southern Argentina: September 1990 to March 1991. J. Geophys. Res. **98**: 8891–8897.
Häder, D.P. (1997) Penetration and effects of solar UV-B on phytoplankton and macroalgae. Plant Ecol **128**: 4–13.
Häder, D.P., Lebert, M. and Helbling, E.W. (2000) Photosynthetic performance of the chlorophyte *Ulva rigida* measured in Patagonia on site. Recent Res. Dev. Photochem. Photobiol. **4**: 259–269.
Häder, D.P., Lebert, M. and Helbling, E.W. (2001a) Effects of solar radiation on the Patagonian macroalgae *Enteromorpha linza* (L.) J. Agardh – Chlorophyceae. J. Photochem. Photobiol. B Biol. **62**: 43–54.
Häder, D.P., Lebert, M. and Helbling, E.W. (2001b) Photosynthetic performance of marine macroalgae measured in Patagonia on site. Trends Photochem. Photobiol. **8**: 145–152.
Häder, D.P., Lebert, M., Sinha, R.P., Barbieri, E.S. and Helbling, E.W. (2002) Role of protective and repair mechanisms in the inhibition of photosynthesis in marine macroalgae. Photochem. Photobiol. Sci. **1**: 809–814.
Häder, D.P., Lebert, M. and Helbling, E.W. (2003) Effects of solar radiation on the Patagonian Rhodophyte *Corallina officinalis* (L.). Photosynth. Res. **78**: 119–132.
Häder, D.P., Lebert, M. and Helbling, E.W. (2004) Variable fluorescence parameters in the filamentous Patagonian Rhodophytes, *Callithamnium gaudichaudii* and *Ceramium* sp. under solar radiation. J. Photochem. Photobiol. B Biol. **73**: 87–99.
Häder, D.P., Kumar, H.D., Smith, R.C. and Worrest, R.C. (2007) Effects of solar UV radiation on aquatic ecosystems and interactions with climate change. Photochem. Photobiol. Sci. **6**: 267–285.
Hanelt, D. (1998) Capability of dynamic photoinhibition in Arctic macroalgae is related to their depth distribution. Mar. Biol. **131**: 361–369.

Hanelt, D., Melchersmann, B., Wiencke, C. and Nultsch, W. (1997) Effects of high light stress on photosynthesis of polar macroalgae in relation to depth distribution. Mar. Ecol. Prog. Ser. **149**: 255–266.

Hargreaves, B.R. (2003) Water column optics and penetration of UVR, In: E.W. Helbling and H.E. Zagarese (eds.) *UV Effects in Aquatic Organisms and Ecosystems*. The Royal Society of Chemistry, Cambridge, pp. 59–105.

Helbling, E.W., Menchi, C.F. and Villafañe, V.E. (2002) Bioaccumulation and role of UV-absorbing compounds in two marine crustacean species from Patagonia, Argentina. Photochem. Photobiol. Sci. **1**: 820–825.

Helbling, E.W., Barbieri, E.S., Sinha, R.P., Villafañe, V.E. and Häder, D.P. (2004) Dynamics of potentially protective compounds in Rhodophyta species from Patagonia (Argentina) exposed to solar radiation. J. Photochem. Photobiol. B: Biol. **75**: 63–71.

Helbling, E.W., Barbieri, E.S., Marcoval, M.A., Gonçalves, R.J. and Villafañe, V.E. (2005) Impact of solar ultraviolet radiation on marine phytoplankton of Patagonia, Argentina. Photochem. Photobiol. **81**: 807–818.

Holm-Hansen, O., Lubin, D. and Helbling, E.W. (1993) Ultraviolet radiation and its effects on organisms in aquatic environments, In: A.R. Young, L.O. Björn, J. Moan and W. Nultsch (eds.) *Environmental UV Photobiology*. Plenum, New York, pp. 379–425.

Kirchhoff, V.W.J.H., Schuch, N.J., Pinheiro, D.K. and Harris, J.M. (1996) Evidence for an ozone hole perturbation at 30° south. Atmos. Environ. **30**: 1481–1488.

Korbee Peinado, N., Abdala Díaz, R.T., Figueroa, F.L. and Helbling, E.W. (2004) Ammonium and UV radiation stimulate the accumulation of mycosporine like amino acids in *Porphyra columbina* (Rhodophyta) from Patagonia, Argentina. J. Phycol. **40**: 248–259.

Korbee Peinado, N., Figueroa, F.L. and Aguilera, J. (2006) Acumulación de aminoácidos tipo micosporina (MAAs): biosíntesis, fotocontrol y funciones ecofisiológicas. Rev. Chil. Hist. Nat. **79**: 119–132.

Madronich, S. (1993) The atmosphere and UV-B radiation at ground level, In: A.R. Young, L.O. Björn, J. Moan and W. Nultsch (eds.) *Environmental UV Photobiology*. Plenum Press, New York, pp. 1–39.

Martin, J.P. and Cuevas, J.M. (2006) First record of *Undaria pinnatifida* (Laminariales, Phaeophyta) in Southern Patagonia, Argentina. Biol. Inv. **8**: 1399–1402.

Menchi, C.F. (2001) Bioacumulación de compuestos potencialmente protectores de la radiación ultravioleta (RUV) en crustáceos herbívoros del mesolitoral, Puerto Madryn, Chubut, Argentina.

Molina, M.J. and Molina, L.T. (1992) Stratospheric ozone, In: D.A. Dunnette and R.J. O'Brien (eds.) *The science of global change: The impact of human activities on the environment*. American Chemistry Society, Washington DC, pp. 24–35.

Niyogi, K.K., Grossman, A.R. and Björkman, O. (1998) *Arabidopsis* mutants define a central role for the xanthophyll cycle in the regulation of photosynthetic energy conversion. Plant Cell. **10**: 1121–1134.

Orce, V.L. and Helbling, E.W. (1997) Latitudinal UVR-PAR measurements in Argentina: extent of the "ozone hole". Global Planet. Change **15**: 113–121.

Piriz, M.L., Eyras, M.C. and Rostagno, C.M. (2003) Changes in biomass and botanical composition of beach-cast seaweeds in a disturbed coastal area from Argentine Patagonia. J. Appl. Phycol. **15**: 67–74.

Richter, P., Gonçalves, R.J., Marcoval, M.A., Helbling, E.W. and Häder, D.P. (2006) Diurnal changes in the composition of Mycosporine-like Amino Acids (MAA) in *Corallina officinalis*. Trends Photochem. Photobiol. **11**: 33–44.

Siegenthaler, U. and Sarmiento, J.L. (1993) Atmospheric carbon dioxide and the ocean. Nature **365**: 119–125.

Villafañe, V.E., Barbieri, E.S. and Helbling, E.W. (2004) Annual patterns of ultraviolet radiation effects on temperate marine phytoplankton off Patagonia, Argentina. J. Plankton Res. **26**: 167–174.

PART 5:
BIOFUEL – SEAWEEDS AS A SOURCE OF FUTURE ENERGY

Notoya
Rhodes

Biodata of **Masahiro Notoya**, author of *"Production of Biofuel by Macroalgae with Preservation of Marine Resources and Environment"*

Masahiro Notoya is currently a Professor in the Laboratory of Applied Phycology in the Tokyo University of Marine Science and Technology, Tokyo, Japan. He obtained his Ph.D. from Hokkaido University in 1978. Professor Notoya's scientific interests are in the areas of ecology of macroalgal and seagrass communities, i.e., "Moba ecology," integrated multitrophic aquaculture, algal bioremediation, biotechnology of useful algae, algal breeding technology, taxonomy, phylogeny and physiology of algae, and the life history of *Porphyra*.

E-mail: **notoya@kaiyodai.ac.jp**

PRODUCTION OF BIOFUEL BY MACROALGAE WITH PRESERVATION OF MARINE RESOURCES AND ENVIRONMENT

MASAHIRO NOTOYA
Notoya Research Institute of Applied Phycology, Mukojima-4, Sumida-ku, Tokyo 131-8505, Japan

1. Introduction

Biofuel production and environment are issues of concern in the world. First, the author describes the real needs of biofuel, and what kind of materials can serve this purpose. This is followed by the argument that under the present global circumstances, macroalgae are the most effective raw material for biofuel production. Seaweeds are the most important in the marine ecosystem for the preservation of marine bioresources and seawater quality by preventing pollution and eutrophication, and also in the absorption and fixation of CO_2 aided by solar energy. The validity of macroalgae is also explained by various additional useful substances found in their tissues, and by having high productivities compared with terrestrial plants and commercial crops. Algae can be produced in the coast and unused vast ocean area within the exclusive economic zone. Finally, the author's idea for the construction of an effective production system of macroalgae is explained.

2. Environmental Destruction and Bioenergy Production

Threats to human life on a global scale in the near future is thought to include environmental destruction, shortage of drinking water and food, water pollution by the contamination of chemical substances or radioactivity, and energy problems. Most of these problems are due to unjustified destruction of natural environments, and they originated by spendthrift economy of mass production/consumption. It is also considered that a key factor of present global warming is CO_2 emissions and other greenhouse gases emitted artificially by the excessive use of fossil fuel (fourth IPCC report). The need of energy production (bioenergy, physical energy from solar light, wind) with environmental conservation approaches without discharging CO_2 is required. Therefore, in biofuel production, the technology using a lot of energy and discharging a lot of waste for producing the raw materials and for the conversion process for fuel is not suitable. Furthermore, neither the technology of changing food into energy nor the technology that uses a life place should be used.

3. Biofuel Production and Global Environment

Recent increases in atmospheric CO_2 levels are also caused by the anthropogenic environmental destruction, such as the excessive consumption of fossil fuels, deforestation, and development of farmland. Thus, to enhance the accumulation of CO_2 in a forest, stopping deforestation and developing farmland have been recommended among immediate measures to be taken in every corner of the world, until now. However, recently measures are moving to control the consumption of the fossil fuel and production of carbon-neutral energy.

Wood, weeds, corn, sugarcane, palm, sunflower, and rapeseed have all been evaluated as raw materials for carbon-neutral energy, such as alcohol or diesel engine oil. Physical energies have also been considered, such as solar, wind, geothermal, tidal, and current power. However, the energy for transportation that can replace petroleum should be liquid, or gas fuel. Land crop resources for carbon-neutral energy have also been used. However, land comprises only about 30% of the surface of the earth, and this includes mountains, deserts, and areas close to lakes and rivers; and besides using it as a region of economic activity, land also serves as a human being's region of livelihood, such as the city, farmland, and pasture. There was a feeling that not enough area has been allotted for the production of biofuel resources. Moreover, the shortage of food material in the world at present should be taken into consideration as well as the rapid increase in global population in the near future. Therefore, using up land space for the production of biofuel resources is considered a problem given the expected food crisis in the near future; thus all land space should be solely used for food production.

On the basis of the above-mentioned facts, production of biofuel resources should use marine plants rather than terrestrial plants. Especially in Japan, the small islands with a large exclusive economic zone require the development of technologies for large-scale culture of macroalgae on the coastal and offshore areas, such that the production of biofuel from macroalgae does not compete with that of food and does not destroy the environment. From our experimental trials, it was estimated that the annual bio-ethanol production was 20 million kiloliters from 10,000 ha (or 100 km^2). This corresponds to about one third the amount of petrol used annually in Japan. Our project has estimated that a production of biohydrogen of about 4.7 m^3/t wet weight of *Ecklonia stolonifera* Okamura is also possible.

4. Necessity of Preservation of the Coastal Environment and the Marine Bioresources

Generally, production of marine bioresources in coastal areas is greatly influenced by the ocean current and water temperature. Macroalgae grow well in coasts having lower temperatures generated by cold currents. Good fishing areas are formed at the boundaries of warm and cold currents.

In the coasts of Japan affected by cold currents such as those in the Pacific, from the north-east of Tohoku to Hokkaido, large amounts of brown macroalgae like *Laminaria* spp. can be found. Along the Pacific coast of Japan, the cold current flows from northeast to south and warm current flows from south to north. Both currents collide at the offing of the Tohoku region, and both make whirlpools where good fishing areas are found. However, the fish resources are continuously decreasing every year from these fishing areas. Moreover, on a global scale, large-sized wide-ranging fish resources are decreasing because of artificial destruction of the environment from various origins. In a certain shocking report, it has been predicted that the natural fish resources in the seas will almost disappear by 2050, addressing already exterminated fishes as well (Worm et al., 2006). On a global scale, it is shown that the coast and the ocean as well as the land and atmosphere are exposed to environmental destruction, and the preservation of the coastal and marine environment is needed.

About 48% of the coastline of Japan has been modified artificially until now, and the natural seashore and useful large macroalgal communities for fisheries have been destroyed by the construction of shore-protecting artificial seawalls from the viewpoint of seashore preservation (Ministry of the Environment, Government of Japan 1994). Generally, various algal species grow in this shallow coastal area, especially communities of large macroalgae such as *Laminaria* spp., *Sargassum* spp., and *Ecklonia* spp., together with sea-grass developing in the sand and muddy shallow bottoms. These areas are used as the spawning ground of fishes and shellfishes, the growing area of larval fish, and the feeding area of large-sized fish. Thus, this is an important area for the preservation of the environment as well as of useful animals and alga resources for fishery. Moreover, these areas in which macroalgae and sea-grasses grow also serve for water purification. Therefore, the shallow coastal area and the communities of marine organisms are important, as they receive the benefits of ecosystem services and environmental preservation.

In Japan, these useful areas of macroalgal communities and ecosystems for fisheries production have been specially called "Moba" since ancient times (Fig. 1). The definition of "Moba" in Japanese is "a useful and important economic area and/or ecosystem dominated by over-sized-seaweed and sea-grass communities in which marine resources are produced and preserved, and the environment has balanced functioning." "Moba" has been protected, managed, preserved, restored, and developed. It has also been advanced by the public works of country and local government. Besides the "Moba," there are various kinds of high productive areas and important ecosystems in tidal areas. There are various organisms involved in

Figure 1. "Moba" of *Sargassum macrocarpum* at a depth of about 6 m, and the algal frond length reaching 10 m, at Toyoda port, Nakanoshima Island, Oki Islands, Shimane Prefecture, Japan.

water quality purification in tidal flats, sandy regions, and salt marshes. However, in recent years these "Moba" and other natural tidal areas are greatly decreased by the coast development, and as a part of the project of "natural reproduction." In addition, the "Isoyake" phenomenon, which is covered with crustose coralline algae and the disappearing communities of the useful macroalgae, is spreading and progressing in the shallow waters along the coasts of Japan. There have also been eutrophication and pollution due to outflows of wastes from areas of fish and shellfish culture, and various economic activities of human beings along the coastal area. Occurrence of blooms of various photosynthetic organisms has also been observed. From the above-mentioned facts, invasion of the environment and its destruction can be seen from the near shore shallow waters to the offing, and we need to take immediate conservation measures both locally and globally. The restoration and preservation of seaweed communities, such as a "Moba" as a foundation of the coastal ecosystem, are needed.

5. Productivity of the Macroalgae Compared with Other Photosynthetic Organisms

The life cycles and major accumulation of carbon in most macroalgae is relatively short, taking about 1 year, or a few years. It is very short compared with tens to hundreds of years it takes for a tree in a forest to reach maturity.

Therefore, macroalgae, in which a lot of CO_2 accumulates in a short time with high productive capacity, are more effective as a recycling resource for fuel than wood, in which CO_2 accumulation and holding takes a much longer time. On the other hand, it is considered that the global production by the phytoplankton of ocean and terrestrial plants is almost equivalent. And the natural production by the macroalgae does not compare to it at all. However, the macroalgal production is limited by only the narrow attached area on the coastal line.

Yearly net production of macroalgae from the coast of Japan was reported to be 1.3 kg/m² for *Laminaria angutata* (Fuji and Kawamura, 1970), 8.3 kg/m² for *Sargassum macrocarpum*, 5.5 kg/m² for *Sargassum patensi* (Taniguchi and Yamada, 1978), 3 kg/m² for *Ecklonia cava* (Yokohama et al., 1987), and 1.9 kg/m² for *Sargassum yezoensis* (Okamura, 2003). It was estimated that the terrestrial crop plants of the average productivity was approximately 2.3–10 kg/m². Therefore, the productivity of macroalgae and terrestrial crop plants is similar. If it is possible to culture the macroalgae on the unused vast marine area for the biofuel resources, there is no bigger production of photosynthetic organisms of terrestrial crop plants or marine plants. On the other hand, with easy techniques and management of culture, the use of unicellular algae on land as a biofuel resource may also be considered. However, the filtration cost and cultivation area for the same amount of harvest of *Laminaria* or *Sargassum* are not comparatively equivalent. From the above-mentioned facts, since macroalgae (such as *Laminaria* spp. or *Sargassum* spp.) have high productivities and the growing periods are short (from 1 to few years), they should be considered as optimal resources for biofuel production.

6. *Sargassum* spp. and *Laminaria* spp. Algae as Raw Materials of Biofuel

Sargassum spp. grows in the shape of a tree with the help of a holdfast, and since the branches float on air vesicles, solar energy can be used efficiently. *Laminaria* spp. is heavier than seawater, and the leaf-like frond grows downward, or lies on the water surface, and more information is needed to devise good culture techniques for their optimal growth. The floating fronds of the *Sargassum* spp. are convenient, as they could be harvested just by separating their holdfasts from the culture system, which is very easy compared with the *Laminaria* spp.

Most species of *Sargassum* are grown in regions that have comparatively warm currents. Large communities are built on coastal areas, thus allowing the growth of useful fish and shellfishes as well as maintenance and preservation of marine resources (i.e., "Moba" are formed). Moreover, a few of the branches are floated out and they grow, and many fronds are accumulated and move with the current as the "flowing algae." It is well established that the "flowing algae" are used for spawning and as a habitat for various useful fish and shellfishes, such as yellowtail, Pacific saury, and rock fish, and also as a growing area for larval fish. Moreover, they are used as seeds for coastal fish farming. Therefore, artificial propagation and culture of *Sargassum* spp. is effective for the increase and preservation of marine fish and shellfish resources. In addition, owing to the macroalgae's nutrient uptake during their growth in the growing area, they can also be used for the function of restoring the eutrophication at the coast.

Laminaria species grow along the coasts of Japan, in the regions of Tohoku and Hokkaido, and they produce big quantities of fronds and form large community areas. Among the Laminariales, *Ecklonia*, *Eisenia*, and *Undaria* are grown on the warmer coasts of Honshu, Shikoku, and Kyushu. These species are also part of "Moba" and are useful for important marine resources and for their preservation. Therefore, it is necessary to consider whether *Sargassum* spp. or *Laminaria* spp. should be used according to the environmental characteristics of the coastal area.

Some years ago, the author considered and proposed the construction of an algal biofuel production system, which involved the cultivation of large amounts of macroalgae and the uptake and fixation of CO_2 through solar energy in the vast unused offing ocean area. At the same time, various marine organisms of fish and shellfish would be attracted and preserved, and the marine ecosystem could be formed focusing on growing macroalgae, by which the production of marine resources would be greatly increased. This system is capable of using deep seawater, ocean power, wave power, wind power, and solar light energy. It will operate as a self-reliant energy production and consumption system, in which there are various self-reliant, zero-emission types of floating production that are complex, and can also be used for exploration, collection, and utilization of useful industrial raw materials contained in the sea and the seabed (Fig. 2).

General seaweed cultivation, such as *Laminaria*, *Undaria*, and *Porphyra*, has been performed in very calm regions of the inner bay. However, it is necessary to install a large-sized culture construction in the open sea for the production of biofuel resources. Therefore, new possibilities such as technology for large-sized culture construction and installation, production of seaworthiness facility, and harvesting technology for big amount of products will emerge. Then, the production and harvesting of algal biofuel resources, which do not use a culture facility, may be considered. That is, it involves the development of "artificial flowing algae" of *Sargassum* spp., harvest of grown algae at the points of flow and reach, and their use in biofuel production (Fig. 3).

For example, in the case of the western coast in the Japan Sea, if bits of *Sargassum* spp. branches are stocked in large quantities on the northern part of Kyushu, it can be "artificial flowing algae." These algae are flow along with the Tsushima warm current, and they go north, accumulating and growing. It takes at least around 2 weeks to go from the northernmost end of Kyushu to the entrance of Tsugaru Strait according to the reported movement of Nomura's jellyfish (*Stomolophus nomurai*). The northing route and attainment time of these algae change with the stock point and time. Although some algae drift along the shore in the direction of the flow, each flock of algae concentrates and reaches Tsugaru Strait or Soya Strait, and the "artificial flowing alga" does not disperse in the Japan Sea.

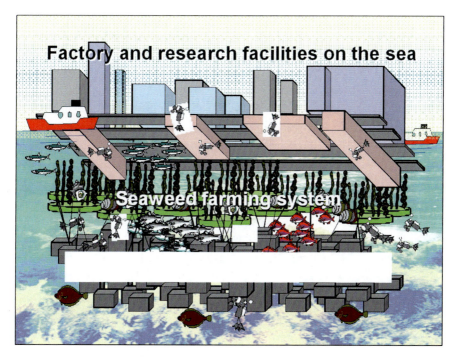

Figure 2. Image of the synthesic construction of a huge raft, self-reliant energy production and consumption, and zero-emission system on the ocean. The construction of each facility for the production, processing, research, as well as management, circulation, and a segment of the economy have been arranged in the upper and lower parts of the float, respectively.

In fact, as far as the route or attainment time is concerned, if the highly precise simulation program developed by the group headed by Professor Yamagata of Tokyo University is used, it could also be said that it is possible to predict the correct position, which changes every moment, and the attainment time, and the course and time can also be specified arbitrarily.

Since all the culture facility and energy on the cultivation management of "artificial flowing alga" depends on natural ocean current and solar light, it is possible to produce low-cost resources for biofuel in large quantities.

In recent years, there have been global problems of the eutrophication and contamination of ocean water. We are also anxious about the same thing in the Japan Sea. Moreover, there are eutrophication problems by the loaded nutrients from the fish and shellfishes culture and from human activities on land, and its roads to blooming, such as red tide and green tide (Fig. 4) in coastal areas of each country.

To take up nutrients from large coastal or marine areas, it is important to cultivate or propagate useful macroalgae in large quantities in those areas, and

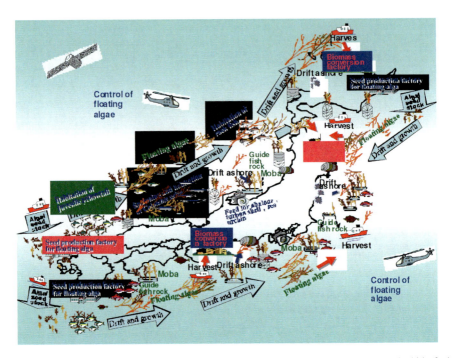

Figure 3. Images of the "artificial flowing alga" and "Moba" development for the macro-algal bio-fuel production with the preservation of global environment and marine resources.

Figure 4. Green Tide blooming of *Ulva* spp. at Park of Sea in Yokohama, Kanagawa Prefecture, Japan.

harvest and utilize them for our life. These macroalgae have also been used as foods since ancient times in some countries. Moreover, the substance is also used in various health foods and supplements, cosmetics, and fertilizers, and phycocolloid has been used as a food additive, recently. In addition, it is also used for integrated multitrophic aquaculture with fish and shellfishes. The author mentions that it is not proper to carry out burial or incineration disposal of the propagated plant. The extensive growing algae on a coast should be effective when utilized as "natural recycled resources." Therefore, these "artificial flowing alga" designs are multipurpose systems, such as a technology for sea water purification with the preservation of useful marine resources and the production of bioenergy resources.

7. Production of Algal Bio-fuel and Use of the Exclusive Economic Zone

The author has already spent about 6–7 years examining the biofuel production from macroalgae, about which several scientific reports were published and various scientific meetings (XIXth International Seaweed Symposium, March 28, 2007, Kobe, Japan and "The Challenges of Seaweed Bio-fuel Production and Preservation of Environment and Fisheries Resources.") were held.

Development of biofuel production technology is fundamental and the preservation of the prospective global environment should be considered. Therefore, as mentioned earlier, biofuel resources and their production area should not be compared with food and the living regions of a human being. Destruction of the environment and its ecosystems should not occur with the process of biofuel production. Natural energy should be used as much as possible, and the resources should be used without futility, at the same time avoiding waste output. The above-mentioned points are required for biofuel production. The type of development in the old capitalist profit-seeking system does not comply with the original meaning of biofuel production.

For that purpose, it may be necessary to establish a new field, for example, "biological energy," which converts steam gasification and synthesises gas with the help of an inorganic catalyst to directly produce useful and good-quality carbide from the fermentation system. As far as the biofuel production from the resources of algal organic matter is concerned, in addition to the former technology of fermentation of methane, alcohol, hydrogen, etc., various synthetic production systems should be developed, such as synthesis of gas by partial combustion. Effective technologies may also be developed using the analogy of pulp, refining of petroleum, or the production process of petrochemicals. About 40% residual substance waste is removed through several types of fermentation. These substances should also be used for the production of the high industrial commodity, ethanol, which is of added value for the substance of various cellulose systems.

In this design, the extensive stable supply of biomass materials is realized with a first premise. The way of thinking and conditions at the unique location of the vast exclusive economic zone of Japan is suited for algal culture and subsequent biofuel production. Views of new, eco-friendly synthetic technology development in global and marine resource preservation will help in Japan's global contribution.

8. References

Ministry of the Environment, Government of Japan (1994) National survey on the natural environment. 1–400 (in Japanese).

Murata, Y. (1980) Photosynthesis and production, In: S. Miyachi and Y. Murata (eds.) *Photosynthesis and Dry Matter Production.* Rikougakusha, Tokyo, pp. 475–510 (in Japanese).

Okamura, D. (2003) *Sargassum yezoensis*, In: M. Notoya (ed.) *Seaweed Marine Forest and Its Developmental Technology.* Seizando-Shoten, Tokyo, pp. 75–81 (in Japanese).

Taniguchi, K. and Yamada, Y. (1978) Ecological study on *Sargassum patens* C. Agardh and *S. serratifolium* C. Agardh in the sublittoral zone at Iida Bay of Noto Peninsula in the Japan Sea. Bull. Jpn. Sea Reg. Fish. Res. Lab. **29**: 239–253.

Taniguchi, K. and Yamada, H. (1988) Annual variation and productivity of *Sargassum horneri* population in Matsushima Bay on the Pacific Coast of Japan Sea. Bull. Tohoku Reg. Fish. Res. Lab. **50**: 59–65 (in Japanese).

Worm, B., Barbier, E.B., Beaumont, N., Duffy, J.E., Folke, C., Halpern, B.S., Jackson, J.B.C., Lotze, H.K., Micheli, F., Palumbi, S.R., Sala, E., Selkoe, K.A., Stachowicz, J.J. and Watson, R. (2006) Impacts of biodiversity loss on ocean ecosystem services. Science **314**: 787–790.

Yokohama, Y., Tanaka, J. and Chihara, M. (1987) Productivity of the *Ecklonia cava* community in a bay of Izu Peninsula on the Pacific Coast of Japan. Bot. Mag. Tokyo **100**: 129–141.

Biodata of **Christopher J. Rhodes**, author of *"Biofuel from Algae: Salvation from Peak Oil?"*

Professor Chris J. Rhodes is currently a visiting professor at the University of Reading and Director of Fresh-Lands consulting. He was awarded a D.Phil from the University of Sussex in 1985 and a D.Sc in 2003. He has wide scientific interests (www.fresh-lands.com) which cover radiation chemistry, catalysis, zeolites, radioisotopes, free radicals, and electron spin resonance spectroscopy, and more recently have developed into aspects of environmental decontamination and the production of artificial fuels. He has published more than 200 peer-reviewed papers and three books.

E-mail: **chris.rhodes@rba.co.uk**

BIOFUEL FROM ALGAE: SALVATION FROM PEAK OIL?

CHRISTOPHER J. RHODES
Fresh-Lands, P.O. Box 2074, Reading, Berkshire, RG4 5ZQ, UK

1. Introduction

There is practically nothing in the modern world that does not depend on the resource of plentiful, cheap oil. The majority of crude oil is refined into fuel for transportation, but it also provides a feedstock for a myriad of industries, producing products ranging from plastics to pharmaceuticals. In total, it is reckoned that worldwide some 86 million barrels of oil are consumed daily, which amounts to just over 31 billion barrels a year. Around one quarter of all oil is used in the USA, which was formerly the world's main oil-producing nation. Now that accolade is with Saudi Arabia, which delivers an almost ten million barrel daily aliquot to the world oil markets, while Russia exhumes an almost equal quantity. In 1999, the price of a barrel of oil was $12, but reached almost $150 in the summer of 2008 preceding a world stock market crash and a fallback to $25 a barrel (Rhodes, 2008). The price rose during the following year and, writing in August 2009, it is now around $70 a barrel. There are many factors held culpable for such frenetic activity in the marketplace, including a seemingly inexorable demand for oil (and indeed all kinds of energy resources) from rising economies such as China and India, a weakening US dollar, and that oil is becoming harder to produce, as a general principle. Over all of this looms the specter of peak oil, which is the point at which production meets a geological maximum, and beyond which it must relentlessly fall. The combination of these factors must culminate in a gap between rising demand and ultimately falling supply. Within a decade or less, the world economies will no longer be able to depend on some limitless growth in oil output. For these reasons, attention is being turned toward "Alternatives," which ideally are also "Renewables," but the issue of biofuels is more complex than is generally realized, and it is at best a partial solution bearing its own attendant environmental costs (Rhodes, 2005).

In addition to the simple fact that growing fuel crops must inevitably compete eventually for limited arable land on which food-crops are to be grown, there are vital differences in the properties of biofuels, e.g., biodiesel and bioethanol, from conventional hydrocarbon fuels such as petrol and diesel, which will necessitate the adaptation of engine designs to use them; for example, in regard to viscosity at low temperatures, e.g., in planes flying in the frigidity of the troposphere. Raw ethanol needs to be burned in a specially adapted engine to recover more of its energy in terms of tank to wheels miles, otherwise it could deliver only about 70% of the energy content of petrol, pound for pound in accordance with its lower enthalpy of combustion (29 MJ/kg) than is typical for an oil-based fuel like petrol (gasoline) or diesel (42 MJ/kg) (Rhodes, 2005).

Of the various means that are being considered, at least in terms of growing our way out of the oil crunch, is making biofuel from algae. There are many advantages claimed as we shall see, but in particular, the quoted yields of oil that might be derived from algae per hectare are high, even when compared with those from high-oil-yielding plants such as jatropha and palm, which translate to around 6 t of diesel per hectare (see Section 6). Most biofuel in Europe is biodiesel and is made, for example, from rapeseed oil, which yields perhaps 1 t of diesel per hectare. In contrast, it is reckoned that some species of high-oil-yielding algae might furnish more than 100 t of diesel per hectare – an attractive prognosis indeed, since an area, say the size of the southern UK, could fuel the entire world (Rhodes, 2005). Algae offer further advantages that they can be used to absorb CO_2 from smokestacks at fossil-fuel-fired power stations, they can be grown on saline waters or wastewaters (cleaning the latter in the process), and furthermore there is no necessity to use arable land for algae production since the tanks they are grown in can be placed anywhere, including brownfield land or on the open ocean. Thus, the competition between fuel and food production is avoided.

The author attempts an overview of some specific aspects of a subject that is, however, not quite as straightforward as it first appears.

2. The Peak Oil Problem

The prediction of peak oil was made in 1956 by Marion King Hubbert (Hubbert, 1956), a geophysicist working for the Shell Development Corporation. Hubbert predicted that the lower 48 states in the USA would peak (hit maximum production) during 1965–1970, depending on his estimate of the total volume of the reserve. At that time, the USA was awash with oil and his prediction was not taken seriously. Hubbert's analysis is based on the logistic function, the first derivative of which gives a peak. Mathematically, this kind of curve can be represented by the logistic differential equation (1).

$$dQ/dt = P = k(1-Q/Q_t)Q \qquad (1)$$

In Eq. (1), P is the production rate, as shown by its equality to the rate of change of cumulative production Q (i.e., the sum quantity of oil recovered from a given source to date), Q_t is the total amount of oil that will ever be recovered from it, and k is the logistic growth rate (a sort of % compound interest). In Hubbert's original paper (Hubbert, 1956), he assumed two scenarios for the lower 48 states in the USA: (1) there were 150 billion barrels worth of oil and (2) there were 200 billion barrels as a total recoverable reserve, Q_t. In those days before computers, he simply reckoned the amount of oil represented by each square on his sheet of graph paper, and drew a curve by hand that enclosed an area equal to the estimated volume of the reserve. For case (1), he predicted that the peak in production would arrive in 1965 and for (2) around 1970. Thus, the method was not predictive of the volume of oil that would be recovered in total; this had to be reckoned first. In fact, US production peaked in 1971, so establishing both his fame and credibility in the basic method. In a later paper (Hubbert, 1982), Hubbert surveyed the mathematics behind all this, from which an alternative and predictive method coined the *Hubbert Linearization* (Deffeyes, 2005) was derived. The basis of this is that Eq. (1), which is a quadratic, can be rewritten in the form of Eq. (2), which is a linear equation of familiar form $y = mx + c$.

$$P/Q = k(1 - Q/Q_t) \qquad (2)$$

By plotting annual production divided by cumulative production (i.e., P/Q) versus cumulative production alone (Q), a straight-line plot is obtained, with a y-axis intercept equal to k and a slope of $-k/Q_t$. Thus, both essential parameters of the logistic peak, k and Q_t, may be estimated without prior assumptions, an improvement on the original approach (Hubbert, 1956). The method is still used extensively in the oil industry, although now with modern PCs, it is easy to fit directly logistic and all other kinds of functions to oil production data using programs like *Origin*. Having established values for k and Q_t, they can be used to construct the logistic curve with considerable accuracy. Because of the symmetry of the curve, when the peak is reached, half the reserve has been extracted, beyond which production falls inexorably. For the entire world, a value for Q_t of around two trillion barrels is estimated, of which we have used close to half. It is expected that once the peak is reached, there will be a decline in world oil production by 3% per annum. This approach is not without its critics, however. Some maintain that it is an oversimplification, and does not allow for future discoveries of oil or the production of unconventional oil and that it is more likely that the postpeak outcome will be a more steady plateau followed by a gradual depletion in supply rather than a mirror-fall of the growth phase. The oil industry actually uses a number of methods of geophysics, e.g., seismic measurements, to estimate the volume of a reserve, and their final predictions are often based on a combination of physical data and various kinds of mathematical and numerical modeling procedures, including "Hubbert."

Various estimates have been given for the time of arrival of peak oil. If all production, ignoring tar sands and natural gas liquids (NGL), is considered, the peak famously reckoned by Kenneth Deffeyes (2005) to arrive on November 24, 2005 (thanksgiving day!), is predicted. If all production including NGL is included, the peak shifts forward to mid-2006. All studies that place peak oil in 2010 and beyond use other methods, but generally consider the rates of depletion of existing oil fields and projections about developing fields. Such studies are termed "bottom-up analysis." Chris Skrebowski, a researcher for the Energy

Institute in Britain, told delegates at a recent oil conference that the world oil supply will peak in 2011 or 2012 at around 93 million barrels a day (it is presently 84 million barrels a day), in line with a general consensus among industry experts there that the peak will arrive by 2012 (Low, 2007). Key pieces of evidence for this include the falling rate of discoveries of new oil fields; the age of the largest fields; geopolitical threats to future oil supplies; and the sustained high price of crude oil. Coincidentally, the CEO of Shell stated earlier this year that he expects to see a gap between supply and demand for oil sometime during 2010–2015. However, the truth is that we will only know retrospectively exactly when peak oil did occur: its effect being to pull down supply while demand continues to rise, thus enlarging the gap from both sides. There is much speculation as to how high oil prices will go. Goldman Sachs predicted in May that we may see the new psychological benchmark $200 barrel within 6–24 months (Rhodes, 2008). In the wake of the record $139 barrel, Morgan Stanley predicted $150 by the beginning of July. This did not happen, in fact, and the price has fallen below $120 a barrel. More alarmingly, the CEO of the Russian Gazprom, Alexey Miller, is talking about a barrel of oil costing $250, which would mean an increase in fuel prices at the pump of 60 p/l and that is before fuel taxes are applied on top of this. Fuel would then cost in excess of £2.00/l (Rhodes, 2008).

3. Conventional Biofuels

Most biofuels produced in Europe are made from plant oils (biodiesel) with a smaller amount of bioethanol that is produced from sugar beet (Duffield et al., 2006). In the USA, the situation is reversed and huge amounts of corn are turned over to the production of "corn ethanol." The ethanol industry in Brazil is mature, as it is made from sugarcane, which grows well there, with the USA as its major customer for exports. As it is not thought that the Brazilian ethanol industry compromises land on which food crops could be otherwise grown, there is a strong objection made with increasing volume to the diversion of corn grown in the USA from the world food markets to making ethanol. Indeed, part of the huge increases in the price of basic staple foods has been blamed on the use of arable land to produce biofuels rather than to grow food (Elgood and Eastham, 2008). There are consequently shortages of rice and wheat, and a significant reduction in the market stockpile of corn, all of which contribute to a potential food crisis particularly in developing nations, including China and India. The yields of biodiesel that can be produced from a hectare of land suitable for different "fuel crops" are shown in Table 2.

4. Biofuel from Algae

There are some truly astounding figures about the amount of biodiesel that might be obtained from farming algae, rather than from growing crops. For example, whereas the yield of biodiesel from soybean is 357 kg/ha/year and 1,000 kg/ha/year from rapeseed, it is 79,832 kg/ha/year from algae, i.e., about 80 t/ha. There are some algae that yield around 50% of their own weight of oil, and from one study it might be deduced that the yield is around 125 t/ha on the basis that 200,000 ha of land could produce 7.5 billion gallons (one quad) of biodiesel (Maio and Wu 2006).

(Since there are 3.875 l to the US gallon, it equals $7.5 \times 10^9 \times 3.875 = 2.91 \times 10^{10}$ l. Since there are 159 l to the barrel and 7.3 barrels to the ton (accepted average), it amounts to $2.91 \times 10^{10}/(159 \times 7.3) = 2.51 \times 10^7$ t of biodiesel produced from 200,000 ha, i.e., $2.51 \times 10^7/200,000 = 125.5$ t/ha).

Some rough calculations are presented to indicate some estimates of scale. In the UK, around 57 million tons of oil are used to run transport – cars, planes, and the whole lot. Another 16 million tons are used as a chemical feedstock for industries etc. However, only the fuel budget is described here. Diesel engines are more efficient in their tank-to-wheel use of fuel than spark-ignition engines, which burn gasoline (petrol), meaning that we could reduce that total by 30%, i.e., to 40 million tons, by switching all forms of transport to run on Diesel "compression" engines. If we take the optimistic 125 t/ha figure for the yield of biodiesel from algae, it implies a crop area of $40 \times 10^6/125 = 320,000$ ha, or 3,200 km^2.

Now this is only 1.3% of the area of the UK mainland, which does look feasible, especially in comparison with values of up to five times the entire area of arable land present that has been deduced, which would be necessary to provide sufficient biofuels derived from land-based crops.

There is no need to use arable land in any case, since the algae would be grown in ponds, and these could be installed essentially anywhere, even in offshore locations, i.e., growing the material on seawater, because salt concentration appears to assist the algal growth.

We can make some guess as to the thickness of the algae too. One hectare = 100×100 m = 10,000 m^2. Hence, 320,000 ha = 3.2×10^9 m^2. The volume of 40 million tons of biodiesel at a specific gravity of 0.84 (based on 79,832 kg = 95,000 l; so, 80 t = 95 m^3) = 4.76×10^7 m^3. Hence, spread over 3.2×10^9 m^2 gives a thickness of 4.76×10^7 m^3/3.2×10^9 m^2 = 0.015 m = 1.5 cm. So, assuming that 50% of it is "oil," it gives a thickness of around 3 cm, which seems reasonable.

How much water would be needed to fill the tanks? Let us assume they are 1-m deep (with the algae floating on top of that), i.e., 3.2×10^9 m$^2 \times 1$ m = 3.2×10^9 m^3. Since this amounts to 3.2 km^3, it is a significant volume of water, and if freshwater would account for about 2% of the UK's total. However, as it has been indicated, seawater can be used instead, or the "ponds" could be fashioned from floating "boon" structures offshore. Closed ponds might be better, since that would permit a much closer control of nutrient supply, and if they were covered, it would restrict the potential for invasion by algae with a lower oil yield.

This might be the only way to provide significant amounts of "oil" post peak-oil (other than by coal liquefaction), and large-scale production should be attempted as soon as possible – well before the world's supply of naturally occurring petroleum begins to wane significantly, which gives us just a few years. The "crop" would take up CO_2 from the atmosphere, thus reducing the burden of greenhouse gas that many are worried about, and that amount of carbon would be re-emitted once the fuel was burned, but with a continual crop production working in symbiosis with the CO_2 content of the atmosphere, taking it up through photosynthesis. There would be no additional CO_2 emitted, other than in the production of an alcohol, methanol, or maybe ethanol, which is needed to transesterify the initial oil into the final biofuel. This would be in a proportion of about 10% of the oil yield, and could be provided from agricultural waste, e.g., wheat grass, some minor compromise of food crops, say to grow sugar beet to ferment into ethanol, and ultimately by hydrolysis and fermentation of cellulosic material once that technology is underway, thought to be by 2015.

What about costs? If we assume a cost per hectare of $80,000, that would equate to $80,000 × 320,000 = $25.6 billion, or around £13 billion. Annual maintenance/operating costs have been estimated at $12,000 per hectare, which is about £2 billion. Assuming a price of $60 a barrel, that may be compared with an annual cost for 40 million tons of oil of $60 × 40 million × 7.3 (barrels/t) = $17.5 billion or about £9 billion. This would mean money that is not going out of the country to unstable regions of the world, and it would break completely UK's dependence on imported oil. It would also reduce the nation's CO_2 emissions by probably 30%. Biodiesel could even be used to substitute for coal in power stations and cut the UK's dependence on coal imports, while reducing CO_2 emissions still further.

5. Chemical Composition of Algae

Algae are made up of eukaryotic cells. These are cells with nuclei and organelles. All algae have plastids, which are chlorophyll-containing species that can perform photosynthesis. However, different algal types have different combinations of chlorophyll molecules. Some have only Chlorophyll A, some A and B, whereas other lines have A and C.

All algae primarily contain proteins, carbohydrates, fats, and nucleic acids, but in varying proportions. As the percentages differ with the type of algae, some algae contain up to 40% of their overall weight in the form of fatty acids. It is this fatty acid component (oil) that can be extracted and converted into biodiesel (Table 1).

The interest in algal oil is not recent, though the widespread interest in making biodiesel from algal oil is. Algae oil is produced for the cosmetic industry, principally from macroalgae (larger-sized algae) such as oarleaf seaweed etc. Most current research on oil extraction from algae is, however, focused on microalgae.

Table 1. Chemical composition of algae expressed on a dry matter basis (%) (Becker, 1994).

Strain	Protein	Carbohydrates	Lipids	Nucleic acid
Scenedesmus obliquus	50–56	10–17	12–14	3–6
Scenedesmus quadricauda	47	–	1.9	–
Scenedesmus dimorphus	8–18	21–52	16–40	–
Chlamydomonas rheinhardii	48	17	21	–
Chlorella vulgaris	51–58	12–17	14–22	4–5
Chlorella pyrenoidosa	57	26	2	–
Spirogyra sp.	6–20	33–64	11–21	–
Dunaliella bioculata	49	4	8	–
Dunaliella salina	57	32	6	–
Euglena gracilis	39–61	14–18	14–20	–
Prymnesium parvum	28–45	25–33	22–38	1–2
Tetraselmis maculata	52	15	3	–
Porphyridium cruentum	28–39	40–57	9–14	–
Spirulina platensis	46–63	8–14	4–9	2–5
Spirulina maxima	60–71	13–16	6–7	3–4.5
Synechoccus sp.	63	15	11	5
Anabaena cylindrica	43–56	25–30	4–7	–

Algal oil is very high in unsaturated fatty acids. Some UFAs found in different algal species include arachidonic acid, eicospentaenoic acid, docosahexenoic acid, gamma-linolenic acid, and linoleic acid.

Table 2. Yield of various plant oils (http://oilgae.com).

Crop	Oil (l/ha)
Castor	1,413
Sunflower	952
Safflower	779
Palm	5,950
Soy	446
Coconut	2,689
Algae	100,000

6. Comparison of Average Oil Yields from Algae with That from Other Oilseeds

Table 2 presents indicative oil yields from various oilseeds and algae. Please note that there are significant variations in yields even within an individual oilseed depending on where it is grown, the specific variety/grade of the plant, etc. Similarly, for algae there are significant variations between oil yields from different strains of algae. The data presented here are indicative in nature, primarily to highlight the order-of-magnitude differences present in the oil yields from algae when compared with other oilseeds (Becker, 1994).

7. Oil Content of Fixed Oils

Cereals only contain about 2% by weight oil, compared with oilseeds that contain much higher levels. The oil content of oilseeds varies widely from one type to the other. It is about 20% in soybeans and as high as 50% in some new Australian varieties of canola (the oil content of canola seed varies from 35% to 50%, and is usually considered to be averaging 40%). Sunflower has one of the highest oil contents among oilseeds – about 55%. Castor seeds have about 45–50% comprising oil. Safflower has about 40% of its contents as oil, and cottonseed has about 20% of its weight as oil. Hemp has 30–35% oil content. Copra, the dried coconut meat, has about 60% (sometimes close to 65%) oil content. Peanuts contain approximately 50% oil on a dry weight basis. Palm kernel has about 50% oil. Corn has only 5–10% of its dry weight as oil. The average oil content of mustard is about 40% – yellow mustard have only about 27% whereas brown mustard have about 36% oil; some oriental mustard have up to 50% oil. Flaxseed has about 45% oil content. For jatropha, the oil content is 35–40% in the seeds and 50–60% in the kernel (Becker 1994) (Table 3).

8. Extraction of Oil from Algae (http://oilgae.com)

Oil extraction from algae is currently a hotly debated topic because this process is one of the more costly features, which can determine the sustainability of algae-based biodiesel. In terms of the concept, the idea is quite simple: Extract the algae from its growth medium (using an appropriate separation process), and use the wet algae to extract the oil (Note: It is not necessary to dry the algae prior to extracting the oil from them). There are three well-known methods to extract the oil from oilseeds, and these methods should apply equally well for algae too:

Table 3. Oil content (% of dry weight) – average values.

Soy	20
Canola/rapeseed	40
Sunflower	55
Castor	45
Safflower	40
Hemp	30
Copra (dry coconut)	60
Peanuts/groundnuts	50
Palm kernel	50
Corn	7
Mustard	40
Flaxseed	45
Jatropha seed	40
Jatropha kernel	55
Algae (for comparison)	15–40

Expeller/Press, Hexane solvent oil extraction, Supercritical Fluid extraction, along with some less familiar procedures which are summarized here.

8.1. EXPELLER PRESS

When algae are dried, they retain the oil content, which then can be "pressed" out with an oil press. Many commercial manufacturers of vegetable oil use a combination of mechanical pressing and chemical solvents in extracting oil. Though more efficient processes are emerging, a simple process is to use a press to extract a large percentage (70–75%) of the oils out of algae.

8.2. SOLVENT EXTRACTION

Algal oil can be extracted using organic solvents. Benzene and ether have been used, but a popular chemical for solvent extraction is hexane, which is relatively inexpensive. The downside to using solvents for oil extraction is the inherent dangers involved in working with particular chemical materials. Benzene is classified as a carcinogen. Chemical solvents also present the problem of causing a potential explosion hazard. Hexane solvent extraction can be used in isolation or it can be used along with the oil press/expeller method. After the oil has been extracted using an expeller, the remaining pulp can be mixed with cyclohexane to extract the remaining oil contents. The oil dissolves in the cyclohexane, and the pulp is filtered out from the solution. The oil and cyclohexane are separated by distillation. These two stages (cold press and hexane solvent) can yield in excess 95% of the total oil present in the algae.

8.3. SUPERCRITICAL FLUID EXTRACTION

This can extract almost 100% of the oils. It does nonetheless require special equipment to contain the working fluid under pressure. When supercritical CO_2 is used, the gas is liquefied under pressure and heated to the point that it has the properties of both a liquid and gas. This liquefied fluid then acts as the solvent in extracting the oil.

8.4. ENZYMATIC EXTRACTION

Enzymatic extraction uses enzymes to degrade the cell walls with water acting as the solvent; this makes fractionation of the oil much easier. The costs of this extraction process are estimated to be much greater than hexane extraction.

8.5. OSMOTIC SHOCK

Osmotic shock is a sudden reduction in osmotic pressure which can cause cells in a solution to rupture. Osmotic shock is sometimes used to release cellular components, in this case oil.

8.6. ULTRASONIC-ASSISTED EXTRACTION

Ultrasonic extraction can greatly accelerate extraction processes. Using an ultrasonic reactor, ultrasonic waves are used to create cavitation bubbles in a solvent material: when these bubbles collapse near the cell walls, they create shock waves and liquid jets that cause the cell walls to break and release their contents into the solvent.

9. Conversion of Algal Oil to Fuel

The underlying chemistry required to convert algal oil to biodiesel can be represented as in Eq. (3), which is a standard transesterification reaction, in which a triglyceride (a long-chain fatty acid ester of glycerol with three fatty acid groups) is reacted with an alcohol such as methanol in the presence of a catalyst to form the fatty acid methyl ester (biodiesel) and free glycerol:

$$\begin{array}{c} CH_2OCOR \\ | \\ CHOCOR \\ | \end{array} + CH_3OH \longrightarrow 3RCOOCH_3 + \begin{array}{c} CH_2OH \\ | \\ CHOH \\ | \end{array} \qquad (3)$$

The same chemical process is involved in the conversion of plant oils, since they are also triglycerides, to biodiesel (Maio and Wu, 2006; Becker, 1994).

10. Oil from Algae: Photosynthetic Efficiencies?

There is much communicated on the prospect of growing oil-rich algae to convert into diesel as a replacement for the inevitably declining supply from crude oil. Now, PetroSun Inc. has announced that their Rio Hondo algae farm in Texas will begin operations at its pilot commercial algae-to-biofuels facility (Fraser, 2008). The farm includes 1,100 acres of salt-water ponds from which it is thought that 4.4 million US gallons of algal oil will be produced along with 110 million pounds of biomass, annually. Expansion of the farm is intended to provide fuel to run existing or putative biodiesel and bioethanol refineries, owned or part-owned by PetroSun. Such an open-pond design effectively consists of a nutrient-loaded/fed aqueous culture medium, in which algae will grow close to the surface, absorbing CO_2 from the atmosphere in the process through photosynthesis – hence the other vital ingredient is sunlight.

Of interest too is the likely photosynthetic efficiency (PSE) of such processes, and of crop-agriculture; therefore some illustrative sums regarding their PSE viability have been attempted:

4.4×10^6 gal (US) of oil = 4.4×10^6 gal × 3.875 l/gal = 1.7×10^7 l.
1.7×10^7 l × 0.84 kg/l = 1.43×10^7 kg = 1.43×10^4 t of oil.

If 110 million pounds of biomass is also produced, this amounts to 110×10^6 lb/2,200 lb/t = 5×10^4 t. Hence, the oil is 22% of the total "mass" produced.

1,100 acres/2.5 ha/acre = 440 ha.

Therefore, the oil yield = 1.43×10^4 t/440 ha = 32.6 t/ha. To "grow" this amount of oil, it would take 7.19×10^8 t/32.6 t/ha = 2.21×10^7 ha.

If the total area of the US is 9.8×10^8 ha (that is land plus water), and the US fuel consumption may be estimated as 0.25 (one quarter of the world's total) × 0.7 (proportion of oil used for transport) × 30×10^9 barrels/7.3 barrels/t = 7.19×10^8 t; to grow enough algae to produce an equivalent amount of algal oil, it would take:

2.21×10^7 ha/9.8×10^8 ha = 2.3% of total US area.

For the entire world, the requisite land area is ×4 that equals 8.84×10^7 ha, to be compared with a total surface area of around 500 million square kilometers (about 147 million square kilometers being land), or <0.2% of it.

If the UK switched all its transportation over to diesel engines, which are more efficient in terms of tank-to-wheel miles than spark-ignition engines that burn petrol (gasoline), we would need 40 million tons of oil (assuming oil = diesel, since the actual yields of diesel from oil are uncertain).

This amounts to 40×10^6 t /32.6 t/ha = 1.23×10^6 ha, which is about 5% of the total area of the UK mainland.

So, what about the photosynthetic efficiencies incurred?

1. A good mean of 5 kWh/m^2/day is taken into consideration. Quite often, values of W/m^2 are quoted over the whole year, which is a bit misleading, since the solar radiation hitting the Earth (insolation) varies from region to region and according to the time of the day and the changing seasons. However, dividing 5 kwh/m^2/day by 24 h gives around 200 W/m^2, to be compared with around 350 W/m^2 hitting the top of the atmosphere, as an annual mean. This would correspond to a sunny clime, e.g., Arizona, Australia, or central Africa; it is probably less than half of this for northern Europe.

Expressing the insolation as kilowatt hour per day is more realistic, however. Hence,
5×10^3 Wh/day × 365 days × 3,600 s/h = 6.57×10^9 W/m^2/year. Since W = J/s, we have:

6.57×10^9 W/m²/year × 10,000 m²/ha = 6.57×10^{13} J/ha/year (i.e., the amount of solar energy falling on each hectare of land).

If we assume that "diesel" uniformly contains 42 GJ/t of energy, each hectare yields:

$$32.6 \times 42 \times 10^9 = 1.37 \times 10^{12} \text{ J}.$$

Thus, the PSE on oil is $1.37 \times 10^{12}/6.57 \times 10^{13} = 2.1\%$.

But only 22% of the total is oil and the rest is (other) "biomass." If we assume that 1 t of biomass contains 15 GJ of energy (about the same as for glucose), we have:

32.6 t × 15 × 10⁹ × (100−22/22) = 1.73×10^{12} J/ha, which gives a PSE of:

$1.73 \times 10^{12}/6.57 \times 10^{13} = 2.6\%$, and so the overall PSE for the PetroSun process is 2.1% + 2.6% = 4.7%.

2. Now this seems quite reasonable, and is in the 6% "ballpark" usually given for the amount of the total solar radiation that is used by plants in photosynthesis, as has been described previously. However, at 32.6 t of oil per hectare, the yield is rather less than the 5,000–20,000 gal/acre quoted in Wikipedia (2009a) for algal oil production.

5,000 gals/acre = 12,350 gals/acre = 12,350 × 3.875 = 47,856 l/ha, and assuming a density of 0.84 t/m², this amounts to 40.2 t/ha, as the lower estimate.

(Assuming, again, a sunny 5 kWh/day or 6.57×10^{13} J/ha/year), if 50% of the algae is oil with an energy content of 42 GJ/t, it contains 40.2 × 42 × 10⁹ = 1.69×10^{12} J, which equates to a PSE of $1.69 \times 10^{12}/6.57 \times 10^{13} = 2.6\%$.

Then there must be another 50% of "other" biomass, which at 15 GJ/t, amounts to:

40.2 × 15 × 10⁹ = 6.03×10^{11} J (PSE = 0.9%), and so the total energy of this high-oil crop = $1.69 \times 10^{12} + 6.03 \times 10^{11} = 2.29 \times 10^{12}$ J, giving a total PSE of 3.5%, which is also reasonable.

However, the upper Wikipedia limit of "20,000 gal/acre" is worrisome. At ×4 the above, this implies a PSE of 3.5% × 4 = 14%, which seems far too high, being above the theoretical PSE limit, at which around 12.7% of the entire insolation is usefully absorbed by a photosynthetic entity. It is conceivable that special methods are employed to increase the yields, e.g., CO_2-injection or forced-UV conditions to increase the PAR of the solar spectrum. At any rate, it could not be achieved in standard open-pond systems. If someone can prove this conclusion wrong, it would be interesting to know the details.

11. Large-Scale Biodiesel Production from Algae

Theoretically, biodiesel produced from algae appears to be the only feasible solution today for replacing petro-diesel entirely. No other feedstock has the oil yield high enough for it to be in a position to produce such large volumes of oil. To elaborate, it has been calculated that for a crop such as soybean or palm to yield enough oil capable of replacing petro-diesel completely, a very large percentage of the current land available needs to be utilized only for biodiesel crop production, which is quite infeasible. For some small countries, in fact it implies that all land available in the country to be dedicated to biodiesel crop production. However, if the feedstock were to be algae, owing to its very high yield of oil per acre of cultivation, it has been found that about ten million acres of land would need to be used for biodiesel cultivation in the USA to produce biodiesel to replace all the petro-diesel used currently in that country. This is just 1% of the total land used today for farming and grazing together in the US (about 1 billion acres). Clearly, algae are a superior alternative as a feedstock for large-scale biodiesel production (Rhodes, 2009).

In practice, however, biodiesel has not yet been produced on the very large scale from algae, though large-scale algae cultivation and biodiesel production appear likely in the near future (4–5 years). To produce biodiesel from algae on a large scale, the following conditions need to be met:

1. The sustainable growth of high-oil-yielding algae strains on a large scale
2. Extraction of the oil from the algae on a large scale
3. Large-scale conversion of algal oil into biodiesel

The first two aspects are specific to algae, whereas the third is a generic aspect for biodiesel production from all plant oils. On the basis of current research, it appears that the real concern would be the following condition: Capability to sustainably produce high-oil-yielding algae strains on a large scale. Though the other two conditions need to be addressed as well, those two are primarily engineering considerations over which we have more control than over the first condition. Hence, this needs to be given more focus.

The capability to sustainably produce high-oil-yielding algae strains on a large scale can again be thought to contain two distinct aspects: (1) Identifying the high-yielding algal strains and (2) optimizing methods to cultivate them. There is currently considerable research effort being made in both areas, and it is hoped that there will be significant developments in the near future.

11.1. BIOFUEL FROM ALGAE USING HIGH-DENSITY VERTICAL REACTORS (HDVR)

The "High-Density Vertical Bioreactor," or HDVB stands in contrast to open-pond systems. The HDVB system is marketed by Valcent Products, Inc. and consists of growing plants or algae in plastic pockets on clear vertical panels that move on a conveyor-belt arrangement. The strategy is designed to maximize the amount of sunlight and to provide an ideal balance of nutrients to achieve optimum growth. It is proposed that such vertical growth systems might provide a solution to the problem of feeding urban populations so that urban living becomes sustainable (Valcent, 2009; Global green Solutions, 2006).

It might be argued that once transportation begins to fail as a consequence of cheap oil supplies waning in 5–10 years, humanity will relocalize into relatively small communities far less dependent on transport. The lack of urban growing space is counteracted by the very high efficiency of crop production in HDVBs. This form of agriculture is also soil-free, and uses perhaps 5% of the amount of water that is required to grow crops by conventional means, since the whole constitutes a closed system with far less evaporation. Since these reactors can be placed anywhere (as can open ponds), there is no necessity to compromise arable land which can still be used for standard agriculture.

However, the HDVB offers the potential for producing fuel as well as food, since algae can be grown in these systems too, and a higher yield is claimed than that in open ponds. Thus, in principle, food is grown locally, thus eliminating much of the fuel costs borne in the carriage of crops from one part of the country to another or even across the world, by air or by ship, and also biodiesel can be made from oil extracted from algae grown using the technology, by transesterification with methanol or ethanol, as is done with plant-derived oils. Growing food both efficiently and locally also averts much of the spoilage that occurs on long hauls, during which as much as 50% of it is thus rendered inedible.

It is claimed that 100,000 gal (US) of diesel can be produced per acre of HDVB area, which does seem very high. Regarding photosynthetic efficiencies, the figure of 20,000 gal/acre quoted in the wikipedia entry on "permaculture" seems well above the theoretical efficiency for a horizontal open-pond/algal system, but higher surface areas could be attained using vertical reactor arrangements; however, to install this paraphernalia on the very large scale is going to take a lot of plastic (derived from oil) and a lot of engineering, especially since the HDVB systems are more intricate than the basis that has been indicated. Irrespective of whether the algae are grown in open ponds or HDVB systems, there will also be need for a massive construction of transesterification plants and a source of methanol or ethanol must be found, in an amount equal to perhaps 10% of the diesel that is produced.

It sounds like a great idea and sits comfortably with most of the values mentioned here and projections as to what precisely we need to achieve to form a stable, sustainable society. However, the scale-up will be a gargantuan task. If we can cut our fuel use to say 25% in relocalized communities, we still need to produce around 700 million tons of biodiesel annually (15 million tons just for the UK and 175 million tons for the USA) and convert most vehicles to run on diesel engines; the question is: Can this be done quickly enough to breach the demand/supply gap facing conventional oil production?

12. Could Peak Phosphate be Algal Diesel's Achilles' Heel?

The depletion of world phosphate reserves will impact on the production of biofuels, including the potential wide-scale generation of diesel from algae. The world population has risen to its present number of 6.7 billion in consequence of cheap fertilizers, pesticides, and energy sources, particularly oil. Almost all modern farming has been engineered to depend on phosphate fertilizers, and those made from natural gas, e.g., ammonium nitrate, and on oil to run tractors, etc. and to distribute the final produce. Worldwide production of phosphate has now peaked (in the USA, the peak came in the late 1980s), which lends fears as to how much food the world will be able to grow in the future, against a rising number of mouths to feed (Phillpott, 2008). Consensus of analytical opinion is that we are close to the peak in world oil production too.

The algae route sounds almost too good to be true. Having set up these ponds, albeit on a large scale, i.e., they would need an area of 3,200 km^2 to produce 40 million tons of diesel, which is enough to match the UK's transportation demand for fuel, if all vehicles were run on diesel engines (the latter are more efficient in terms of tank-to-wheels miles by about 40% than petrol-fuelled spark-ignition engines), one could ideally leave them to absorb CO_2 from the atmosphere (thus simultaneously solving another little problem) by photosynthesis, driven only by the flux of natural sunlight. The premise is basically true; however, for algae to grow, vital nutrients are also required, as a simple elemental analysis of dried algae will confirm. Phosphorus, though present in under 1% of that total mass, is one such vital ingredient, without which algal growth is negligible. Two different methods of calculation have been used here to estimate how much phosphate would be needed to grow enough algae, first to fuel the UK and then to fuel the world:

1. The analysis of dried Chlorella (Wikipedia, 2009b) has been taken as an illustration, which contains 895 mg of elemental phosphorus per 100 g of algae.

UK Case: To make 40 million tons of diesel, 80 million tons of algae would be required (assuming that 50% of it is oil and this can be converted 100% to diesel).

The amount of "phosphate" in the algae is $0.895 \times (95/31) = 2.74$ %. (MW PO_4^{3-} is 95, that of P = 31).

Hence this amount algae would contain: 80 million $\times$ 0.0274 = 2.19 million tons of phosphate.

World case: The world gets through 30 billion barrels of oil a year, of which 70% is used for transportation (assumed). Since 1 t of oil is contained in 7.3 barrels, this equals $30 \times 10^9/7.3 = 4.1 \times 10^9$ t and 70% of that = 2.88×10^9 t of oil for transportation.

So, this would need twice that mass of algae = 5.76×10^9 t of it, containing: $5.76 \times 10^9 \times 0.0274 = 158$ million tons of phosphate.

2. To provide an independent estimate of these figures, it has been noted that growth of this algae is efficient in a medium containing a concentration of 0.03–0.06% phosphorus; the lower part of the range, i.e., 0.03% P is used here. "Ponds" for growing algae vary in depth from around 0.6 to 1.5 m and so a depth of 1 m could be assumed for simplicity.

UK case: Previously, the author has worked out (Rhodes, 2009) that producing 40 million tons of oil (assumed equal to the final amount of diesel, to simplify the illustration) would need a pond/tank area of 3,200 km². An area of 3,200 km² is equal to 320,000 ha, and at a depth of 1 m, this amounts to a volume of: 320,000 × (1 × 10⁴ m²/ha) × 1 m = 3.2 × 10⁹ m³.

A concentration of 0.03 % P = 0.092% phosphate, and so each cubic meter (1 m³ weighs 1 t) of volume contains 0.092/100 = 9.2 × 10⁻⁴ t (920 g) of phosphate. Therefore, we need:

3.2 × 10⁹ × 9.2 × 10⁻⁴ = 2.94 million tons of phosphate, which is in reasonable accord with the amount of phosphate taken up by the algae (2.19 million tons), as deduced earlier.

World case: The whole world needs 2.88×10^9 t of oil, which would take an area of $2.88 \times 10^9/125$ t/ha $= 2.30 \times 10^7$ ha of land to produce it.

2.3×10^7 ha × $(10^4$ m²/ha$) = 2.3 \times 10^{11}$ m² and at a pond depth of 1 m they would occupy a volume $= 2.30 \times 10^{11}$ m³. Assuming a density of 1 t = 1 m³, and a concentration of $PO_4^{3-} = 0.092\%$, we need:

$2.30 \times 10^{11} \times 0.092/100 = 2.13 \times 10^8$ t of phosphate, i.e., 213 million tons.

This is also in reasonable accordance with the figure deduced from the mass of algae accepting that not all of the P would be withdrawn from solution during the algal growth. Indeed, the ratio of algal phosphate to that present originally in the culture medium (i.e., 158/213) suggests that 74% of it is absorbed by the algae.

Now, world phosphate production amounts to around 140 million tons (noting that we need 213 million tons to grow all the algae), and food production is already being thought to be compromised by phosphate resource depletion. The USA produces less than 40 million tons of phosphate annually, but would require enough to produce around 25% of the world's total algal diesel, in accord with its current "share" of world petroleum-based fuel, or 53 million tons of phosphate. Hence, for the USA, security of fuel supply could not be met by algae-to-diesel production using even all its indigenous phosphate rock output, and imports (of phosphate) are still needed. The world total of phosphate is reckoned at 8,000 million tons and that in the USA at 2,850 million tons (by a Hubbert Linearization analysis). However, as is true of all resources, what matters is the rate at which they can be produced.

13. Conclusions

Some aspects of the practicalities of algae-to-fuel conversion, including field trials, have been discussed previously (Gao and McKinley, 1994; Miyamoto, 1997; Christi, 2007; Sheehan et al., 1998). The author remains optimistic over algal diesel, but clearly if it is to be implemented on a serious scale, its phosphorus has to come from elsewhere than phosphate rock mineral. There are regions of the sea that are relatively high in phosphates and could in principle be concentrated to the desired amount to grow algae, especially as salinity is not necessarily a problem. Recycling phosphorus from manure and other kinds of plant and animal waste appears to be the only means to maintain agriculture at its present level, and certainly if its activities will be increased to include growing algae. In principle too, the phosphorus content of the algal waste left after the oil-extraction process could be recycled into growing the next batch of algae. These are all likely to be energy-intensive processes, however, requiring "fuel" of some kind, in their own right.

It is salutary that there remains a competition between growing crops (algae) for fuel and those for food, even if not directly in terms of land, for the fertilizers that both depend on. This illustrates for me the complex and interconnected nature of, indeed Nature, and that like any stressed chain, will ultimately converge its forces onto the weakest link in the "it takes energy to extract energy" sequence.

A Hubbert analysis of human population growth indicates that rather than rising to the putative "9 billion by 2050" scenario quoted from WHO figures, it will instead peak around the year 2025 at 7.3 billion, and then fall (Phillpott, 2008). It is probably significant too that that population growth curve fits very closely both with that for world phosphate production and another for world oil production (Phillpott, 2008). It seems to be highly indicative that it is the decline in resources that will underpin our demise in numbers as is true of any species: from a colony of human beings growing on the Earth, to a colony of bacteria growing on agar nutrient in a petri dish.

14. References

Becker, E.W. (1994) In: J. Baddiley et al. (eds.) *Microalgae: Biotechnology and Microbiology*. Cambridge University Press, Cambridge/New York, p. 178.
Christi, Y. (2007) Biodiesel from microalgae. Biotechnol. Adv. **25**: 294–306.
Deffeyes, K.S. (2005) *Beyond Oil*. Hill & Wang, New York.
Duffield, J.A., Shapouri, H. and Wang, M. (2006) Assessment of biofuels, In: J. Dewulf and H. Van Langenhove (eds.) *Renewables-Based Technology*. Wiley, Chichester.
Elgood, G. and Eastham, T. (2008) Biofuels blamed for food price crisis, http://uk.reuters.com/article/businessNews/idUKL0340750020080704.
Fraser, J. (2008). http://thefraserdomain.typepad.com/energy/2008/03/fyi-petrosun-to.html.
Gao, K. and Mckinley, K.R. (1994) Use of macroalgae for marine biomass production and CO_2 remediation – a review. J. Appl. Phycol. **6**: 45–60.
Global Green Solutions (2006). http://www.globalgreensolutionsinc.com/s/VertigroFAQ.asp.
Hubbert, M.K. (1956) Nuclear energy and the fossil fuels. Presented before the Spring meeting of the Southern District, American Petroleum Institute, Plaza Hotel, San Antonio, TX, March 7–9, 1956.
Hubbert, M.K. (1982) Techniques of production as applied to oil and gas, In: Glass SI (ed.) *Oil and Gas Supply Modelling, Special Publication 631*. National Bureau of Standards, Washington DC, pp. 16–141.
Low, D. (2007) ASPO conference confirms a peak in global oil production by 2012. http://www.energybulletin.net/35127.html
Maio, X. and Wu, Q. (2006) Biodiesel production from heterotropic microalgal oil. Bioresour. Technol. **97**: 841–847.
Miyamoto, K. (ed.) (1997). Renewable biological systems for alternative sustainable energy production (FAO Agricultural Services Bulletin – 128).
Phillpott T. (2008) Biofuels and the fertilizer problem. http://www.energybulletin.net/print.php?id=40300.
Rhodes, C.J. (2005) *Energy Balance*: http://ergobalance@blogspot.com.

Rhodes, C.J. (2008) Oil calculator. *Chemistry and Industry,* July 7, p. 12.
Rhodes, C.J. (2009) Oil from algae; salvation from peak oil? Sci. Prog. **92**: 39–90.
Sheehan, J., Dunahay, T., Benemann, J.R. and Roessler, P. (1998) A Look Back at the U.S. Department of Energy's Aquatic Species Program – Biodiesel from Algae, NREL/TP-580-24190.
Valcent (2009) http://www.valcent.net/s/TomorrowGarden.asp.
Wikipedia (2009a) Algaculture, http://en.wikipedia.org/wiki/Algaculture.
Wikipedia (2009b) Chlorella, http://en.wikipedia.org/wiki/Chlorella.

PART 6:
CULTIVATION OF SEAWEEDS IN GLOBALLY CHANGING ENVIRONMENTS

Hayashi	**Nagata**
Hurtado	**Issar**
Msuya	**Neori**
Bleicher-Lhonneur	**Mantri**
Critchley	**Reddy**
Tang	**Jha**
Taniguchi	**Pereira**
Zhou	**Yarish**

Biodata of **Leila Hayashi, Anicia Q. Hurtado, Flower E. Msuya, Genevieve Bleicher-Lhonneur,** and **Alan T. Critchley**, authors of *"A Review of Kappaphycus Farming: Prospects and Constraints"*

Dr. Leila Hayashi is currently a Postdoc of the Universidade Federal de Santa Catarina, Brazil. She obtained her Ph.D. from Universidade de São Paulo, Brazil, in 2007, and continues her studies and research at the Universidade Federal de Santa Catarina. She gives classes to undergraduate and postgraduate students and has participated in projects regarding the commercial viability of *Kappaphycus alvarezii* in Brazil. Her scientific interests are in the areas of seaweed cultivation and commercial processing, extraction of value-added products, and integrated cultivation.

E-mail: **leilahayashi@hotmail.com**

Anicia Q. Hurtado is currently a Visiting Scientist of the Aquaculture Department, Southeast Asian Fisheries Development Center, Tigbauan, Iloilo, Philippines. She has been there since 2006 after serving the Center as Senior Scientist for 20 years. She finished her Doctor of Agriculture at Kyoto University, Kyoto, Japan. She works mainly on the aquaculture of *Kappaphycus* as a Consultant to international and local, nongovernment organizations directly involved with seaweed farmers. At present, she is developing "new strains" of *Kappaphycus* using tissue culture techniques for possible sources of propagules for commercial farming.

E-mail: **aqhurtado@gmail.com**

Leila Hayashi   **Anicia Q. Hurtado**

Flower E. Msuya is currently working as a Senior Researcher at the Institute of Marine Sciences of the University of Dar es Salaam in Zanzibar, Tanzania. She obtained her Ph.D. from Tel Aviv University, Israel, in 2004 and continued her research at the Institute of Marine Sciences. Dr. Msuya's scientific interests are in seaweed farming, physiology, ecology, and value addition; Integrated mariculture; and socio-economic studies in marine science and associated fields.

E-mail: **flowereze@yahoo.com**

Genevieve Bleicher-Lhonneur is currently working as a senior Strategic Raw Materials Procurement manager for Cargill. She has spent her entire career in procurement, putting her expertise at the disposal of the industry. Her contribution to many seaweed farming projects and fair support to the farmers are widely recognized. She is still involved in developing new sources and improving farming conditions for the benefit of the farmers and the whole industry. To reach this objective, Cargill is financing specific studies related to strain selection or environmental impact on cultivated seaweed properties in close cooperation with outside scientists.

E-mail: **Genevieve_Bleicher-Lhonneur@cargill.com**

Flower E. Msuya   **Genevieve Bleicher-Lhonneur**

Alan T. Critchley is a reformed academic. He graduated from Portsmouth Polytechnic, UK, and had a university career in Southern Africa teaching phycology, marine ecology, and botany (KwaZulu Natal, Wits and Namibia). He moved to the "dark side" in 2001 and took up a position in a multinational industry with Degussa Texturant Systems (now Cargill TS), where he was responsible for new raw materials for the extraction of the commercial colloid carrageenan. It was here he began a love affair with *Kappaphycus* and its cultivation. Since 2005, he has worked as head of research for Acadian Seaplants Limited working on value addition to seaweed extracts and on-land cultivation of seaweed for food and bioactive compounds. Not able to turn his back on the academic world entirely, he is currently adjunct professor at the Nova Scotia Agricultural College. It has been his absolute priviledge and pleasure to work with such excellent scientists and friends as Leila, Anne, Flower, and Genevieve on research into the production and improved quality of carrageenan-producing seaweeds.

E-mail: **Alan.Crithley@acadian.ca**

A REVIEW OF *KAPPAPHYCUS* FARMING: PROSPECTS AND CONSTRAINTS

LEILA HAYASHI[1], ANICIA Q. HURTADO[2], FLOWER E. MSUYA[3], GENEVIEVE BLEICHER-LHONNEUR[4], AND ALAN T. CRITCHLEY[5]

[1] Depto. BEG, Centro de Ciências Biológicas, Universidade Federal de Santa Catarina, Trindade, 88040-900 Florianópolis, Santa Catarina, Brazil
[2] Aquaculture Department, Southeast Asian Fisheries Development Center, Tigbauan, Iloilo, 5021, Philippines
[3] Institute of Marine Sciences, University of Dar es Salaam, P.O. Box 668, Zanzibar, Tanzania
[4] Raw Material Procurement, Cargill Texturizing Solutions, 50500 Baupte, France
[5] Acadian Seaplants Limited, 30 Brown Ave., Dartmouth, NS, Canada B3B 1X8

1. Introduction

Global warming is of increasing concern worldwide. The question of how to mitigate the CO_2 released into the atmosphere is the most topical issue, and sustainable solutions are constantly being sought. Aquaculture has been proposed as one method for the sequestration or immobilization of CO_2 through filtration or mechanical/chemical processes for long-term storage (Carlsson et al., 2007). However, the development of new sustainable technologies are but in their infancy, as the aquaculture sector moves to becoming more efficient and sustainable.

In the above context, Chopin (2008) suggested that if limitations to nutrient emission are to be put in place, extractive species such as seaweeds could be considered as suitable for nutrient credits (e.g. for the extraction of nitrogen, phosphorus, etc.), in a similar system to that of carbon credits, which are becoming increasingly acceptable to be traded in the global economy.

Kappaphycus alvarezii, commercially recognized as "cottoni," is considered the main source of *kappa* carrageenan, while *Eucheuma denticulatum*, known as "spinosum," is the main source of *iota* carrageenan. Both species are responsible for approximately 88% of raw material processed for carrageenan production, yielding about 120,000 dry tons year^{-1} mainly from the Philippines, Indonesia, and Tanzania (Zanzibar) (McHugh, 2003). Considered as two of the most commercially successful species, they could be potential candidates for carbon and nutrient credits, if research and technology are developed, since the farms have the potential to increase the area occupied for seaweed cultivation, thus improving the carbon dioxide sequestration and acting as a nutrient sink when cocultivated with other organisms, thereby improving water quality to some extent.

A review of *Kappaphycus* (broadly including *Eucheuma*) farming is presented, including the current possibilities and challenges with the goal of contributing to sustainable mariculture management practices.

2. Worldwide Trends of *Kappaphycus* Production

The world's geographical area for the *Kappaphycus* farming lies within ±10° latitude (Fig. 1), notably from the Southeast Asian countries extending to East Africa and Brazil. However, the Southeast Asian region, primarily the Brunei–Indonesia–Malaysia–Philippines (East Association of Southeast Asian Nations (ASEAN) Growth Area – BIMP-EAGA – integrated countries) has by far the greatest potential for expanded tropical seaweed cultivation, consisting 60% of the sites in the world. In particular, Indonesia, Malaysia, and the Philippines provide sheltered areas that are favorable for cultivation (IFC, 2003).

The current and estimated increase in *Kappaphycus* production in the BIMP-EAGA region is shown in Fig. 2. Although the production of the southern Philippines and Sabah combined is about 100,000 t dry weight year^{-1}, with the potential to increase by approximately 50%, the shared projected capacity of West, Central, and East Indonesia is huge (approximately 450,000 t) (IFC, 2003). The high potential of Indonesia can be attributed to its extensive coastline, which fits 100% within the tenth parallel latitude, where tropical seaweeds grow abundantly and robustly and, most importantly, where typhoons seldom occur. On the other hand, the single largest area of *Kappaphycus* production in the Philippines (Sitangkai, Tawi-Tawi) offers considerable potential for expansion since it has 60,000 ha available for mariculture purposes, even though only an estimated 10,000 ha are presently used for cultivating *Kappaphycus* (PDAP, 2007). Sabah (Malaysia) had a total production of 50,000 t fresh weight (around 6,250 t dry weight), which was grown on 1,000 ha (= 50 t ha^{-1} year^{-1}) in 2005 (Neish, 2008). Projection for this area aims an expansion to 12,000 ha in 2010, with a target production of 250,000 t fresh weight.

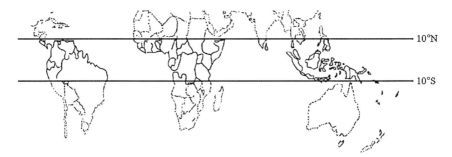

Figure 1. The world's potential geographical area for the *Kappaphycus* farming.

A REVIEW OF *KAPPAPHYCUS* FARMING: PROSPECTS AND CONSTRAINTS

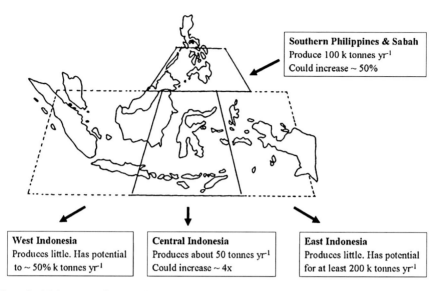

Figure 2. Major areas of *Kappaphycus* cultivation and potential dry weight production in the BIMP-EAGA region (IFC, 2003).

Table 1. World production of *Kappaphycus* (cottonii) in 2006 (Hurtado, 2007).

Country	Volume (ton dry weight)	Total (%)
Philippines	89,000	55.5
Indonesia	61,000	38.0
Malaysia	4,000	2.5
Cambodia/Vietnam	2,200	1.4
China	800	0.5
Kiribati	1,100	0.7
India	400	0.2
Tanzania/Madagascar	1,500	0.9
Brazil	500	0.3
Total	160,500	100

The *Kappaphycus* (cottonii) world production for commercial extraction of kappa carrageenan shows that the BIMP-EAGA region produces 96.5% of the total production, of which 55% comes from the Philippines, followed by Indonesia (38%) and Malaysia (2.5%). The rest of the producing areas contribute relatively very small volumes (Table 1).

3. Cultivation Techniques and Post-harvest Management

Since the first successful farming of *Kappaphycus* in the Philippines in 1970, the cultivation technique has undergone many modifications. The main commercial varieties are presented in Fig. 3 (Hurtado et al., 2008a). The most traditional type of farming is the fixed off-bottom method usually practiced in shallow reef areas. Innovations in the deeper water cultivation areas include using the hanging long line (fixed and swing) and the multiple-raft long line. The fixed hanging long line technique has both ends tied to an anchor bar or block, while in the swing hanging

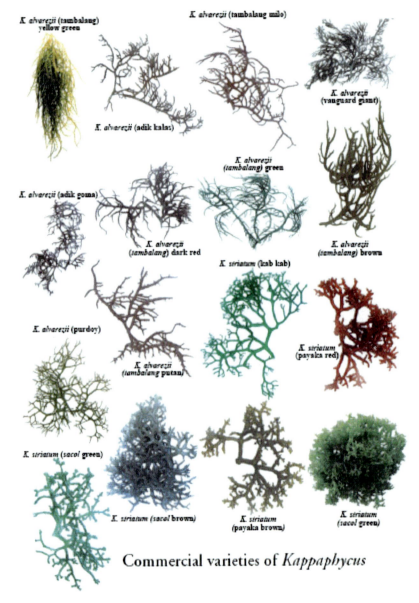

Figure 3. The main commercial varieties of *Kappaphycus* spp. cultivated worldwide (Photos retrieved from Hurtado et al., 2008a).

long line, only one end is tied to an anchor bar or block, and the other end is allowed to swing freely with the current. Figure 4a–d shows examples of the different farming methods commonly used.

Table 2 shows the summary of growth rates of *Kappaphycus* grown by different methods and in different areas. The growth rates varied widely depending on the cultivation system adopted but, in general, plants cultivated in the usual fixed off-bottom system or in rafts showed high values (between 0.2% and 5.3% day^{-1} in the first case and between 0.5% and 10.7% day^{-1} in the last case). When plants were co-cultivated with other organisms in tanks, acting as biofilters, the growth rates were notably lower (0.7–2.7% day^{-1}).

Both the quantity and quality of carrageenan is influenced by the maturity of the thallus. The work of Mendonza et al. (2006) suggested that young (apical) segments of *K. striatum* var. *sacol* yielded higher gel strength, cohesiveness, viscosity properties, and lower average molecular weight than old (basal) segments. However, the latter yielded greater amounts of carrageenan in total. Furthermore, they emphasized that as the thalli aged, the content of *iota* (and precursor) carrageeenan decreased, which may be related to certain physiological and structural functions during the growth and structural maturation of the alga.

The work of Hurtado et al. (2008b) on *Kappaphycus striatum* var. sacol showed that a lower stocking density (e.g. 500 g m^{-1}) and a shorter period of culture (30 days) yielded a higher growth rate than a higher stocking density (1,000 g m^{-1}) and longer period of culture (45 and 60 days) when grown vertically on rafts. Furthermore, the results revealed a higher yield in carrageenan at a lower stocking density (500 g m^{-1} line^{-1}) than at 1,000 g m^{-1} line^{-1}; and the molecular weight was greater in plants growing at 50–100 cm depth both for 30 and 45 days of cultivation, when compared with those that grow at 150–200 cm depth and 60-day cultivation period. These results simply indicate that "sacol" prefers a shorter period of culture and at a lower depth to synthesize more carrageenan. It is surprising how little such relatively simple studies have been employed given the value of the seaweed biomass to national economies and the carrageenan yield and quality to the extraction/processors.

With regard to *Kappaphycus alvarezii* cultivated in Ubatuba, Brazil, Hayashi et al. (2007) observed that plants cultivated at high density (24 plants of 40 g each in 1 m^2), in PVC pipe multiple raft, showed higher productivity when cultivated at surface for 44 and 59 days. Higher carrageenan yields were obtained from plants cultivated for 28 days; in fact, in this study, the *iota* carrageenan content was highest at 59 days of cultivation; and unlike the observations for *K. striatum*, the older tissues of *K. alvarezii* showed the higher molecular weight and gel strength. Clearly, there is a need for much greater site-specific studies on methods of cultivation and duration of the commonly practiced farming period.

The molecular weight has become an important criterion for the food regulatory authorities since the Scientific Committee on Food (SCF) of the European Commission endorsed in 2003 a molecular weight distribution limit on carrageenan as precautionary approach. This regulation was created based on some suspects that high consumption of low molecular weight carrageenan could

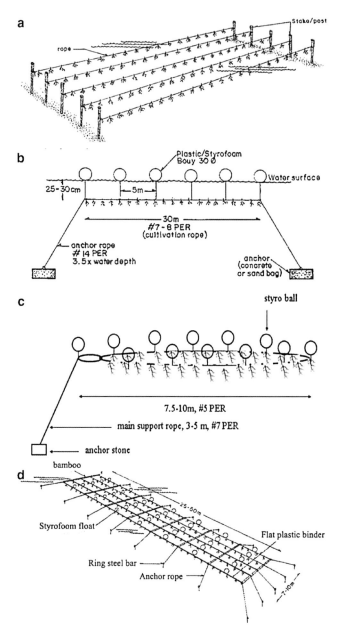

Figure 4. Culture techniques commonly used (**a**) fixed-type (Hurtado et al., 2008a); (**b**) hanging long line (fixed-type) (Hurtado et al., 2008a); (**c**) hanging long line (swing type) (Hurtado et al., 2008c); (**d**) multiple-raft long line (Hurtado and Agbayani, 2002).

Table 2. Reported daily growth of *Kappaphycus alvarezii* using different culture techniques and sites.

Culture technique	Growth rate (% day^{-1})	Reported by	Country
Fixed off-bottom	5.7	Lim and Porse, 1981	Bohol, Philippines
	3–4	Adnan and Porse, 1987	Indonesia
	3.5–3.7	Luxton et al., 1987	Fiji
	2.04	Samonte et al., 1993	Western Visayas, Philippines
	0.2–3.2	Hurtado et al., 2001	Antique, Philippines
	2.0–8.1	Muñoz et al., 2004	Yucatan, Mexico
	4.2–4.3	Wakibia et al., 2006	Southern Kenya
	1.6–4.6	Hung et al., 2009	Camranh Bay, Vietnam
Bamboo raft (single)	1.24	Samonte et al., 1993	Western Visayas, Philippines
	0.5–5.6 (Unfertilized)	Msuya and Kyewalyanga, 2006	Zanzibar, Tanzania
	0.5–7.6 (Fertilized)	Msuya and Kyewalyanga, 2006	
	4.4–8.9	Dawes et al., 1994	Zanzibar, Tanzania
	6.5–10.7	Paula et al., 2002	
	4.5	Gerung and Ohno, 1997	Philippines
	4–5	Bulboa and Paula, 2005	São Paulo, Brazil Southern Japan
	5.2–7.2	Hayashi et al., 2007	São Paulo, Brazil
PVC pipe raft (multiple)	0.8–1.3	Lombardi et al., 2006	São Paulo, Brazil
	3.7–7.1	Hurtado-Ponce, 1992	São Paulo, Brazil
Cages	0.9–3.8	Hurtado-Ponce, 1995	Guimaras, Philippines
	3.7	Hurtado-Ponce, 1994	Guimaras, Philippines
Cages: polyculture	3.8		
vertical lines	5.3		
cluster			
horizontal	0.2–4.2	Hurtado et al., 2001	Antique, Philippines
Hanging long line (fixed)	2.8–3.0	Hurtado et al., 2008b	Tawi-Tawi, Philippines
Hanging long line (swing)	1.8–10.86	Wu et al., 1989	China
Vertical lines	1.9–6.3	Glenn and Doty, 1990	Hawaii
Pens (loose thalli)	9–11	Ohno et al., 1996	Vietnam
Lagoon	7–9	Ohno et al., 1996	Vietnam
Inlet	5–6	Ohno et al., 1996	Vietnam
Pond	2–2.7	Rodrigueza and Montaño, 2007	Philippines
Biofilter in tanks	0.7–0.8	Hayashi et al., 2008	São Paulo, Brazil
Polyculture with oyster	1.9–6.1	Qian et al., 1996	China
Polyculture with grouper	3.7–5.3	Hurtado-Ponce, 1994	Antique, Philippines

provoke peptic ulcer. However, until nowadays, several animal studies have been made and supported the safety of carrageenan for use in foods; thus, regulatory authorities saw no reason to question the safety of carrageenan as long as the average molecular weight was 100,000 Da or higher (Watson, 2008).

4. Moisture Content of Raw Dried Seaweed

Moisture content (MC) of the harvested seaweed biomass plays a major role in its market acceptance and commercial value. The more water content the seaweed has, the lower the farm gate price. Frequently, seaweed farmers have no capacity to determine the MC properly, so most of the time they are dependent on decisions of the traders. In turn, these traders rely on the final measurement by the processors, often after some considerable time of transportation. It would be prudent if one small testing laboratory could be installed at major producing areas in order to accurately determine MC of the seaweed, so that the farmers could be paid accordingly. MC is a critical factor along the value chain.

Generally, fresh seaweeds from far-flung islands of the Philippines are sun-dried for 1–2 days only due to very limited availability of space for drying. These are sold immediately to collectors (first trader) with an estimated MC of 45–50% (given the local conditions and humidity; obtaining lower levels without mechanical drying are not practical). The collector will in turn sell to a larger trader (second trader) who will sell either to local processors or to exporters of Raw Dried Seaweed (RDS) (Fig. 5). In other areas, e.g. Tanzania, MC varies from 15% to 20% after 2–3 days of sun drying.

There are advantages when the farmers are organized. The bargaining power of a farmers' association is much stronger. The possibility of selling directly either to a local or off-shore processor is increased and bypasses two to three layers of traders, thus providing a higher profit margin to the association.

5. Social and Economic Aspects of *Kappaphycus* Cultivation

Seaweed farming is an industry that can contribute significantly to the economy of the producing countries, bringing a foreign income and improving the socio-economic situation of the coastal people involved. Apart from the governments of the respective countries taking a key role in helping their communities through seaweed farming, other institutions have also taken actions on such activities. In Tanzania, NGOs like the USAID-funded Agricultural Cooperative Development International and Volunteers in Overseas Cooperative Assistance (ACDI-VOCA) have implemented different programs on seaweed farming for the benefit of coastal communities. The successful cultivation practices have increased the economic purchasing power and social empowerment of women seaweed farmers (Pettersson-Löfquist, 1995; Bryceson, 2002; Msuya, 2006a). In this country,

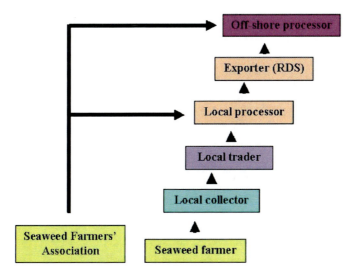

Figure 5. Trade chain of *Kappaphycus alvarezii* crop.

the significance of the industry as a foreign income was documented in 2006 when it had contributed 14.7% and 27.3% of the Zanzibar's marine resources exports between 1993 and 1994 (Msuya, 2006a). In Indonesia, it has been shown that 7,350 families source their livelihood from seaweed farming (Watson, 2000), while in the Philippines it was shown that seaweed farming became the main source of income to village communities where seaweed was farmed (Quiñonez, 2000).

6. Current Trends

While seaweed farming has contributed significantly to the economies of many countries, there have been recent changes that have significantly affected the industry and consequently farmers and countries at large. First of all, is the preference by the world market for one species *Kappaphycus alvarezii* over *Eucheuma denticulatum* because of its stronger gel (*kappa* carrageenan) compared with the latter's *iota* carrageenan, which is a weaker gel? As a result, the price of *K. alvarezii* is higher than that of *E. denticulatum*. Examples are in Tanzania where the price of the former in 2008 was US$0.2 kg^{-1} of dry weight, almost double that of the latter, which is US$0.1. In 2002, the price was US$0.09 (Bryceson, 2002). In the Philippines, the current farm gate price (as of 30 March 2009) is US$0.67–0.73. This picture directly reflects the farmers payment, which can vary among the countries, e.g. US$49–91 month^{-1} in Pacific Islands (Bergschmidt, 1997), and US$2,000 in the Philippines (Murphy, 2002).

Seaweed sales can also be influenced by the distance from the farming sites to the export point and the farmer's efforts. In Tanzania, for example, farmers are currently receiving US$50–500 month^{-1}. Other aspects coupled with the market change is the proliferation of "diseases," as ice-ice, and die-offs of the higher-priced *K. alvarezii*, experienced in a number of countries.

A recent study showed that while seaweed farming in Songosongo Island in the South of Tanzania had proved to be an alternative livelihood to the people, the farming is failing in many cultivation sites in the shallow intertidal areas where it used to grow. The main cause observed during a short survey by Msuya (2009) was the water temperature, which was higher than had been observed previously in seaweed farms in other areas and other years in Tanzania. Water temperatures had been less or equal to 33°C. In Msuya (2009), surface seawater temperature values taken between 11:00 and 13:00 h along the seaweed farming area showed values ranging from 33°C to 38°C. The seaweed showed signs of being burnt with bleached thalli; sometimes, the whole thallus was bleached. Farmers have abandoned these sites to concentrate in a smaller area. The number of farmers decreased by 60% between 2003 (seaweed production peak) and 2008. Production decreased by 94% during the same period (Msuya, 2009).

In Zanzibar, in an area where *K. alvarezii* used to be purchased for seed during experimental studies, farmers have stopped cultivating the species due to die-offs. Seed is now taken from either South or North of the study site that has been used since 2005. In a continuing study in the area, highest temperatures in January–February 2008 were 36–37°C (Msuya, 2009). Further studies are required to ascertain if the area in general is suitable for *Kappaphycus* and or *Eucheuma* spp.

In an effort to farm the higher-priced seaweed, farmers spend the growing seasons (up to 6 months) only to end up losing the seedlings during bad seasons. This has frustrated them who have been wasting their time, energy, and resources with no returns. In so doing, the production of even lower value *Eucheuma* species has decreased. Equally affected by the situation are the middlemen who buy and export the seaweeds as they cannot get enough materials for their business. Such events cause dismay and farmers stop the cultivation activity. Pollnac et al. (1997) observed several examples of other communities in North Sulawesi that had taken up and then abandoned seaweed farming in the past. A drop in seaweed prices in 2001 led to a reduction in seaweed farming in the Bentenan–Tumbak area of North Sulawesi (Unsrat, 2001).

7. *Kappaphycus alvarezii* Diseases and Epiphytes

"Ice-ice disease" and epiphyte infection are two factors occurring as "outbreaks" that can be frequently observed in high-density commercial farming (Vairappan et al., 2008). "Ice-ice" is a symptom that *Kappaphycus alvarezii* presents after suffering stresses such as abrupt changes in temperature and/or salinity. The term "ice-ice" was coined by Philippine farmers using the English term "ice" to describe the senescent tissue devoid of pigments that causes healthy branches to break off

(Doty, 1987). The "ice-ice" problem was first reported in 1974 during the start of commercial seaweed farming in Tawi-Tawi, the Philippines (Barraca and Neish, 1978, Uyengco et al., 1981) and severe instances have reportedly wiped out entire farms (Largo et al., 1995). According to Doty (1987), the onset is a sharp loss of thallus pigmentation until it becomes white; the segment may remain for a day or two before it dissolves away, separating the two adjacent parts of the thallus, which seem to be otherwise unaffected (Fig. 6).

Although most of the publications consider "ice-ice" as a disease, Doty (1987) and Ask (2006) affirmed that it is not a disease since there was little evidence that it was caused by pathogenic bacteria. However, Largo et al. (1995) presented convincing evidence that bacterial groups from the complex *Cytophaga-Flavobacterium* and the *Vibrio-Aeromonas* could be the causative agents in the development of the symptom. The numbers of bacteria on and in "ice-ice"-infected branches were 10–100 times greater than from normal, healthy plants. In other work, Largo et al. (1999) noted that the combined effect of stress and biotic agents, such as opportunistic pathogenic bacteria, were primary factors in the development of "ice-ice," and that the different bacterial groups have different strategies of infection. Mendonza et al. (2002) observed de-polymerization of carrageenan from "ice-ice"-infected portions of the *K. striatum* thallus as well as lowered levels of *iota* carrageenan and methyl-constituents, which consequently lowered the average molecular weight (30 kDa) of the colloid, which could be extracted. Appreciable decreases in carrageenan yield, gel strength, and viscosity, with a combined increase in the syneresis index were also noted. The authors recommended complete removal of the infected portions of thallus prior to sun drying to prevent contamination of yields by low molecular weight carrageenan.

Another problem more recently described is the occurrence of epiphyte outbreaks in *Kappaphycus alvarezii* farms. Epiphytic and endophytic filamentous algae (EFA) are a serious threat to the health of the seaweeds and to overall farm productivity (Ask, 2006) (Fig. 7).

The development of problems caused by epiphytic algae has been described in the Philippines, Malaysia, Tanzania, and India (Hurtado et al., 2006, Msuya and Kyewalyanga, 2006; Muñoz and Sahoo, 2007, Vairappan, 2006). Hurtado

Figure 6. "Ice-ice" symptoms in some commercial *Kappaphycus alvarezii* samples (Photos retrieved from Hurtado et al., 2008a).

Figure 7. Examples of epiphytic and endophytic filamentous algae (EFA). From *left* to *right*, *Ceramium*, *Boodlea composite*, *Neosiphonia*, and *Ulva* (Photos retrieved from Hurtado and Critchley (2006) and courtesy of A. Hurtado).

Figure 8. *Neosiphonia* infection on *Kappaphycus alvarezii* thallus (Courtesy of C. Vairappan).

and Critchley (2006) used the terminology of Kloareg, who studied epiphytes of *Gracilaria*, in which he classified epiphytes into five types: Type I – epiphytes weakly attached to the surface of the host and with no evidence of host tissue damage; Type II – epiphytes strongly attached to the surface of the host but host tissue damage still absent; Type III – epiphytes that penetrate the outer layer of host cell wall without damaging the cortical cells; Type IV – epiphytes that penetrate the outer layer of the host cell wall, associated with host's cortical disorganization and Type V – epiphytes that invade the tissues of the host, growing intercellularly, and associate with destruction of cortical and (in some cases) medullary cells (i.e. forming a parasitic relationship). Among these types, the most harmful is certainly the last one: the red filamentous alga first attributed to the genus *Polysiphonia* and subsequently described as *Neosiphonia*, which causes great losses in the crops (Fig. 8).

According to Hurtado and Critchley (2006), *Neosiphonia* (= *Polysiphonia*) infestations were first observed in a dense population by an American Peace Corp Volunteer, Jesse Shubert, in Calaguas Island, the Philippines, in 2000. He observed that there was widespread infestation of this epiphyte in this specific locality and this was brought to the attention of the international seaweed community through the internet. This epiphyte caused a distortion of the *K. alvarezii* thallus in the site of penetration, from the cortical to the medullary layers, and was named "goose bumps" by D. Largo (Hurtado et al., 2006) (Fig. 9).

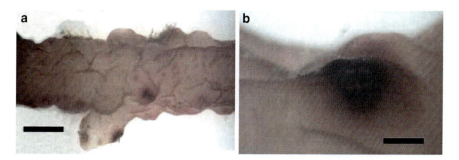

Figure 9. "Goose bump" formation caused by *Neosiphonia* infections on *Kappaphycus* thalli: (**a**) at the end of epiphyte infection phase and (**b**) epiphyte infected "mounts" with the onset of secondary bacterial infection. Scale bar = 300 μm (Courtesy of C. Vairappan).

Hurtado et al. (2006) affirmed that the occurrence of *Neosiphonia* (= *Polysiphonia*) epiphytes in Calaguas Is. resulted in tremendously reduced biomass production of *Kappaphycus* in the formerly productive cultivation area. Even now, only a few people continue to farm the species in the area since the *Neosiphonia* (= *Polysiphonia*) outbreak. The infestation is persistent rather than periodic, unlike the observations of Msuya and Kyewalyanga (2006) in Jambiani, Zanzibar, where seasonal presence of reddish filaments on the thalli of almost all seaweed cultivated after 6 weeks was noted, besides some additional signs of "ice-ice."

In Malaysia, Vairappan (2006) isolated a total of five epiphytic species from outbreaks: *Neosiphonia savatieri*, *N. apiculata*, *Ceramium* sp., *Acanthophora* sp., and *Centroceras* sp. The author observed the first emergence in late February 2006, with the appearance of tiny black spots on surface epidermal layer, which then became rough and the vegetative epiphyte surfaced after 3–4 weeks. Epiphytes were observed as solitary plants growing on the algal surface with rhizoids penetrating into the tissue of the cortical cell layers.

In the peak season, the dominant epiphytes, *N. savatieri*, were seen to grow close to each other at a maximum density of 40–48 epiphytes cm^{-2} (Vairappan, 2006). The emergence of epiphytes coincided with drastic changes in salinity and temperature; the author (op. cit.) suggests that there could be a possible correlation between the fluctuations in the abiotic factors and the emergence of epiphytes. In fact, Hurtado et al. (2006) found a strong correlation between the percentage cover of "goose bumps"- *Neosiphonia* (= *Polysiphonia*), light intensity, and water movement. According to these last authors, if these factors were limiting (leading to crop stress), then problems with epiphyte infestations increased. It seems that infestation by EFA in *Kappaphycus alvarezii* has a direct implication for careful selection of appropriate farm sites and selection of noninfected seedlings.

Other species of epiphytes had been identified by Hurtado et al. (2008a), e.g., red macro-epiphytes (*Actinotrichia fragilis*, *Acanthophora spicifera*, *A. muscoides*, *Amphiroa foliacea*, *A. dimorpha*, *A. fragilissima*, *Ceramium* sp., *Gracilaria arcuata*,

Hydropuntia edulis, Hypnea musciformis, H. spinella, H. valentiae, H. pannosa, Champia sp., *Chondrophycus papillosus*), brown macro-epiphytes (*Hydroclathrus clathratus, Dictyota divaricata, D. cervicornis, Padina australis, P. santae-crucis*), and green macro-epiphytes (*Boodlea composita, Chaetomorpha crassa, Ulva clathrata, U. compressa, U. fasciata, U. media, U. pertusa, U. reticulata*).

Vairappan et al. (2008) in a collaborative study including Filipino, Indonesian, Malaysian, and Tanzanian researchers identified the causative organism of epiphyte infestation in these countries, analyzed the infection density, the stages of infection, and observed the occurrence of secondary bacterial infection after the epiphytes dropped off. In all countries, the causative organism was identified as *Neosiphonia apiculata*, which presented on the host seaweed as follows: the Philippines (88.5 epiphytes cm^{-2}), Tanzania (69.0 epiphytes cm^{-2}), Indonesia (56.5 epiphytes cm^{-2}), and Malaysia (42.0 epiphytes cm^{-2}). According to these authors (op. cit.), the "goose-bump" is a characteristic feature of the epiphyte infection, with a formation of a pit in the middle where the epiphyte's basal primary rhizoid was loosely attached (Fig. 10a). After the drop off of the secondary rhizoids and their upper main thalli, tissue degradation began with the formation of tiny pores on the "goose-bumps" (Fig. 10b), followed by their disintegration and the establishment of secondary bacterial infection, mainly *Alteromonas* sp., *Flavobacterium* sp., and *Vibrio* sp. (Fig. 10c). C. Vairappan (2008, personal communication) observed in *N. apiculata* infected plants 25.6% lower carrageenan yield, 74.5% lower viscosity, 54.2% lower gel strength, 22.4% higher syneresis than healthy plants from commercial farms of *K. alvarezii* in Sabah, Malaysia, and a reduction of carrageenan size from 800 kDa in healthy specimens to 80 kDa in infected plants.

To try to minimize the occurrence of epiphytes, the use of uninfected, clean, and healthy "seedlings" of *Kappaphycus* is strongly recommended besides the careful selection of a farming site with clean and moderate to fast water movement, which has less siltation (Hurtado and Critchley, 2006). However, if a proliferation of *Neosiphonia* (= *Polysiphonia*) is noted, the cultivated seaweed must be totally harvested; attempts to select young branches of the infected plant for "seedling" purposes should not be made to prevent the transfer of epiphytes from one crop

Figure 10. Scanning electron microscopy (SEM) micrographs showing phases of cellular decomposition of the epiphyte-infected site. (**a**) Epiphytes rhizoids drop off from the "goose-bump"; (**b**) tissue degradation of the "goose-bumps"; and (**c**) secondary bacterial infection (Photos retrieved from Vairappan et al., 2008).

to another. If possible, it is recommended not to use the same farming site for the next cropping season (for other details, consult Hurtado and Critchley, 2006).

The impact of increasing surface seawater temperatures and the noted stress on cultivated seaweeds remains largely unknown. However, it could be speculated that the cultivated seaweeds are stressed and their vigor compromised, which then leaves them susceptible to colonization by epiphytes.

Other contamination, still not identified, was observed in Brazil, in *Kappaphycus alvarezii* cultivated in vitro. Despite the "optimum" conditions of the laboratory, some strains developed black spots on the thallus (within the cortical tissue) where after some days signs of "ice-ice" began (Fig. 11) (L. Hayashi, 2008, personal communication). An unsubstantiated report also exists of a graying of cortical tissue of plants from the Philippines, which resulted in even the normally white, finished, refined carrageenan powder having a gray discoloration (A.T. Critchley, 2008, personal communication). It would not be surprising if these signs are caused by fungal attack, especially given the history of the pathology of terrestrial crops. It is not "rocket science" to expect that carrageenophyte cultivars, asexually selected, and grown as monocrops will become increasingly the target of marine pathogens. The outcome needs to be seen, but much closer attention is required to the pathology of seaweed crops.

Considering all these problems that are possibly correlated with environmental change, and to avoid future problems with "ice-ice," epiphytes and even the presence of undesirable pathogenic organisms (such as fungal infestation), it is essential, now more than ever, to choose the best cultivation areas. Moreover, only the very best management practices available should be employed since *K. alvarezii* is a crop with limited genetic variability, i.e., cultivated plants are propagated only vegetatively and have no sexual reproduction, so that the species is very vulnerable to pathogenic agents. Vairappan et al. (2008) suggest that the outbreak of *N. apiculata* in Malaysian farms was caused by the negligent introduction of already "infected" *K. alvarezii* seedlings from the Philippines, without sufficient monitoring

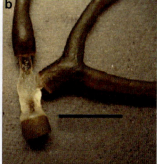

Figure 11. Unknown infection observed in *Kappaphycus alvarezii* in vitro cultivation in Brazil. (**a**) Three infected thalli. (**b**) Details of one infected part. Scale bar = 1 cm (Photos by L. Hayashi).

and quarantine procedures, after the farms were badly infected by "ice-ice disease" and epiphytes.

Adequate quarantine protocols and strict adherence are very important to minimize the risk of importing associated species or any diseased plants (Sulu et al., 2006). According to Ask et al. (2003), among all of the worldwide introductions of *K. alvarezii*, appreciative quarantine procedures were made only in the Solomon Islands and Brazil. In the first case (Solomon Islands), the seaweed was placed in raceways for 14 days, with the initial aim to remove invertebrates rather than to prevent the spread of infectious disease (and/or superficial algal epiphytes). In Brazil, a branch of 2.5 g was isolated and propagated in unialgal conditions in the laboratory, with seawater sterilized for 10 months (Paula et al., 1999). These procedures are quite different from each other, the first too imprecise and other (perhaps) too stringent. Sulu et al. (2006) recommend washing the seaweeds with filtered seawater to remove macro- and microbiota and to reduce the incidence of organisms on the transplanted fragments over a 2-week quarantine period. In Brazil, successive washes with distilled water and sterilized seawater, and then drying with tissue towel has been effective; in addition, the *K. alvarezii* branches were kept in quarantine for at least 2 months (L. Hayashi, 2008, personal communication).

8. The Issue of *Kappaphycus alvarezii* as an Invasive Organism

According to Zemke-White and Smith (2006), the genus *Kappaphycus* was introduced in 19 tropical countries versus *Eucheuma* into at least 13 tropical countries. Despite the polemics of the exotic species introduction, after more than 30 years from the beginning of *Kappaphycus* commercial cultivation in the Philippines, it is only in recent years that cases of bioinvasion have been reported. In the 19 tropical countries, two presented effective cases of invasion, causing serious environmental damage, mainly in coral reefs. Another good example of the edict is that just because a species will grow in a new environment is not a sufficient case for its introduction. Several people involved in seaweed cultivation (including Ask, 2008, personal communication) have stated that *Kappaphycus* and *Eucheuma* (while eminently suitable for cultivation) should not be relocated outside their natural range of distribution.

The most studied case of the impacts of *K. alvarezii* cultivation in the tropics is from Hawaii. The first study conducted by Russell (1983) 2 years after the introduction of the species (which he refers as *Eucheuma striatum*) in Kaneohe Bay, attested that *Eucheuma* did not establish over deep water or out of depressions, hollows, or channels, and was unable to colonize neighboring reefs without human help. He observed that the greatest accumulation of *Eucheuma* (23 t) occurred on the reef edge, but this was not a permanent or established population.

In 1996, Rodgers and Cox went back to the same bay and observed that *Kappaphycus* spp. (mentioned as *K. alvarezii* and *K. striatum*) had spread 6 km away from the initial site of introduction, at an average rate of 250 m year^{-1} (Rodgers and Cox, 1999).

After a further 5 years, Smith et al. (2002) concluded that *Kappaphycus* spp. had still not spread outside of Kaneohe Bay but had continued to spread northward in the bay since the Rodgers and Cox (1999) study.

Conklin and Smith (2005) found that in just 2 years, the alga had increased from less than 10% to over 50% cover on some patch and fringing reefs. According to this work and their account to Zemke-White and Smith (2006), in many cases the alga occupied over 80% cover of the benthos and generally grew in large three-dimensional mats in Kaneohe Bay, eventually overgrowing or interacting with reef-building coral (Fig. 12a and b). The removal of these plants has been made using a new weapon specially developed to this case, named "Super Sucker": an underwater vacuum that sucks the algae right off the reef (for more details, please consult: www.nature.org/wherewework/northamerica/states/hawaii/projectprofiles/art22268.html).

Bioinvasion is just one of the major problems in places where *Kappaphycus* is introduced. However, a fundamental issue involves the identification of the problem genus: *Kappaphycus* or *Eucheuma*? Recently, the problem seems to receive a highlight. Molecular analyses indicated that plants introduced to Hawaii, presumably from the Philippines, are distinct from all other *Kappaphycus* worldwide cultivated samples and their unique genotypes, as expressed in their unique haplotypes, may explain their invasive nature in Hawaii (Zuccarello et al., 2006). The work of Conklin et al. (2009) confirms that the species is *Kappaphycus alvarezii*, but the Hawaiian strain forms a separated grouping of the strains from Venezuela, Tanzania, and Madagascar.

Between 2007 and 2008, *Kappaphycus* bioinvasion problems received yet further, serious media attention. This time in India and as such received much attention even in journals as *Nature and Science*. Chandrasekaran et al. (2008) observed that *K. alvarezii* had successfully invaded and established on both dead and live corals in Kurusadai Island, India. The species had specifically invaded

Figure 12. *Kappaphycus alvarezii* (**a**) and *Eucheuma denticulatum* (**b**) overgrowing coral reefs in Kaneohe Bay, Hawaii (Photos from Zemke-White and Smith, 2006).

Figure 13. *Kappaphycus alvarezii* overgrowing corals in Kurusadai Island, India (Photos from Chandrasekaran et al., 2008).

the *Acropora* sp. and destroyed them by shadowing and smothering effects. The authors (op. cit.) noted an extraordinary phenotypic plasticity in terms of distinct variations in color and shape of the thallus, thickness of its major axis, morphological features, and frequency of primary and secondary branching (Fig. 13). The shadowing was due to smothering effect in which the major axis extends like an elastic rubber sheet and covers the maximum surface area of the corals. The invaded corals lost their skeletal integrity, stability, and rigidity and could be easily detached from the reef matrix.

The seaweed colonies could have been established from vegetative fragments of past and present trials of *Kappaphycus* cultivation. Many fragments can be generated and if not recollected, can be dispersed through wave action and settle on coral substrata. According to Chandrasekaran et al. (2008), other factors could include the relatively long duration (1 year) of cultivation experiments conducted in 1997 at different depths, as well as ongoing cultivation and ideal environmental conditions such as water temperature and availability of nutrients. Detachment of the thalli from the open and raft cultures during rough weather conditions, especially during the southwest and northeast monsoon seasons and their dispersal to other areas, cannot be ruled out.

In a public response to the publications concerning the invasion in India, scientists of Central Salt and Marine Chemical Research Institute (CSMCRI) contested the fact, arguing that they surveyed the region from 2005 to 2008 and had not observed the occurrence of nonfarmed populations of *Kappaphycus*. Furthermore, they stated that there was little possibility of the original germplasm in Kurusadai be the source of the reported outcome. The activities were discontinued in November 2003 due to inferior growth and heavy grazing (in this document however, they refer the plants as *Eucheuma* and not *Kappaphycus* – a common misrepresentation which has to be borne in mind when each genus is referred to, particularly in popular literature).

According to a review by Pickering et al. (2007), many Pacific Island countries attempted to cultivate *Kappaphycus* with no success. In some places, abandoned *K. alvarezii* aquaculture trials resulted in the thalli dying out, while in other places there were small residual populations. Zemke-White and Smith (2006), in their review, related the spread of *Kappaphycus alvarezii* and *Eucheuma denticulatum* from farms to neighboring reefs in Zanzibar, Fiji, and Venezuela, which had proceeded without bioinvasion into their coral reef environments. According to Pickering et al. (op. cit.) in this oligotrophic reef environments, *K. alvarezii* may be said to "persist" but it is not "proliferating."

It is clear that the invasive potential of the *Kappaphycus* or *Eucheuma* in environments is a prevailing danger but remains unpredictable. Zemke-White and Smith (2006) presupposed that given enough time, *Kappaphycus* and *Eucheuma* used in commercial cultivation could have the ability to spread from farm sites and establish independent populations. However, both the extent of this spread and the effects on local species may differ between locations and species. The introduction of *Eucheuma/Kappaphycus* from the Phillipines into Hawaii, and later into the Pacific Islands, demonstrates the complexity of this subject. Confusion in identifications between the genera is also a point of contention; the original report of *K. alvarezii* is regarded as a serious pest at one location in Hawaii. However, the cultivar *K. alvarezii* has in fact failed to establish in many localities in the Pacific Ocean where it was deliberately introduced for commercial cultivation (Pickering et al., 2007). Eucheumatoids are normally cultivated in relatively oligotrophic conditions, and the impacts of anthropogenic eutrophication of the bays in Hawaii have not been fully investigated, given sufficient association to changes in environmental conditions that may have tipped the balance in favor of successful development of eucheumatoids as invasive species.

Craige (2007) affirmed that properties of the invaded community could determine the success of the invader. Shifts from coral to algal domination on reefs have been found to be associated with the loss of biodiversity, reduction in the numbers of fish, decrease in the intrinsic value of the reef, and ultimately the erosion of the physical structure of the reef (Cocklin and Smith, 2005).

According to Pickering et al. (2007), there are many cases of *K. alvarezii* having been introduced, often without quarantine protocols for aquaculture projects that were underfinanced, or commercially, socio-economically or institutionally unsound, even without guaranteed markets. These ventures were essentially set up to fail from the outset. By so doing, these states expose themselves to the full extent of potential environmental risk, without capturing any of the anticipated socio-economic benefits against which they had balanced this risk.

Whether the bioinvasion is caused by *Kappaphycus* or *Eucheuma* is not the main concern. The main issue is what is the cause and how will the situation be controlled. Effective control options for current cultivations and establishment of quarantine protocols before any new introductions are even envisioned and would prevent similar problems in other reefs around the world.

9. Integrated Multitrophic Aquaculture

Aquaculture is probably the fastest growing food-production sector worldwide, being responsible for almost 50% of the fish for food and with greatest potential for satisfying the growing demand for aquatic food (FAO, 2006). According to

FAO, in the last 50 years, production increased from 20 million tons in the early 1950s to 157 million tons in 2005 (FishStat Plus – version 2.3). However, concerns over the impacts of these growing, intensive aquaculture activities has prompted the development of Integrated Multi-Trophic Aquaculture (IMTA) as one of the most sustainable alternative to minimize these impacts and increase the income of the organisms involved (Chopin et al., 2001).

According to Neori et al. (2007), the sustained expansion of intensive seafood production inescapably requires "trophic diversification" – an ecologically balanced, combined culture of organisms of high trophic levels (carnivores) with "service crops" from lower trophic levels (mainly seaweeds and filter feeders) to perform these tasks while adding income (Figs. 14 and 15). Based on this, the application

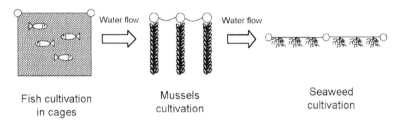

Figure 14. IMTA design in open sea and in Hainan Island cultivation (China) (Photos retrieved from Neori and Shpigel, 2006).

Figure 15. IMTA design in tanks and in Seaor Marine Enterprises in Israel (Photos retrieved from Neori and Shpigel, 2006).

of IMTA principles has been proposed as an eco-technological alternative to the optimization of productivity and utilization of energy, for removal and recycling toxic metabolites with the use of re-circulation systems. Adoption of such principles works towards the minimization and mitigation of the ecological impact of the activity (Folke and Kautsky, 1992; Buschmann et al., 2001; Chopin et al., 2001; Troell et al., 2003).

The best IMTA farms – whether intensive or extensive and whether in coastal waters, ponds, or tanks – are those in which there is a balance of waste production and extraction, thus becoming environmentally benign minibiospheres (Neori et al., 2007).

Seaweeds with commercial value were considered to be more advantageous, since they may offer high bioremediation efficiency and the nutrients absorbed are converted to biomass products (i.e. a nutrient sink), used for their growth. Apart from acting as agents for nutrient removal, seaweeds are promising sources of high-value biochemicals (such as antibiotics, cosmetics, and nutritional additives), phycocolloids (agar, carrageenans, and alginates), and nutritious food for other cultivated marine animals (such as abalone), and serve as popular and healthy foods, particularly in Southeast Asia. Together with shellfish, they are sources of healthy marine omega-3 fatty acids. With further research and development, and strain selection, many new uses could be found for various seaweeds, and they might even replace a significant proportion of the fishmeal employed to supply carnivorous fish diets with balanced proteins and oils (Neori et al., 2007). Ask and Azanza (2002) suggested that polyculture systems with commercial eucheumatoid species might improve the economic potential for cultivation if their production costs decreased and/or the production increased due to benefits derived from co-cultivation.

The potential for *Kappaphycus alvarezii* to remove N-compounds from seawater has been proven. Doty (1987) reported reductions of 24% in nitrate and nitrite concentrations, and a reduction of 6% in phosphate concentration by the commercial cultivation of *Eucheuma* (*Kappaphycus*) in the Philippines. Li et al. (1990) observed that plants fertilized for 1 h in 5–25 mM of ammonium had higher growth rates than plants without fertilization; however, plants fertilized with 35–50 mM of ammonium decreased their growth rates, probably because of the toxic effects of higher concentrations. These authors also observed that plants fertilized and transferred to open sea cultivation resulted in higher growth rates and carrageenan yield and quality than the control. According to them, the coloration of plants changed from yellowish-brown to dark brown, possibly indicating N storage. Mairh et al. (1999) tested other species, i.e., *K. striatum* and observed that cultures supplied with 1–3 mg N L^{-1} from (NH$_4$)$_2$SO$_4$ or 3–5 mg N L^{-1} NH$_4$NO$_3$ showed a significant increase in wet weight and bioaccumulation of total nitrogen content. The conclusions of this study proposed that the alga could accumulate NH$_4^+$ ions from low concentrations to maintain its growth. Dy and Yap (2001) observed that a surge uptake of ammonium by *K. alvarezii* usually occurred within the first 30 min of fertilization and ranged from 15 to 35 μmol NH$_4^+$ g^{-1} dry weight h^{-1}. The authors forecast that surge uptake might be an important strategy for the alga to survive in areas with low nitrogen concentrations and take advantage of the naturally occurring nitrogen pulses in the field. Msuya and Salum (2007) observed higher carrageenan yield and gel strength in fertilized than unfertilized *K. alvarezii*, with TAN (total ammonia nitrogen) uptake efficiency of 81% after fertilization with 12:10:8 NPK in small natural pools that remained when the tide was out, in Zanzibar, Tanzania. However, in contrast, Msuya and Kyewalyanga (2006) working in a similar setup observed an average of 67 ± 15% TAN removal efficiency of *K. alvarezii* and no significant differences were observed in dry matter ($p = 0.347$), carrageenan yield ($p = 0.059$), *iota* content ($p = 0.767$), and gel viscosity ($p = 0.968$) between fertilized and unfertilized treatments.

Taking all this evidence into account, the efficiency of *K. alvarezii* as a biofilter has been tested in IMTA with oyster, shrimp, and fish. Qian et al. (1996) observed that *K. alvarezii* fertilized for 1 h with pearl oyster (*Pinctada martensi*) wastes grew faster than those without fertilization in sea cultivation. Lombardi et al. (2006) estimated a production of 17.5 kg of fresh *Kappaphycus* m^{-2} year^{-1} co-cultivated with shrimps (*Litopeneaus vannamei*) in floating cages $1 \times 1 \times 1$ m^3. Despite the promising results, unfortunately, not one of the previous authors quantified the nutrients removed by the species. They only verified the effects of nutrient pulses on *K. alvarezii* growth (in the first case) and the productivity of the species cultivated in cages with the shrimps (in the second case).

The first attempts to evaluate the biofilter potential were made by Rodrigueza and Montaño (2007) and Hayashi et al. (2008), with an IMTA project including *K. alvarezii* with fish (*Chanos chanos* and *Trachinotus carolinus*, respectively). Despite the low growth rates obtained in these studies, the removal of ammonium from fish effluents was similar (e.g. 66–70%), in both studies. Rodrigueza and Montaño (op. cit.) observed a considerable increase in carrageenan yield from the plants cultivated in effluents; the rheological properties and colloid quality did not vary. Hayashi et al. (op. cit.) also observed maximal removal efficiency of 18.2% nitrate, 50.8% nitrite, and 26.8% phosphate, but did not distinguish any differences in carrageenan yield between plants cultivated in effluent or in the open sea. They also noted that after the period nutrient preconditioning by tank cultivation (10 days), healthy seedlings transferred to the sea cultivation presented higher growth rates than seedlings without previous tank cultivation, suggesting the possibility of nutrient storage.

As Troell et al. (2003) suggested, for future research it is essential to consider some important points listed below:

- Be familiar with biological/biochemical processes in closed recirculating and open seaweed cultivation systems
- Optimize the production of the extractive organisms as well as the fed organisms
- Try to establish the best water flow and circulation
- Conduct research into technologies at scales relevant to commercial implementation or suitable extrapolation

- Attain a detailed understanding of the temporal variability in seaweed-filtered mariculture systems and study the influences of location-specific parameters, such as latitude, climate, and local seaweed strains/species on seaweed filter performance, among other things

For details, see Troell et al. (2003).

The first published results obtaining by Rodrigueza and Montaño (2007) and Hayashi et al. (2008) are promising and require further study, searching the best site-specific system(s) that optimize the biofilter potential of the species, the seaweed growth, and (importantly) improve the carrageenan yield and quality to increase the economic benefits of the operational efforts.

According to Neori et al. (2007), the introduction of a nutrient emission tax or its exemption through the implementation of biomitigative practices would make the economic validity of the IMTA approach even more obvious. Moreover, by adopting the IMTA approach, the mariculture industry could increase its social acceptability. Although it is very difficult to assign a monetary value to such a sociopolitical variable, gaining public acceptance is imperative for the development of the industry's full potential. Also, reducing environmental and economic risk in the long term should make financing easier to obtain.

10. Efforts Taken to Combat the Problems

Apart from the efforts to combat the "ice-ice" problem, only limited efforts have been made by stakeholders, especially scientists and processing companies, to combat the die-off problem and increase seaweed production. One such effort was the development of the floating methods in contrast to the off-bottom method (Msuya, 2006b Msuya et al., 2007). Whereas the off-bottom method is used during low tides only, the deep-water floating lines technique can be used all the time. However, boats or canoes are required to go into deep waters for the floating lines technique. This may be a challenge to the poor coastal communities, but with more effort, entrepreneurship training, and encouragement, the farmers are able to possess at least one boat/canoe per family or per group.

Another example of efforts applied is the use of natural fertilization with seaweeds extracts to mitigate the proliferation of epiphytes, reduce die-offs, and increase productivity. Experiments utilizing Acadian Marine Plant Extract Powder (AMPEP) have been published; viz. Loureiro et al. (2009) tested apical branches of *Kappaphycus alvarezii* cultivated in vitro in different concentrations of AMPEP normally used in horticultural or agricultural applications for land plants. These authors observed considerable reduction of epiphytes and an increase in *Kappaphycus* growth rates. Other experiments could be made to test the effects of the extracts in open sea farms.

The growth of the three color morphotypes of *Kappaphycus alvarezii* var. *tambalang* was also enhanced using AMPEP. Furthermore, there was no incidence of *Neosiphonia* infestation during the 3-month culture period and there was a limited occurrence of macro-epiphytes (A. Hurtado, 2008, personal communication). The presence of filamentous green *Ulva* did not hamper the robust and healthy growth of the three color morphotypes of *K. alvarezii* var. *tambalang*. The results of the study indicate the capacity of AMPEP to reduce the incidence of epiphytes. In another study, the use of AMPEP in tissue culture of *K. alvarezii* var. *tambalang* was successful in the micropropagation of plantlets for outplanting purposes (Hurtado et al., 2008).

Buyers of seaweeds who are also the exporters from producing countries have tried to solve the problem of die-off by introducing different varieties (or species) of *Kappaphycus*. In Tanzania, the buyers of the seaweed introduced a species of the seaweed, *Kappaphycus striatum* (locally known as "kikarafuu") about 5 years ago. This was done in collaboration with the government due to import restrictions on introducing new species. The species got its local name from its fronds that look like the clove, which is locally called karafuu in Kiswahili, the local language of Tanzania. The species is now farmed in a number of villages in the Zanzibar Islands. The buyers also introduced a species that is thought to be a variety of *Kappaphycus alvarezii* known locally as "kikorosho," which has not been as successful as the "kikarafuu." The additional species and variety had lower growth rates than the commonly farmed *K. alvarezii* (variety "cottonii"). However, they are not as prone to environmental problems as the cottonii, although *K. striatum* does get black spots (of unknown cause) during bad growth seasons.

Farmers have been practicing what we can call "shift cultivation." When they find that the seaweed is failing, the farmers have moved the cultivation site to a different area within the shallow lagoons. In Tanzania, for example, farmers in the area of Bagamoyo, on the mainland, have been doing this for sometime and it is possible for them to tell which area they can farm at what time of the year, although the weather has been very unpredictable in the most recent 2–3 years. Having found it difficult, due to the unpredictable weather, the farmers are now also using the deepwater floating lines technique following the efforts of scientists in Tanzania (Msuya et al., 2007).

11. Issues To Be Considered for the Future of *Kappaphycus* Farming

It has been shown that the farming of the genus has been failing in many cultivation sites where it used to grow well. It may well be that changes in environmental conditions over the cultivation periods have affected the seaweed negatively. Unfortunately, the actual causes of the die-offs are not studied extensively nor well documented, but possible causes such as abrupt changes in surface seawater temperature, salinity, pH, have been mentioned by several scientists as contributing factors, e.g. Bryceson (2002), Mmochi et al. (2005). Other negative impacts are suggested having anthropogenic causes, e.g. eutrophication from runoff, agrochemicals used in agriculture, etc. Whether these changes are related to global climate change in surface seawater temperatures or direct anthropogenic causes must remain speculative until the issues are studied in much greater detail.

Floating cages are a good way of producing more *Kappaphycus* whether in polyculture or monoculture (Hurtado-Ponce, 1992). The fact that polyculture of the seaweed with animals such as shrimps (Lombardi et al., 2001, 2006) or fish (Rodrigueza and Montaño, 2007; Hayashi et al., 2008) leads to more production provides a bright future for production of the seaweed when demand for raw materials is high.

There is also need to further study the genetics of eucheumatoids and to cross-breed the *Kappaphycus* varieties currently cultivated to develop more resistant strains that can be farmed in the areas where the current strains have failed. This is a challenge to scientists as well as other stakeholders worldwide. The genetic modification of eucheumatoids should be rejected outright, since there will be significantly reduced demand for the carrageenan products if labeling (as practiced in EU countries) were required.

As seaweed farming has already been taken as a viable economic activity of many communities worldwide and as a tool for economic empowerment of coastal women in developing countries (Pettersson-Löfquist, 1995; Quiñonez, 2000; Bryceson, 2002; van Ingen et al., 2002; Msuya, 2006a), there is need to take more action to sustain the activity. The world preference of *K. alvarezii* as a source of *kappa* carrageenan as opposed to *E. denticulatum* (as the dominant source of *iota* carrageenan) is a challenge that should be taken by all stakeholders to find ways of producing more of the higher-valued, but environmentally challenged, species. Without concentrated, sustained additional efforts, the future of *Kappaphycus* monocrop cultivation could be at stake. All the stakeholders have a challenge of providing ways of solving the problem for the benefit of coastal communities, the producer countries, and the industry dependent on the reliable supply of high-quality, sustainably produced raw materials.

12. References

Adnan, H. and Porse, H. (1987) Culture of *Eucheuma cottonii* and *Eucheuma spinosum* in Indonesia. Hydrobiologia **151/152**: 355–358.

Ask, E.I. (2006) Cultivating cottonii and spinosum: a "how to" guide, In: A.T. Critchley, M. Ohno and D.B. Largo (eds.) *World Seaweed Resources – An Authoritative Reference System*. ISBN 90 75000 80 4, ETI Information Services, UK. DVD-ROM.

Ask, E.I. and Azanza, R.V. (2002) Advances in cultivation technology of commercial eucheumatoid species: a review with suggestions for future research. Aquaculture **206**: 257–277.

Ask, E.I., Batibasaga, A., Zertuche-Gonzalez, J.A. and de San, M. (2003) Three decades of *Kappaphycus alvarezii* (Rhodophyta) introduction to non-endemic locations, In: A.R.O. Chapman, R.J. Anderson, V.J. Vreeland and I.R. Davison (eds.) *Proceedings of the 17th International Seaweed Symposium*. Oxford University Press, pp. 49–57.

Barraca, R.T. and Neish, I.C. (1978) A survey of *Eucheuma* farming practices in Tawi-Tawi. I. The Sitangkai, Sibutu, Tumindao Region. Report, Marine Colloids (Phils.) Inc. 65 pp.

Bergschmidt, H. (1997) Seaweed production in Kiribati: a new cash crop. Pacific Bull. **10**: 9–12.

Bryceson, I. (2002) Coastal aquaculture developments in Tanzania: sustainable and non-sustainable experience. Western Indian Ocean J. Mar. Sci. **1**(1): 1–10.

Bulboa, C.R. and Paula, E.J. (2005) Introduction of non-native species of *Kappaphycus* (Rhodophyta) in sub-tropical waters: comparative analysis on growth rates of *Kappaphycus alvarezii* and *Kappaphycus striatum in vitro* and in the seas in south-eastern Brazil. Phycol. Res. **53**: 183–188.

Buschmann, A.H., Troell, M. and Kautsky, N. (2001) Integrated algal farming: a review. Cah. Biol. Mar. **42**: 83–90.

Carlsson, A.S., van Beilen, J.B., Möller, R. and Clayton, D. (2007) Micro- and macro-algae: utility for industrial applications, In: D. Bowles (ed.) *Outputs from the EPOBIO Project*. CPL Press, United Kingdon, September, 86 pp.

Chandrasekaran, S., Nagendran, N.A., Pandiaraja, D., Krishnankutty, N. and Kamalakannan, B. (2008) Bioinvasion of *Kappaphycus alvarezii* on corals in the Gulf of Mannar, India. Cur. Sci. **94**: 1167–1172.

Chopin, T. (2008) Opinion: let's not make IMTA an afterthought in open ocean development. Integrated multi-trophic aquaculture (IMTA) will also have its places when aquaculture moves to the open ocean. Fish farmer. **31**(2): 40–41.

Chopin, T., Buschmann, A.H., Halling, C., Troell, M., Kautsky, N., Neori, A., Kraemer, G.P., Zertuche-González, J.A., Yarish, C. and Neefus, C. (2001) Integrating seaweeds into marine aquaculture systems: a key toward sustainability. J. Phycol. **37**: 975–986.

Cocklin, E.J. and Smith, J.E. (2005) Abundance and spread of the invasive red algae, *Kappaphycus* spp., in Kane'ohe Bay, Hawai'i and an experimental assessment of management options. Biol. Inv. **7**: 1029–1039.

Conklin, K.Y., Kurihara, A. and Sherwood, A.R. (2009) A molecular method for identification of the morphologically plastic invasive algal genera *Eucheuma* and *Kappaphycus* (Rhodophyta, Gigartinales) in Hawaii. J. Appl. Phycol. **21**(6):691.

Craige, R.J. (2007) Seaweeds invasions: a synthesis of ecological, economic and legal imperatives. Bot. Mar. **50**: 321–325.

Dawes, C.J., Lluisma, A.O. and Trono, G.C. (1994) Laboratory and field growth studies of commercial strains of *Eucheuma denticulatum* and *Kappaphycus alvarezii* in the Philippines. J. Appl. Phycol. **6**: 21–24.

Doty, M.S. (1987) The production and use of *Eucheuma*, In: M.S. Doty, J.F. Caddy and B. Santelices (eds.) *Case Studies of Seven Commercial Seaweed Resources*. FAO Fish Tech. Pap. **281**: 123–161.

Dy, D.T. and Yap, H.T. (2001) Surge ammonium uptake of the cultured seaweed, *Kappaphycus alvarezii* (Doty) Doty (Rhodophyta, Gigartinales). J. Exp. Mar. Biol. Ecol. **265**: 89–100.

FAO (2006) *State of World Aquaculture. FAO Fish Tech Pap 500*, Rome, 134 pp.

Folke, C. and Kautsky, N. (1992) Aquaculture with its environment: prospects for sustainability. Ocean Shor. Manag. **17**: 5–24.

Gerung, G.S. and Ohno, M. (1997) Growth rates of *Eucheuma denticulatum* (Burman) Collins et Harvey and *Kappaphycus striatum* (Schmitz) Doty under different conditions in warm waters of Southern Japan. J. Appl. Phycol. **9**: 413–415.

Glenn, E.P. and Doty, M.S. (1990) Growth of the seaweeds *Kappaphycus alvarezii, K. striatum* and *Eucheuma denticulatum* as affected by environment in Hawaii. Aquaculture **84**: 245–255.

Hayashi, L., Oliveira, E.C., Bleicher-Lhonneur, G., Boulenguer, P., Pereira, R.T.L., von Seckendorff, R., Shimoda, V.T., Leflamand, A., Vallée, P. and Critchley, A.T. (2007) The effects of selected cultivation conditions on the carrageenan characteristics of *Kappaphycus alvarezii* (Rhodophyta, Solieriaceae) in Ubatuba Bay, São Paulo, Brazil. J. Appl. Phycol. **19**: 505–511.

Hayashi, L., Yokoya, N.S., Ostini, S., Pereira, R.T.L., Braga, E.S. and Oliveira, E.C. (2008) Nutrients removed by *Kappaphycus alvarezii* (Rhodophyta, Solieriaceae) in integrated cultivation with fishes in re-circulating water. Aquaculture **277**: 185–191.

Hung, L.D., Hori, K., Nang, H.Q., Kha, T. and Hoa, L.T. (2009) Seasonal changes in growth rate, carrageenan yield and lectin content in the red alga *kappaphycus alvarezii* cultivated in Camranh Bay, Vietnam. J. Appl. Phycol. **21**: 262–272.

Hurtado-Ponce, A.Q. (1992) Cage culture of *Kappaphycus alvarezii* var. *tambalang* (Gigartinales, Rhodophyceae). J. Appl. Phycol. **4**: 311–313.

Hurtado-Ponce, A.Q. (1994) Cage culture of *Kappaphycus alvarezii* (Doty) Doty and *Epiphenelus* sp. Proc. Nat. Symp. Marine Sci. **2**: 103–107.
Hurtado-Ponce, A.Q. (1995) Carrageenan properties and proximate composition of three morphotypes of *Kappaphycus alvarezii* Doty (Gigartinales, Rhodophyta). Bot. Mar. **38**: 215–219.
Hurtado, A.Q. (2007) Establishment of seaweed nurseries in Zamboanga City, Philippines. Terminal Report submitted to IFC-ADB, 24 September 2007, 17 pp.
Hurtado, A.Q. and Critchley, A.T. (2006) Epiphytes, In: A.T. Critchley, M. Ohno and D.B. Largo (eds.) *World Seaweed Resources – An Authoritative Reference System*. ISBN 90 75000 80 4, ETI Information Services, UK. DVD-ROM.
Hurtado, A.Q., Agbayani, R.F., Sanares, R. and Castro-Mallare, M.T.R. (2001) The seasonality and economic feasibility of cultivating *Kappaphycus alvarezii* in Panagatan Cays, Caluya, Antique, Philippines. Aquaculture **199**: 295–310.
Hurtado, A.Q. and Agbayani, R.F. (2002) Deep-sea farming of *Kappaphycus alcvarezii* using the multiple raft, long-line method. Bot. Mar. **45**: 438–444.
Hurtado, A.Q., Critchley, A.T., Trespoey, A. and Bleicher-Lhonneur, G. (2006) Occurrence of *Polysiphonia* epiphytes in *Kappaphycus* farms at Calaguas Is., Camarines Norte, Phillippines. J. Appl. Phycol. **18**: 301–306.
Hurtado, A.Q., Bleicher-Lhounner, G. and Critchley, A.T. (2008a) *Kappaphycus "cottonii"* farming. Cargill Texturizing Solutions, 26 pp.
Hurtado, A.Q., Critchley, A.T., Trespoey, A. and Bleicher-Lhonneur, G. (2008b) Growth and carrageenan quality of *Kappaphycus striatum* var. sacol grown at different stocking densities, duration of culture and depth. J. Appl. Phycol. **20**: 551–555.
Hurtado, A.Q., Yunque, D., Tibubos, K. and Critchley, A.T. (2009) Use of acadian marine plant extract from *Ascophyllum nodosum* in tissue culture of *Kappaphycus* varieites. J. Appl. Phycol. **21**: 633–639.
IFC (2003) *PENSA Seaweed Report for EAST*, Makassar, Indonesia, 39 pp.
Largo, D.B., Fukami, K. and Nishijima, T. (1995) Occasional pathogenic bacteria promoting *ice–ice* disease in the carrageenan-producing red algae *Kappaphycus alvarezii* and *Eucheuma denticulatum* (Solieriaceae, Gigartinales, Rhodophyta). J. Appl. Phycol. **7**: 545–554.
Largo, D.B., Fukami, K. and Nishijima, T. (1999) Time-dependent attachment mechanisms of bacterial pathogen during ice-ice infection in *Kappaphycus alvarezii* (Gigartinales, Rhodophyta). J. Appl. Phycol. **11**: 129–136.
Li, R., Li, J. and Wu, C.Y. (1990) Effect of ammonium on growth and carrageenan content in *Kappaphycus alvarezii* (Gigartinales, Rhodophyta). Hydrobiologia **204/205**: 499–503.
Lim, J.R. and Porse, H. (1981) Breakthrough in the commercial culture of *Eucheuma spinosum* in Northern Bohol, Philippines. Int. Seaweed Symp. **10**: 602–606.
Lombardi, J.V., Marques, H.A. and Barreto, O.J.S. (2001) Floating cages in open sea water: an alternative for promoting integrated aquaculture in Brazil. World Aquacult. September: 47–50.
Lombardi, J.V., Marques, H.L.A., Pereira, R.T.L., Barreto, O.J.S. and Paula, E.J. (2006) Cage polyculture of the Pacific white shrimp *Litopenaeus vannamei* and the Philippine seaweed *Kappaphycus alvarezii*. Aquaculture **258**: 412–415.
Loureiro, R.R., Reis, R. and Critchley, A.T. (2010) *In vitro* cultivation of three *Kappaphycus alvarezii* (Rhodophyta, Areschougiaceae) variants (green, red and brown) exposed to a commercial extract of the brown alga *Ascophyllum nodosum* (Fucaceae, Ochrophyta). J. Appl. Phycol. **22**: 101–104.
Luxton, D.M., Robertson, M. and Kindley, M.J. (1987) Farming of *Eucheuma* in the south Pacific islands of Fiji. Hydrobiologia **151/152**: 359–362.
Mairh, O.P., Zodape, S.T., Tewari, A. and Mishra, J.P. (1999) Effect of nitrogen sources on the growth and bioaccumulation of nitrogen in marine red alga *Kappaphycus striatum* (Rhodophyta, Solieriaceae) in culture. Indian J. Mar. Sci. **28**: 55–59.
McHugh, D.J. (2003) *A Guide to the Seaweed Industry*. FAO Fisheries Technical Paper. Food and Agriculture Organization of the United Nations, Rome, 105 pp.
Mendonza, W.G., Montaño, N.E., Ganzon-Fortes, E.T. and Villanueva, R.D. (2002) Chemical and gelling profile of *ice-ice* infected carrageenan from *Kappaphycus striatum* (Schmitz) Doty "sacol" strain (Solieriaceae, Gigartinales, Rhodophyta). J. Appl. Phycol. **14**: 409–418.

Mendonza, W.G., Ganzon-Fortes, E.T., Villanueva, R.D., Romero, J.B. and Montano, M.N.E. (2006) Tissue age as a factor affecting carrageenan quantity and quality in farmed *Kappaphycus striatum* (Schmitz) Doty ex Silva. Bot. Mar. **49**: 57–64.

Mmochi, A.J., Shaghude, Y.W. and Msuya, F.E. (2005) Comparative Study of Seaweed Farms in Tanga, Tanzania. Submitted to SEEGAAD Project, August 2005, 37 pp.

Msuya, F.E. (2006a) The impact of seaweed farming on the social and economic structure of seaweed farming communities in Zanzibar, Tanzania, In: A.T. Critchley, M. Ohno and D.B. Largo (eds.) *World Seaweed Resources*, Version: 1.0, ISBN: 90-75000-80-4. 27 p. (www.etiis.org.uk).

Msuya, F.E. (2006b). The seaweed cluster initiative in Zanzibar, Tanzania, In: B.L.M. Mwamila and A.K. Temu (eds.) *Proceedings of the Third Regional Conference on Innovation Systems and Innovative Clusters in Africa*. Dar es Salaam, Tanzania, pp. 246–260.

Msuya, F.E. (2009) Impacts of environmental changes on the farmed seaweed and seaweed farmers in Songosongo Island, Tanzania. Report submitted under a Collaborative Project on Sustaining Coastal Fishing Communities, Memorial University of Newfoundland-University of Dar es Salaam, 15 pp.

Msuya, F.E. and Kyewalyanga, M.S. (2006) Quality and quantity of phycocolloid carrageenan in the seaweeds *Kappaphycus alvarezii* and *Eucheuma denticulatum* as affected by grow out period, seasonality, and nutrient concentration in Zanzibar, Tanzania. Report submitted to Cargill Texturizing Solutions, 46 pp.

Msuya, F.E. and Salum, D. (2007) Effect of cultivation duration, seasonality, nutrients, air temperature and rainfall on carrageenan properties and substrata studies of the seaweeds *Kappaphycus alvarezii* and *Eucheuma denticulatum* in Zanzibar, Tanzania. WIOMSA/MARG I n° 2007-06. 36 pp.

Msuya, F.E., Shalli, M.S., Sullivan, K., Crawford, B., Tobey, J. and Mmochi, A.J. (2007) A comparative economic analysis of two seaweed farming methods in Tanzania. The Sustainable Coastal Communities and Ecosystems Program. Coastal Resources Center, University of Rhode Island and the Western Indian Ocean Marine Science Association, 27 p. (www.crc.uri.edu, www.wiomsa.org).

Muñoz, J. and Sahoo, D. (2007) Impact of large scale *Kappaphycus alvarezii* cultivation in coastal water of India, XIXth International Seaweed Symposium, Kobe, Japan, Program and Abstracts, 121 pp.

Muñoz, J., Freile-Pelegrin, Y. and Robledo, D. (2004) Mariculture of *Kappaphycus alvarezii* (Rhodophyta, Solieriaceae) color stains in tropical waters of Yucatan, Mexico. Aquaculture **239**: 161–177.

Murphy, D. (2002) Philippines swap guns for rakes: Christian Science Monitor, March 4.

Neish, I. (2008) First Indonesia Seaweed Forum, Makassar, Indonesia, Oct 27–Nov 30.

Neori, A. and Shpigel, M. (2006) Algae: key for sustainable mariculture – profitably expanding mariculture without pollution, In: A.T. Critchley, M. Ohno and D.B. Largo (eds.) *World Seaweed Resources – An Authoritative Reference System*. ISBN 90 75000 80 4, ETI Information Services, UK. DVD-ROM.

Neori, A., Troell, M., Chopin, T., Yarish, C., Critchley, A. and Buschmann, A.H. (2007) The need for a balanced ecosystem approach to blue revolution aquaculture. Environment **49**: 36–43.

Ohno, M., Nang, H.Q. and Hirase, S.T. (1996) Cultivation and carrageenan yield and quality of *Kappaphycus alvarezii* in the waters of Vietnam. J. Appl. Phycol. **8**: 431–437.

Paula, E.J., Pereira, R.T.L. and Ohno, M. (1999) Strain selection in *Kappaphycus alvarezii* var. *alvarezii* (Solieriaceae, Rhodophyta) using tetraspore progeny. J. Appl. Phycol. **11**: 111–121.

Paula, E.J., Pereira, R.T.L. and Ohno, M. (2002) Growth rate of the carrageenophyte *Kappaphycus alvarezii* (Rhodopyta, Gigartinales) introduced in subtropical waters of Sao Paolo State, Brazil. Phycol. Res. **50**: 1–9.

Pettersson-Löfquist, P. (1995) The development of open-water algae farming in Zanzibar: reflections on the socioeconomic impact. Ambio **24**(7–8): 487–491.

PDAP (2007). *Sitangkai Seaweed Industry Master Plan*. Quezon City, Philippines, 78 pp.

Pickering, T.D., Skelton, P. and Sulu, R.J. (2007) Intentional introductions of commercially harvested alien seaweeds. Bot. Mar. **50**: 338–350.

Pollnac, R.B., Sondita, F., Crawford, B., Mantjoro, E., Rotinsulu, C. and Siahainenia, A. (1997) Baseline assessment of socioeconomic aspects of resources use in the coastal zone of Bentenan and

Tumbak. Proyek Pesisir Technical Report No: TE-97/02-E. Coastal Resources Center, University of Rhode Island, Narragansett, Rhode Island, USA, 79 pp.

Qian, P.Y., Wu, C.W.Y., Wu, M. and Xie, Y.K. (1996) Integrated cultivation of the red alga *Kappaphycus alvarezii* and the pearl oyster *Pinctada martensi*. Aquaculture **147**: 21–35.

Quiñonez, N.B. (2000) Change in behaviour and seaweed farming lead to improved lives. Centre for empowerment and resource development, Inc. (CERD), Hinatuan, Surigao del Sur, Mindanao, Philippines.

Rodgers, S.K. and Cox, E.F. (1999) Rate of spread of introduced rhodophytes *Kappaphycus alvarezii, Kappaphycus striatum*, and *Gracilaria salicornia* and their current distributions in Kane'ohe Bay, O'ahu, Hawai'i. Pac. Sci. **53**: 232–241.

Rodrigueza, M.R.C. and Montaño, M.N.E. (2007) Bioremediation potential of three carreegenophytes in tanks with seawater from fish farms. J. Appl. Phycol. **19**: 755–762.

Russell, D.J. (1983) Ecology of the imported red seaweed *Eucheuma striatum* Schmiz on Coconut Island, Oahu, Hawaii. Pac. Sci. **37**: 87–107.

Samonte, G.P.B., Hurtado-Ponce, A.Q. and Caturao, R. (1993) Economic analysis of bottom line and raft monoline culture of *Kappaphycus alvarezii* var. *tambalang* in Western Visayas, Philippines. Aquaculture **110**: 1–11.

Smith, J.E., Hunter, C.L. and Smith, C.M. (2002) Distribution and reproductive characteristics of nonindigenous and invasive marine algae in the Hawaiian Islands. Pac. Sci. **56**: 299–315.

Sulu, R., Kumar, L., Hay, C. and Pickering, T. (2006) *Kappaphycus* seaweed in the Pacific: review of introductions and field testing proposed quarantine protocols, In: A.T. Critchley, M. Ohno and D.B. Largo (eds.) *World Seaweed Resources – An Authoritative Reference System*. ISBN 90 75000 80 4, ETI Information Services, UK. DVD-ROM.

Troell, M., Halling, C., Neori, A., Chopin, T., Buschmann, A.H., Kautsky, N. and Yarish, C. (2003) Integrated mariculture: asking the right questions. Aquaculture **226**: 69–90.

Unsrat (2001) Budidaya laut dan pengembangan mata pencharian tambahan. Konsultan Fakultas Perikanan dan Ilmu Kelautan Universitas Sam Ratulangi. Technical Report TE-01/08-I Coastal Resources Center, University of Rhode Island, USA, 77 pp.

Uyengco, F.R., Saniel, L.S. and Jacinto, G.S. (1981) The 'ice-ice' problem in seaweed farming. Proc. Inter. Seaweed Symp. **10**: 625–630.

Vairappan, C.S. (2006) Seasonal occurrences of epiphytic algae on the commercially cultivated red alga *Kappaphycus alvarezii* (Solieriaceae, Gigartinales, Rhodophyta). J. Appl. Phycol. **18**: 611–617.

Vairappan, C.S., Chung, C.S., Hurtado, A.Q., Soya, F.E., Bleicher-Lhonneur, G. and Critchley, A. (2008) Distribution and symptoms of epiphyte infection in major carrageenophyte-producing farms. J. Appl. Phycol. **20**: 477–483.

Van Ingen, T., Kawau, C. and Wells, S. (2002) Gender Equity in Coastal Zone Management: Experiences from Tanga, Tanzania. UCN Eastern Africa Regional Programme. Tanga Coastal Zone Conservation and Development Programme, 26 pp.

Wakibia, J.G., Bolton, J.J., Keats, D.W. and Raitt, L.M. (2006) Factors affecting the growth rates of three commercial eucheumoids at coastal sites in southern Kenya. J. Appl. Phycol. **18**: 565–573.

Watson, D.B. (2000) Seaweed cultivation in Indonesia; social, environmental and Public health and carrageenan regulation: a review and analysis. J. Appl. Phycol. **20**: 505–513.

Wu, C., Li, J., Xia, E., Peng, Z., Tan, S., Li, J., Wen, Z., Huang, X., Cai, Z. and Chen, G. (1989) On the transplantation and cultivation of *Kappaphycus alvarezii* in China. Chin. J. Oceanol. Limnol. **7**: 327–334.

Zemke-White, W.L. and Smith, J.E. (2006) Environmental impacts of seaweed farming in the tropics, In: A.T. Critchley, M. Ohno and D. Largo (eds.) *World Seaweed Resources*. Expert Centre for Taxonomic Identification (ETI), University of Amsterdam (CD-ROM series).

Zuccarello, G.C., Critchley, A.T., Smith, J.E., Sieber, V., Bleicher-Lhonneur, G. and West, J.A. (2006) Systematics and genetic variation in commercial *Kappaphycus* and *Eucheuma* (Solieriaceae, Rhodophyta). J. Appl. Phycol. **18**: 643–651.

Biodata of Associate **Professor Jing-Chun Tang, Hideji Taniguchi, Professor Qixing Zhou,** and **Professor Shinichi Nagata,** authors of *"Recycling of the Seaweed Wakame Through Degradation by Halotolerant Bacteria"*

Jing-Chun Tang is an Associate Professor at Nankai University (China). He received his Doctor's degree from Nagoya University of Japan (2004) specializing in environmental microbiology and biological solid waste disposal. He analyzed the microbial community of composting by quinone profile and 16S rDNA DGGE method. Now, he is undertaking a national key research project (863 Projects) on bioremediation of petroleum contaminated soil in oil field of China.

E-mail: **tangjch@nankai.edu.in**

Hideji Taniguchi is a Researcher at the General Testing Research Institute of Japan, Oil Stuff Inspectors Corporation. After graduation of Chiba University at 1986, he received master degree of Maritime Science and Technology from Kobe University of Japan in 2009. Present field of research is the effective use of oceanic wastes.

E-mail: **taniguchi2046@nykk.or.jp**

Jing-Chun Tang   Hideji Taniguchi

Qixing Zhou is the Dean of College of Environmental Science and Engineering, Nankai University (China). He graduated from Chinese Academy of Sciences (Ph.D., 1992) and is a famous scientist in the area of pollution ecology and environmental toxicity in China and worldwide.

E-mail: **Zhouqx@nankai.edu.cn**

Shinichi Nagata is the Professor at the Research Center for Inland Seas, Organization of Advanced Science and Technology, Kobe University, Japan. He received his doctoral degree in 1976 from Kyoto University of Japan. He is engaged in the osmo-adaptation mechanism of halophilic and halotolerant bacteria and its application, focusing on the synthesis and uptake of compatible solutes under high salinity.

E-mail: **nagata@maritime.kobe-u.ac.jp**

Qixing Zhou **Shinichi Nagata**

RECYCLING OF THE SEAWEED WAKAME THROUGH DEGRADATION BY HALOTOLERANT BACTERIA

JING-CHUN TANG[1], HIDEJI TANIGUCHI[2],
QIXING ZHOU[1], AND SHINICHI NAGATA[2]
[1]*Key Laboratory of Pollution Processes and Environmental Criteria, Ministry of Education, College of Environmental Science and Engineering, Nankai University, Tianjin 300071, China*
[2]*Environmental Biochemistry Division, Research Center for Inland Seas, Organization of Advanced Science and Technology, Kobe University, Kobe 658-0022, Japan*

1. Introduction (Seaweed Recycling, Composting, and the Marine Environment)

With the increase in population and the development of industries, organic pollution has become a great problem of worldwide concern. N and P in organic wastes often enter into the water body and accumulate in the sea, especially in inland sea areas, in which the water body is often confined within an enclosed area. The high N and P contents in the water of inland seas may have great impact on the marine ecological system. For example, some species of microorganisms will grow rapidly and result in the occurrence of red tide. Decaying algal blooms may accumulate as a deposit and become a source of N and P pollutants in inland seas.

Removing the excess amounts of N and P in seawater is important to improve the seashore environment in inland areas. Seaweed planting is one of the options. Seaweed cultivation could be a possible solution to the problem of eutrophication since it reduces the nutrients in seawater, as cultivated seaweeds grow much better in those areas where N and P are abundant (Ohno and Critchley, 1993; Fei, 2004). For example, *Porphyra yezoensis* Ueda grows well in the regions where N concentrations (NO_3-N + NH_4-N) in seawater are above 100 mg/m³. N, P, and other pollutants in the water of inland seas can thus be accumulated into seaweed biomass and the seawater might be cleaned.

Table 1 shows the growth of the seaweed *Gracilaria lemaneiformis* within about 1 year, using a rope cultivation technique. The weight of the seaweed increased during this year 100 times, with a mean daily growth rate of 3.09%. As calculated by Fei (2004), the removal rate of N resulted in 6,600 mgN/m³ in every cultivation season, 16.5 times higher than 400 mgN/m³, which is the indication level of N eutrophication. Cultivation of the brown seaweed *Undaria pinnatifida* (wakame) in Kobe University yields about 30–60 t/ha in every cultivation season, but the harvested seaweeds are wastes that are not suitable to be

Table 1. Growth of *Gracilaria lemaneiformis* in transplantation experiment (Fei, 2004).

Culture rope length (m)	Start (kg)	Finish (kg)	Increase (times)	Growth (% day^{-1})
92	3.22	356.5	110.7	3.14
273	8.19	814.4	99.4	3.06
365	11.4	1,170.9	102.6	3.09

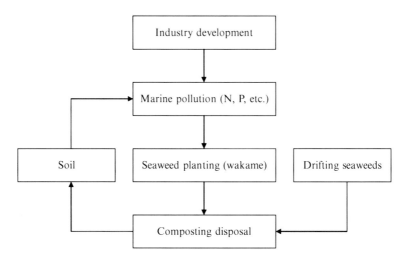

Figure 1. Diagram of seaweed recycling for the preservation of the marine environment.

consumed as food. On the other hand, drifting seaweeds are also collected to avoid in situ rotting and to reduce pollution that would interfere with the recreational use of the beach. For example, on the shores of Puerto Madryn in northeastern Patagonia, Argentina, about 8,000 t of seaweed have been collected every year (Eyras et al., 1998). Recycling and disposal of seaweeds has become a key problem that must be solved for the sustainable development of seawater and the marine environment.

Most seaweeds are edible, but harvested wakame may contain a variety of pollutants such as heavy metals, herbicides, etc., and cannot be used as a food source for human beings (Castlehouse et al., 2003). Taking the economy, technology, and environmental protection into consideration, composting is one of the best choices to dispose seaweed wastes which contain high concentrations of salts (Eyras et al., 1998). Further, the compost produced from seaweed can be reused as fertilizer on farmland (Vendrame and Moore, 2005), which is important for the recycling of C, N, and P (Fig. 1).

Massive use of seaweed biomass – resulting from the eutrophication of coastal ecosystems – for composting purposes is now developed as a recycling strategy (Eyras et al., 1998). As seen from the chemical composition of wakame (*U. pinnatifida*) (Yamada, 2001) shown in Table 2, the content of proteins and

Table 2. Chemical composition of marine algae *Undaria pinnatifida* (g/100 g) (Yamada, 2001).

	Water	Protein	Lipid	Carbohydrate	Ash	Food fiber	Salt
Fresh	89.0	1.9	0.2	5.6	3.3	3.6	1.5
Dried	12.7	13.6	1.6	41.3	30.8	32.7	16.8
Refresh	90.2	2.0	0.3	5.9	1.6	5.8	0.7

carbohydrates is relatively high, indicating that there exist enough organic substrates for microbial growth. However, it seems to be difficult to decompose seaweeds with bacteria, since they contain both carbohydrates that are hard to degrade and high salt concentrations (1.5% and 16.8% in fresh and dried wakame, respectively). If the water content is adjusted to 70% during wakame composting, the resulting salt concentration will be about 6%, which is about twice higher than seawater. Salts and alginate, which reach concentrations as high as 50% in dried wakame samples (Skriptsova et al., 2004), may act as the limiting factor for effective decomposition of wakame. Until now, studies have been conducted on the decomposition of alginate by chemical (Matsushima et al., 2005) as well as biological methods (Moen et al., 1997; Moen and Ostgaard, 1997). To our knowledge, the decomposition of alginate during seaweed composting has not been studied. Alginate can be decomposed into oligosaccharides by microbial alginate lyase. The resulting oligosaccharides have been proven to be effective in promoting the growth of plant roots (Iwasaki and Matsubara, 2000). Thus, the decomposition of alginate is important and is a key factor for the applicability of wakame compost products.

In this chapter, we introduce the application of *Bacillus* sp. HR6 isolated from soil (Mimura et al., 1999; Mimura and Nagata, 1999) for the disposal of wakame with high salinity (Tang et al., 2007). Since marine microorganisms isolated from seaweed are, in general, more suitable to degrade wakame under high salinity than general bacteria (Tang et al., 2008), we attempted to search alginate-degrading bacteria and to use them for wakame composting.

2. Composting of Wakame Using Halotolerant Bacteria

2.1. EFFECT OF SALINITY ON THE COMPOSTING BY HR6

In this experiment, sun-dried wakame was used. To obtain washed wakame, dried wakame was cleaned with fresh water and redried. The chemical composition of dried and washed wakame is shown in Table 3. Higher N and C contents were detected in the latter than in the former. For the degradation of wakame, we first used *Bacillus* sp. HR6 that can grow in a wide range of pH (5.7–8.5), temperature (20–50°C), and salinity (2–10% NaCl) (Mimura et al., 1999). The time necessary to reach the maximum temperature, through 19, 25, 35, and 40 h of incubation, is dependent on the salinity in wakame composting, which indicates that lower salinity brought about a shorter lag time (Fig. 2). Maximum temperatures were

Table 3. Chemical composition of wakame used for the experiment.

	Water (%)	N (%)	C (%)	C/N	NaCl (mg/g)[a]
Dried wakame	8.22	3.64	29.82	8.19	89.0
Washed wakame	13.72	4.14	36.95	8.92	46.2

[a]NaCl concentration was determined for aqueous layer after mixing of water and wakame with 10:1.

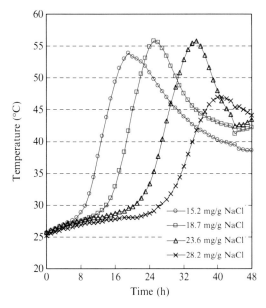

Figure 2. Effect of salinity on the temperature changes during wakame composting by strain HR6.

53–55°C at a salinity level of 28.2 mg/g NaCl at 15.2°C, 18.7°C, and 23.6°C, and 48°C. The reduction of wet weight during 24–48 h of incubations was the highest in the presence of 15.2 mg/g NaCl (Table 4). In the samples containing 15.2–28.2 mg/g NaCl, there existed an insignificant difference in maximum temperature, which proved that composting can be conducted in salinity as high as 28.2 mg/g of NaCl with the inoculation of HR6, although a lower salinity is better for the initialization time and activity of the composting process. This is important for the field treatment of wakame when the salinity content is high. In addition, the salinity of 15.2 mg/g of NaCl is also relatively high compared with that of the common composting materials, suggesting that HR6 is superior to other bacteria for the disposal of waste with high salinity. The presence of Na^+ sometimes stimulates the decomposition of some components in wakame. For example, alginate is degraded in the form of Na-alginate (Moen et al., 1997) and a Ca-cross-linked alginate matrix may present a limiting factor when microbes decompose

Table 4. Relative changes of wet weight during composting of wakame with strain HR6 under different salinities.

Incubation time(h)	Wet weight (%)			
	15.2 mg/g NaCl	18.7 mg/g NaCl	23.6 mg/g NaCl	28.2 mg/g NaCl
0	100.0	100.0	100.0	100.0
24	98.7	99.2	99.2	99.6
48	97.2	97.4	97.6	98.5

brown algal tissue (Moen and Ostgaard, 1997). An appropriate level of salinity may therefore be required for the composting and disposal of wakame.

2.2. ISOLATION AND CHARACTERIZATION OF HALOTOLERANT BACTERIA FOR SEAWEED COMPOSTING

An attempt was made to examine the screening of marine bacteria for the degradation of wakame. As a result, an effective strain *Halomonas* sp. AW4 was isolated and selected for further work. Strain AW4 initiated the wakame composting process smoothly, showing a high decomposition rate within a short time. Dominant species in the agar plate (5 g polypeptone, 3 g yeast extract, 30 g NaCl, 15 g agar in 1,000 mL distilled water, pH 7.5) were further isolated and finally six predominant strains, AW1–AW6, were selected for the present study. The characteristics of these strains are shown together with those of HR6 for comparison (Table 5). The colonies of these bacteria showed different colors: light green (AW2), light brown (AW1), brown (AW5), light white (AW3, AW4), and white (AW5). All the strains grew well in the presence of 0.5–1 M NaCl, but strains AW1 and AW2 could not grow unless NaCl was supplemented. Morphologically, AW3 and AW6 were long rods with low motility and cocci, respectively, while other strains were ellipse-shaped. In small-scale composting, AW4 showed the highest decomposition rate when compared with other species and HR6. In addition, AW4 revealed an optimal growth rate in the presence of 0.5–1 M NaCl. Surprisingly, this strain still proliferated at a high salinity of 3 M NaCl. Inoculation of AW4 resulted in a more rapid increase of the composting temperature than inoculation of HR6.

For 16S rDNA analysis, strain AW4 was first incubated in modified LB medium for 24 h. After extraction of DNA, we amplified 16S rDNA fragments using the primers EUB27F (5′-AGAGTTTGATCCTGGCTCAG-3′) and EUB 533r (5′-TTACCGCGGCKGCTGRC-AC-3). Partial 16S rDNA analysis suggested that AW4 belongs to the genus *Halomonas*. The phylogenetic allocation and 16S rRNA gene sequence similarities of AW4 and related bacterial strains are presented in Fig. 3. AW4 was most closely related to the strain *Halomonas venusta*, with 99.6% similarity in partial 16S rDNA analysis. Higher similarity with AW4 was also found in *H. hydrothermalis*, *H. variabilis*, *H. magadiensis*, *H. boliviensis*,

Table 5. Characterization of six marine bacteria isolated from seaweeds for wakame composting.

	Source	Color of colony	30°C growth rate (max) (h⁻¹) 0 M NaCl	0.5 M NaCl	1 M NaCl	2 M NaCl	Growth at 30°C (3 M NaCl)	Growth at 50°C (0.5 M NaCl)	Microscopic observation Shape	Size (m)	Motility	Degradation of wakame (%)[a]
AW1	Awaji island	Light brown	0	0.66	0.54	0.21	−	+	Ellipse	<1	+	11.2
AW2	Awaji island	Light green	0.47	0.34	0.17	0	−	+	Ellipse	<1	−	8.45
AW3	Awaji island	Light white	0	0.4	0.27	0.13	+	+	Rod	<25	+	11
AW4	Awaji island	White	0	0.88	0.64	0.37	+	+	Ellipse	<1–2	+	16.5
AW5	Awaji island	Brown	0.21	0.64	0.54	0	−	−	Ellipse	<1	+	12.5
AW6	Awaji island	Light white	0.66	0.64	0.59	0.03	−	+	Coccus	<1–2	+	8.8
HR6	Soil	White	0.61	0.41	0.2	0.09	−	+	Rod	<25	+	8.9

[a] Data by small scale composting.

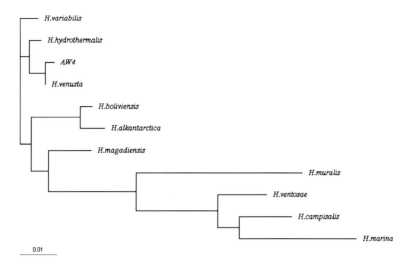

Figure 3. Phylogenetic tree based on partial 16S rRNA gene of strain AW4. Evolutionary distances were calculated using the "neighbor-joining" method. The scale bar represents 1% sequence divergence.

and *H. alkantarctica* with 98.7%, 98.5%, 97.4%, 97.2%, and 97.0%, respectively. The genus *Halomonas* mainly consists of halotolerant, halophilic, and haloalkaliphilic species that were isolated from salterns, saline soil and lakes, and the Dead Sea (Ventosa et al., 1998). Some of these species have already proven to be effective bacteria for waste degradation at high salinity. *H. campisalis* showed a high capacity for degradation of saline and alkaline wastes owing to denitrification, broad carbon utilization, and pH as well as salinity tolerance (Mormile et al., 1999). *H. muralis* was isolated from alkaline two-phase olive mill wastes (Ntougias et al., 2006) and *H. marina* was found to be one of the main microorganisms that degrade the brown alga *Fucus evanescens* (Ivanova et al., 2002).

2.3. THE COMPOSTING PROCESS OF WAKAME BY AW4

The initial moisture content of wakame, which was cut into pieces smaller than 1 cm, was adjusted to about 70%. About 1.5 kg of wakame was composted in a 5-L container (Fig. 4) with an isolation film outside to keep the temperature constant. The container was aerated with an air pump at a flow rate of 0.5 L/min. Each bacterial strain was added as suspension. Changes of O_2 and CO_2 concentrations in exhaust gas as well as the temperature inside the container were measured. Changes of the water content, pH, C, as well as N contents and viable cell numbers were also measured during composting. Figure 5 shows the changes of temperature as well as of O_2 and CO_2 in exhaust gas during composting with strains HR6 or AW6. Temperature changes corresponded to changes of the O_2 and CO_2

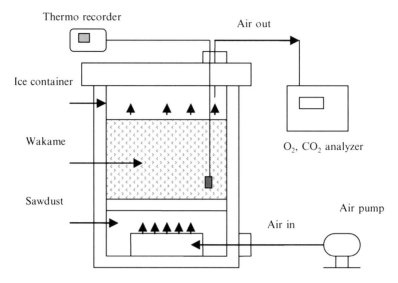

Figure 4. Composting system of wakame in a 5-L container.

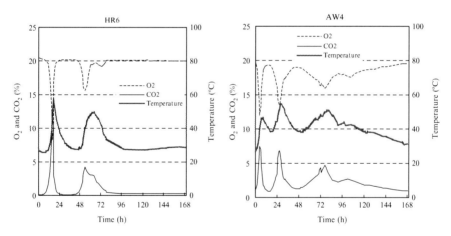

Figure 5. Changes of O_2 and CO_2 in exhaust gas and corresponding temperature during wakame composting with strains HR6 and AW4.

concentration in exhaust gas, although a lag time of about 2 h was observed, which suggests that the microbial activity was not always stable but variable during wakame composting. Higher temperatures and more O_2 consumption and CO_2 production, however, were found during wakame composting with AW4 when compared with composting with HR6. The strain AW4 seems, therefore, to be more adapted to the compost environment of wakame than strain HR6.

Table 6 shows the changes of dry weight, pH, and chemical composition during the composting process of AW4 and HR6. Decomposition rates of dry

Table 6. Changes of dry weight, pH, and chemical composition during wakame composting with strains AW4 and HR6.

Time (h)	Time (h)	Dry weight (%)	pH	N (%)	C (%)	C/N
AW4	0	100.0	7.16	3.09	31.34	10.1
	72	87.8	8.22	3.01	30.19	10.1
	168	73.6	9.00	3.15	28.95	9.2
HR6	0	100.0	6.98	2.28	32.84	14.4
	72	93.4	8.24	2.10	31.16	14.9
	168	87.8	8.48	1.91	32.37	16.9

Values of C and N are average from two independent experiments.

Table 7. Changes of cell viability in wakame composting with strains AW4 and HR6 ($\times 10^8$ CFU/g).

Time (h)		0	8	24	48	55	72	120	168
AW4	AW4	6	25.3	23.3	1.5	–	4.3	1.5	1.8
	Total	6	25.3	23.3	12.7	–	26.3	22.5	35.0
HR6	HR6	1.1	2.8	1.2	0.4	0.1	0.0	0.0	0.0
	Total	1.1	2.8	1.2	18.5	183.0	6.5	9.8	10.8

weight wakame were 12.2% and 26.4% after 3 and 7 days in the experiment of AW4, respectively, which were much higher than that of HR6 under the same condition, 6.6% and 12.2%. pH values during the decomposition with AW4 increased to 8.2 and 9.0 after 72 and 168 h of incubations, respectively, and high NH_3 content, about 500 ppm, was detected in exhaust gas showing that an alkaline environment was caused by the production of NH_3 in the exhaust gas. C and N contents decreased and increased slightly after 168 h, respectively, leading to a reduction of the C/N ratio, 9.2. Since the N content was relatively high compared with other composts (Tang et al., 2003), the C/N ratio was lower than that of general compost, >15 (Bernal et al., 1998).

Changes of viable cell numbers of strain HR6, AW4, and total bacteria were observed during the incubation process (Table 7). The viability of strain HR6 increased within 24 h of composting. Then it decreased gradually, although a sharp increase of total bacteria to 1.83×10^{10} CFU/g was observed at 55 h of incubation. After 3 days of composting, total viable cell numbers decreased to the level of 6.5 $\times 10^8$ CFU/g. On the other hand, the cell viability of AW4 increased during 24–72 h of composting, which corresponded with temperature and respiratory changes. The cell viability remained at a high level until the end of composting, although the temperature and respiratory rates gradually decreased. The appearance of the wakame biomass showed great change after composting, that is, its color changed from brown-green to grey-black and the original shape of the wakame was gradually destroyed. It became fine granule as that of matured compost, which suggests that a diverse microbial community was induced by inoculating AW4, which was effective in decomposing wakame during composting.

3. Role of Alginate-Degrading Bacteria in the Recycling of Wakame

Alginate is a linear copolymer of β-1,4-D-mannuronic acid and α-1,4-L-guluronic acid with the residues organized in blocks of not only polymannuronate and polyguluronate but of heteropolymeric sequences of both uronic acids (Moen and Ostgaard, 1997). Alginate can be degraded by radiation or thermal treatment (Nagasawa et al., 2000). Biological degradation, on the other hand, has been generally conducted by alginate lyase that acts on the 4-O-linked glycosidic linkage of alginate (Iwamoto et al., 2001). Alginate lyases have been isolated from a wide range of organisms such as algae, marine invertebrates, marine microorganisms, etc. (Wong et al., 2000). Since the degradation of alginate is not easily preceded by most microorganisms, which is mainly because of its complicated molecular structure, there has been no report on the isolation of alginate-degrading bacteria and their application in the decomposition of seaweed wastes. However, isolation of alginate-degrading bacteria and their use is interesting and indispensable for the effective degradation of these wastes.

3.1. ISOLATION AND IDENTIFICATION OF ALGINATE-DEGRADING BACTERIA

In the primary screening, four strains from a total of 56 isolated microorganisms were selected based on the ability of alginate degradation. Four bacteria designated as A7, N7, N10, and N14 were Gram positive, long rods in shape, and of white color colony on agar plates. All these bacteria were unable to grow in the medium without alginate supplemented. Quantitative changes of reducing and unsaturated polysaccharides that were produced by each bacterium in a medium containing initially 5 g/L alginate are shown in Fig. 6. The reducing sugar was detected by 3,5-dinitrosalicylic acid (DNS) method (Miller, 1959). The supernant of culture medium was mixed with DNS solution and boiled for 5 min. Absorbance of the reaction solution was analyzed at 540 nm after dilution by distilled water at the rate of 1:10, and the result was expressed as mg glucose/mL culture. The production of unsaturated sugar through β-elimination reaction by alginate lyase was determined by measuring the absorbance at 235 nm (Iwamoto et al., 2001). Both saccharides were lower than in the control after inoculation of the marine bacterium AW4, but alginate-degrading bacteria showed an increase of both saccharides after 48–96 h of incubations. An increase of OD_{235} at 48 h of incubation compared with that of reducing sugars might be caused by the existence of endo-alginate hydrolase, which was reported in fungi (Schaumann and Weide, 1990). Strain A7 showed the most rapid degradation of alginate among the strains that were examined.

A7 was long rod-shaped, 5–10 μm in size, and filamentous, and its colonies were of circular morphology and creamy white (data not shown). Partial 16S

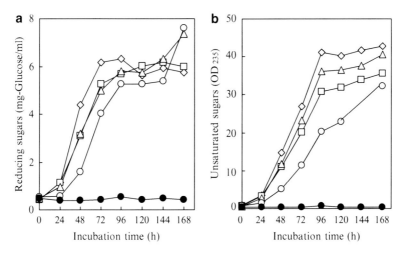

Figure 6. Changes of reducing and unsaturated sugars by four alginate-degrading bacteria and marine bacterium AW4 as reference. Changes of reducing sugars (**a**) and unsaturated sugars (**b**) were followed for A7 (*open diamonds*), N7 (*open squares*), N10 (*open triangles*), N14 (*open circles*), and AW4 (*closed circles*).

rDNA analysis showed that A7 was most closely related to *Gracilibacillus halotolerans*, at a similarity of 99% (Waino et al., 1999). In the phylogenetic tree based on partial 16S rRNA genes (Fig. 7), some alginate-degrading bacteria such as *Vibrio* sp. O_2 (Kawamoto et al., 2006) or *Bacillus algicola* (Ivanova et al., 2002) have been described, but this is the first time to report alginate-degrading bacteria within the genus *Gracilibacillus*.

3.2. CHARACTERIZATION OF ALGINATE-DEGRADING BACTERIUM A7

The growth of strain A7 in the presence of low concentrations of nutrients, 0.1 g polypeptone and 0.06 g yeast extract in 1 L, was characterized under different pH, NaCl concentrations, temperature, and nutrient contents (Fig. 8). Rapid growth was observed at pH 8.5–9.5, suggesting that an alkaline environment is favorable for growth. The bacterium grew well in the presence of 0.5–2 M NaCl, whereas no growth was found in the absence of NaCl. At a temperature of 30°C, the highest growth rate was observed after 48 h of incubation, although a sharp increase of growth within 24 h was observed at 45°C and a much slower rate was observed at 20°C. The optimal condition for growth of A7 was observed when the strain was incubated in the presence of 5 g/L of sodium alginate. Thus, the best overall condition for growth was as follows; pH 8.5–9.5, NaCl 0.5 M, temperature 30°C, and nutrient content of 2–5 g/L of polypeptone.

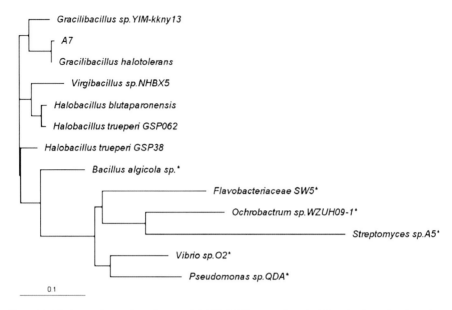

Figure 7. Phylogenetic tree based on partial 16S rRNA gene of strain A7 and other related species.

3.3. DECOMPOSITION OF SEAWEED ALGINATE BY THE STRAIN A7 AND COMPOSTING PROCESS

In medium containing 1% or 3% wakame, similar changes of reducing and unsaturated sugars were observed as those in the presence of 5 g/L of sodium alginate (Fig. 9). Three percent wakame resulted in lower reducing sugars at the initial stage of incubation, but they became much higher at later periods of incubation when compared with 1% wakame. A7 therefore appears as effective in degrading alginate constituents of wakame, and it might be possibly used for the disposal of wakame waste. As shown in Fig. 10, alginate in wakame compost was about 32–36% initially and decreased to 14.3% after 72 h of composting with A7. A similar level of degradation was observed for AW4 after 168 h. On the basis of the present results, strain A7 might be more effective for alginate decomposition during wakame composting than others examined.

As an example, changes in the concentration of different alginate components during the composting process with AW4 are shown in Table 8 (TA, total alginate; WSA, water soluble alginate; WIA, water insoluble alginate). During composting, a decomposition rate of alginate of about 17% and 56% was reached after 72 and 168 h of composting, respectively. In contrast, dry weights were almost unchanged during the same period. The present experimental results indicate that easily available substances such as protein and other carbohydrates were degraded at the initial period of composting and decomposition of alginate

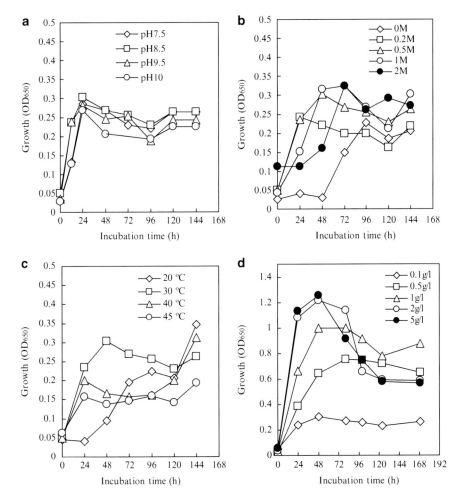

Figure 8. Growth of strain A7 under different pHs (**a**: 0.5 M NaCl, 30°C, 0.1g/L polypeptone), salinities (**b**: pH 8.5, 30°C, 0.1 g/L polypeptone), temperatures (**c**: pH 8.5, 0.5 M NaCl, 0.1 g/L polypeptone) and polypeptone (**d**: pH 8.5, 30°C, 0.5 M NaCl). Alginate content was fixed to 5 g/L and yeast extract was added at a ratio of 3:5 with polypeptone.

was performed at the following stage, after 72–168 h of composting. The diverse microbial community that developed until the late period of composting was therefore more effective in degrading alginate in wakame than the initial community. In comparison with strain HR6, the marine bacterium AW4 showed better degradation ability for wakame, probably due to the fact that marine bacteria are suitable to induce the complex microbial community that is responsible for the decomposition of alginate.

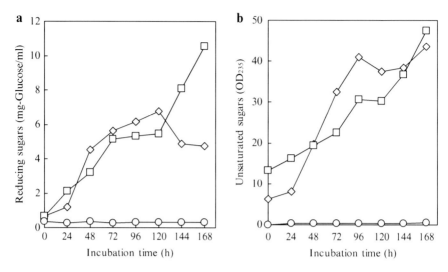

Figure 9. Changes of reducing sugars (**a**: OD_{540}) and unsaturated sugars (**b**: OD_{235}) during the incubation of strain A7 in the presence of 0% (*circles*), 1% (*diamonds*), and 3% wakame (*squares*).

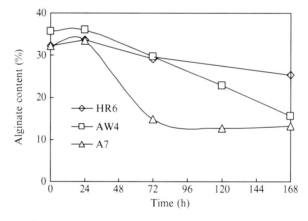

Figure 10. Changes of alginate content during composting processes using different bacteria.

4. Effect of Seaweed Compost on Plant Growth and Its Potential Ecological Toxicity

The products of alginate degradation are polysaccharides with different carbon chain length, some of which have been proven to be effective in promoting the growth of plant roots (Iwasaki and Matsubara, 2000; Cao et al., 2007). Hence, they may possibly be used as fertilizers in the agriculture of plants. Application of alginate-degrading bacteria instead of alginate lyase is more preferable for degradation of seaweed wastes. A germination experiment was carried out by the method

Table 8. Changes of alginate contents during composting process of wakame by the strain AW4.

Time (h)	WSA (%)	WIA (%)	TA (%)	WSA/WIA
0	3.18 ± 0.14	32.42 ± 0.14	35.60	0.10
24	3.88 ± 0.29	32.20 ± 0.58	36.08	0.12
72	2.61 ± 0.32	26.98 ± 0.74	29.60	0.10
120	5.31 ± 0.71	17.54 ± 0.80	22.85	0.30
168	4.83 ± 0.06	10.80 ± 1.26	15.63	0.45

Values of WSA and WIA are averages ± SD from two independent experiments.

introduced by Tiquia and Tam (1998) using Komatsuna (*Brassica campestris*). Fresh samples of wakame compost were extracted with distilled water at a compost to water ratio of 1:9 and then filtrated. Diluted extract was applied onto a filter paper in a Petri dish at 5 mL. Twenty seeds placed in the dish were incubated in the dark. All germination experiments were conducted in triplicates with distilled water as control. The root length was measured after 3 days and the relative root length (RRL) was calculated as follows:

$$\text{RRL (\%)} = \frac{\text{Root Length in extract}}{\text{Root Length in control}} \times 100 \tag{1}$$

Changes of RRL after the incubation with different bacterial strains are shown in Table 9. The medium was diluted 10–50 times with distilled water, since the salt present in the medium may reduce the germination of plants. RRL values with 1% wakame were 263.2% and 120.4% for strains A7 and AW4 after 25-fold dilution, respectively. Among different dilution rates, 25-fold resulted in the highest value of RRL, 194.5%. Higher RRL values for wakame and A7 culture medium suggest that oligosaccharides produced during incubation with A7 might enhance the elongation of plant roots, different from nonalginate-degrading bacteria like AW4. The products of A7 treatment of alginate or seaweeds containing alginate may therefore be used as a kind of fertilizer or enhancing material for phytoremediation.

The phytotoxicity of the compost samples after AW4 treatment was analyzed through a germination test in which the changes of RRL values of Komatsuna (*B. campestris*) during culture with wakame extracts collected after different composting times were monitored. Since salts inhibit plant growth, we tried to examine the germination using compost extracts under different dilution ratios. RRL values were lower than 50% under the dilution rate of 2 and increased to more than 60% and 80% at dilutions of 10 and 100, respectively. Increasing RRL values were caused by oligosaccharides that had been produced during alginate decomposition in wakame (Iwasaki and Matsubara, 2000; Xu et al., 2003). The present result demonstrates that the promotion effect of composted wakame stimulated the plant growth, which is substantially important for the application of the wakame compost.

Table 9. Changes of relatively root length (RRL) by using culture medium after incubation of strains A7 and AW4.

	AW4	A7 (1% wakame)	A7 (culture medium)		
Dilution rate	25	25	10	25	50
RRL_{CM}	120.4 ± 8.3	263.2 ± 5.1	106.7 ± 11.2	194.5 ± 14.4	161.6 ± 5.4
RRL_{DW}	66.0 ± 4.6	144.2 ± 2.8	58.5 ± 6.1	106.6 ± 7.9	88.6 ± 3.0

Values are average ± SD from three independent experiments. RRL_{CM} and RRL_{DW} indicate relative values of root length that were calculated on the basis of culture medium and distilled water as control, respectively.

The ecological toxicity of seaweed compost was examined using a bioluminescence assay as described by Nagata and Zhou (2006) and Zhou et al. (2006). A luminescent bacterium, *V. fischeri* DSM 7151, was grown in luminescence medium (LM) containing 3 g/L of glycerol for 18 h. The change of luminescence intensity after 30 min of incubation was monitored. INH% values (percentage of inhibition efficiency) were calculated according to the following equation:

$$INH\% = \left[1 - \left(\frac{IT_t \times IC_0}{IT_0 \times IC_t}\right)\right] \times 100\% \qquad (2)$$

where INH%: percentage of inhibition efficiency, IT_t: bioluminescence of sample after contact time t, IT_0: initial bioluminescence of sample, IC_t: bioluminescence of control after contact time t, IC_0: initial bioluminescence of control.

For all the samples collected at different composting time, INT% values were lower than 20. Especially, samples collected after 0, 1, and 3 days of composting showed 0% toxicity, suggesting that the compost extracts stimulate, in turn, the activity of *V. fischeri*. INT% values increased after 5–7 days of composting, that is, the toxicity increased, probably because of the decrease of nutrient amounts with composting time and intermediate metabolic substances with toxicity produced during this period. INT% values of toxicity obtained were different from those obtained with the germination assay, in which a decrease of phytotoxicity was found with the succession of composting. Multiple approaches and further related studies should be a prerequisite for a proper evaluation of ecological effect of composting products.

5. References

Bernal, M.P., Paredes, C., Sanchez-Monedero, M.A. and Cegarra, J. (1998) Maturity and stability parameters of composts prepared with a wide range of organic wastes. Bioresource Technol. **63**: 91–99.

Cao, L.X., Xie, L.J., Xue, X.L., Tan, H.M., Liu, Y.H. and Zhou, S.N. (2007) Purification and characterization of alginate lyase from *Streptomyces* species strain A5 isolated from Banana Rhizosphere. J. Agric. Food Chem. **55**: 5113–5117.

Castlehouse, H., Smith, C., Raab, A., Deacon, C., Meharg, A.A. and Feldmann, J. (2003) Biotransformation and accumulation of arsenic in soil amended with seaweed. Environ. Sci. Technol. **37**: 951–957.

Eyras, M.C., Rostagno, C.M. and Defosse, G.E. (1998) Biological evaluation of seaweed composting. Compost Sci. Util. **6**: 74–81.

Fei, X.G. (2004) Solving the coastal eutrophication problem by large scale seaweed cultivation. Hydrobiologia **512**: 145–151.

Iwamoto, Y., Araki, R., Iriyama, K., Oda, T., Fukuda, H., Hayashida, S. and Muramatsu, T. (2001) Purification and characterization of bifunctional alginate lyase from *Alteromonas* sp. strain no. 272 and its action on saturated oligomeric substrates. Biosci. Biotechnol. Biochem. **65**: 133–142.

Iwasaki, K. and Matsubara, Y. (2000) Purification of alginate oligosaccharides with root growth-promoting activity toward lettuce. Biosci. Biotechnol. Biochem. **64**: 1067–1070.

Ivanova, E.P., Bakunina, I.Y., Sawabe, T., Hayashi, K., Alexeeva, Y.V., Zhukova, N.V., Nicolau, D.V., Zvaygintseva, T.N. and Mikhailov, V.V. (2002) Two species of culturable bacteria associated with degradation of brown algae *Fucus evanescens*. Microb. Ecol. **43**: 242–249.

Kawamoto, H., Horibe, A., Miki, Y., Kimura, T., Tanaka, K., Nakagawa, T., Kawamukai, M. and Matsuda, H. (2006) Cloning and sequencing analysis of alginate lyase genes from the marine bacterium *Vibrio* sp. O2. J. Mar. Biotechnol. **8**: 481–490.

Matsushima, K., Minoshima, H., Kawanami, H., Ikushima, Y., Nishizawa, M., Kawamukai, A. and Hara, K. (2005) Decomposition reaction of alginic acid using subcritical and supercritical water. Ind. Eng. Chem. Res. **44**: 9626–9630.

Miller, G.L. (1959) Use of dinitrosalicylic acid reagent for determination of reducing sugars. Anal. Chem. **31**: 426–428.

Mimura, H., Maeda, K. and Nagata, S. (1999) Chromatographic analysis of bean curd refuse decomposed by *Bacillus* sp. HR6. Biocontrol Sci. **4**: 23–26.

Mimura, H. and Nagata, S. (1999) Physiological characteristics of *Bacillus* sp. HR6 in the process of decomposing bean curd refuse. Biocontrol Sci. **4**: 105–108.

Moen, E., Horn, S. and Ostgaard, K. (1997) Alginate degradation during anaerobic digestion of *Laminaria hyperborea* stipes. J. Appl. Phycol. **9**: 157–166.

Moen, E. and Ostgaard, K. (1997) Aerobic digestion of Ca-alginate gels studied as a model system of seaweed tissue degradation. J. Appl. Phycol. **9**: 261–267.

Mormile, M.R., Romine, M.F., Garcia, T., Ventosa, A., Bailey, T.J. and Peyton, B.M. (1999) *Halomonas campisalis* sp nov., a denitrifying, moderately haloalkaliphilic bacterium. Syst. Appl. Microbiol. **22**: 551–558.

Nagasawa, N., Mitomo, H., Yoshii, F. and Kume, T. (2000) Radiation-induced degradation of sodium alginate. Polym. Degrad. Stab. **69**: 279–285.

Nagata, S. and Zhou, X. (2006) Analyses of factors to affect the bioassay system using luminescent bacterium *Vibrio fischeri*. J. Health Sci. **52**: 9–16.

Ntougias, S., Zervakis, G.I., Ehaliotis, C., Kavroulakis, N. and Papadopoulou, K.K. (2006) Ecophysiology and molecular phylogeny of bacteria isolated from alkaline two-phase olive mill wastes. Res. Microbiol. **157**: 376–385.

Ohno, M. and Critchley, A.T. (1993) *Seaweed Cultivation and Marine Ranching*, Kanagawa International Fisheries Training Center, Japan International Cooperation Agency (JICA), Tokyo.

Schaumann, K. and Weide, G. (1990) Enzymatic degradation of alginate by marine fungi. Hydrobiologia **204/205**: 589–596.

Skriptsova, A., Khomenko, V. and Isakov, V. (2004) Seasonal changes in growth rate, morphology and alginate content in Undaria pinnatifida at the northern limit in the Sea of Japan (Russia). J. Appl. Phycol. **16**: 17–21.

Tang, J.C., Inoue, Y., Yasuta, T., Yoshida, S. and Katayama, A. (2003) Chemical and microbial properties of various compost products. Soil. Sci. Plant Nutr. **49**: 273–280.

Tang, J.C., Wei, J.H., Maeda, K., Kawai, H., Zhou, Q., Hosoi-Tanabe, S. and Nagata, S. (2007) Degradation of seaweed wakame (*Undaria pinnatifida*) by composting process with inoculation of *Bacillus* sp. HR6. Biocontrol Sci. **12**: 47–54.

Tang, J.C., Xiao, Y., Oshima, A., Kawai, H. and Nagata, S. (2008) Disposal of seaweed wakame (*Undaria pinnatifida*) in composting process by marine bacterium *Halomonas* sp. AW4. Int. J. Biotechnol. **10**: 73–85.

Tiquia, S.M. and Tam, N.F.Y. (1998) Elimination of phytotoxicity during co-composting of spent pig-manure sawdust litter and pig sludge. Bioresource Technol. **65**: 43–49.

Vendrame, W. and Moore, K.K. (2005) Comparison of herbaceous perennial plant growth in seaweed compost and biosolids compost. Compost Sci. Util. **13**: 122–126.

Ventosa, A., Nieto, J.J. and Oren, A. (1998) Biology of moderately halophilic aerobic bacteria. Microbiol. Mol. Biol. Rev. **62**: 504–544.

Waino, M., Tindall, B.J., Schumann, P. and Ingvorsen, K. (1999) *Gracilibacillus* gen. nov., with description of *Gracilibacillus halotolerans* gen. nov., sp. nov.; transfer of *Bacillus dipsosauri* to *Gracilibacillus dipsosauri* comb. nov., and *Bacillus salexigens* to the genus *Salibacillus* gen. nov., as *Salibacillus salexigens* comb. nov. Int. J. Syst. Bacteriol. **49**: 821–831.

Wong, T.Y., Preston, L.A. and Schiller, N.L. (2000) Alginate lyase: review of major sources and enzyme characteristics, structure–function analysis, biological roles, and applications. Ann. Rev. Microbiol. **54**: 289–340.

Xu, X., Iwamoto, Y., Kitamura, Y., Oda, T. and Muramatsu, T. (2003) Root growth-promoting activity of unsaturated oligomeric uronates from alginate on carrot and rice plants. Biosci. Biotechnol. Biochem. **67**: 2022–2025.

Yamada, N. (2001) *Science of Seaweed Utilization*. Seizando Press, Tokyo, Japan.

Zhou, X., Okamura, H. and Nagata, S. (2006) Applicability of luminescent assay using fresh cells of *Vibrio fischeri* for toxicity evaluation. J. Health Sci. **52**: 811–816.

Biodata of **Arie S. Issar** and **Amir Neori**, authors of *"Progressive Development of New Marine Environments – IMTA (Integrated Multi-Trophic Aquaculture) Production"*

Professor Arie S. Issar is a Professor Emeritus at the Jacob Blaustein Institutes for Desert Research (BIDR), and the Geological Department of Ben-Gurion University of the Negev. He founded, and was the head of, the Water Resources Center in the BIDR from 1975 until his retirement in 1998. Professor Issar's current research focuses on the impact of climate change on the hydrological cycle and socio-economic systems, with the aim of developing conceptual models that can mitigate the negative impact of global change. Professor Issar has published about 120 papers, edited five books of collections of papers and seven books in the fields of geology, hydrogeology, climate change, and philosophy of science.

E-mail: **issar@bgu.ac.il**

Dr. Amir Neori is a Senior Scientist at the Israel Oceanographic & Limnological Research, Ltd., The National Center for Mariculture, Eilat, Israel. He obtained his Ph.D. from the University of California San Diego – Scripps Institution of Oceanography in 1986 in Marine Biology and continued his research in sustainable mariculture and algae in his present capacity. Dr. Neori's scientific interests are in the area of environmentally friendly aquaculture, algal aquaculture, reduction in aquaculture environmental impact, integrated multi-trophic aquaculture (IMTA), and biofuel from algae. He has published over 70 peer-reviewed publications.

E-mail: **neori@ocean.org.il; aneori@gmail.com**

Arie S. Issar   Amir Neori

PROGRESSIVE DEVELOPMENT OF NEW MARINE ENVIRONMENTS – IMTA (INTEGRATED MULTI-TROPHIC AQUACULTURE) PRODUCTION

ARIE S. ISSAR[1] AND AMIR NEORI[2]
[1]Ben Gurion University of the Negev, J. Blaustein Institutes for Desert Research, Zuckerman Institute for Water Resources Sede Boker Campus, 84990, Israel
[2]Israel Oceanographic & Limnological Research Ltd, National Center for Mariculture, P.O. Box 1212, Eilat 88112, Israel

1. Introduction

The impact of the accelerating global warming on natural and human environments of arid and semi-arid zones is forecasted to be catastrophic. It is therefore doubtful whether adherence to the principles of Sustainable Development can avert the forthcoming catastrophes, especially in developing societies. The unavoidable conclusion is that a different policy of development has to be drawn up, which will ensure progress toward a safer way of life and at the same time alleviate the consequences of environmental catastrophes.

The suggested name of this policy is "Progressive Development," because it involves, first and foremost, profound and sweeping changes in the human and natural environmental resources of arid and semi-arid zones. These changes will utilize (in contrast to some interpretations of "sustainability") the still undeveloped resources of human and natural environments (Issar, 2008).

In a nutshell, Progressive Development aims to pave a new road to the survival and well-being of future generations, especially in arid and semi-arid zones of the Third World, by giving priority to investment in the planning and development of new environments, while advancing the local populations in the dimension of knowledge by education.

2. Reduced Nutrient Supply to the Sea due to the Suppression of Desert Dust Storms by Global Warming

Storms from the Sahara transport about 184 (Ginoux et al., 2001) to 259 (Tegen et al., 2004) million tons of dust annually to the North Atlantic. The dust supplies the marine life with nitrogen, iron, phosphorus, and micronutrients – some

---

A new policy is proposed for the mitigation of impacts of climate change on coastal regions, particularly in arid and semi-arid climates.

of which are common on land but scarce in the open ocean. This fertilization stimulates oceanic primary production (Jickells et al., 2005). Since 1993, ten international research teams have completed relatively small-scale ocean trials demonstrating this effect (Carbo et al., 2005).

Algae, in particular phytoplankton but also seaweed, as well as aquatic higher plants sequester carbon dioxide (CO_2) from the atmosphere and the sea to supply the ocean's food chain with organic carbon and oxygen. A part of the carbon taken up by the phytoplankton ends up on the bottom as dead organic matter or in the carbonaceous skeletons of many organisms, including calcareous algae, zooplankton (such as foraminifera), shellfish, and corals. This process captures large quantities of atmospheric carbon for long periods of time.

Investigations into the Quaternary paleo-climatology on a global scale show that during cold periods, high-pressure systems over the Gobi and Sahara deserts caused heavy dust storms, which led to iron oxide-rich and phosphate-rich "red rains" in other regions (Issar, 2003a). Such rains are also documented from historical cold periods, when cyclonic storms passed the Sahara on their way to southern Europe (Bücher, 1986). Once the global cooling process stopped, when its prime driving forces disappeared or diminished and, as the atmosphere and the oceans became warmer, carbon sequestration by phytoplankton dropped. These conditions, characterizing postglacial periods, changed in response to the warming of the atmosphere and oceans (Issar, 2003a).

Evidence for the enhancement of dust storms by global cooling, especially during the Last Glacial Period, has been found in both marine and continental sediments. Loess deposits some tens of meters thick over the Sinai and Negev (Issar and Bruins, 1983) and the clay component of the red soils in Jordan were brought by rainstorms associated with intensive cyclonic lows (Lucke, 2007). These storms moved into the Levant after crossing the Libyan and Egyptian deserts. In the coastal plain of Israel, the continental deposits during humid periods were characterized by reddish silt or clay, whereas deposition during arid periods was characterized by sand (Issar, 2001).

During cold periods, aeolian contribution from the Sahara to marine sediments in the SE Mediterranean reached 65%, while during warm periods about 70% of the deposits were "Nile particulate matter" (Schilman et al., 2001). A well-dated core from 200 km off the Atlantic African coast near Mauritania revealed a sudden reduction in velocity (or strength) of the trade winds above North Africa, synchronous with the onset of global deglaciation and a decrease in primary plankton productivity (Koopmann, 1981).

In eastern Asia, loess formations arose from dust accumulated during the dry glacial periods, while during interglacial warm humid periods, soil layers were characterized by higher iron concentration and low loess accumulation. A correlation was observed between the deposition of loess layers in continental China and greater accumulation of aeolian material in the adjacent deep sea. During the ice melt at the end of the Last Glacial Period, characterized by a reduction in aeolian flux, ocean water also became lighter in its delta O^{18} composition because of water influx from the melting glaciers (Hovan et al., 1989).

It is now generally accepted that the leading factors responsible for the glaciation and deglaciation processes during the Quaternary were the Milankovitch cycles, variations in solar output, changes in the chemical composition of the atmosphere and oceans, and volcanic activity (Issar, 2003b). Interactions of several of these factors compounded their influence on climate change. When a combination of these factors brought about a warm period, the glaciers melted, the ocean water became lighter, and there was a reduction in aeolian dust coming from the deserts. It is thus rather plausible that with the present warming of the climate due to, or amplified by the greenhouse effect, i.e., warming that is dependent on increased CO_2, the number and intensity of the dust storms and with them, the supply of nutrient-rich dust to the oceans, will diminish. This is likely to start a vicious cycle, as the oceans' rise in temperature will cause a decrease in the energy of the desert dust storms, thus reducing the supply of fertilizers to the oceans and with it marine primary production. The concomitant decrease in the marine food chain will reduce the marine sequestration of CO_2. The inevitable increased rate of rise in atmospheric levels of CO_2 will accelerate global warming.

3. Other Effects of Global Climate Change on Marine Environments – Negative and Positive

The global climate change can be anticipated to have negative, but also positive socio-economic and environmental–biological effects on the marine and coastal regions. Chief among them are enhancement of and interference with the functioning and population size of marine biota, rising sea level, rising sun radiation at the surface of land and sea, changes in hydrography and in marine-current regimes.

3.1. DAMAGE TO MARINE LIFE AND FISHERIES – A NEGATIVE EFFECT

A drop in global primary production will inevitably lead to a proportional drop in the production of marine organisms, birds, and seafood, with its global implications in a highly populated dry world.

3.2. RISE IN SEA LEVEL – A NEGATIVE EFFECT

A rise in sea level will harm if not ruin the livelihood and economy of many coastal countries. This rise is a function of the fact that climate warming causes the gradual melting of land-based glaciers. Estimates of the predicted rise in sea level differ among scientists. According to the latest report (February 2007) of the

Intergovernmental Panel on Climate Change (IPCC, 2007), a rise in sea level of 0.2–0.6 m is predicted by the year 2100. However, this excludes the possibility of future rapid changes in the Greenland and West Antarctic glaciers. Melting of all of the Greenland glaciers would add 7 m to the level of the sea, but the timing of such a catastrophe remains controversial. Although no one expects complete glacier disintegration in the twenty-first century, there are worrisome signs that the process has already begun. Eventual disintegration of the West Antarctic ice sheet also remains controversial; nevertheless, if it does happen, it will add another 7 m to sea level (IPCC, 2007).

According to David Wheeler from the Center for Global Development, a rapid rise in sea level would threaten millions with inundation. A 3-m rise in sea level in this century is no longer beyond the bounds of informed discussion in the scientific literature. Such a rise threatens many low-lying countries: it would drown Egypt's Nile Delta, and even a 1-m rise would inundate much of this delta's fertile land. The Vietnamese high-risk "red zone," less than 5 m above sea level, encompasses 38% of the country's population, 36% of its GDP (gross domestic product), and 87% of its wetlands. As the level of the sea rises, progressive inundation, high tides, and storm surges will take an increasing toll (Wheeler, 2007).

3.3. HIGH CO_2 IN THE ATMOSPHERE CAN HAVE POSITIVE EFFECTS

A high content of CO_2 in the atmosphere benefits photosynthesis (Griffin and Seemann, 1996). Algae and higher plants whose primary photosynthetic products have three carbon atoms per molecule (C3) respond positively to increases in CO_2 levels. At the current atmospheric levels of CO_2, up to half of the photosynthate in C3 plants is returned to the atmosphere by photorespiration, which occurs simultaneously with photosynthesis in sunlight. High levels of CO_2 in the air reduce photorespiration and thus increase net primary production. However, a rise in photosynthesis that is not matched by a balanced rise in the supply of other nutrients can lead to blooms of cyanobacteria, some of which are harmful.

3.4. HIGHER SEA TEMPERATURES

A change in the pattern of sea currents may become an important factor in changing temperatures of oceans and inner seas, such as the Caribbean, the Baltic, and the Mediterranean, among others. Nevertheless, the character, potency, and impact of these changes are still under investigation. Higher temperatures and lower wind velocities, expected particularly at the equatorial and mid-latitudes, may initially promote the growth of phytoplankton and other aquatic plants; however, at the same time, the increased stability of the water column, owing to warming and reduced salinity in surface water, may reduce the vertical mixing and supply of nutrients from deeper waters and thus limit productivity or lead to changes in the structure of the entire aquatic biota.

3.5. LESS CLOUDS – MORE RADIATION

Another positive factor for phytoplankton in the mid-latitude zones, especially where precipitation regimes are governed by westerly rainstorms, is the decrease in cloud cover associated with global warming. Higher radiation rates are beneficial to plant growth and promote carbon sequestration, if matched by a proportional increase in nutrient supply. However, at the same time, increased UV radiation can damage life on land and in the sea.

4. Progressive Development: A Blueprint for Countering the Negative Impacts of Global Warming on the Marine Environment by Capitalizing on Its Positive Effects

As the continuation of climate warming is quite probable, a rise in sea level should be regarded as an impending catastrophe on a global scale. As already discussed, such a rise, thanks to the advancement of science, can be forecasted, though still not in detail. For some regions, engineering devices, for example dikes such as those of the Delta Works in the Netherlands, may be of help, while for others, such measures can provide only partial relief.

The Delta Works are an example of how the conceptual model of Progressive Development can be applied, using positive characteristics of global warming to mitigate its negative effects. The main approach is to create artificial lagoons behind dikes built against the rising sea level, and to use the new lagoons for intensive mariculture of seaweed, shellfish, and fish. The dike and lagoon protective infrastructure will thus sustain a new superstructure of food and energy production. The produced seaweed can serve as a commodity in the food, feed, chemical, and health product industries, and can also support aquaculture of herbivorous and omnivorous shellfish, fish, and shrimp using IMTA technologies (Chopin et al., 2001, 2008; Troell et al., 2003; Neori et al., 2004; Buschmann et al., 2008). Recently, the use of seaweeds for biofuel has also been gaining interest (Anonymous – Royal Society, 2008). Moreover, while seaweed may benefit from the consequences of global warming, their growth can actually reduce several of its negative impacts by taking up CO_2 and waste nutrients and oxygenating the water. The other major group of extractive species – shellfish, will consume microalgae that will grow in the lagoons and thus sequester much CO_2 as carbonaceous shells and edible meat. Of course, such lagoons can support a certain quantity of cultured fish, balanced ecologically by seaweed and shellfish within the framework of IMTA (Tournay, 2006).

It is beyond the scope of the present chapter to discuss the building of dikes for the protection of low-lying coastal areas from flooding by the rising sea level. However, needless to say that such an activity demands huge investments in planning and construction, as well as in maintenance, not to speak of the need to develop novel engineering and coastal oceanographic know-how. The dikes and polders of the Netherlands are the best pilot project for studying the many issues involved in the protection of regions threatened by future sea-level rises.

A choice that some countries may face is whether to follow the Dutch people's method of stopping the sea or to yield to the forces of nature, retreat before the rising sea, and pay the price in loss of land and habitat. In any case, avoiding a decision will mean loss of life, as well as of national and private assets on a tremendous scale.

Once a decision is made to follow the Dutch model and build a series of dikes, it becomes necessary to choose between placement of the dikes along the existing shorelines and reclamation of additional land that is presently under water. Reclamation of continental shelf areas into diked lagoons opens up new prospects for the production of seafood, as well as new sources of income connected to the production of biofuel; last but not the least, it will create new sinks for sequestering atmospheric carbon and waste nutrients.

4.1. GROWING SEAWEED IN ARTIFICIAL LAGOONS BEHIND DIKES

Dikes constructed offshore, at a distance that depends on bathymetric and coastal oceanographic data, will create large shallow coastal lagoons, with relatively warm, quiet, highly lit, and clear water, loaded with CO_2 and low in concentrations of other nutrients. In many cases, wastewater from household, industrial, and agriculture sources will be available from nearby urban and agricultural regions. Such wastewater, when spread prudently over the surface of the lagoons, can supplement the level of nutrients and promote high rates of primary production per unit of water area (Ryther et al., 1975), with little impact on the natural ecosystems outside the dikes.

Primary production values from untended hypertrophic warm-water lagoons have been reported at over 12.5 t C ha^{-1} year^{-1} (Hung and Hung, 2003). Similar values have been reported from modern, warm-water coastal seaweed farms in China (Yang et al., 2004). Semi-intensive IMTA systems in Israel and elsewhere have reported yields of nearly 400 t ha^{-1} year^{-1} of *Ulva* sp. (fresh weight) comparable with 20 t C ha^{-1} year^{-1} in two-species combination culture systems, while *Ulva* sp. monoculture production in tanks reached even higher values (Neori et al., 2004). Reported yields of *Ulva* sp. in commercial seaweed farms in South Africa lie between those two values (Bolton et al., 2008). It is anticipated that seaweed production values of 10–20 t C ha^{-1} year^{-1} can be achieved in any Mediterranean lagoon that will be enriched with wastewater and managed for maximal seaweed growth.

Demand for seafood is on the rise. This trend is expected to continue according to FAO Fisheries Reports (FAO, 2006). While global capture of fish in the oceans has leveled off and many fish stocks have essentially collapsed, the demand for seafood has led to the rapid and sustained expansion of aquaculture. Moreover, an even greater demand for seafood may be anticipated if the desertification of agricultural land and exhaustion of freshwater reserves continues, because marine aquaculture, or mariculture, requires neither arable land nor fresh water. Intensive mariculture should advance according to the general outlines of the conceptual model of Progressive Development (Issar, 2008). Current high-volume-feeding monospecies aquaculture, particularly of carnivorous fish in pens or shrimp and fish in ponds, receive high-protein fish-based diets, but part of this feed becomes waste. Constraints resulting from their environmental impact and rising feed costs hamper further growth of such farms (FAO, 2006). As in certain traditional polyculture schemes, plants can drastically reduce feed use and environmental impact of industrialized mariculture, and at the same time add to its income. These nutrient-assimilating photoautotrophic plants, mostly seaweed and microalgae, use solar energy to turn nutrient-rich effluents into profitable resources. Plants counteract the environmental effects of the heterotrophic fish and shrimp and restore water. This is the basis of ecologically balanced IMTA farms, which are quite suitable for implementation in lagoons of the kind envisaged in the Progressive Development proposal.

The algal form that develops in a lagoon can be influenced by man to a large degree: maintaining a balanced supply of nutrients will promote eukaryotic algae over cyanobacteria (Anderson, 1995). Enrichment with silicates will promote diatom microalgae, which are a good food for bivalves (Lee et al., 2004). Shallow lagoons, where ample light reaches the bottom, and structures that are submerged at shallow depth, will promote seaweed growth (Sahoo and Ohno, 2000). Most of the world's seaweed culture makes use of ropes or rafts, submerged at shallow depths in coastal waters and in shallow ponds (McHugh, 2003).

The main practical use for microalgae is the in situ feeding of bivalves, whose production quantitatively shares with seaweeds the top position in world mariculture (FAO, 2006). All other uses of marine microalgae involve expensive processes for separation of the cells from the water (Shumway et al., 2003). Seaweeds, on the other hand, are marine algae that can be easily (and cheaply) harvested. Different species of seaweed can be produced and processed at relatively low cost and have many uses in the food, natural gel, cosmetics, and medicinal industries, which is why seaweed culture has become a leading sector in mariculture (FAO, 2006). Thus, the envisaged lagoons can become a source of seaweed biomass, shellfish, and other seafood, while taking up CO_2 and waste nutrients and producing oxygen.

A promising product of seaweed is biofuel. Albeit much research and development is still needed to determine which species are the most efficient in converting atmospheric and marine sources of nutrition into biomass, and which can be converted into CO_2-neutral biofuels; scientists claim that algae, grown in either artificial lakes or lagoons, may help in solving the problems of future supply of automobiles' fuels (Sheenan et al., 1998). Research is now focusing on methods by which algae can thrive to a level which will allow the realization of these industrial and ecological possibilities. Auxiliary benefits will involve wastewater treatment and the reduction of CO_2 emissions from fossil-fuel power plants, which contain about 10–30 times as much CO_2 as normal air (University of Virginia 2008). A 1,000 km² area of seaweed culture lagoon (such as 10-km wide along 100 km of coastline) can sequester about one million ton CO_2 year⁻¹ and produce a biomass with a net combustion energy content equal to a couple percents of Israeli annual coal use.

Drawing some benefit from the negative impact of power plants that use fossil fuels – oil, gas, or coal, is within the framework of the general philosophy underlying the artificial lagoon idea. One must take into consideration that these conventional power plants will continue to be power sources, and thus will continue to emit their gases, especially CO_2. In many countries, these plants are built near the sea, from which water is pumped for cooling and steam production. The warm water as well as the gas emitted after burning the fossil fuel can enhance the production of seaweed in nearby lagoons.

4.2. ADDITIONAL BENEFITS FROM SEA DIKES BUILT OFFSHORE

As already mentioned, the main benefit from sea dikes is the prevention of coastal flooding by the rising sea. Prefeasibility studies will help determine whether to construct the dikes on or away from the shore, forming artificial lagoons. The benefit of linking the dikes to a series of artificial islands off-shore needs to be considered. A prefeasibility study on artificial islands off the Mediterranean coast of Israel, carried out by a Dutch-Israeli team,[1] has identified environmental, technological, and economical feasibilities, especially for airports: because of the ever-increasing population density in the coastal plain of Israel, airports impose restrictions on the surrounding environments and vice versa. Artificial islands offshore may prove inevitable for the facilitation of the anticipated increase in air traffic. Moreover, the construction of hotels and artificial beaches along these dikes may benefit the growing world tourist industry, particularly tourism geared toward the sunny beaches of the Mediterranean and other warm seas.

5. Pilot Projects Along the Coastal Shelves

Progressive Development is aimed at the earliest possible implementation, so as to limit and check the catastrophic effects of climate change in the regions and environments most susceptible to inundation and a drop in marine production. It is therefore suggested that pilot projects be started that can evaluate and improve the capacity of the proposed infrastructures to fulfill their intended functions.

---

[1] "Artificial Islands off the Mediterranean Coast of Israel" Feasibility R&D study (Phase 1), Final Report by The Dutch/Israeli Steering Committee. Submitted to the Ministry of National Infrastructures, Jerusalem and The Ministry of Transportation, Public Works and Water Management. Den Haag. February 2000. The cooperation of Dr. Michael Beyth, the Israeli co-chairman of the Steering Committee, is herewith acknowledged.

An investigation of the technical feasibility of pilot experimental projects along the coasts of southern Israel, the Gaza Strip, and the Nile Delta is suggested as a first step for the following reasons:

1. The continental shelf in these areas, in particular off the Delta, is shallow.
2. The severe damage expected from a rise in sea level to these relatively low-lying and highly populated shorelines.
3. The availability of large quantities of fertilizer-rich water effluents produced in nearby countries.
4. The shorelines location of most fossil-fuel power stations in these countries.
5. The extensive know-how on seaweed and IMTA culture in Israel.

6. Summary and Conclusions

Accelerating global warming is forecasted to have a negative impact on natural and human environments in arid and semi-arid zones. This includes damage to agriculture and flooding of the coastal plains. However, an important and often overlooked anticipated impact of global warming is a significant decrease in aquatic productivity, because of an expected drop in nutrient transport by dust storms from deserts to the marine environment. Significantly lower oceanic production will lead to socioeconomic crises, which will be particularly devastating in developing countries. Adoption by policymakers of the proposed conceptual model "Progressive Development" can ameliorate the adverse impact of global climate change on already thirsty and hungry societies by combining the prevention of flooding with an increased supply of seafood and water resources. "Progressive Development" entails profound and sweeping changes in the management of the natural, environmental, and human resources of impacted regions. It recognizes the fact that our entire civilization has been based on modified ecosystems, both on land and in water, that flooding will have huge impacts on both land and sea sides of the coastline and that further modifications are required in view of the negative impacts of global climate change as well as the needs of the expanding human population. It questions the interpretation of the "sustainability principle," which insists on preserving prairies, deserts, seas, and oceans in their present state, or on reverting them to their premodification status. Underutilized natural resources of native human societies – nondeveloped land and water (fresh, brackish, and seawater) – should be developed according to long-range planning. Seaweed culture and IMTA (integrated multitrophic aquaculture) can play fundamental roles in the application of Progressive Development to marine and coastal environments.

The following considerations justify setting up seaweed and IMTA farms according to the principles of Progressive Development of marine environments:

1. Dike construction along much of the world's coastline is nearly inevitable to forestall catastrophic coastal flooding.

2. By simply pushing the dikes several km offshore, the proposed lagoons and islands with their increased economic activities will recuperate at least some of the huge investment involved.
3. Proximity to coastal power stations allows easy enrichment of the lagoons by flue gas CO_2.
4. Primary productivity can benefit from CO_2-enriched lagoon water.
5. The mariculture products will capture large quantities of CO_2. The produced crops have proven profitable large markets as food, chemicals, and even biofuel.
6. The lagoons can function as sinks for agricultural and urban wastewater and rain runoff.
7. The lagoons may function as a buffer zone between the open and high seas and the coastal population centers against dike breaches.
8. Artificial islands integrated with the dikes can serve as an infrastructure for airports and industry.
9. Hotels and tourist attractions on the dikes and on artificial islands built along them can benefit the tourist industry.

7. References

Anderson, D.M. (1995) Toxic red tides and harmful algal blooms: a practical challenge in coastal oceanography. U.S. National Report to the IUGG American Geophysical Union, pp. 1189–1200.

Bolton, J.J., Robertson-Andersson, D.V., Shuuluka, D. and Kandjengo, L. (2008) Growing *Ulva* (Chlorophyta) in integrated systems as a commercial crop for abalone feed in South Africa: a SWOT analysis. J. Appl. Phycol. doi:10.1007/s10811-008-9385-6.

Bücher, A. (1986) Recherches sur les Poussieres Minerales d'Origine Saharienne. These de Doctorat d'Etat. University of Reims-Champagne-Ardenne, France, 165 pp.

Buschmann, A.H., Hernández-González, M., Aranda, C., Chopin, T., Neori, A., Halling, C. and Troell, C. (2008) Mariculture waste management, In: S.E. Jørgensen (ed.) and B.D. Fath (Editor-in-Chief) *Encyclopedia of Ecology*, Vol. 3: Ecological Engineering. Elsevier, Oxford, pp. 2211–2217.

Carbo, P., Herut, B., Krom, M. and Homoky, W. (2005) Impact of Saharan dust on N and P geochemistry in the South Eastern Levantine basin. Deep Sea Research CYCLOPS dedicated volume.

Chopin, T., Buschmann, A.H., Halling, C., Troell, M., Kautsky, N., Neori, A., Kraemer, G., Zertuche-Gonzalez, J., Yarish, C. and Neefus, C. (2001) Integrating seaweeds into aquaculture systems: a key towards sustainability. J. Phycol. **37**: 975–986.

Chopin, T., Robinson, S.M.C., Troell, M., Neori, A., Buschmann, A.H. and Fang, J. (2008) Multi-trophic integration for sustainable marine aquaculture, In: S.E. Jørgensen (ed.) and B.D. Fath (Editor-in-Chief) *Encyclopedia of Ecology*, Vol. 3: Ecological Engineering. Elsevier, Oxford, pp. 2463–2475.

FAO (2006) State of World Aquaculture, FAO Fisheries Technical Paper No. 500. Rome, 134 p.

Ginoux, P., Chin, M., Tegen, I., Prospero, J., Holben, B., Dubovik, O. and Lin, S.-J. (2001) Global simulation of dust in the troposphere: model description and assessment. J. Geophys. Res. **106**: 20,255–20,273.

Griffin, K.L. and Seemann, J.R. (1996) Plants, CO2 and photosynthesis in the 21st century. Chem. Biol. **3**: 245–254.

Hovan, S.A., Rea, D.K., Pisias, N.G. and Shackleton, N.J. (1989) A direct link between the China loess and marine $d^{18}O$ records: aeolian flux to the north Pacific. Nature **340**: 296–298.

Hung, J.J. and Hung, P.Y. (2003) Carbon and nutrient dynamics in a hypertrophic lagoon in southwestern Taiwan. J. Mar. Syst. **42**: 97–114.

IPCC (2007) Summary for policymakers, In: S. Solomon, D. Qin, M. Manning, Z. Chen, M. Marquis, K.B. Averyt, M. Tignor and H.L. Miller (eds.) *Climate Change 2007: The Physical Science Basis*. Contribution of Working Group I to the Fourth Assessment Report of the Intergovernmental Panel on Climate Change. Cambridge University Press, Cambridge/New York. The full report is available at http://www.ipcc.ch/ipccreports/ar4-wg1.htm

Issar, A.S. (2001) Paleo-environments of the uppermost Quaternary in Israel. Curr. Res. GSI **12**: 235–238.

Issar, A.S. (2003a) *Climate Changes During the Holocene and Their Impact on Hydrological Systems*. Cambridge University Press, Cambridge, UK.

Issar, A.S. (2003b) The driving force behind the cold climate spells during the holocene, In: A. Kotarba (ed.) *Holocene and Late Vistulian Paleogeography and Paleohydrology*. Polska Akademia Nauk, Prace Geograficzne No. 189, pp. 291–297.

Issar, A. (2008) Progressive development in arid environments: adapting the concept of sustainable development to a changing world. Hydrogeol. J. **16**(6): 1431–2174.

Issar, A.S. and Bruins, J. (1983) Special climatological conditions in the deserts of Sinai and the Negev during the latest Pleistocene. Palaeogeogr. Palaeocl. **43**: 63–72.

Jickells, T.D., An, Z.S., Andersen, K.K., Baker, A.R., Bergametti, G., Brooks, N., Cao, J.J., Boyd, P.W., Duce, R.A., Hunter, K.A., Kawahata, H., Kubilay, N., laRoche, J., Liss, P.S., Mahowald, N., Prospero, J.M., Ridgwell, A.J., Tegen, I. and Torres, R. (2005) Global iron connections between desert dust, ocean biogeochemistry, and climate. Science **308**: 67–71.

Koopmann, B. (1981) Sedimentation von saharastaub im subtropischen Atlantik während der letzten 25,000 Jabre. "Meteor" Forsch. Ergeb. **C35**: 23–59.

Lee, J.J., Rodriguez, D., Zmora, O., Neori, A., Symons, A. and Shpigel, M. (2004) Nutrient study for the transition from earthen sedimentation ponds in integrated mariculture systems to ones lined with PVC, what needs to be done? J. Appl. Phycol. **16**: 341–353.

Lucke, B. (2007) *Demise of the Decapolis Past and Present Desertification in the Context of Soil Development, Land Use and Climate*. Faculty of Environmental Sciences and Process Engineering, Brandenburg University of Technology Cottbus, Germany.

McHugh, D.J. (2003) A guide to the seaweed industry. FAO Fisheries Technical paper 441. FAO, Rome, 105 p.

Neori, A., Chopin, T., Troell, M., Buschmann, A.H., Kraemer, G.P., Halling, C., Shpigel, M. and Yarish, C. (2004) Integrated aquaculture: rationale, evolution and state of the art, emphasizing seaweed biofiltration in modern mariculture. Aquaculture **231**: 361–391.

Anonymous – Royal Society (2008) Sustainable biofuels: prospects and challenges. Science Policy. The Royal Society, London, 90 p. Available at http://royalsociety.org/

Ryther, J.H., Goldman, J.C., Gifford, J.E., Huguenin, J.E., Wing, A.S., Clarner, J.P., Williams, L.D. and Lapointe, B.E. (1975) Physical models of integrated waste recycling—marine polyculture systems. Aquaculture **5**: 163–177.

Sahoo, D. and Ohno, M. (2000) Rebuilding the ocean floor—construction of artificial reefs around the Japanese coast. Curr. Sci. **78**: 228–230.

Schilman, B., Almogi-Labin, A., Bar-Matthews, M., Labeyrie, L., Paterne, M. and Luz, B. (2001) Long- and short-term carbon fluctuations in the Eastern Mediterranean during the late Holocene. Geology **29**(12): 1099–1102.

Sheenan, J., Dunahay, T., Benemann, J. and Roessler, P. (1998) A look back at the US Department of Energy's aquatic species program – biodiesel from algae. National Renewable Energy Laboratory, USA. Available online at www1.eere.energy.gov/biomass/pdfs/biodiesel_from_algae.pdf

Shumway, S.E., Davis, C., Downey, R., Karney, R., Kraeuter, J., Parsons, J., Rheault, R. and Wikfors, G. (2003) Shellfish aquaculture—in praise of sustainable economies and environments. World Aquacult. **34**: 15–17.

Tegen, I., Werner, M., Harrison, S.P. and Kohfeld, K.E. (2004) Relative importance of climate and land use in determining present and future global soil dust emission. Geophys. Res. Lett. **31**: L05105, doi:10.1029/2003GL019216.

Tournay, B. (2006) IMTA: template for production? Fish Farming Int **33**(5): 27.

Troell, M., Halling, C., Neori, A., Chopin, T., Buschmann, A.H., Kautsky, N. and Yarish, C. (2003) Integrated mariculture: asking the right questions. Aquaculture **226**: 69–90.

University of Virginia (2008) Algae: biofuel of the future? ScienceDaily October 28, 2008.

Wheeler, D. (2007) *The IPCC Debate on Sea-Level Rise: Critical Stakes for Poor Countries*. Center for Global Development, February 2007. http://blogs.cgdev.org/globaldevelopment.

Yang, Y.F., Li, C.H., Nie, X.P., Tang, D.L. and Chung, I.K. (2004) Development of mariculture and its impacts in Chinese coastal waters. Rev. Fish Biol. Fisher. **14**: 1–10.

Biodata of **Vaibhav A. Mantri, C.R.K. Reddy**, and **Professor Bhavanath Jha**, authors of *"Reproductive Processes in Red Algal genus Gracilaria and impact of Climate Change"*

Mr. Vaibhav A. Mantri is currently Scientist in the Discipline of Marine Biotechnology and Ecology, Central Salt and Marine Chemicals Research Institute (Council of Scientific and Industrial Research), Bhavnagar, India. He received his M.Sc. in the year 1999 from Pune University, India, with Phycology as specialization, subsequently obtained PhD in 2010 from Bhavnagar University, India. His doctoral thesis work was on 'Biology of *Gracilaria dura* from north cost of India' He has been recently awarded the Raman Research Fellowship for the post doctoral work at Ghent University, Belgium. Dr. Mantri's scientific interests are in the areas of: seaweed biodiversity, population studies, life cycle, and bioprospecting of seaweeds.

E-mail: **vaibhav@csmcri.org**

Dr. C.R.K. Reddy is currently Group Leader for Seaweed Biology and Cultivation in the Discipline of Marine Biotechnology and Ecology, Central Salt and Marine Chemicals Research Institute (Council of Scientific and Industrial Research), Bhavnagar, India. He obtained his Ph.D. from Nagasaki University, Nagasaki, Japan, in 1992 and worked as Teaching Assistant for 1 year in Faculty of Fisheries, Nagasaki University. Dr. Reddy's scientific interests are in the areas of: genetic improvement of seaweeds through biotechnological interventions; seaweed biodiversity and bio-prospecting, seaweed biology and cultivation, and nutritional aspects of edible seaweeds.

E-mail: **crk@csmcri.org**

Vaibhav A. Mantri   **C.R.K. Reddy**

Professor Bhavanath Jha is currently the Head and Co-ordinator of Discipline of Marine Biotechnology and Ecology, Central Salt and Marine Chemicals Research Institute (Council of Scientific and Industrial Research), Bhavnagar, Gujarat, India. He obtained his Ph.D. from Jawaharlal Nehru University, New Delhi, in 1983 and did Post-Doctoral research at the University of Cambridge, England, and J.W. Goethe University, Frankfurt. Professor Jha's scientific interests are in the areas of: molecular phylogeny of seaweeds, genomics and metagenomics of extremophiles, halotolerant plant growth promoting rhizobacteria, quorum sensing and biofilm formation, stress genomics and proteomics, and development of transgenics.

E-mail: **bjha@csmcri.org**

REPRODUCTIVE PROCESSES IN RED ALGAL GENUS *GRACILARIA* AND IMPACT OF CLIMATE CHANGE

VAIBHAV A. MANTRI, C.R.K. REDDY, AND BHAVANATH JHA
Discipline of Marine Biotechnology and Ecology, Central Salt and Marine Chemicals Research Institute, Council of Scientific and Industrial Research (CSIR), Bhavnagar, 364002, India

1. Introduction

The red algae are eukaryotic autotrophs, the majority of which are reported from the marine environment. The marine macroscopic red algae, commonly known as red seaweeds, are distantly related to two other groups, namely green and brown seaweeds. The independent evolutionary lineage of red algae is characterized by a combination of morpho-anatomical, metabolic, and physiological features that include complex life-history patterns (Hawkes, 1990; Coelho et al., 2007). About 5,800 distinct species have been described in this class (Brodie and Zuccarello, 2007) with a few having immense commercial potential. Some red algal species are edible (Dippakore et al., 2005; Wang et al., 2008), while others are used for the extraction of industrially important phycocolloids such as agar and carrageenan (Zemke-White and Ohno, 1999; Meena et al., 2007). Although in most of the cases, naturally occurring biomass has been harvested for industrial use, the cultivation practice is considerably on the rise for the last couple of decades to meet the surging global market demand (Hanisak, 1998; Ganesan et al., 2006; Subba Rao and Mantri, 2006). In 2004, ca. 4 million wet tons of red seaweed have been cultivated worldwide valuing ca. US$1.9 billion (FAO, 2006).

The development of a new cultivation method for many types of seaweed relies on the effective control of their reproduction, and, therefore, the knowledge about their life-history patterns and reproductive processes is imminent. The world witnessed the tremendous benefits from the understanding of the reproductive process when Professor Drew discovered the "conchocelis-phase" in the life history of red alga *Porphyra umbilicalis* (Drew, 1949). The conchocelis has proven to be another stage in the life history of Bangiales rather than a new species as described earlier. Thus, based on the full illustration of the life history of this commercially important genus, the cultivation technique of *Porphyra* has been established. Considerable progress has been made in the cultivation of *Porphyra* since then and an average of 400,000 t (wet wt.) *Porphyra* per year is being produced with a market value of over US$1,500 million (Dippakore et al., 2005).

Another example is van den Meer and Tood's (1980) discovery of extremely small female gametophyte of the commercial red algal species *Palmaria palmata* (dulse). Field phycologists had found only males and tetrasporophytes, so the plant was believed to have no sexual reproduction (see Bold and Wynne, 1978). The discovery of female gametophyte has helped in the development of a suitable cultivation procedure for this commercially utilized genus (Gall et al., 2004). Apart from culture and cultivation aspects, the knowledge of reproductive patterns and processes has also been essential to understand the ecological process, such as early recruitment to determine the organization of the seaweed communities (Santelices and Aedo, 2006; Lamote and Johanson, 2008), dispersion (Norton, 1992), and resource management (Subbarangaiah, 1984).

The red algal genus *Gracilaria* that forms the raw material for the industrial production of agar recorded more than 110 species from the tropical shores (Rueness, 2005). The genus *Gracilaria* is characterized by the presence of tubular nutritive cells and nonsuperficial spermatia. The annual global harvest of *Gracilaria* has been in excess of 37,000 dry tons, of which about one third accounts for aquaculture (Ye et al., 2006). Despite increasing interest in agar as a commercial product, relatively little is known about the overall reproductive processes in genus *Gracilaria*. Generally, it is assumed that most of the species of *Gracilaria* are characterized by triphasic sexual life history with an alternation of generations. As in the case of most red algae, the life cycle is diplohaplontic, with haploid gametophytes alternating the diploid sporophytes. However, deviations to this have also been reported. The variation in reproductive success in relation to haploid and diploid stages has also been reported. Studies have also confirmed the differences in the dispersal abilities in haploid and diploid life-cycle stages. The different survival strategies under artificial environmental conditions have been reported by both haploid and diploid in *Gracilaria*. It has been shown by using microsatellite DNA markers that the variation in male fertilization success depends on the distance traveled by spermatia, male–male competition, and female choice. Thus, the fertilization is not sperm-limited as it was earlier thought. The knowledge about reproductive biology would be helpful not only for cultivation but also for resource management. This paper briefly appraises the information on the different reproductive processes in the genus *Gracilaria* and the possible effects of the changing environment.

2. Life-Cycle Patterns

The sexual life cycle in the seaweeds involves a cyclic alternation between diploid and haploid phases, with meiosis mediating the transition from the diploid to the haploid state, while syngamy reconstitutes a diploid genome. The typical life history of *Gracilaria* follows a basic pattern known as "*Polysiphonia* type," which is triphasic in nature (Fig. 1). The gametophyte as well as the sporophyte is morphologically identical and independent, while the gametophytes are dioecious. The

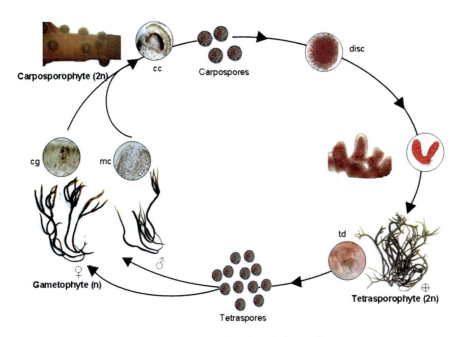

Figure 1. Life cycle of *Gracilaria dura* (C. Agardh) J. Agardh from Indian waters.

haploid male and female gametophytic plants in nature are presumed to be in equal number; however, their identity can be confirmed under the microscope only at maturity. The female gamete is fertilized in situ by the single fertilization event. The diploid zygote is then formed involving very complex cytological events. The zygote germinates within the mother tissue forming the distinct new phase called "carposporophyte." This stage can easily be identified in the field with the naked eye, as hemispherical, protruding cystocarps are formed all over the surface of the female gametophyte. The formation of cystocarp is regarded as a separate phase that grows parasitically on the female gametophyte. The numerous carpospores are produced in the cystocarp by mitotic division, hence genetically identical and diploid. The carpospore production could be compared with that of polyembryony, where the embryo derived from bi-parental fusion splits into many genetically identical embryos that are different from the mother (Engel et al., 2004). Each carpospore germinates into a new diploid tetrasporophyte that is morphologically indistinguishable from the gametophytic plants. The tetraspore mother cell upon maturity undergoes meiosis, resulting in genetically variable haploid tetraspores. Half of the tetraspores would germinate and develop into male and the remaining half into female gametophytic plants, re-establishing the sequence of cyclic pattern. Numerous field as well as laboratory studies have confirmed the above-mentioned life-cycle pattern in a majority of *Gracilaria* species (Ogata et al., 1972; Plastino and Oliveira, 1988; Engel et al., 2001, Mantri, 2010). This type of sexual life cycle has also been reported in other Florideophyceae members including economically important genera such as *Gelidium, Chondrus, Eucheuma,* and *Kappaphycus* (Coelho et al., 2007). In the laboratory, the life history could be completed within 5–12 months of culture, depending on the species (Kain and Destombe, 1995; Oza and Gorasia, 2001).

However, deviations from the typical *Polysiphonia* type of life cycle in *Gracilaria* have also been reported. The in situ germination of tetraspores has resulted in the formation of a very interesting feature called "mixed-phase" individuals, who exhibit both sporophyte and gametophyte characters in the same single generation. This type of variation has been reported in *G. tikvahia, G debilis,* and *G. verrucosa* (Bird et al., 1977; Oliveira and Plastino, 1984; Destombe et al., 1989). This variation effectively allows the population to skip the haploid microscopic phase of life cycle to some degree. Oliveira and Plastino (1984) reported the aberrant sequence of development in *G. chilensis*. The in vitro culture of carpospore has produced spermatangia at maturity in few cultures instead of tetrasporangia. The authors have further reported the formation of only spermatangia in nonaerated culture of tetraspores; however, the aerated culture of tetraspores produced male and female gametophytes in normal 1:1 ratio. The study inferred that the development of spermatangia on putative female gametophytes, or on putative tetrasporophytes, could be due to the different culture conditions under which these have been grown. This study also postulates the environmentally controlled sex determination in red algae. The development of female gametophyte has been reported from the carpospores in *G. edulis* (Krishnamurthy et al., 1969). Monocious condition bearing both male and female structures along with tetrasporangia has also been recorded (Kain and Destombe, 1995). The mitotic recombination and mutation involving male or female repression could be the genetic explanation for such a deviation (van der Meer, 1977, 1981; van der Meer and Todd, 1977).

Although reproduction is often linked with sex, this is not always the rule. The reproduction involving propagation through asexual spore, specialized vegetative structures, etc. can exist either instead of, or in addition to, a sexual cycle. Asexual cycles produce a succession of genetically identical individuals with the same level of ploidy. The vegetative multiplication is known to occur in many natural populations of *Gracilaria* species through fragmentation of the thalli, but only few instances have reported the formation of specialized vegetative structures called propagules in culture (Plastino and Oliveira, 1988; Xing-Hong and Wang, 1993). More recently, alternative life-cycle pattern has been reported in *G. gracilis* from north-eastern Italy (Polifrone et al., 2006). The formation of spherical agglomerates of cells has been reported that upon release were able to proliferate and develop into new thalli. The propagules have been formed from the germinating carpospore and not from the modified branchlets of the tetrasporophytic plants as reported earlier in *G. chilensis* (Plastino and Oliveira, 1988). Such kind of asexual reproduction could help one to ensure the multiplication of species to recover the population growth when it faces sudden environmental changes such as elevated temperatures (Polifrone et al., 2006). Zhao et al. (2006) have also reported the development of filamentous structures during laboratory culture of carpospore of *G. asiatica*. The development of filamentous fronds is in addition and independent of normal young sporelings. The single cell detached from the filament develops into a new filamentous frond. However, the formation of normal sporeling from filamentous frond is not reported. However, filament formation during carpospore germination could be considered as a new sexual pathway in *G. asiatica*. Recently, asexual propagules have also been observed in *G. dura* collected from north-western coast of Indian peninsula (Mantri, 2010). In old stock culture of tetrasporophyte (nearly 13–14 months old), many thin proliferations arising from the common base have been formed all over the mother thallus. These structures have the capacity to produce the complete mature thallus, if excised from the mother plant. This finding, however, is limited only for laboratory culture. If these propagules are also present in the natural population of *G. dura*, they would represent a unique means of propagation and distribution.

3. Phenology and Differences in Life-History Stages

Phenology is the study of reproductive stages in the life cycle of an organism in accordance with the changes in climatic season. The phenology of species of *Gracilaria* has been studied widely from different geographical areas. At equivalent latitudes, the phonologies are remarkably alike (Kain and Destombe, 1995). At higher latitudes, the peak growth has been reported in late summer or autumn (Destombe et al., 1988); however, at lower latitudes, it takes place in winter as it is controlled by the monsoon (Kaliaperumal et al., 1986). Only one growth season has been reported in *G. cornea* from Mexico. The biomass production has been significantly correlated with seawater temperature (Orduña-Rojas and Robledo, 2002). The wet season recorded the peak biomass in *G. cornea* followed by cold and dry season with lesser production of plants. Similarly, the higher biomass values have been recorded in *G. heteroclada* during winter in Central Philippines (Luhan, 1996). The bimodal growth has been reported in *G. edulis, G. arcuata* var. *arcuata, G. corticata* var. *cylindrica, G. verrucosa* from India (Umamaheswara Rao, 1973; Kaliaperumal et al., 1986; Oza et al., 1989). Similarly, *G. bursa-pastoris* and *G. gracilis* from Mediterranean lagoon, Thau, France, have shown two peaks in the biomass production (Marinho-Soriano et al., 1998).

It is well evident from the literature that induction of reproduction in *Gracilaria* is environmentally controlled (Ye et al., 2006). Hay and Norris (1984) reported that reproduction in subtidal *Gracilaria* species has been associated with the onset of dry season along the Caribbean coast. They have further mentioned the dominance of gametophytic population over tetrasporophytes in three species and vice versa in two other species. In most of the perennial red alga, the sporophytic phase in a population dominates the gametangial generation. The maximum occurrence of 65–33% has been reported for tetrasporophytic phase in *G. cornea* throughout the year from Mexico with only 21–17% carposporophytes and 12%

male gametophytes (Orduña-Rojas and Robledo, 2002). The highest proportion of tetrasporophytes in the natural population has also been reported in *G. gracilis* from north-eastern Italy (Polifrone et al., 2006). However, Destombe et al. (1989) reported equal proportion of diploid tetrasporophytes and haploid gametophytes in *G. verrucosa* from Northern France. Recently, the tetrasporophyte predominance in Gelidiales has been attributed to the ecophysiological differences in the phases (Carmona and Santos, 2006). It has also been shown that the diploid tetrasporophytic fronds of species of *Gelidium* have efficient reattaching capacity with more rhizoidal cluster-forming ability than gametophytic fronds (Juanes and Puente, 1993). Similar ecophysiological difference has been also reported in few species of *Gracilaria*. The diploid tetrasporophytes of *G. verrucosa* have been shown to be better in growth, survival, tolerance to heavy metal, and UV radiations (Destombe et al., 1993). The agar yield of diploid tetrasporophytes has been shown to be higher than haploid female gametophyte in *G. bursa-pastoris* (Marinho-Soriano et al., 1999).

A few species of *Gracilaria* have been cultivated extensively; still very little is known about the relationship between the cultivated and natural propagation. Very little attention has been paid to the possible changes in life-history characters in relation to the culture practice, so also to the causes of using sexual verses asexual propagules for farming. Recently, such studies have been undertaken in *G. chilensis*, which has been cultivated extensively in the Chilean water for more than 25 years (Guillemin et al., 2008). The process of domestication has produced certain voluntary or involuntary selection for superior growth performance. The microsatellite DNA markers have been used to study the changes in genetic diversity and life-history traits associated with farming. The results suggested that genetic diversity has been reduced owing to continuous clonal propagation. The predominance of diploid individuals in farm suggested that farming practices had significantly modified the important life-history traits when compared with wild populations. The recruitment of individuals resulting from sexual reproduction is very infrequent in farmed individuals, which strongly supports the fact that the sexual life cycle is not competed under farmed conditions. The dominance of diploid individuals associated with farming practices may have important consequences on the evolution of the haploid–diploid life cycle. The spread of selected genotype at local scale has been reported and attributed to the large scale and continuous farming of specific genotype.

4. Distribution, Development of Reproductive Structures, and Sex- and Phase-Linked Molecular Markers

The reproductive structures in *Gracilaria* are distributed randomly throughout the surface. Few studies have shown that the basal portion of the main axis represents the highest number of both tetrasporangia and cystocarps (Garza-Sanchez et al., 2000). In *G. lemaneiformis*, it has been observed that the distribution of tetraspores on the first-generation branches was significantly higher than the subsequent-generation branches. The first-generation branches possess 80% of the total tetraspores while the remaining 20% are produced on second- and third-generation branches (Ye et al., 2006). Similarly, the germination rate in terms of percentage survival depends on from where they have been released. The spore death rate has been considerably lower (14%) for the tetraspores originated from first-generation branches than the subsequent order (54%) in *G. lemaneiformis* (Ye et al., 2006). Although such a study has not been carried out for the other species of *Gracilaria*, more research in this aspect is desirable to attribute the causes for such discrepancy.

The information related to the development of reproductive structures as a whole in genus *Gracilaria* is still fragmentary, but it has been worked out fairly well in few species (Greig-Smith, 1954; Oza, 1976; Ryan and Nelson, 1991; Bouzon et al., 2000). The female gametes in *Gracilaria* are stationary, called "carpogonia." These are produced profusely all along the surface of female gametophyte at maturity. The male gametes as in case of other red algae are nonmotile, called "spermatia." The spermatia are produced in specialized structure called spermatangial cavity. The arrangement of spermatangial cavity is one of the characteristics of paramount importance, as has been used for delineating the genera as well as species of gracilariod alga. The carpogonium branch has been formed on the supporting cells and has cortical origin. The presence of true auxiliary cell in certain taxa is doubtful. The fusion of two to many vegetative cells to form a "fusion-cell" has been reported. The gonimoblast has been developed from the fusion cells. The spermatangia have been produced in sori and can be of two types: the simple or confluent cavity or patches in shallow depression. The five types of spermatangial cavities have been reported in *Gracilaria*, namely *corda* type, *symmetrica* type, *textorii* type, *verrucosa* type, and *henriquesiana* type (Kain and Destombe, 1995). The location of spore mother cell has been reported in the outer cortex in *G. corticata* (Oza, 1976). The male gametogenesis has been studied in *G. caudata* and *G. mammillaris* (Bouzon et al., 2000). The electron microscopic examination revealed that the spermatangial mother cells differ from the surrounding vegetative cells by having poorly developed chloroplasts and numerous plastoglobuli. The endoplasmic reticulum has been concentrated at the cell periphery, contributing to the formation of the spermatangial vesicle. Each spermatium has two layers in its cell walls, but becomes necked upon its release. The tetraspore mother cell has been differentiated in the cortical cell in *G. corticata* (Oza, 1976). Morphologically, this is the end cell of the lateral system. During the development, the cell accumulates, and the reserve food enlarges and becomes densely pigmented. The spermatium divides and forms four tetraspores, which are either cruciately or cunately arranged.

It is very difficult in *Gracilaria* to identify the sexuality before maturity owing to the isomorphic nature of the life cycle. The phase- and sex-related genetic markers have been identified in *G. lemaneiformis* through amplification of genomic DNA by RAPD method (Xiang et al., 1998) and more recently by improved ISSR and AFLP analysis (Pang et al., 2010).

5. Spore and Gamete Production, Viability, and Fertilization

The population structure is determined by the vital rates of spore mortality, recruitments, and reproduction associated with different stages of life cycle. In *Gracilaria*, the recruitment and the relative frequencies of the gametophyte stage depend on the vital rates of the tetrasporophyte stage, and vice versa. However, the dominance of one generation of the life history in certain species could be explained by phase-specific differences in the above-mentioned factors. However, it has been discovered that in perennial red alga *G. gracilis*, the survival of the gametophyte and tetrasporophyte stages is more important for population persistence and growth than for the fertility aspects (Engel et al., 2004). This indicates that the survival of adults is much more important for population dynamics than is reproductive success. The large numbers of studies have been carried out to estimate the spore output and the factors that control the spore release in *Gracilaria* (Kain and Destombe, 1995). A few studies have also dealt with the gamete production and their survival. The spore-shedding experiments conducted in the laboratory have resulted in a very interesting finding pertaining to the reproductive strategies in *Gracilaria*. It is generally considered that the presence of fertile plants in the natural population guarantee the spore release and their subsequent recruitment. The spore shedding can be more seasonally confined as shown in many species of *Gracilaria* (Kaliaperumal et al., 1986; Mal and Subbaramaiah, 1990). However, some of the species such as *G. corticata* from Mandapam coast, India, showed the spore shedding throughout the year (Umamaheswara Rao, 1976). The carpospores have advantage over tetraspores as released as a mass in mucilaginous sac, which ensures rapid and maximum spore settlement (Polifrone et al., 2006). The maximum spore shedding has been reported within the first 3 days in many of the *Gracilaria* species under laboratory conditions (Oza and Krishnamurthy, 1968; Chennubhotla et al., 1986; Shyam Sundar et al., 1991). Nevertheless, prolonged spore discharge lasting for about 25 days has been reported in *G. corticata* (Joseph and Krishnamurthy, 1977) and for 30 days in *G. edulis* (Rama Rao and Thomas, 1974) and *G. verrucosa* (Lefebvre et al., 1987). The marked diurnal variation has been reported in spore-shedding pattern of different *Gracilaria* species with peak output at different times. *G. corticata* and *G. sjoestedtii* reported maximum carpospore as well as tetraspore production during night and the lengthened darkness has enhanced the spore production (Umamaheswara Rao, 1976; Umamaheswara Rao and Subbarangaiah, 1981; Chennubhotla et al., 1986). In contrast, in *G. textorii*, peak spore shedding has been reported at the end of the day (Subbarangaiah, 1984). Umamaheswara Rao and Subbarangaiah (1981) have further reported that the timing of spore shedding has been affected by temperature alone but not irradiance, desiccation, and salinity. The difference in the carpospore and tetraspore size has been reported by many workers. It has been suggested that there is a positive correlation between spore size and their sedimentation rate. The spore dispersal depends primarily on their viability and longevity in the suspended state. Destombe et al. (1992) showed that in *G. verrucosa* the spore dispersal has not been affected by the size of the spore. Further, they have reported that the haploid spores have better dispersal abilities than the diploid ones because of their higher longevity and buoyancy in the water column under in vitro conditions.

The stationary female gamete in *Gracilaria* fuses with spermatia. The gametic fusion involved complex chain of events described elsewhere in this chapter. The spermatia of *G. verrucosa* are effective for less than 5 h and have the dispersion range of 80 m (Destombe et al., 1990). The spermatia viability has been thus considered the prime cause of fertilization limitation in *Gracilaria*. The performance of nonmotile spermatia has been evaluated under the field conditions in *G. gracils* through microsatellite markers. Interestingly, the results showed that the cystocarp yield has not been sperm-limited. It has been further inferred from the observations that the variation in male fertilization success depends on various factors such as distance traveled by spermatia, male–male competition, and female choice (Engel et al., 1999).

The intrinsic factors such as life cycle and mating system are thought to be responsible for genetic structure within the population. Such factors also govern the abilities of dispersal between different populations. Engel et al. (2004) evaluated the consequences of the haploid–diploid life history and intertidal rocky shore landscape on a fine-scale genetic structure in *G. gracilis* using seven polymorphic microsatellite loci. The reproduction in *G. gracilis* occurs in an allogamous manner. Within single population, no significant difference has been observed in allele frequencies, gene diversities, and mean number of alleles between the haploids and diploids. Although within-population allele frequencies have been similar between haploid and diploid samples, the overall genetic difference among haploid samples has been more than twice that of diploid ones. The weak but significant population differentiation has been detected in both haploids and diploids that varied with landscape features and not with geographic distance.

6. Spore Germination and Coalescence

The spores (both carpospores and tetraspores) in *Gracilaria* that have been attached to the substratum immediately undergo germination without any resting stage. The germination of spore follows the *Dumontia*-type of cell division (Oza and Krishnamurthy, 1967; Oza, 1975; Orduña-Rojas and Robledo, 1999; Polifrone et al., 2006; Mantri et al., 2009). In *G. gracilis*, *G. corticata*, and *G. cornea*, the first division of spore took place in transverse plane to form two-celled sporeling (Oza, 1975; Orduña-Rojas and Robledo, 1999; Polifrone et al., 2006). In *G. corticata*, slightly oblique division took place in carpospore to form three-celled stage, whereas in tetraspore, two resultant cells underwent second division in an oblique plane with respect to the first to form four-celled germling. In *G. gracilis*, the first two divisions have taken place without expanding the cell volume (Polifrone et al.,

2006). After the four-celled stage, each cell in *G. corticata* divides transversely to the first median plane forming an eight-celled primary disc. The cells of the primary disc are arranged in two superimposed tiers of cells. Further, the peripheral cells of the primary disc divide periclinally to form the arched shape dome forming at the center. The establishment of the apical cell takes place at the summit, while the initiation of the rhizoids takes place in the lower half of the sporeling from the outermost cells that are in direct contact of the substratum (Oza, 1975). Similar pattern of spore germination and development as disc, holdfast stage, and apical dome formation has also been reported in *G. changii* (Yeong et al., 2008).

The germinating spores that are in close proximity tend to fuse and form the irregular shaped tissue mass in natural as well as under laboratory conditions from which the chimeric plant develops. This phenomenon is termed as spore coalescence and has been first reported in *G. verrucosa* (Jones, 1956). The spore coalescence produces more and larger shoots developed than the isolated spore. The viability in terms of survival increases manyfold in coalescing spores. In the majority of cases, coalesced spore mass is a mosaic of spore derivatives and thus exhibits the significant differences in growth and other phenotypic characters. In *G. chilensis*, bicolor individuals have been produced owing to spore coalescence (Santelices et al., 1996). The chimeric holdfast produced the red as well as green cells, so also the upright chimeric fronds. The random amplified polymorphic DNA analysis further confirmed the existence of two genetically different phenotypes combined because of coalescence of spores. The coalescence of spores under natural conditions forms the basal crust, which could withstand the adverse environmental conditions owing to its microscopic nature that is physiologically better adapted than the adult thallus. The coalescing macroalgae are thus ecologically important members of intertidal and shallow subtidal communities. Since the number of upright shoot formation is increased due to the spore coalescence, up to a certain extent it governs the population structure. In general, the spore coalescence increases the size of the sporelings, thereby reducing further probability of sporeling mortality.

7. Global Climate Change and Its Predicted Effects on Reproductive Processes in *Gracilaria*

The human population has precipitated irreversible changes in the biosphere since the advent of the industrial era. The increasing deposition of pollutants and high levels of greenhouse gases have elevated the global temperature. In addition to these, the habitat disturbance by anthropogenic activities also caused substantial change in both the terrestrial and aquatic ecosystems. Thus, the climate change is ranking high in scientific and public agendas. The ecological processes in few groups of marine organisms have been alarmingly altered because of global change in the environmental parameters (Edwards and Richardson, 2004; Broitman et al., 2008; Wanless et al., 2008; Gibbons and Richardson, 2009). Certain seaweeds such as members of nongeniculate coralline algae provide excellent material to assess their response to the changing climatic conditions due to the intact fossil deposits available for the study. However, there are too few data to allow any confident statements on the effects of global climate change on the different biological parameters in the seaweeds, including reproduction. It has been generally assumed that the global warming would cause a poleward shift in the distributional boundaries of species with an associated replacement of cold-water species by warm-water species. There are interestingly no concise and confirmatory evidences in support of the above statement available for any of the groups of marine organisms. However, the critical role of temperature in determining algal species distributions has been evident on much smaller spatial scales. A recent such study has revealed that the rise in seawater temperature by 3.5°C, induced by the thermal outfall of a power-generating station, over 10 years along 2 km of rocky coastline in California has resulted in significant community-wide changes in 150 marine species including algae. The communities have been greatly altered in cascading responses to changes in abundance of several key taxa, particularly habitat-forming subtidal kelps and intertidal foliose red algae. The temperature-sensitive algae have been decreased greatly in abundance (Schiel et al., 2004). The fragmentary information is also available to correlate the potential environment-induced changes on the algal distribution, dispersal, and establishment (Sagarin et al., 1999; Edwards and Richardson, 2004; Smayda et al., 2004; Coleman et al., 2008). The study has been conducted along the North Sea coast to understand the possible effects of predicted climate change in the seaweed flora. The results have demonstrated that *G. verrucosa*, which have been conspicuous members of the macro-algal flora of North Sea until the middle of the twentieth century, has completely disappeared along with few other seaweed species by 1997 (Ducrotoy, 1999). The author has related this change to the altered environmental parameters such as increased levels of nutrients and oxygen depletion.

Having the tropical origin, genus *Gracilaria* would be capable of reproducing over a wide range of temperatures. The adult plants in *Gracilaria* have been found to be more sensitive to the elevated temperatures. The Mediterranean population of *G. gracilis* has shown the rapid decline in number of macroscopic thalli at the end of May when the water temperature reaches its peak (25–28°C). Similarly, *G. dura* population from northwestern peninsular India vanishes at the end of June when the seawater temperature reaches its maximum (above 30°C) (Mantri, 2010). However, interestingly a fall in temperature (below 10°C) induces the reproduction in *G. gracilis* (Polifrone et al., 2006). However, the expected rise of average 2°C in global temperature may or may not induce the sporulation in *G. gracilis*. However, correlating reproductive phenology and temperature has been a difficult task because of our incomplete knowledge about how temperature influences reproductive maturity and at what stage of development. Temperature has been also the important factor that controls the spore shedding in many of the tropical species of *Gracilaria*. It has been shown that higher and lower temperatures than the normal would either prepone or postpone the timings of diurnal periodicity of the spores in *G. corticata*, *G. textorri*, and *G. sjoestedtii* (Subbarangaiah, 1985). In *G. edulis*, about 3,180 carpospores cystocarp^{-1} day^{-1} have been produced at 28.5°C in March; however, in May when the seawater temperature reaches 32.5°C the carpospore yield has been only 1,696 spores cystocarp^{-1} day^{-1} at Gulf of Mannar, India (Mal and Subbaramaiah, 1990). The 4°C increase in seawater temperature has significantly reduced the carpospore output in this species by half of the average output. Garza-Sanchez et al. (2000) also reported that carpospore as well as tetraspore release and attachment in *G. pacifica* are temperature- and irradiance-dependent. The maximum tetraspore release has been achieved at 24°C and 140 µM quanta m^{-2} s^{-1} in winter as well as autumn. However, the maximum carpospore release has been achieved at moderate temperature 20°C and 60 µM quanta m^{-2} s^{-1}. The higher temperature of 24°C and higher irradiance of 140 µM quanta m^{-2} s^{-1} has drastically reduced the spore output in *G. pacifica*.

The spore survival in *G. pacifica* has been studied in spring, summer, autumn, and winter. In all the seasons, the moderate temperature of 20°C and moderate irradiance of 60 µM quanta m^{-2} s^{-1} has shown higher percentage survival. In *G. lemaneiformis*, it has been observed that the tetraspore survival rate remains at the same level of about 85% with the temperature variance from 10°C to 30°C. However, the higher temperatures have resulted in significantly lower growth of the spore disc (Ye et al., 2006).

Ozone layer in the stratosphere provides the protection to all the living organisms on the Earth from harmful ultraviolet radiation (UVR). The human-induced ozone loss has resulted in elevated levels of UVR reaching the Earth's surface in recent decades due to the thinning of stratospheric ozone layer. In spite of the fact that the worldwide production of ozone-depleting chemicals has already been reduced by 95%, the environmental disturbances are expected to persist for about the next half a century. Since UVR can penetrate water column, they can cause a range of deleterious effects on aquatic organisms, with early life stages at particular risk (reviewed by Hader et al., 2007). UVR have been found to affect marine macroalgae by affecting key physiological processes including photosynthesis, enzyme metabolism DNA lesions, etc. (reviewed by Franklin and Forster, 1997; Xue et al., 2005). However, the effect of UVR on reproductive process in seaweeds is not frequent in literature. Roleda et al. (2004) have studied the effect of PAR and UVBR on early life stages of *Mastocarpus stellatus* and *Chondrus crispus*. The germination and photosynthesis of the low light adapted carpospores of both species has been inhibited with increase in PAR. The carpospore viability in *C. crispus* has been found to be sensitive to UVR. Similarly, the effect of UVB radiation on early developmental stages, spore survival, and embryo growth has been studied in *Mazzaella laminarioides*, *Gigartina skottsberdii*, and *Macrocystis pyrifera* (Navarro et al., 2007). The survival and growth has been found to be affected in these seaweeds due to UVB. It has been observed that the percentage inhibition of sporulation in cystocarpic plants of *G. corticata* increased with increasing period of exposure to UV radiation. The maximum inhibition of sporulation of about 55% has been observed when the cystocarpic plant has been subjected to 60-min exposure to UVB (CSMCRI, unpublished results). Therefore, it could be concluded that the UVR may affect the reproductive process and recruitment in *Gracilaria* species.

The global climate change has increased the levels of atmospheric carbon dioxide in the last few decades. The abundant growth of *Gracilaria* has been reported in the intertidal area of some topical countries. The few species of *Gracilaria* with high growth rate could be also considered as the important CO_2 sink, although such studies have not yet been carried out. The genus *Gracilaria* has complex life history and varying reproductive strategies. The studies have proved beyond doubt that the differences do exist in different life-cycle stages at physiological, ecological, as well as molecular level. The juvenile stages such as spores, germling disc, and early growth forms are of great importance, as such stages have an essential role to play in dispersal and subsequent recruitment. The survival of these transitional life-history stages is the most critical phase leading to the successful formation of a benthic population. The spermatia viability in *Gracilaria* is highly limited; this could be easily vulnerable to climatic changes. Considering the small size and simple cellular organization of these reproductive as well as early life-history stages, any kind of climatic change may exert stress, which ultimately would affect the biology of the species. However, the response of particular species toward change has been often very subtle and cannot be assessed simply from the results of few studies that subjected different life stages to a range of variable environmental conditions. Each species has its separate and unique strategy to combat the changing environmental conditions. Moreover, all the climatic changes are not sudden but take several years, and thus species might adjust to such changes up to a certain extent by altering the physiological needs. The *G. salicornia* populations collected from Japan and Thailand have shown resilience against the changes in temperature, salinity, and irradiance under in vitro conditions (Phooprong et al., 2007). The different responses in photosynthesis and respiration ability toward the short-term change in irradiance, salinity, and temperature have been recorded. The population from Thailand has shown the adaptability to high irradiance and temperature; however, the one from Japan has shown adaptability to the low submarine irradiance under in vitro conditions. The same species has been intentionally introduced in Hawaii, in the 1970s for experimental aquaculture purpose (Smith et al., 2004). It took 30 years for this species to prove its invasive tendencies. This phenomenon has shown that the species might respond differently to the environmental change over the years by climatic adaptation. The overall understanding of the climatically driven environmental changes is much more complex and includes the species tolerance limit toward the particular change, alternative dispersal or reproductive strategy, physiological adaptation, biotic interactions, etc. The effect of varying environmental conditions on reproductive processes of genus *Gracilaria* is largely unpredicted since the information pertaining to the effects is limited.

8. Acknowledgments

Authors would like to thank Council of Scientific and Industrial Research, New Delhi (NWP 018 and 019), for funding support. We would also like to thank handling editor for constructive suggestions.

9. References

Bird, N., McLachlan, J. and Grund, D. (1977) Studies on *Gracilaria*. V. In-vitro life history of *Gracilaria* sp. from Maritime Provinces. Can. J. Bot. **55**: 1282–1289.

Bold, H.C. and Wynne, M.J. (1978) Introduction to the algae. Prentice-Hall, New Jersey Publication.

Bouzon, Z.L., Miguens, F. and Oliveira, E.C. (2000) Male gametogenesis in red algae *Gracilaria* and *Gracilariopsis* (Rhodophyta, Gracilariales). Cryptogamie. Algol. **21**: 33–47.

Brodie, J. and Zuccarello, G.C. (2007) Systematics of the species rich algae: red algal classification, phylogeny and speciation, In: T.R. Hodkinson and J.A.N. Parnell (eds.) *Reconstructing the Tree of Life: Taxonomy and Systematics of Species Rich Taxa*. CRC Press, Taylor & Francis, Boca Raton, pp. 323–336.

Broitman, B.R., Mieszkowska, N., Helmuth, B. and Blanchette, C.A. (2008) Climate and recruitment of rocky shore intertidal invertebrates in the eastern north Atlantic. Ecology **89**: S81–S90.

Carmona, R. and Santos, R. (2006) Is there an ecophysiological explanation for the gametophyte-tetrasporophyte ratio in *Gelidium sesquipedale* (Rhodophyta). J. Phycol. **42**: 259–269.

Chennubhotla, V.S.K., Kaliaperumal, N., Ramalingam, J.R. and Kalimuthu, S. (1986) Growth, reproduction and spore output in *Gracilaria folifera* (Forsskål) Børgesen *Gracilariopsis sjoestedtii* (Kylin) Dwson around Mandapam. Indian J. Fish. **33**: 76–84.

Coelho, S.M., Peters, A.F., Charrier, B., Roze, D., Destombe, C., Valero, M. and Cock, J.M. (2007) Complex life cycles of multicellular eukaryotes: new approaches based on the use of model organisms. Gene **406**: 152–170.

Coleman, M.A., Kelaher, B.P., Steinberg, P.D. and Millar, A.J.K. (2008) Absence of a large brown macroalga on urbanized rocky reefs around Sydney, Australia, and evidence for historical decline. J. Phycol. **44**: 897–901.

Destombe, C., Godin, J. and Bodard, M. (1988) The decay phase in the life history of *Gracilaria verrucosa*: consequences in intensive cultivation, In: T. Stadler, J. Mollion, M.C. Verdus, Y. Karamanos, H. Morvan and D. Christiaen (eds.) *Algal Biotechnology*. Elsevier Applied Science, London, pp. 287–303.

Destombe, C., Valero, M., Vernet, P. and Couvet, D. (1989) What controls haploid–diploid ratio in the red alga, *Gracilaria verrucosa*. J. Evol. Biol. **2**: 317–338.

Destombe, C., Godin, J. and Remy, J.M. (1990) Viability and dissemination of spermatia of *Gracilaria verrucosa* (Gigartinales, Rhodophyta). Hydrobiologia **204/205**: 219–223.

Destombe, C., Godin, J., Lefebvre, C. and Vernet, P. (1992) Differences in dispersal abilities of haploid and diploid spores of *Gracilaria verrucosa* (Gigartinales, Rhodophyta). Bot. Mar. **35**: 93–98.

Destombe, C., Godin, J., Nacher, M., Richered, S. and Valero, M. (1993) Differences in response between haploid and diploid isomorphic phases of *Gracilaria verrucosa* (Rhodophyta: Gigartinales) exposed to artificial environmental conditions. Hydrobiologia **260/261**: 131–137.

Dippakore, S., Reddy, C.R.K. and Jha, B. (2005) Production and seeding of *Porphyra okhaensis* (Bangiales, Rhodophyta) in laboratory culture J. Appl. Phycol. **17**: 331–337.

Drew, K. (1949) Conchocelis-phase in the life-history of *Porphyra umbilicalis* (L.) Kützing, Nature **166**: 748–749.

Ducrotoy, J.P. (1999) Indications of change in the marine flora of the North Sea in the 1990s. Mar. Pollut. Bull. **38**: 646–654.

Edwards, M. and Richardson, A.J. (2004) Impact of climate change on marine pelagic phenology and trophic mismatch. Nature **430**: 881–884.

Engel, C., Wattier, R., Destombe, C. and Valero, M. (1999) Performance of non-motile male gametes in the sea: analysis of paternity and fertilization success in natural population of red seaweed, *Gracilaria gracilis*. Proc. R. Soc. Lond. B Biol. Sci. **226**: 1879–1886.

Engel, C., Aberg, P., Gaggiotti, O.E., Destombe, C. and Valero, M. (2001) Population dynamics and stage structure in a haploid–diploid red seaweed, *Gracilaria gracilis*. J. Ecol. **89**: 436–450.

Engel, C., Destombe, C. and Valero, M. (2004) Mating system and gene flow in the red seaweed *Gracilaria gracilis*: effect of haploid–diploid life history and intertidal rocky shore landscape on fine scale genetic structure. Heredity **92**: 289–298.

FAO (2006) State of world aquaculture 2006. Food and Agriculture of the United Nations, Rome. Technical paper no. 500.

Franklin, L. and Forster, R. (1997) The changing irradiances environment: consequence for marine macrophytes physiology, productivity, and ecology. Eur. J. Phycol. **32**: 207–237.

Gall, L.L., Pien, S. and Rusig, A.M. (2004) Cultivation of *Palmaria palmata* (Palmariales, Rhodophyta) from isolated spores in semi-controlled conditions. Aquaculture **229**: 181–191.

Ganesan, M., Thiruppathai, S. and Jha, B. (2006) Mariculture of *Hypnea musciformis* (Wulfen) Lamouroux in South east coast of India. Aquaculture **256**: 201–211.

Garza-Sanchez, F., Zertuche-Gonzalez, J.A. and Chapman, D.J. (2000) Effect of temperature and irradiance on the relaease, attachment and survival of spores of *Gracilaria pacifica* Abbott (Rhodophyta). Bot. Mar. **43**: 205–212.

Gibbons, M.J. and Richardson, A.J. (2009) Patterns of jellyfish abundance in the North Atlantic. Hydrobiologia **616**: 51–65.

Greig-Smith, F. (1954) Cytologial observations on *Gracilria multipartata*. Phycol. Bull. **2**: 4.

Guillemin, M.L., Faugeron, S., Destombe, C., Viard, F., Correa, J.A. and Valero, M. (2008) Genetic variation in wild and cultivated populations of the haploid–diploid red alga *Gracilaria chilensis*: how farming practices favor asexual reproduction and heterozygosity. Evolution **62**: 1500–1519.

Hader, D.P., Kumar, H.D., Smith, R.C. and Worrest, R.C. (2007) Effects of solar UV radiation on aquatic ecosystems and interactions with climate change. Photochem. Photobiol. Sci. **6**: 267–285.

Hanisak, M.D. (1998) Seaweed cultivation global trends. World Aquacult. **29**: 18–21.

Hawkes, M.W. (1990) Reproductive strategies, In: K.M. Cole and R.G. Sheath (eds.) *Biology of the Red Algae*. Cambridge University Press, Cambridge, pp. 455–476.

Hay, M.E. and Norris, J.N. (1984) Seasonal reproduction and abundance of six sympatric species of *Gracilaria* Grev. (Gracilariaceae, Rhodophyta) on a Caribbean subtidal sand plain. Hydrobiologia **116/117**: 63–94.

Jones, W.E. (1956) Effect of spore coalescence on the early development of *Gracilaria verrucosa* (Hudson) papenfuss. Nature **178**: 426–427.

Joseph, M.M. and Krishnamurthy, V. (1977) Studies on shedding of carpospores in *Gracilaria corticata*. Seaweed Res. Util. **2**: 1–8.

Juanes, J.A. and Puente, A. (1993) Differential reattachment capacity of isomorphic life history phases of *Gelidium sesquipedale*. Hydrobiologia **260/261**: 139–144.

Kain, J. and Destombe, C. (1995) A review of the life history, reproduction and phenology of *Gracilaria*. J. Appl. Phycol. **7**: 269–281.

Kaliaperumal, N., Chennubhotla, V.S.K., Kalimuthu, S. and Ramalingam, J.R. (1986) Growth, phenology and spore shedding in *G. corticata* var. *arcuuta* (Zanardinii) Umamaheswara Rao and *G. corticata* var. *cylindrica* (J. Agardh) Umamaheswara Rao (Rhodophyta). Indian J. Mar. Sci. **15**: 107–110.

Krishnamurthy, V., Venkataraju, P. and Venugopal, R. (1969) An aberrant life history in *Gracilaria edulis* (Gmel.) Silva and *Gracilaria corticata*. J. Agardh. Curr. Sci. **14**: 343–344.

Lamote, M. and Johnson, L.E. (2008) Temporal and spatial variation in the early recruitment of fucoid algae: the role of microhabitats and temporal scales. Mar. Ecol. Prog. Ser. **368**: 93–102.

Lefebvre, C.A., Destombe, C. and Godin, J. (1987) Le fonctionnement du carposporophyte de *Gracilaria verrucosa* et ses repercussions sur la strategie de reproduction. Cryptogamie Algol. **8**: 113–126.

Luhan, M.R.J. (1996) Biomass and reproductive states of *Gracilaria heteroclada* Zhang et Xia collected from Jaro, Central Philippines. Bot. Mar. **39**: 207–211.

Mal, T.K. and Subbaramaiah, K. (1990) Diurnal periodicity of carpospore shedding in the red alga *Gracilaria edulis* (Gmel.) Silva [Rhodophyta]. Indian J. Mar. Sci. **19**: 63–65.

Mantri, V.A. (2010) Studies on biology of *Gracilaria dura* (C. Agardh) J. Agardh. PhD Thesis, Bhavnagar University, 150 pp.

Mantri, V.A., Thakur, M.C., Kumar, M., Reddy, C.R.K. and Jha, B. (2009) The carpospore culture of industrially important red alga *Gracilaria dura* (Gracilariales, Rhodophyta). Aquaculture **297**: 85–90.

Marinho-Soriano, E., Laugier, T. and de Casabianca, M.L. (1998) Reproductive strategy of two *Gracilaria* species, *G. bursa-pastoris* and *G. gracilis*, in a Mediterranean lagoon (Thau, France). Bot. Mar. **41**: 559–564.

Marinho-Soriano, E., Bourret, E., de Casabianca, M.L. and Maury, L. (1999) Agar from the reproductive and vegetative stages of *Gracilaria bursa-pastoris*. Bioresour. Technol. **67**: 1–5.
Meena, R., Prasad, K., Ramavat, B.K., Ghosh, P.K., Eswaran, K., Thiruppathi, S., Mantri, V.A., Subbarao, P.V. and Siddhanta, A. K. (2007) Preparation, characterization and benchmarking of agarose from *Gracilaria dura* of Indian waters. Carbohydr. Polym. **69**: 179–188.
Navarro, N.P., Mansilla, A. and Palacios, M. (2007) UVB effects on early developmental stages of commercially important macroalgae in southern Chile. J. Appl. Phycol. **20**: 447–456.
Norton, T.A. (1992) The biology of seaweed propagules. Br. Phycol. J. **27**: 217.
Ogata, E., Matsui, T. and Nakamura, H. (1972) The life cycle of *Gracilaria verrucosa* (Rhodophyceae, Gigartinales) in vitro. Phycologia. **11**: 75–80.
Oliveira, E.C. and Plastino, E.M. (1984) The life history of some species of *Gracilaria* (Rhodophyta) from Brazil. Jpn. J. Phycol. **32**: 203–208.
Orduña-Rojas, J. and Robledo, D. (1999) Effect of irradiance and temperature on the release and growth of carpospores from *Gracilaria cornea* J. Agardh (Gracilariales, Rhodophyta). Bot. Mar. **42**: 315–319.
Orduña-Rojas, J. and Robledo, D. (2002) Studies on the tropical agarophyte *Gracilaria cornea* J. Agardh (Rhodophyta, Gracilariales) from Yucatan, Mexico: II. Biomass assessment and reproductive phenology. Bot. Mar. **45**: 459–464.
Oza, R.M. (1975) Studies on Indian *Gracilaria*. I. Carpospore and tetraspore germination and early stages of development in *Gracilaria corticata* J. Ag. Bot. Mar. **18**: 199–201.
Oza, R.M. (1976) Studies on Indian *Gracilaria*. II. The development of reproductive structures of *Gracilaria corticata*. Bot. Mar. **19**: 107–114.
Oza, R.M. and Gorasia, S. (2001) Life history of *Gracilaria corticta* J. Ag. (Gracilariaceae, Rhodophyta) *in vitro*. Seaweed Res. Util. **23**: 27–31.
Oza, R.M. and Krishnamurthy, V. (1967) Carpospore germination and early stages of development in *Gracilaria verrucosa* (Huds.) Papenf. Phykos **6**: 84–85.
Oza, R.M. and Krishnamurthy, V. (1968) Studies on carposporic rhythm of *Gracilaria verrucosa* (Huds.) Papenf. Bot. Mar. **11**: 118–121.
Oza, R.M., Tewari, A. and Rajyagurau, M.R. (1989) Growth and phenology of red alga *Gracilaria verrucosa* (Huds) Papenf. Indian J. Mar. Sci. **18**: 82–86.
Pang, Q., Sui, Z., Kang, K.H., Kong, F. and Zhang, X. (2010) Application of SSR and AFLP to the analysis of genetic diversity in *Gracilariopsis lemaneiformis* (Rhodophyta). J. Appl. Phycol. DOI: 10.1007/s10811-009-9500-3.
Phooprong, S., Ogawa, H. and Hayashizaki, K. (2007) Photosynthetic and respiratory responses of *Gracilaria salicornia* (C. Ag.) Dawson (Gracilariales, Rhodophyta) from Thailand and Japan. J. Appl. Phycol. **19**: 447–456.
Plastino, E.M. and Oliveira, E.C. (1988) Deviation in the life history of *Gracilaria* sp. (Rhodophyta, Gigartinales) from Coquimbo, Chile, under different culture conditions. Hydrobiologia **164**: 67–74.
Polifrone, M., Masi, F.D. and Gargiulo, G.M. (2006) Alternative pathways in the life history of *Gracilaria gracilis* (Gracilariales, Rhodophyta) from north-eastern Sicily (Italy). Aquaculture **261**: 1003–1013.
Rama Rao, K. and Thomas, P.C. (1974) Shedding of carpospores in *Gracilaria edulis* (Gmel.) Silva. Phykos **13**: 54–59.
Roleda, M.Y., van de Poll, W.H., Hanelt, D. and Wiencke, C. (2004) PAR and UVBR effects on photosynthesis, viability, growth and DNA in different life stages of two coexisting Gigartinales: implications for recruitment and zonation pattern. Mar. Ecol. Prog. Ser. **281**: 37–50.
Rueness, J. (2005) Life history and molecular sequences of *Gracilaria vermiculophylla* (Gracilariales, Rhodophyta), a new introduction to European waters. Phycologia **44**: 120–128.
Ryan, K.G. and Nelson, W. (1991) Comparative study of reproductive development in two species of *Gracilaria* (Gracilariales, Rhodophyta) I. Spermatogenesis. Crypto. Bot. **2/3**: 229–233.
Sagarin, R.D., Barry, J.P., Gilman, S.E. and Baxter, C.H. (1999) Climate-related change in an intertidal community over short and long time scales. Ecol. Monogr. **69**: 465–90.

Santelices, B. and Aedo, D. (2006) Group recruitment and early survival of *Mazzaella laminarioides*. J. Appl. Phycol. **18**: 583–589.
Santelices, B., Correa, J.A., Meneses, I., Aedo, D. and Varela, D. (1996) Sporeling coalescence and intraclonal variation in *Gracilaria chilensis* (Gracilariales, Rhodophyta). J. Phycol. **32**: 313–322.
Schiel, D.R., Steinbeck, J.R. and Foster, M.S. (2004) Ten years of induced ocean warming causes comprehensive changes in marine benthic communities. Ecology **85**: 1833–1839.
Shyam Sundar, K.L., Subba Rao, P.V. and Subbaramaiah, K. (1991) Studies on carpospore shedding in the red alga *Gracilaria crassa* (Gigartinales: Rhodophyta). Indian J. Mar. Sci. **20**: 70–71.
Smayda, T.J., Borkman, D.G., Beaugrand, G. and Belgrano, A. (2004) Responses of marine phytoplankton populations to fluctuations in marine climate, In: N.C. Stenseth, G. Ottersen, J.H. Hurrell and A. Belgrano (eds.) *Marine Ecosystems and Climate Variation: The North Atlantic: A Comparative Perspective.* Oxford University Press, Oxford, pp. 49–58.
Smith, J.E., Hunter, C.L., Conklin, E.J., Most, R., Sauvage, T., Squair, C. and Smith, C.M. (2004) Ecology of the Invasive Red Alga Gracilaria salicornia (Rhodophyta) on O'ahu. Hawaii. Pac. Sci. **58**: 325–343.
Subba Rao, P.V. and Mantri, V.A. (2006) Indian seaweed resources and sustainable utilization: scenario at the dawn of a new century. Curr. Sci. **91**: 164–174.
Subbarangaiah, G. (1984) Growth, reproduction and spore shedding in *Gracilaria textorii* (Sur.) J. Ag. along Visakhapatanam coast. Phykos **23**: 246–253.
Subbarangaiah, G. (1985) Influence of temperature on the diurnal periodicity of tetraspores of some Gigartinales (Rhodophyta). Seaweed Res. Util. **8**: 23–28.
Umamaheswara Rao, M. (1973) Growth and reproduction in some species of *Gracilaria* and *Gracilariopsis* in the Palk Bay. Indian J. Fish. **20**: 182–192.
Umamaheswara Rao, M. (1976) Spore libration in Gracilaria corticata J. Ag. growing at Mandapam. J. Exp. Mar. Biol. Ecol. **21**: 91–98.
Umamaheswara Rao, M. and Subbarangaiah, G. (1981) Effects of environmental factors on the diurnal periodicity of tetraspores of some gigartinales (Rhodophyta), In: T. Livring (ed.) X^{th} *International Seaweed Symposium*. Walter de Gruyter, Berlin, pp. 209–214.
van der Meer, J.P. (1977) Genetics of *Gracilaria* sp. (Rhodophyceae, Gigartinales) II. The life history and genetic implications of cytokinetic failure during tetraspore formation. Phycologia **16**: 367–371.
van der Meer, J.P. (1981) Genetics of *Gracilaria tikvahiae* (Rhodophyceae) VII. Further observations on mitotic recombination and construction of polyploids. Can. J. Bot. **59**: 787–792.
van der Meer, J.P. and Todd, E.R. (1977) Genetics of *Gracilaria* sp. (Rhodophyceae, Gigartinales) IV. Mitotic recombination and its relationship to mixed phases in life history. Can. J. Bot. **55**: 2810–2817.
van der Meer, J.P. and Todd, E.R. (1980) The life-history of *Palmaria palmata* in culture. A new type for the Rhodophyta. Can. J. Bot. **58**: 1250–1256.
Wang, W.J., Zhu, J., Xu, P., Xu, J., Lin, X., Huang, C., Song, W., Peng, G. and Wang, G. (2008) Characterization of the life history of *Bangia fuscopurpurea* (Bangiaceae, Rhodophyta) in connection with its cultivation in China. Aquaculture **278**: 101–109.
Wanless, S., Harris, M.P., Lewis, S., Frederiksen, M. and Murray, S. (2008) Later breeding in northern gannets in the eastern Atlantic. Mar. Ecol. Prog. Ser. **370**: 263–269.
Xiang, F.L., Zheng, H.S. and Xue, C.Z. (1998) Application of RAPD in genetic diversity study on *Gracilaria lemaneiformis* III. Phase and sex related markers. Chinese J. Oceanol. Limnol. **16**: 147–151.
Xing-Hong, Y. and Wang, S.J. (1993) Regeneration of whole plants from *Gracilaria asiatica* Chang et Xia protoplast (Gracilariaceae, Rhodophyta). Hydrobiologia **260/261**: 429–436.
Xue, L., Zhang, Y., Zhang, T., An, L. and Wang, X. (2005) Effects of enhanced ultraviolet–B radiation on algae and cyanobacteria. Crit. Rev. Microbiol. **31**: 79–89.
Ye, N., Wang, H. and Wang, G. (2006) Formation and early development of tetraspores of *Gracilaria lemaneiformis* (*Gracilaria*, Gracilariaceae) under laboratory conditions. Aquaculture **254**: 219–226.

Yeong, H.Y., Khalid, N. and Phang, S.W. (2008) Protoplast isolation and regeneration from *Gracilaria changii* (Gracilariales, Rhodophyta). J. Appl. Phycol. **20**: 641–651.

Zemke-White, L.W. and Ohno, M. (1999) World seaweed utilization: end-of-century summary. J. Appl. Phycol. **11**: 369–376.

Zhao, F., Wang, A., Liu, J. and Duan, D. (2006) New phenomenon in early development of sporelings in *Gracilaria asiatica* Chang et Xia (Gracilariaceae, Rhodophyta). Chinese J. Oceanol. Limnol. **24**: 364–369.

Biodata of **Rui Pereira** and **Charles Yarish**, authors of *"The Role of Porphyra in Sustainable Culture Systems: Physiology and Applications"*

Dr. Rui Pereira is currently a Postdoc Researcher at the Centre for Marine and Environmental Research in Porto, Portugal. He obtained his Ph.D. in Biology from the University of Porto in 2004, having developed most of his work at the Seaweed Marine Biotechnology Laboratory at the University of Connecticut, USA. Presently, he continues his studies and research on aquaculture of seaweed in integrated multitrophic systems and its applications, namely as ingredients in fish feed. Dr. Pereira's scientific interests are in the areas of seaweed eco-physiology, applied phycology and seaweed biodiversity.

E-mail: **rpereira@ciimar.up.pt**

Professor Charles Yarish received his Ph.D. from Rutgers University (1976) and then joined the faculty at the University of Connecticut (1976) where he is a professor of biology in the Departments of Ecology and Evolutionary Biology and Marine Sciences. At the University of Connecticut, he developed an internationally known laboratory for seaweed research and has been one of the leaders in the development of integrated multi-trophic aquaculture (IMTA). In 2008, Profesor Yarish was elected to the Connecticut Academy of Science and Engineering. He has published extensively including two co-edited books entitled "Economically Important Marine Plants of the Atlantic: Their Biology and Cultivation" and also "Seaweeds – Their Environment, Biogeography and Ecophysiology." He has also been an adjunct Professor of Marine Sciences at the State University of New York at Stony Brook, visiting scientist at the Biologische Anstalt Helgoland, Germany, visiting professor of Marine Biology at the University of Groningen, The Netherlands, a guest professor at Shanghai Fisheries University, China, and most recently a visiting sea grant scholar at the Rhode Island Sea Grant College Program. He has served with many organisations including the International Executive Service Corps' Aquacultural Project (Kenya), member of the Organising Committee and the Executive Secretariat for the Sixth International Phycological Congress, Qingdao, China, and The Advisory Board of the Institute of BioSciences (Halifax), National Research Council of Canada. Professor Yarish received the 1992 Marinalg Award's First Prize at the XIVth International Seaweed Symposium, Brittany, France, for his work in East Africa (Kenya) judged to be the most useful to the economic development of the world seaweed industry. He has also been a national lecturer, The secretary, member of the Society's Executive Committee, vice president, president (2001) and past president of the Phycological Society of America. He has been an invited participant in many international invited symposia and meetings in Canada, Chile, Cuba, Germany, Ireland, Japan, People's Republic of China, Mexico, Portugal, South Korea, Spain, The Netherlands and the USA, which have dealt with the ecophysiology and biogeography of seaweeds, as well as the development of integrated multi-trophic aquaculture (IMTA) systems. He is the elected co-chairman of the Science Technical Advisory Committee for the US Environmental Protection Agency's Long Island Sound Study and has assisted the Office of Oceanic and Atmospheric Research, NOAA, US Department of Commerce on matters dealing with aquaculture in the People's Republic of China, South Korea and Japan. Among other memberships in many professional societies, he is also a member of the Advisory Board for the City of Bridgeport's Regional Vocational Aquaculture High School. He has been a co-principal investigator numerous grants award by the National and Connecticut Sea Grant College Programmes, as well as The State of Connecticut. In August 2007, he was the recipient of The 2007 Faculty Recognition Award, which recognises sustained outstanding achievements in teaching, research and services benefiting University of Connecticut at Stamford. He is a also a member of the scientific advisory committee of a national panel, "The Sustainable Seafood Forum," Aquarium of the Pacific, Marine Conservation Research Institute, Long Beach, California.

E-mail: **charles.yarish@uconn.edu**

Rui Pereira   **Charles Yarish**

THE ROLE OF *PORPHYRA* IN SUSTAINABLE CULTURE SYSTEMS: PHYSIOLOGY AND APPLICATIONS

RUI PEREIRA[1] AND CHARLES YARISH[2]
[1] *CIIMAR/CIMAR Centre for Marine and Environmental Research, Rua dos Bragas, 289, Porto, Portugal*
[2] *Departments of Ecology and Evolutionary Biology and Marine Sciences, University of Connecticut, One University Place, Stamford, CT 06901-2315, USA*

1. The Genus *Porphyra*

The genus *Porphyra* belongs to the division Rhodophyta, Class Bangiophyceae, Order Bangiales and Family Bangiaceae. It has a global distribution. The genus consists of more than 140 species and is the most speciose genera in the red algae (Yoshida et al., 1997; Silva, 1999; Brodie and Zuccarello, 2007). *Porphyra* is also one of the oldest red algae known. Fossils that closely resemble this genus were found in the 570 million-year-old Doushantuo formation in southern China (Graham and Wilcox, 2000). Butterfield (2000) estimates the lineage of the Order Bangiales to be at least 1.2 billion years old.

Some of the characteristics used to identify species of *Porphyra* are: shape, thickness and color (in fresh and dry conditions) of the blades; distribution of the reproductive cells (monoecious or dioecious species); arrangement of the reproductive cells; shape and size of the vegetative cells; shape of the margins and of the marginal cells; seasonality of the gametophyte phase; position on shore (in relation to the tidal level); and number and size of the chromosomes (Hus, 1902; Coll and Oliveira Filho, 1976; Wilkes et al., 1999; Brodie and Irvine, 2003). Despite its simple morphology, it may have variable morphologies that may ultimately require molecular techniques for positive identification. Recent molecular evidences suggest that the genus is polyphyletic (Nelson et al., 2006). More studies are using molecular techniques that include the sequencing of the SSU rRNA gene, ITS1-5.8s-ITS2 and *rbc*L subunits (Broom et al., 2002; Brodie et al., 1996, 2007; Klein et al., 2003; Neefus et al., 2008). There is a need to find genes capable of distinguishing species, without detecting differences between populations of the same species, which could lead to formation of many 'new' species. As Klein et al. (2003) pointed out, phylogenetic analysis of the Bangiophyceae have previously been conducted based on the sequences of a single or, at most, two examples of each species. This limited sampling is inadequate to assess intraspecific variation or to reveal cryptic variation in morphologically similar taxa. Those authors developed molecular screens that can reliably and objectively sort specimens of northwest Atlantic *Porphyra* into taxa (Klein et al., 2003). It is a DNA-based system that uses partial sequences of the nuclear small subunit ribosomal RNA gene (SSU) or the plastid ribulose-1,5-bisphosphate carboxylase-oxygenase large subunit gene (rbcL). A similar study was carried out for *Porphyra* and *Bangia* species from the northeast Atlantic (Brodie et al., 1998). Nevertheless, morphological and life-history traits are still useful and are used to distinguish species.

The macroscopic phase of *Porphyra* consists of thalli with either one (monostromatic) or two (distromatic) cell layers. This is usually considered a species-specific character, although Brodie et al. (1998) proved that it has no taxonomic significance. Molecular systematic studies suggest that the monostromatic type may be older and that the distromatic type may have arisen at least twice (Oliveira et al., 1995). The blades of *Porphyra* grow by division of the marginal cells and attach to the substrate (rocks, shells or other algae) through numerous thin, colourless rhizoidal cells. The genus is typically divided into three groups, based on the number of cells and on the number of chloroplasts per cell. There are monostromatic species with one chloroplast per cell, monostromatic species with two chloroplasts per cell and distromatic species with one chloroplast per cell (Brodie and Irvine, 2003).

With *Porphyra* being such a unique group of economic species, there has been interest in establishing it as a model organism for basic and applied studies in plant science (Sahoo et al., 2002). There are several advantages to study development, physiology, cytology, genetics and genomics. For example, establishment of several pure lines, the small genome size (consisting of up to five chromosomes) and short generation time (1–3 months) are suitable for genetic analysis. To understand the whole genetic system in red algae, large-scale expressed sequence tag (EST) analysis of *P. yezoensis* has been initiated (Kitade et al., 2008). Of the several thousand EST sequences produced so far, approximately 40% have similarity to those of registered genes from various organisms, including higher plants, mammals, yeast and cyanobacteria. *Porphyra* is currently the focus of a very significant human genome-like study by the US Department of Energy's Joint Genome Institute Community Sequencing Project and the US National Science Foundation (Gantt et al., 2009).

2. Economic Importance of *Porphyra*

The genus *Porphyra* is the most valuable seaweed crop, considering the value per kilogram produced. According to the data made available by the Food and Agriculture Organization of the United Nations (FAO, 2009a), in 2007, more than 1.5 million tons of *Porphyra* were produced valued at over US$1.5 billion (Table 1).

Porphyra, commercially known as nori (its Japanese name) or 'laver' is mainly used for human consumption as ingredient of the Japanese delicacy 'sushi'. The alga contains high levels of protein (25–50%), vitamins (higher vitamin C

Table 1. Main seaweed aquaculture production and value in 2007.

Genera	Production (t)	Value (US$1,000)	US$ t^{-1}
Laminaria	4,613,104	2,894,539	627
Eucheuma/Kappaphycus	3,164,064	550,934	174
Undaria	1,765,492	790,875	448
Porphyra	1,510,634	1,541,934	1,020
Gracilaria	1,066,700	522,159	490
Other	326,577	97,613	299

than in oranges), trace minerals and dietary fibres (Noda, 1993). Nearly 17 types of free amino acids are found in the tissue of *Porphyra*, including taurine, which controls blood cholesterol levels (Tsujii et al., 1983). This alga has yet other biologically active substances beneficial for human health. For instance, Zhang et al. (2003) showed that a sulphated polysaccharide fraction from *P. haitanensis* can be used to compensate for the decline in total antioxidant capacity and the activities of antioxidant enzymes. This suggests that it may be helpful in retarding the aging process. Other works with polysaccharides extracted from *P. yezoensis* showed anticoagulant (Zhou et al., 1990) and immune-stimulating activities (Yashizawa et al., 1995). Saito et al. (2002) showed that *Porphyra* peptides induced a significant reduction of the blood pressure in hypertensive patients.

Porphyra is a preferred source of the red pigment r-phycoerythrin, which is utilised as a fluorescent marker in in situ hybridisation studies, in the medical diagnostic industry (Mumford and Miura, 1988). More recent applications of *Porphyra* are its use as source of mycosporines (Korbee et al., 2005a, b; Sampath-Wiley et al., 2008), compounds with UV protection properties. It is also an ingredient in fish-feed as a partial substitute for fish meal because of its high protein content; however, it is now receiving widespread acceptance and is still not commercially available (Davies et al., 1997; Walker et al., 2009; Pereira et al., 2010).

3. The Life Cycle of *Porphyra*

The life cycle of the genus was first described by Kathleen Drew (1949). She found that the microscopic, shell-inhabiting 'Conchocelis rosea' was in fact the sporophytic phase of the genus *Porphyra*. The life cycle of *Porphyra* is, therefore, biphasic and heteromorphic, with a foliose haploid gametophyte and a filamentous diploid sporophyte (Fig. 1). The sporophyte phase is still commonly referred to as the conchocelis phase (Brodie and Irvine, 2003).

The conchocelis serves as a perennating stage in nature and it can also be maintained in laboratory cultures for long periods through vegetative propagation. Under the appropriate environmental conditions, the conchocelis produces conchosporangial filaments that produce diploid conchospores. The conchospores

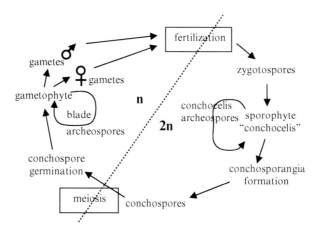

Figure 1. Schematic representation of the life cycle of *Porphyra*.

are then released and, during the germination, the meiosis yields a four haploid celled germling, a chimera. However, confirmation of the position of meiosis is only available for four species: *Porphyra yezoensis* (Ma and Miura, 1984; Tseng and Sun, 1989), *P. torta* (Burzycki and Waaland, 1987), *P. tenera* (Tseng and Sun, 1989) and *P. purpurea* (Mitman and van der Meer, 1994). Some authors suggest that meiosis may not be a fixed event, varying according to species or growing conditions (Burzycki and Waaland, 1987; Guiry, 1991).

The distribution of the reproductive cells in the gametophytes varies between species and can be used as a distinctive character. There are monoecious species and dioecious species. Within the monoecious species, there are also different distribution patterns of the reproductive cells. In some species, the blade presents a longitudinal division with male regions (with spermatangial cells) on one side and female regions (with carpogonial cells) on the other, i.e., sectored. In other species, there are patches of spermatangial and carpogonial cells throughout the blade without a particular order. In any case, after fertilisation, diploid zygotospores are formed, liberated and germinate to form the conchocelis and thus closing the cycle.

In some species of *Porphyra*, alternative forms of reproduction can be found. The conchocelis can reproduce through formation of neutral conchospores and conchocelis archeospores (*sensu* Nelson et al., 1999), in any case originating new conchocelis. As for the gametophytes, in some species these are able to form blade archeospores, a form of asexual spore that germinates to form a new blade (formerly referred to as monospores). Besides archeospores, other kinds of asexual spores (agamospores, neutral spores, endospores) have been described in *Porphyra* (Notoya, 1997; Gantt et al., 2009).

The lack of vegetative propagation capacity of the gametophytes constitutes a disadvantage to the expansion of *Porphyra* aquaculture, when compared with other Rhodophyta. The need to obtain spores from conchocelis, for every crop, makes the process more complex and expensive. In fact, the few species of *Porphyra* cultivated in large scale are the ones that naturally produce archeospores, an episode that allows an exponential increase in the production, leading to up to six to eight harvests per production net (1.8 × 18 m^2).

The recognition of the difficulties in the cultivation of *Porphyra* has led researchers to look for techniques to overcome that limitation, either with new ways of cultivating the conchocelis (He and Yarish, 2006) or with artificial methods to obtain protoplasts and/or asexual spores from the gametophytes (Polne-Fuller et al., 1984; Le Gall et al., 1993; Mizuta et al., 2003; Saito et al., 2008). The work by He and Yarish (2006) is particularly important in the sense that it provided an easier (less complex and less expensive) way of cultivating the conchocelis phase of *Porphyra*, therefore facilitating the production of conchospores and new gametophytes. In terms of protoplast production, the cells obtained with enzyme cocktail not always developed into new gametophytes. In some cases, the protoplasts formed callus-like tissues or cell clumps (Le Gall et al., 1993). The works by Mizuta et al. (2003) and by Saito et al. (2008) are of particular interest because they use a simpler method to obtain protoplasts. In the work by Saito et al. (2008), the method was applied successfully to a species, *Porphyra pseudolinearis*, which does not produce archeospores naturally.

Recently, Pereira et al. (2006) reported on an unusual form of reproduction, or perennation, in *Porphyra dioica*. The individuals of this species are capable of producing 'new' blades from the basal parts of the primary blades obtained from conchospores. Although the importance of this strategy in the perennation of the populations is still not clear, studies in the laboratory showed that regeneration of 'new' blades, once the old ones were cut out, occurred for at least 3 months (Pereira et al., 2010).

4. Physiological Characteristics of *Porphyra*

Environmental conditions such as temperature, salinity, photon flux density (PFD), photoperiod and nutrient availability, among others, are known to affect the physiological mechanisms of seaweed. *Porphyra* is not an exception, and being an economically important genus, its physiology has been the subject of many works in the past 40–50 years.

The genus *Porphyra* is distributed around the world and species of *Porphyra* can be found from cold to warm waters. Some species, like *Porphyra moriensis* (Notoya and Myashita, 1999), *P. dioica* (Pereira et al., 2006) can withstand a wide range of temperatures, although it is common for a species to have strict temperature requirements for specific events in the life cycle. For instance, in *Porphyra linearis*, conchosporangia are formed at 5–20°C but release of conchospores occurs only at 13°C (Bird et al., 1972).

Some species of *Porphyra* occur in the intertidal (e.g., *P. dioica*, *P. umbilicalis*, *P. linearis*) while others are found only in the low-intertidal (e.g., *P. leucosticta*, *P. purpurea*) or the subtidal regions (e.g. *P. amplissima*). Intertidal species are naturally expected to survive salinity variations, but in reality not many studies have looked into the effects of salinity in *Porphyra*. Stekoll et al. (1999) investigated the effects of salinity on the growth of the conchocelis phase of *P. abbottae*, *P. torta* and *P. pseudolinearis*. Those authors found that salinity between 20‰ and 40‰ had little effect on growth, but there was virtually no growth at salinity 10‰ and below. Conitz et al. (2001), on the other hand, did not find significant effects of salinity on growth of juvenile gametophytes of *P. torta*, from 30‰ to 7.5‰.

The effects of photon flux density (PFD) and photoperiod have been studied for many species, especially due to their influence in the life cycle of virtually all species. The gametophytes and sporophytes of *Porphyra dioica*, for instance, grow in short-day, neutral-day and long-day photoperiods. However, formation of conchosporangia and release of conchospores in the laboratory occurred only in short days (Pereira et al., 2006).

In terms of nutrient removal, the works by Chopin et al. (1999), Carmona et al. (2006) and Pereira et al. (2008), among others, showed that some species of *Porphyra* perform well in that function. Chopin et al. (1999) showed that tissue nitrogen (N) and phosphorus (P) of *Porphyra* increased in specimens grown near salmon cages. Of the several species studied, these authors concluded that *Porphyra yezoensis* and *Porphyra purpurea* respond to high nutrient loading in coastal waters resulting from anthropogenic activities (salmon aquaculture and intense scallop dragging).

Studies in the laboratory have also shown that *Porphyra* responds to high concentrations of ammonium (NH_4^+), a common form of N in water enriched by anthropogenic activities. Wu et al. (1984) studied the utilisation of NH_4^+ by *Porphyra yezoensis* and obtained higher growth rates (11.6% day^{-1}) and tissue nitrogen content (4.72% dry weight [dw]) with NH_4^+ concentration ranging from 5 to 10 ppm (approx. 350–700 µM). For the same species, Amano and Noda (1987) reported that the optimal fertilising effect was obtained using 20 ppm NH_4^+ during 48 h. More recently, Carmona et al. (2006) assessed the bioremediation potential of several native northeast American species of *Porphyra* and compared those with well-known Asian species. In that study, growth and tissue N reached maximal levels at inorganic N concentrations of 150–300 µM. Maximum growth rates ranged from 10 to 25% day^{-1}. Pereira et al. (2008) studying a North Atlantic species, *Porphyra dioica*, also reported an interesting capacity to uptake equally well NO_3^- or NH_4^+, in concentrations ranging from 25 to 300 µM, when only one of these N forms is present. If both forms were present, the preference was for NH_4^+ even if available at a much lower concentration than NO_3^-.

Still in terms of nutrient physiology, considerably fewer studies have focused on aspects related to phosphorus (P) and carbon (C). Carmona et al. (2006) showed that P biofiltering efficiency was higher when NH_4^+ was supplied in *P. amplissima*, *P. purpurea*, *P. umbilicalis* and *P. yezoensis*, but not in *P. katadai* and *P. haitanensis*.

These authors also determined that P uptake was not saturated up to 30 µM P in *P. umbilicalis* and *P. haitanensis*, while uptake rate by *P. purpurea* was saturated at 15 µM P, regardless of the N source, and at even lower P concentrations in the other species. Pereira et al. (2008) showed that, for *P. dioica*, P enrichment up to 400 µM P did not translate into higher P uptake. In fact, despite the increasing P concentration, the total amount of P removed from the medium was the same. Both studies confirm what was first pointed out by Hafting (1999) for *P. yezoensis*, a lack of capacity to store P. Interestingly, Chopin et al. (2004) described the presence of polyphosphate granules in the cells of *P. purpurea*. Polyphosphate granules are a form of P storage in yeast, microalgae and some macroalgae Sommer and Both, 1938; Lundberg et al., 1989; Chopin et al., 1997). Chopin et al. (2004), however, also point out that these granules do not seem to have the same storage function in *P. purpurea*.

In terms of C metabolism, there are seaweed species capable of using only CO_2 and others that can use CO_2 or HCO_3^-. Bicarbonate utilisation has been suggested for *P. leucosticta* (Mercado et al., 1997), *P. umbilicalis* (Maberly, 1990) and, although with a limited capacity, for *P. linearis* (Israel et al., 1999). For *P. leucosticta* and *P. linearis*, those authors detected the activity of intra- and extracellular carbonic anhydrase (CA). In *P. yezoensis*, on the other hand, Gao et al. (1992) found evidences for active HCO_3^- transport, in that CO_2 uptake was extremely slow compared with the photosynthesis, and that external CA was never found in that species.

5. *Porphyra* in IMTA

According to the FAO (2009b), in 2006, 47% of the world's fish food supply was produced in aquaculture. This activity continues to grow more rapidly than all other animal food-producing sectors and in 2006 accounted for 36% of the world total aquatic animal production by weight. In turn, in 2007, aquatic plants aquaculture represents nearly 23% of the world aquaculture production, based principally in Asian countries. The seaweeds that are produced are for human consumption or for extraction of hydrocolloids (agar, alginates or carrageenan).

There are two main biological techniques for treatment of animal aquaculture waters: bacterial dissimilation into gases and plant assimilation into biomass. Bacterial biofilters allow effective and significant aquaculture water recirculation (van Rijn, 1996), but the technology is not simple and such systems usually accumulate nitrate and sludge that need to be disposed, and are expensive to operate. Biofiltration by algae is assimilative (Krom et al., 1989) and they use excess nutrients (particularly C, N, and P) to produce new biomass that can easily be removed from the water and that can have an economic value. In China, the annual production of over 7.4 Mt of seaweed (FAO, 2009a) may be responsible for removal of more than 40,000 t of nitrogen (N) from coastal waters. It is clear that seaweed production can help mitigate the potential environmental impacts of animal production and contribute to the development of an environmentally and economically sustainable aquaculture.

The principle of Integrated Multi-trophic Aquaculture (IMTA) systems is to build a simplified ecosystem where the resources provided, mainly feed and water, will be used by two to three other extractive aquaculture organisms: molluscs and seaweeds. This allows a system to virtually use much of the nutrients minimising the production of wastes. The best examples of IMTA systems operating in different parts of the world were described by Chopin et al. (2008).

Because of its morphological characteristics, high surface area to volume ratio (SA/V), *Porphyra* is one of the most promising species to be used as biofilter in IMTA. The thallus of *Porphyra* is a thin blade with one or two layers of cells, all potentially involved in nutrient absorption. It can be argued that a thallus with high surface area to volume ratio does not allow storage of nutrients in reserve tissues like those of brown algae (e.g. Laminariales and Fucales). The advantage of *Porphyra* is its rapid growth, over 25% day^{-1} (Pereira et al., 2006; Carmona et al., 2006), which can allow repeated harvests and continuous removal of nutrients from the water. There are yet other factors supporting the use of *Porphyra* in IMTA. When compared with most other seaweeds, besides its rapid growth rate, *Porphyra* has high nutrient uptake rates, is capable of coping with high NH_4^+ concentrations and is able to store N in its tissue up to 6% dry weight (dw). Furthermore, the biomass produced has several valuable applications.

The best growth rates and nutrient removal capacities are usually found in species with high surface area to volume ratios as explained by the functional-form model (Hanisak et al., 1990; Littler and Littler, 1980). For that reason, a lot of work has been done using thin blade-like species of *Ulva* (Chlorophyta). Martínez-Aragón et al. (2002) compared phosphate removal, from sea bass cultivation effluents, by *Ulva rotundata*, *Ulva intestinalis* (formerly *Enteromorpha intestinalis*) and *Gracilaria gracilis*. The maximum P uptake rate (2.86 µmol PO_4^{-3} g^{-1} dw h^{-1}) was found in *U. rotundata*. In a follow-up study with the same species (Hernández et al., 2002), *U. rotundata* also showed the highest NH_4^+ uptake rate, 89.0 µmol g^{-1} dw h^{-1}. Chung et al. (2002) recorded an uptake rate of 114.6 µmol NH_4^+ g^{-1} dw h^{-1} for *U. pertusa*. Fujita (1985) obtained, for *U. lactuca*, uptake rates between 2 × 10^3 and 3.6 × 10^3 µg N g^{-1} dw h^{-1}. The same author obtained, for *U. intestinalis*, a maximal uptake of 14 × 10^3 µg N g^{-1} dw h^{-1}. Mata et al. (2006) reported what is likely to be the highest N uptake rate for a seaweed in integrated aquaculture environment. In that work, the authors report a total ammonia nitrogen (TAN) uptake rate of 90 µmol l^{-1} h^{-1} at a TAN flux of about 500 µmol l^{-1} h^{-1} and with 5 g fresh weight (fw) l^{-1} stocking density. This is the equivalent to approximately 4.5 mg N day^{-1} g^{-1} fw of the alga.

The results of N uptake by *Porphyra*, although lower than those reported for *A. armata*, are very interesting and promising. In laboratory conditions, Pereira et al. (2006) obtained an N uptake rate of 1.5 mg N day^{-1} g fw of *P. dioica*. Also in the laboratory, Carmona et al. (2006) obtained a maximum removal of 1.75 mg N day^{-1} g^{-1} fw of *P. purpurea* and *P. haitanensis*. In both cases, tissue N recorded was over 6% dw. As mentioned earlier, *Porphyra* has other characteristics that constitute important advantages when choosing a biofilter. Particularly relevant, in terms of the potential role of *Porphyra* in ecosystem sustainability, is the ability of some species to uptake N equally well in the form of NH_4^+ or NO_3^- as shown by Pereira et al. (2006) and Carmona et al. (2006).

The results obtained so far show that *Porphyra* has an important role in the nutrient balance in the ecosystems. If we consider a 10:1 dw:fw ratio and average values of 5.5% N, 0.7% P and 38% C in tissue dw, we realise that the aquaculture of *Porphyra* removes, annually, a significant amount of these nutrients from water. In fact, considering FAO last production figures (FAO, 2009a), approximately 1.51 Mt of *Porphyra* were produced in 2007. This may equal to a removal of 9,063 t of N, 1,057 t of P and 57,404 t of C, every year.

These results are not just theoretical but supported by field studies of production of *Porphyra* and its nutrient removal capacity (e.g. Chopin et al., 1999; He et al., 2008). In the most recent paper, He et al. (2008) showed the bioremediation capability and efficiency of large-scale *Porphyra* cultivation in the removal of inorganic nitrogen and phosphorus from open sea area. The study took place in 2002–2004, in a 300 ha nori farm along the Lusi coast, Qidong County, Jiangsu Province, China. Nutrient concentrations were significantly reduced by the seaweed cultivation. Compared with a control area, *Porphyra* farming resulted in the reduction of NH_4–N, NO_2–N, NO_3–N and PO_4–P by 50–94%, 42–91%, 21–38% and 42–67%, respectively. Nitrogen and phosphorus contents in dry *Porphyra* thalli harvested from the Lusi coast averaged 6.3% and 1.0%, respectively. The authors concluded that the annual biomass production of *P. yezoensis* (about 800 kg dw ha^{-1} from a 300 ha cultivation) equaled to an average of 14,708 kg of tissue N and 2,373 kg of tissue P harvested. These results indicate clearly that *Porphyra* can efficiently remove excess nutrient from nearshore eutrophic coastal areas.

The nutrient uptake capacity of *Porphyra* can also be useful in inland IMTA systems, providing biofiltration for intensive fish aquacultures. The effluents from these systems can reach high concentrations of both NH_4–N and NO_3–N. *Porphyra* is capable of coping with high concentrations N in either form whether for uptake or growth.

The main constraint in the case of land-based tank cultivation is the area of cultivation needed to significantly reduce the nutrient loads of the effluents. For instance, for an effluent with 143 µM N and 10 µM P at a flux of 190 l min^{-1}, Carmona et al. (2006) calculates that 282 m^2 of *Porphyra amplissima* cultivation are needed to remove 90% of the N. Pereira et al. (2006), for *Porphyra dioica*, calculated that 179 m^2 of culture area would be needed to remove 50% of the N. This is considering that approximately 600 g of N are released per day by ton of fish (data from a 50 fish pond, 36 m^3 each, turbot and sea bass farm, as described by Matos et al., 2006). Those authors refer, however, that in an IMTA system, the performance of *P. dioica* is likely to improve. Under those tank cultivation conditions, the algae would have more nutrients and CO_2 available, owing to the constant water flux. This would also allow to experiment with higher stocking densities,

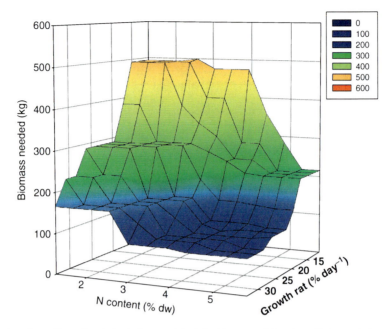

Figure 2. Three-dimensional simulation model of the biomass of *Porphyra dioica* needed to achieve an 80% reduction in a N load (during 12 h light day) of 600 g N day^{-1} (195 l min^{-1} × 150 µM-N).

possibly increasing nutrient removal. Ultimately, the best nutrient removal performance would be the result of the conditions that could yield the optimal combination of growth and tissue N content, for a given stocking density (Fig. 2).

It is clear that *Porphyra* can play an important role in the sustainable development of intensive aquaculture. Species of *Porphyra* have high ammonia uptake rates, high ammonia uptake efficiencies, high yields and high protein contents (N content). Detailed economic studies are now needed to confirm the feasibility of such systems.

In conclusion, we think *Porphyra* is a very promising marine organism for biotechnological exploitation, one of the best seaweed for application in sustainable aquaculture (IMTA) and an interesting organism to be used as a model in biological studies.

6. References

Amano, H. and Noda, H. (1987) Effect of nitrogenous fertilizers on the recovery of discoloured fronds of *Porphyra yezoensis*. Bot. Mar. **30**: 467–473.
Bird, C.J., Chen, L.C.-M. and McLachlan, J. (1972) The culture of *Porphyra linearis* (Bangiales, Rhodophyta). Can. J. Bot. **50**: 1859–1863.

Brodie, J., Hayes, P.H., Baker, G.L., Irvine, L.M. and Bartsch, I. (1998) A reappraisal of *Porphyra* and *Bangia* (Bangiophycidae, Rhodophyta) in the northeast Atlantic based on the rbcL-rbcS intergenic spacer. Journal of Phycology **34**: 1069–1074.

Brodie, J. and Irvine, L.M. (2003) *Seaweeds of the British Isles*, Vol. 1 Part 3B: Bangiophycidae. Intercept, Hampshire, 167 pp.

Brodie, J. and Zuccarello, G.C. (2007) Systematics of the species-rich algae: red algal classification, phylogeny and speciation, In: T.R. Hodkinson and J.A.N. Parnell (eds.) *Reconstructing the Tree of Life: Taxonomy and Systematics of Species Rich Taxa*. Systematics Association Special Volume Series 72. CRC Press, Boca Raton, pp. 323–336.

Brodie, J., Bartsch, I., Neefus, C., Orfanidis, S., Bray, T. and Mathieson, A. (2007) New insights into the cryptic diversity of the North Atlantic–Mediterranean '*Porphyra leucosticta*' complex: *P. olivii* sp. nov. and *P. rosengurttii* (Bangiales, Rhodophyta). Eur. J. Phycol. **42**: 3–28.

Broom, J.E., Nelson, W.A., Yarish, C., Jones, W.A., Aguilar Rosas, R. and Aguilar Rosas, L.E. (2002) A reassessment of the taxonomic status of *Porphyra suborbiculata*, *Porphyra carolinensis* and *Porphyra lilliputiana* (Bangiales, Rhodophyta) based on molecular and morphological data. Eur. J. Phycol. **37**: 227–236.

Burzycki, G.M. and Waaland, J.R. (1987) On the position of meiosis in the life history of *Porphyra torta* (Rhodophyta). Bot. Mar. **30**: 5–10.

Butterfield, N.J. (2000) *Bangiomorpha pubescens* n. gen., n. sp.: implications for the evolution of sex, multi-cellularity and the Mesopreterozoic–Neoproterozoic radiation of eukaryotes. Paleobiology **26**: 386–404.

Carmona, R., Kraemer, G.P. and Yarish, C. (2006) Exploring Northeast American and Asian species of *Porphyra* for use in an Integrated Finfish-Algal Aquaculture System. Aquaculture **252**: 54–65.

Chopin, T., Lehmal, H. and Halcrow, K. (1997) Polyphosphates in the red macroalga *Chondrus crispus* (Rhodophyceae). New Phytol. **135**: 587–594.

Chopin, T., Yarish, C., Wilkes, R., Belyea, E., Lu, S. and Mathieson, A. (1999) Developing *Porphyra*/salmon integrated aquaculture for bioremediation and diversification of the aquaculture industry. J. Appl. Phycol. **11**: 463–472.

Chopin, T., Robinson, S.M.C., Troell, M., Neori, A., Buschmann, A.H. and Fang, J. (2008) Multitrophic integration for sustainable marine aquaculture, In: S.E. Jørgensen and B.D. Fath (Editor-in-Chief) *Ecological Engineering*, Vol. 3 of Encyclopedia of Ecology, 5 vols. Elsevier, Oxford, pp. 2463–2475.

Chopin, T., Morais, T., Belyea, E. and Belfry, S. (2004) Polyphosphate and siliceous granules in the macroscopic gametophytes of the red alga *Porphyra purpurea* (Bangiophyceae, Rhodophyta). Botanica Marina **47**: 272–280.

Chung, I., Kang, Y.H., Yarish, C., Kraemer, G. and Lee, J. (2002) Application of seaweed cultivation to the bioremediation of nutrient-rich effluent. Algae **17**(3): 187–194.

Coll, J. and Oliveira Filho, E. (1976) The genus *Porphyra* C. Ag. (Rhodophyta, Bangiales) in the American South Atlantic. II. Uruguaian species. Botanica Marina **19**: 191–196.

Conitz, J.M., Fagen, R., Lindstrom, S.C., Gerald Plumley, F. and Stekoll, M.S. (2001) Growth and pigmentation of juvenile *Porphyra torta* (Rhodophyta) gametophytes in response to nitrate, salinity and inorganic carbon. J. Appl. Phycol. **13**: 423–431.

Davies, S.J., Brown, M.T. and Camilleri, M. (1997) Preliminary assessment of the seaweed *Porphyra purpurea* in artificial diets for thick-lipped grey mullet (*Chelon labrosus*). Aquaculture **152**: 249–258.

Drew, K.M. (1949) Conchocelis phase in the life history of *Porphyra umbilicalis* (L.) Kütz. Nature **164**: 748–749.

FAO Fisheries and Aquaculture Information and Statistics Service (2009a) Aquaculture production 1950-2007. Rome.

FAO (2009b) *The State of the World Fisheries and Aquaculture 2008*. FAO Fisheries and Aquaculture Department. Food and Agriculture Organization of the United Nations, Rome.

Fujita, R. M. (1985) The role of nitrogen status in regulating transient ammonium uptake and nitrogen storage by macroalgae. J. Exp. Mar. Biol. Ecol. **92**: 283–301.

Gantt, E., Berg, G.M., Bhattacharya, D., Blouin, N.A., Brodie, J.A., Chan, C.X., Collén, J., Cunningham, F.X., Gross, J., Grossman, A.R., Karpowicz, S., Kitade, Y., Klein, A.S., Levine, I.A., Lin, S., Lu, S., Lynch, M., Minocha, S.C., Müller, K., Neefus, C.D., de Oliveira, M.C., Rymarquis, L., Smith, A., Stiller, J.W., Wu, W.K., Yarish, C., Zhuang, Y.Y. and Brawley, S.H. (2010) *Porphyra*: complex life histories in a harsh environment. In: J. Seckbach and D. Chapman (eds.) *Cellular Origins, Life in Extreme Habitats and Astrobiology*, Vol 13: Red Algae in the Genomic Age, Springer, New York, 495 pp.

Gao, K., Aruga, Y., Asada, K., Ishihara, T., Akano, T. and Kiyohara M. (1992) Photorespiration and CO_2 fixation in the red alga *Porphyra yezoensis* Ueda. Jpn. J. Phycol. **40**: 373–3377.

Graham, L.E. and Wilcox, L.W. (2000) Algae. Prentice Hall, NJ, USA. 640 pp.

Guiry, M.D. (1991) Sporangia and spores, In: K.M. Cole and R.G. Sheath (eds.) *Biology of the Red Algae*. Cambridge University Press, Cambridge, pp. 347–376.

Hafting, J.T. (1999) Effect of tissue nitrogen and phosphorus quota on growth of *Porphyra yezoensis* blades in suspension cultures. Hydrobiologia **398/399**: 305–314.

Hanisak, M.D., Littler, M.M. and Littler, D.S. (1990) Application of the functional-form model to the culture of seaweeds. Hydrobiologia **204/205**: 73–77.

He, P. and Yarish, C. (2006) The developmental regulation of mass cultures of free-living conchocelis for commercial net seeding of *Porphyra leucosticta* from Northeast America. Aquaculture **257**: 373–381.

He, P., Xua, S., Zhanga, H., Wena, S., Daia, Y., Lin, S. and Yarish, C. (2008) Bioremediation efficiency in the removal of dissolved inorganic nutrients by the red seaweed, *Porphyra yezoensis*, cultivated in the open sea. Water Res. **42**: 1281–1289.

Hernández, I., Martínez-Aragón, J.F., Tovar, A., Pérez-Lloréns, J.L. and Vergara, J.J. (2002) Biofiltering efficiency in removal of dissolved nutrients by three species of estuarine macroalgae cultivated with sea bass (*Dicentrarchus labrax*) waste waters 2. Ammonium. Journal of Applied Phycology **14**: 375–384.

Israel, A., Katz, S., Dubinsky, Z., Merrill, J.E. and Friedlander, M. (1999) Photosynthetic inorganic carbon utilization and growth of *Porphyra linearis* (Rhodophyta). J. Appl. Phycol. **11**: 447–453.

Kitade, Y., Asamizu, E., Fukuda, S., Nakajima, M., Ootsuka, S., Endo, H., Tabata, S. and Saga, N. (2008) Identification of genes preferentially expressed during asexual sporulation in *Porphyra yezoensis* gametophytes (Bangiales, Rhodophyta). J. Phycol. **44**: 113–123.

Korbee, N., Huovinen, P., Figueroa, F.L., Aguilera, J. and Karsten, U. (2005a) Availability of ammonium influences photosynthesis and the accumulation of mycosporine-like amino acids in two *Porphyra* species (Bangiales, Rhodophyta). Mar. Biol. **146**: 645–654.

Korbee, N., Figueroa, F.L. and Aguilera, J. (2005b) Effect of light quality on the accumulation of photosynthetic pigments, proteins and mycosporine-like amino acids in the red alga *Porphyra leucosticta* (Bangiales, Rhodophyta). J. Photochem. Photobiol. B **8**: 71–78.

Krom, M.D., Erez, J., Porter, C.B. and Ellner, S. (1989) Phytoplankton nutrient uptake dynamics in earthen marine fishponds under winter and summer conditions. Aquaculture **76**: 237–253.

Le Gall, Y., Chiang, Y.-M. and Kloareg, B. (1993) Isolation and regeneration of protoplasts from *Porphyra dentata* and *Porphyra crispata*. Eur. J. Phycol. **28**: 277–283.

Littler, M.M. and Littler, D.S. (1980) The evolution of the thallus form and survival strategies in benthic marine macroalgae: field and laboratory tests of a functional form model. Am. Nat. **116**: 25–44.

Lundberg, P., Weich, R.G., Jensen, P. and Vogel, H.J. (1989) Phosphorus-31 and nitrogen-14 studies of the uptake of phosphorus and nitrogen compounds in the marine macroalga *Ulva lactuca*. Plant Physiology **89**: 1380–1387.

Ma, J.H. and Miura, A. (1984) Observations of the nuclear division in the conchospores and their germlings in *Porphyra yezoensis* Ueda. Jpn. J. Phycol. **32**: 373–378.

Maberly, S.C. (1990) Exogenous sources of inorganic carbon for photosynthesis by marine macroalgae. J. Phycol. **26**: 439–449.

Martínez-Aragón, J.F., Hernández, I., Pérez-Lloréns, J.L., Vázquez, R. and Vergara, J.J. (2002) Biofiltering efficiency in removal of dissolved nutrients by three species of estuarine macroalgae cultivated with sea bass (*Dicentrarchus labrax*) waste waters 1. Phosphate. Journal of Applied Phycology **14**: 365–374.

Mata, L., Silva, J., Schuenhoff, A. and Santos, R. (2006) The effects of light and temperature on the photosynthesis of the *Asparagopsis armata* tetrasporophyte (*Falkenbergia rufolanosa*), cultivated in tanks. Aquaculture **252**: 12–19.

Matos, J., Costa, S., Rodrigues, A., Pereira, R. and Sousa Pinto, I. (2006) Experimental integrated aquaculture of fish and red seaweeds in Northern Portugal. Aquaculture **252**: 31–42.

Mercado, J.M., Niell, F.X. and Figueroa, F.L. (1997) Regulation of the mechanism for HCO_3^- use by inorganic carbon level in *Porphyra leucosticta* Thur. in Le Jolis (Rhodophyta). Planta **201**: 319–325.

Mitman, G. and van der Meer, J.P. (1994) Meiosis, blade development, and sex determination in *Porphyra purpurea* (Rhodophyta). J. Phycol. **30**: 147–159.

Mizuta, H., Yasui, H. and Saga, N. (2003) A simple method to mass produce monospores in the thallus of *Porphyra yezoensis* Ueda. J. Appl. Phycol. **15**: 345–349.

Mumford, T.F. and Miura, A. (1988) *Porphyra* as food: cultivation and economics, In: C.A. Lemby and J.R. Walland (eds.) *Algae and Human Affairs*. Cambridge University Press, Cambridge, pp. 87–117.

Neefus, C., Mathieson, A.C., Bray, T.L. and Yarish, C. (2008) The occurrence of three introduced Asiatic species of *Porphyra* (Bangiales, Rhodophyta) in the northwestern Atlantic. J. Phycol. **44**: 1399–1414.

Nelson, W.A., Brodie, J. and Guiry, M. (1999) Terminology used to describe reproduction and life history stages in the genus *Porphyra* (Bangiales, Rhodophyta). J. Appl. Phycol. **11**: 407–410.

Nelson, W.A., Farr, T.J. and Broom, J.E.S. (2006) Phylogenetic relationships and generic concepts in the red order Bangiales: challenges ahead. Phycologia **45**: 249–259.

Noda, H. (1993) Health benefits and nutritional properties of nori. J. Appl. Phycol. **5**: 255–258.

Notoya, M. (1997) Diversity and life history in the genus *Porphyra*. Nat. Hist. Res. Special Issue No. **3**: 47–56.

Notoya, M. and Miyashita, A. (1999) Life history, in culture, of the obligate epiphyte *Porphyra moriensis* (Bangiales, Rhodophyta). Hydrobiologia **398/399**: 121–125.

Oliveira, M.C., Kurniawan, J., Bird, C.J., Rice, E.L., Murphy, C.A., Singh, R.K., Guttel, R.G. and Ragan, M.A. (1995) A preliminary investigation of the order Bangiales (Bangiophycidae, Rhodophyta) based on sequences of the nuclear small-subunit ribosomal RNA genes. Phycological Research **42**: 71–79.

Pereira, R., Yarish, C. and Sousa-Pinto, I. (2006) The influence of stocking density, light and temperature on the growth, production and nutrient removal capacity of *Porphyra dioica* (Bangiales, Rhodophyta). Aquaculture **252**: 66–78.

Pereira, R., Kraemer, G.P., Yarish, C. and Sousa-Pinto, I. (2008) Nitrogen uptake by gametophytes of *Porphyra dioica* (Bangiales, Rhodophyta) under controlled culture conditions. Eur. J. Phycol. **43**(1): 107–118.

Pereira, R., Valente, L., Sousa-Pinto, I. and Rema, P. (2010) Apparent digestibility coefficients of four species of seaweeds by rainbow trout (*Oncorhynchus mykiss*) and nile tilapia (*Orecochromis niloticus*). Aquaculture 2010. World Aquaculture Society Meeting, San Diego, March 1–5, 2010.

Polne-Fuller, M., Biniaminov, M. and Gibor, A. (1984) Vegetative propagation of *Porphyra perforata*. Hydrobiologia **116/117**: 308–313.

Sahoo, D., Tang, X. and Yarish, C. (2002) *Porphyra* – the economic seaweed as a new experimental system. Curr. Sci. **83**(11): 1313–1316.

Saito, M., Kawai, M., Hagino, H., Okada, J., Yamamoto, K., Hayashida, M. and Ikeda, T. (2002) Antihypertensive effect of nori-peptides derived from red alga *Porphyra yezoensis* in hypertensive patients. American Journal of Hypertension, vol 15, No4, Part 2, 210A. Published by Elsevier Science Inc.

Saito, A., Mizuta, H., Yasui, H. and Saga, N. (2008) Artificial production of regenerable free cells in the gametophyte of *Porphyra pseudolinearis* (Bangiales, Rhodophyceae). Aquaculture **281**: 138–144.

Sampath-Wiley, P., Neefus, C.D. and Jahnke, L.S. (2008) Seasonal effects of sun exposure and emersion on intertidal seaweed physiology: fluctuations in antioxidant contents, photosynthetic pigments and photosynthetic efficiency in the red alga *Porphyra umbilicalis* Kützing (Rhodophyta, Bangiales). J. Exp. Mar. Biol. Ecol. **361**: 83–91.

Silva, P.C. (1999) *Porphyra from the Index Nominum Algarum.* http://ucjeps.berkeley.edu/rlmoe/porphyra.html. University of Herbarium, UC Berkeley. Rev: 12/20/99.

Sommer, A.L. and Booth, T.E. (1938) Meta- and pyro-phosphate within the algal cell. Plant Physiol. **13**: 199–205.

Stekoll, M.S., Lin, R. and Lindstrom, S.C. (1999) *Porphyra* cultivation in Alaska: conchocelis growth of three indigenous species. Hydrobiologia **398/399**: 291–297.

Tseng, C.K. and Sun, A. (1989) Studies on the alternation of the nuclear phases and chromosome numbers in the life history of some species of *Porphyra* from China. Bot. Mar. **32**: 1–8.

Tsujii, K., Kchikawa, T., Matusuura, Y. and Kawamura, M. (1983) Hypercholesterolemic effect of taurocyamine or taurine on the cholesterol metabolism in white rats. Sulphur Amino Acids **6**: 239–248.

van Rijn, J. (1996) The potential for integrated biological treatment systems in recirculating fish culture – a review. Aquaculture **139**: 181–201.

Walker, A.B., Fournier, H.R., Neefus, C.D., Nardi, G.C. and Berlinsky, D.L. (2009) Partial replacement of fishmeal with Laver (*Porphyra* spp.) in diets of Atlantic cod (Gadus morhua). N. Am. J. Aquacult. **71**: 39–45.

Wilkes, R.J., Yarish, C. and Mitman, G.G. (1999) Observations on the chromosome numbers of *Porphyra* (Bangiales, Rhodophyta) populations from Long Island Sound to the Canadian Maritimes. Algae **14**(4):219–222.

Wu, C.-Y., Zhang, Y.-X., Li, R.-Z., Penc, Z.-S., Zhang, Y.-F., Liu, Q.-C., Zhang, J.-P. and Fan, X. (1984) Utilization of ammonium-nitrogen by *Porphyra yezoensis* and *Gracilaria verrucosa*. Hydrobiologia **116/117**: 475–477.

Yashizawa, Y., Ametani, A., Tsunehiro, J., Numera, K., Itoh, M., Fukui, F. and Kaminogawa, S. (1995) Macrophage stimulation activity of the polysaccharide fraction from a marine alga (*Porphyra yezoensis*): structure-function relationships and improved solubility. Bioscience Biotechnology and Biochemistry **59**: 1933–1937.

Yoshida, T., Notoya, M., Kikuchi, N. and Miyata, M. (1997) Catalogue of species of *Porphyra* in the world, with special reference to the type locality and bibliography. Nat. Hist. Res. Special Issue **3**: 5–18.

Zhang, Q., Li, N., Zhou, G., Lu, X., Xu, Z. and Li, Z. (2003) In vivo antioxidant activity of polysaccharide fraction from *Porphyra haitanensis* (Rhodophyta) in aging mice. Pharmacological Research **48**: 151–155.

Zhou, H.P. and Chen, Q.H. (1990) Anticoagulant and antihyperlipedemic effects of polysaccharide from *Porphyra yezoensis*. J. China Pharm. Univ. **21**: 358–360.

PART 7:
BIOTECHNOLOGICAL POTENTIAL OF SEAWEEDS

Turan
Neori
Reisser
Cavas
Pohnert

Biodata of **Gamze Turan** and **Amir Neori,** authors of *"Intensive Seaweed Aquaculture: A Potent Solution Against Global Warming"*

Dr. Gamze Turan is a faculty member at Ege University, Fisheries Faculty, Aquaculture Department, Izmir, Turkey. She obtained her Ph.D. from Ege University Institute of Natural and Applied Sciences in 2007 in Aquaculture and continued her research on seaweeds, their cultivation, and usage. Dr. Turan's scientific interests cover the areas of seaweed cultivation, and their usage as phycocolloids, human food, animal feed, fish feed, fertilizers and soil conditioners, biomass for fuel, cosmetics, integtated multitrophic aquaculture (IMTA), wastewater treatment, etc. She has published over 20 peer-reviewed publications.

E-mail: **gamze.turan@ege.edu.tr**

Dr. Amir Neori is a senior scientist at the Israel Oceanographic & Limnological Research, Ltd. The National Center for Mariculture, Eilat, Israel. He obtained his Ph.D. from the University of California San Diego – Scripps Institution of Oceanography in 1986 in Marine Biology and continued his research in sustainable mariculture and algae in his present capacity. Dr. Neori's scientific interests are in the area of environmentally-friendly aquaculture, algal aquaculture, reduction in aquaculture environmental impact, integrated multitrophic aquaculture (IMTA), and biofuel from algae. He has published over 70 peer-reviewed publications.

E-mail: **neori@ocean.org.il; aneori@gmail.com**

Gamze Turan   **Amir Neori**

INTENSIVE SEA WEED AQUACULTURE: A POTENT SOLUTION AGAINST GLOBAL WARMING

GAMZE TURAN[1] AND AMIR NEORI[2]
[1]*Fisheries Faculty, Aquaculture Department, Ege University, Bornova, Izmir, 35100, Turkey.*
[2]*Israel Oceanographic & Limnological Research Ltd, National Center for Mariculture, P.O. Box 1212, Eilat, 88112, Israel*

1. Introduction

On the basis of current understanding of the relationship between climate change and energy policy, development of an effective and multistructured renewable energy sector is crucial, as acknowledged in the United Nations Framework Convention on Climate Change (UNFCCC) and the fourteenth Conference of the Parties (COP-14), held in December 2008 in Ponzan, Poland. The worldwide energy demand is increasing rapidly as many industries and populations are rapidly expanding. Since fossil fuels are finite resources and their combustion leads to a further increase of greenhouse gases, such as CO_2, SO_2, and NO_x, their continued use is not sustainable. Today, renewable energy sources supply 14% of the total global energy demand. Some expect that in 2040, 50% of the world energy supply will come from renewable sources (Demirbas, 2008). Additional efforts and further research and development on biofuels, toward environmentally and economically sustainable processes, are essential for the full exploitation of this given market opportunity.

The substitution of conventional fuels by biofuels can reduce pollution and support sustainability. First-generation biofuels, such as biodiesel and bioethanol derived from biomass, have their environmental benefits related to carbon-neutral energy. However, increasing biofuel production from land crops strains the global food supply. Owing to these limitations, second-generation (bio) fuels – from biomasses that generate carbon neutral energy without competing with food production – have been developed. These can be produced from the residual nonfood parts of current crops, as well as novel energy crops, such as seaweeds.

The culture of seaweeds has unique characteristics, which make it different and in many ways attractive in comparison with other biofuel sources. Seaweeds, also known as "marine macroalgae," "aquatic plants," or "sea vegetables," are autotrophic organisms that produce biomass using sunlight and extracting from the water-dissolved inorganic nutrients, including carbon. Several seaweed species are perhaps the most attractive of all CO_2 removal and biofuel aquatic crops, thanks to high yields and low cost of production. Therefore, efficient production of biodiesel and bioethanol from seaweeds has been considered (Hanisak and Ryther, 1986; Bird and Benson, 1987; Flowers and Bird, 1987; Morand et al., 1991; Gao and Mckinley, 1994; Kelly and Dworjanyn, 2008). Seaweeds can be viewed as miniature biochemical factories, photosynthetically efficient, and effective CO_2 fixers. Many species of seaweed are rich in oil or sugars that can be converted into biofuels; as a result, the biofuel productivity of seaweeds per unit area can be much higher than the conventional farm crops, such as wheat and maize. Producing a ton of dry algal biomass utilizes approximately 360 kg carbon, 63 kg nitrogen, and 8.6 kg phosphorus (Sinha et al., 2001). Utilization of anthropogenic CO_2 as an industrial by-product for seaweed production holds great promise not only as a carbon sink, but also as a source of food, fodder, fuel, and pharmaceutics. The recent Algal Biomass Summit (Seattle, October 2008) stressed the importance of algae to deliver such a mix of energy, feed, and industrial products. From an ecological point of view, generation of biomass should not aim to a single application, treating the remainder as a "waste," but toward a comprehensive solution to several challenges, including biofuel, carbon sequestration, waste remediation, and natural production of food and biochemicals. The goal therefore should be the integration of the processes. Mass balance and energy balance, complemented by energy analysis, can guide the optimization of the technologies and economics of using seaweeds, regarding carbon sequestration, waste remediation, biofuel production, and generation of seaweed products. Additional attractive characteristics of seaweed-based biofuels include the following: (a) some seaweeds are rich in oil, others in processable carbohydrates and proteins, (b) seaweeds of different species can be grown anywhere, in marine, brackish, and freshwater and in most climates, (c) seaweed can grow well on liquid domestic and industrial wastewaters and on streams polluted by agriculture, reducing pollution as they grow, and (d) seaweed biomass is desirable and valuable for a diverse array of commercial purposes, depending on species, quality, and quantity.

This chapter analyzes the current production of both commercially grown and wild-grown seaweeds, as well as their capacity for photosynthetically driven CO_2 assimilation and growth. It is suggested that CO_2 uptake by seaweeds can represent a considerable sink for anthropogenic CO_2 emissions and that harvesting and appropriate use of seaweed primary production is a commercially viable approach for the amelioration of greenhouse gas emissions and biofuel production.

2. Seaweed Culture Utilization

Seaweeds form a large group of amazing aquatic plants, widely used for human food in China, Japan, and Korea. It is also an ingredient in animal feeds, cosmetics, and fertilizers (Critchley and Ohno, 1997, 1998, 2001; Critchley et al., 2006; FAO, 2003; McHugh, 2003). Various seaweeds are also used to produce the hydrocolloids (or phycocolloids) alginate, agar, and carrageenan as thickening and gelling agents in the food and biomedical industries. Seaweed extracts are to be found in a wide range of common products we use daily such as toothpaste, shaving foam, ice cream, cheeses, candy, beer, shower gels, bacteriological agar, and paper. Increasingly, seaweeds are being investigated for the biological activity of their extracts, which are finding applications in pharmaceuticals, biotechnology, and food preservatives (Schuenhoff et al., 2006). In addition, seaweed products are used in textiles and printing to achieve the desired consistency in dyes, paint, and ink. Several companies produce a range of seaweed-based therapy (Thalassotherapy) products, which include seaweed bath salts, bubble bath, shampoos, shower gels, soaps, facial scrubs, body masks, moisturizers, and foot bath salts (De Roeck-Holtzhauer, 1991; Turan and Cirik, 2008). Developments in the paper pulp industry have made seaweeds practical alternatives to the use of wood (De Poli et al., 1994; You, 2008).

In recent decades, several studies have investigated the possible use of seaweeds as a source of biofuel (Hanisak and Ryther, 1986; Bird and Benson, 1987; Flowers and Bird, 1987; Morand et al., 1991; Gao and Mckinley, 1994; Kelly and Dworjanyn, 2008). Potential uses for seaweeds in wastewater treatment due to their ability to absorb nutrients, heavy metal ions such as zinc and cadmium from polluted water have also been developed (Ryther et al., 1975; Schramm, 1991a). The effluent water from fish farms usually contains high levels of nutrients that can cause problems to other aquatic life in adjacent waters. Use of seaweeds as biofilters for mariculture effluents has been developed (Krom and Neori, 1989; Neori et al., 1989, 1996, 2004; Cohen and Neori, 1991; Neori and Shpigel, 1999; Nunes et al., 2003; Fei, 2004; Kang et al., 2008; Xu et al., 2008).

Seaweeds are now being advanced in polyculture systems as an integral component of integrated multitrophic aquaculture (IMTA) (Shpigel et al., 1993; Buschmann et al., 1994, 2001, 2008; Neori et al., 1996, 2000, 2004; Shpigel and Neori, 1996; Troell et al., 1997, 1999, 2003, 2006; Chopin et al., 1999, 2001, 2008; Neori and Shpigel, 1999; Schuenhoff et al., 2003, 2006; Yang et al., 2004; Chopin, 2006; Cirik et al., 2006; Turan et al., 2006, 2007a,b,c; Whitmarsh et al., 2006; Neori, 2007, 2008). The IMTA approach, besides being a form of ecologically balanced aquaculture management, diminishes possible environmental impacts from aquaculture (Edwards, 2004; Fei, 2004; Yang et al., 2006).

3. Intensive Seaweed Aquaculture

Seaweed aquaculture already represents about a quarter of the world's aquaculture production, but its potential is far from being fully exploited. Seaweed cultivation techniques are standardized, routine, and economical. Despite the variety of life forms and the thousands of seaweed species described, seaweed aquaculture currently uses only about 100 taxa. The genera *Laminaria*, *Undaria*, *Porphyra*, *Eucheuma/Kappaphycus*, and *Gracilaria* account for about 98% of global production

(Sahoo and Yarish, 2005; Yarish and Pereira, 2008). However, cultivation of the commercially important species *Asparagopsis/Falkenbergia, Caulerpa, Chondrus, Cystoseira, Ecklonia, Gelidium Gigartina, Hypnea, Macrocystis, Palmaria, Monostroma, Pterocladia, Sargassum*, and *Ulva* takes place in many different places (Kain-Jones, 1991; Schramm, 1991b; Critchley and Ohno, 1997, 1998, 2001; Cirik and Cirik, 1999; FAO, 2003; McHugh, 2003; Critchley et al., 2006; Troell et al., 2006). The world seaweed industry produces over 15 million tons (fresh weight) annually with a total value of over US$7 billion (FAO, 2006b). Value of food products for human consumption contribute US$4–5 billion of this, with the single most valuable crop being *Porphyra* ("Nori" and "Laver"), worth over US$1.3 billion. Substances that are extracted from seaweeds – hydrocolloids – account for a large part of the remaining value. Smaller, miscellaneous uses, such as fertilizers and animal feed additives, make up the rest. The farming of seaweed has expanded rapidly as demand has outstripped the supply available from natural resources. Over 90% of the market is supplied by cultivation.

China is the largest producer of seaweeds with 10.9 million tons (wet weight) followed by the Philippines (1.5 million tons), Indonesia (0.91 million tons), the Republic of Korea (0.77 million tons), and Japan (0.49 million tons) (FAO, 2009). While the bulk of China's contribution comes from the cultivation of *Laminaria japonica*, 50% of Korea production is contributed by *Undaria pinnatifida* and 75% of Japan's contribution comes from the cultivation of *Porphyra* sp. The Philippines and Indonesia are involved mainly in the cultivation of *Kappaphycus alvarezii* and *Eucheuma denticulatum* (carragenophytes) as well as *Gracilaria* species (agarophytes) (Table 1).

Seaweed growth rate and yield depend on species, the site of cultivation, the season, and the cultivation methodology. For example, the daily growth rate

Table 1. World seaweed aquaculture production in 2006 (FAO, 2006a).

Country	Seaweed production (ton)	Species	Utilization	Value (US$'000)
China	10,867,410	*Laminaria japonica* (more than 70% contribution)	Alginates, food, and other industries	5,240,817
Phillippines	1,468,905	*Kappaphycus alvarezii* *Eucheuma denticulatum*	Carrageenan	173,953
Korea	1,209,895	*Undaria pinnatifida* (nearly 50% contribution)	Food	514,022
Indonesia	910,636	*Kappaphycus alvarezii* *Eucheuma denticulatum*	Carrageenan	127,489
Japan	490,052	*Porphyra yezoensis* *P. tenera* (75% of total production)	Food	1,051,361
World total	15,075,612	*Laminaria, Undaria, Kappaphycus, Eucheuma, Porphyra, Gracilaria*	Phycocolloid (agar–agar, alginate, carrageenan), food, and other industries	7,187,125

(DGR) of *Kappaphycus alvarezii* usually varies between 2% and 3% and that of *Gracilaria* spp. between 3.3% and 8.4%, depending on factors such as CO_2 level of surrounding seawater (Chung et al., 2007). Temperature increase may also affect the ability of seaweeds to perform in particular geographic areas (Breeman, 1990), whereas both increased storm events and runoff from land are likely to impact seaweed growth (Dayton and Tegner, 1984; Nielsen, 2003).

Large-scale seaweed culture is attractive because of low-cost technologies that have been in operation for decades, and the multiple uses of the product. Yields of seaweeds can be as high as 80 mt dw ha^{-1} year^{-1} in modern intensive pond farms, whereas extensive low technology coastal farms regularly get yields above 20 mt dw ha^{-1} year^{-1} (Neori et al., 2004). Seaweeds can take up 29 mt carbon ha^{-1} year^{-1} in modern intensive farms and 7.3 mt carbon ha^{-1} year^{-1} in low technology farms (Sinha et al., 2001). Possibilities also exist for promoting intensive growth of seaweeds in integrated aquaculture (IMTA).

IMTA seaweed farming provides exciting new opportunities for valuable crops of seaweeds with higher production. IMTA practice combines the cultivation of fed finfish with extractive shellfish and seaweeds for an ecologically balanced aquaculture. IMTA can increase the long-term and overall sustainability and profitability per cultivation unit as the wastes of the main cultured species are biomitigated through conversion into fertilizer, food, and energy through additional commercially valuable species. In this way, otherwise costly waste mitigation processes become revenue-generating cultivation components, which, by their harvest, export nutrients outside of the coastal ecosystem. It is important to note that 830 tons of CO_2 year^{-1} can be thus exported by an IMTA farm that produces 1,000 mt fish, 2,000 mt shellfish, and 500 mt seaweed. An IMTA farm that produces 1,000 mt fish and 7,000 mt seaweed can export 1,230 tons of CO_2 year^{-1} (Neori, 2008).

The seaweed IMTA component may include species of *Gracilaria, Porphyra, Eucheuma/Kappaphycus, Laminaria, Undaria, Ecklonia, Macrocystis, Ulva,* and *Caulerpa*. However, other commercially important species, such as *Palmaria, Chondrus, Gigartina, Hypnea, Sargassum, Cystoseira, Asparagopsis/Falkenbergia,* etc. have also high potential in IMTA systems. Today, several IMTA projects are being conducted in different parts of the world. The goal is to develop profitable modern IMTA seaweed farming components for different aquaculture environments. IMTA technologies with seaweed culture are bound to play a major role worldwide in sustainable expansions of the aquaculture operations within balanced ecosystem (Neori, 2007). Similarly, seaweeds culture is ready to respond to the worldwide increasing demands for renewable fuel and efficient CO_2 removal.

4. CO_2 Utilization by Seaweeds

Marine photosynthesis accounts for 50% of the total primary productivity of the planet (54–59 Pg (a petagram = 10^{15} g) C year^{-1} from a total of 111–117 Pg C year^{-1}). Of this, marine macrophytes (seaweeds and seagrasses) in the coastal regions account for ~1 Pg C year^{-1}. However, natural beds of seaweeds, such as the kelps genera *Macrocystis* and *Laminaria* are capable of very high rates of photosynthesis and productivity of ≥3,000 gC m^{-2} year^{-1} (Gao and McKinley, 1994; Muraoka, 2004). A range of other seaweed species have high productivities and can contribute significantly to the annual biological drawdown of CO_2 and the global C cycle (Ritschard, 1992; Gao and McKinley, 1994; Muraoka, 2004).

Seaweeds utilize CO_2 that diffuses in through cellular membranes from the surrounding seawater. Increased CO_2 concentration can increase the rate of diffusion and through it the rate of photosynthesis. High CO_2 concentrations increased growth of the macroalga *Porphyra yezoensis* (Gao et al., 1991). Other species are essentially CO_2 saturated under present-day CO_2 levels in the sea and are not expected to show increased performance in the future (Beardall et al., 1998). In their natural habitats, photosynthesis and growth of *Gracilaria* species are likely to be CO_2-limited, especially when the population density is high and water movement is slow. When the CO_2 content continues to rise, these marine seaweeds are expected to grow better (Surif and Raven, 1989; Maberly, 1990; Gao et al., 1991; Levavasseur et al., 1991). In culture studies of the red macroalgae *Gracilaria* sp. and *G. chilensis* in vessels continuously aerated with normal air containing 350 ppm CO_2, air enriched with additional 650 ppm CO_2, and air enriched with additional 1,250 ppm CO_2 for a period of 19 days, the relative growth enhancements in the +650-ppm and +1,250-ppm CO_2 treatments were 20% and 60%, respectively, in *G. chilensis*, and 130% and 190%, respectively, in the *Gracilaria* sp. (Gao et al., 1993). In another study, the red seaweed *Lomentaria articulata* was grown for 3 weeks in hydroponic culture vessels, where various atmospheric CO_2 and O_2 concentrations were applied. Oxygen concentrations ranging from 10% to 200% of ambient had no significant effect on either the seaweed's daily net carbon gain or its total wet biomass production rate. However, CO_2 concentrations ranging from 67% to 500% of ambient had highly significant effects on these parameters (Kubler et al., 1999). When the brown seaweed *Hizikia fusiforme* was grown in aquaria in a wide range of substrate nitrate concentrations, continuous aeration with air enriched with 700 ppm CO_2 enhanced growth, nitrogen assimilation rate, and nitrate reductase activity. The CO_2 enrichment increased the seaweed's growth rate by 50%, nitrate uptake rate during the light periods by 200%, and nitrate reductase activity by approximately 20% (Zou, 2005).

Among 15 seaweed species examined, the red alga *Pachymeniopsis lanceolata* showed the highest carbon removal rate on an areal basis, 2,500–6,000 g C m^{-2} year^{-1}, which is five times higher than that in tropical forests. *Undaria pinnatifida, Porphyra seriata, Laminaria japonica, Ulva pertusa, Ecklonia stolonifera,* and *Grateloupia lanceolata* were also found to be highly efficient in removing aquatic inorganic carbon. The uptake process depends on seasonal variations. For example, the highest carbon uptake rate on fresh-weight basis occured in *Ulva pertusa* collected in autumn, but the carbon uptake rate on area basis was the highest in *Grateloupia lanceolata* in winter, and twice the value in tropical rainforest (Chung, 2008).

5. Global Contribution of Seaweed Culture to the Carbon Cycle

Today, the world uses 15 million tons of fresh (2.3 million tons dry) seaweeds (Table 1). Seaweed dw contains above 30% carbon. Reported carbon content in *Porphyra* is 39%, and in *Undaria* and *Laminaria* 30% (Chung, 2008). Thus, the data in Table 1 imply that nearly 0.7 million tons C year^{-1} are removed from the sea in commercially harvested seaweeds. Consideration of the scale of seaweed production in some selected countries (Table 2) indicates the extent to which their current production might contribute to any offset against their CO_2 emissions. With current levels of production, this represents, for most countries, only a small proportion of C emissions. However, it should be noticed that most countries produce little seaweed given the length and character of their coastline. China produces ~116 tons dry seaweed year^{-1} km^{-1}. If other countries were to match this level, then at least some of them would be in a position where seaweed utilization could offset significant fractions of their C emissions (Table 3). Of course, culture of seaweeds on huge rafts in the open ocean (Notoya, 2010, this volume) can dwarf even this estimate.

Seaweed carbon could make an important contribution to the global production of biofuel. The lipid content of most seaweed species is less than 7%. However, the contents of soluble and structural carbohydrates often surpass 30%

Table 2. A comparison between CO_2 emissions of selected countries, their current seaweed production from aquaculture and wild harvest, and C sequestration in relation to national C emissions. Harvest data from FAO, 2006a; emission data from GHG (greenhause gas) data from UNFCCC (http://unfccc.int).

Country	Seaweed culture (t fw year^{-1})	Seaweed capture (t fw year^{-1})	Total production (t dw year^{-1})	C in production (t year^{-1})	C emissions (10^3 t year^{-1})	C in production/C in emissions (%)
China	10,867,410	308,380	1,676,369	502,911	947,347	0.053086
Phillipines	1,468,905	298	220,381	66,114	20,536	0.321940
Korea	1,209,895	15,212	183,767	55,130	45,905	0.120095
Indonesia	910,636	7,730	137,755	41,327	78,007	0.052978
Japan	490,052	104,893	89,242	26,773	334,020	0.008015
Chile	33,586	409,851	66,516	19,955	14,942	0.133543
Norway	–	153,906	23,086	6,926	9,630	0.071917
France	45	76,633	11,502	3,451	99,132	0.003481
Russia	818	50,262	7,662	2,299	420,098	0.000547
USA	–	35,922	5,388	1,617	1,571,471	0.000103
Ireland	–	29,500	4,425	1,328	11,676	0.011373
Canada	–	20,756	3,110	933	142,201	0.002187
Australia	–	14,167	2,125	638	90,648	0.000704
Italy	–	1,600	240	72	121,799	0.000069
Turkey	–	45	6.75	2	61,053	0.000003
World Total	15,075,612	1,305,803	2,457,212	737,163	6,226,228	0.011840

Table 3. A comparison between current and potential (see text) seaweed production from aquaculture and seaweed wild harvest and cultivation in relation to the coastline in selected countries.

Country	Coastline (km)	Production (t dw year^{-1})/ coastline km	Potential production if increased to match China dw year^{-1})/ coastline km	Production increase (factor)	% of C emissions captured by biomass at the increased coastline usage	potential C sequestration (t year^{-1}) by seaweed production
China	14,500	115, 612	1,676,369	1	0.05	502,910.7
Phillipines	36,289	6,073	4,195,444	19	6.12	1,258,633.2
Korea	2,413	76,157	278,972	1.5	0.18	83,682.6
Indonesia	54,716	2,518	6,325,826	28.6	1.52	1,897,747.8
Japan	29,751	3,000	3,439,573	38.5	0.31	1,031,871.9
Chile	6,435	10,337	743,963	11.2	1.50	223,188.9
Norway	25,148	918	2,907,411	125.9	9.05	872,223.3
France	3,427	3,356	396,202	34.5	0.12	118,860.6
Russia	37,653	204	4,353,139	568.2	0.31	1,305,941.7
USA	19,924	270	2,303,454	427.5	0.05	691,036.2
Ireland	1,448	3,056	167,406	37.8	0.43	50,221.8
Canada	202,080	15	23,362,873	7,512.2	16.43	7,008,861.9
Australia	25,760	8	2,978,165	13.5	0.01	893,449.5
Italy	7,600	3	878,651	3,661.5	0.25	263,595.3
Turkey	8,333	0.09	963,395	142,725.2	0.43	289,018.5
World Total	356,000	6,902	41,157,873	16.8	0.20	12,347,361

(Renaud and Luong-Van, 2006). Seaweed lipid can be directly converted to biodiesel with basically the same technology as other biomass feedstocks. Seaweed carbohydrates and proteins can also be processed to useful fuels and chemical feedstocks (Petrus and Noordermeer, 2006).

Seaweed oil is an interesting sustainable feedstock for biofuel/biodiesel manufacturing. It is a next-generation alternative to land-based biodiesel sources, like soybean, canola, and palm. Seaweed oil can be extracted, processed, and refined for various uses, including transportation, using currently available technology. Other benefits of seaweeds as a potential feedstock are their availability. Seaweeds can be grown nearly everywhere. One of the technical challenges for algae-based biodiesel is the matching of seaweed growth requirements, performance, and chemistry to each potential cultivation site and industrial use. Developing cost-effective engineering of very large-scale farms and their operation are additional keys to success in seaweed-based biofuel.

Seaweed farms are expected to locate mainly in two environments, one being on the high seas (see Notoya, 2010) and in low value coastal lands and waters close to CO_2-emitting industrial centers and power plants. Typical coal-fired power plants emit flue gas with up to 13% CO_2. This high concentration of CO_2 can enhance uptake of CO_2 by some seaweeds. Therefore, where the necessary low-value area is available, the concept of coupling a coal-fired power plant with an algae farm provides an elegant approach to the immediate recycling of CO_2 from coal combustion into a useable liquid fuel.

6. Summary and Conclusions

The characteristics of the seaweed industry that make it suitable to contribute to the abatement of climatic change and supply of biofuel are their widespread availability, diversity of species, high content of practical fuel biochemicals, independence from competition with food agriculture, high yields and CO_2 uptake rates, and low-cost technologies for large-scale production.

Seaweed aquaculture is expected to increase dramatically, both in monoculture and in IMTA. Seaweed coastal CO_2 removal belts can be anticipated to be developed in coastal waters together with sustainable seafood production (see in Issar and Neori and in Notoya, 2010, this volume). Seaweeds can be a considerable sink for anthropogenic CO_2 emissions and excess nutrients in the coastal waters of some countries. Increasing seaweed production to its full potential, and appropriate use of the biomass could thus play a significant role in the amelioration of greenhouse gas emissions and ocean nutrification and the production of C-neutral fuels. The produced biomass can contribute to some extent in meeting the global food, fodder, and pharmaceuticals requirements. It is clear that seaweed fuels will be produced. The timely transition to renewable fuel from seaweed critically depends on scientific progress in three areas, which are:

- *Seaweed biology and biotechnology*: identification and development of seaweed varieties that maximize the production of biodiesel and other fuels
- *Engineering*: development of culture technologies and farm designs that sustainably produce large quantities of seaweed feedstock at low cost
- *Social sciences*: communication to the relevant parties of the social, socioeconomic, and environmental benefits that large-scale seaweed fuel production can have; overcoming opposition to innovation, gaining of socio-political support, and public involvement

Critical to this and other innovative developments is the acceptance by the public financial system of its obligation to support long-term research and innovation programs, ensuring that emerging ideas that are explored consider overall societal needs. Only the public financial systems, e.g., governments and the World Bank, have the resources and time to fund the required investments. Such efforts need a strong political will for their active steering in the right direction.

Preconditions for an accelerated development for industrial applications will be:

- *Government regulations*: shaping a competitive environment for the private sector
- *Public–private partnerships*: prototyping new processes by collaborations between research and industry
- *An educational infrastructure*: drawing together the best possible human resources

When vigorously pursued, direct conversion of seaweed into fuel represents one of the very few major options that humankind has to provide socially, economically, and environmentally robust and resilient renewable fuel, whose production answers additional major human necessities rather than creating them, and with energy security that is guaranteed in a humanitarian instead of confrontational manner. With scalable seaweed-fuel conversion technology, nations can become sustainable producers and exporters at the level of regions, cities, communities, and individual citizens. This may well give rise to a paradigm shift, from the current model where fuel is provided at the lowest possible direct cost by large-scale industries, with a considerable disregard for environmental and societal concerns, to an energy system where fuel is a sustainable source of economic growth for the public that principally owns its source of energy and the benefits that come with it.

7. References

Beardall, J., Beer, S. and Raven, J.A. (1998) Biodiversity of marine plants in an era of climate change: some predictions on the basis of physiological performance. Bot. Mar. **41**: 113–123.

Bird, K.T. and Benson, P.H. (eds.) (1987) *Seaweed Cultivation for Renewable Resources*. Elsevier Press, New York, 381 pp.

Breeman, A.M. (1990) Expected effects of changing seawater temperatures on the geographic distribution of seaweed species, In: J.J. Beukema, W.J. Wolff and J.J.W.M. Brouns (eds.) *Expected Effects of Climate Change on Marine Coastal Ecosystems*. Kluwer, The Netherlands, pp. 69–76.

Buschmann, A.H., Mora, O.A., Gomez, P., Bottger, M., Buitano, S., Retamales, C., Vergara, P.A. and Gutierrez, A. (1994) *Gracilaria chilensis* outdoor tank cultivation in Chile: use of land-based salmon culture effluents. Aquacult. Eng. **13**: 283–300.

Buschmann, A.H., Troell, M. and Kautsky, N. (2001) Integrated algal farming: a review. Can. Biol. Mar. **42**: 83–90.

Buschmann, A.H., Hernandez-Gonzales, M.C., Aranda, C., Chopin, T., Neori, A., Halling, C. and Troell, M. (2008) Mariculture waste management, In: S.E. Jorgensen and B.D. Fath (Editor-in-chief) *Ecological Engineering*, Vol. 3 of Encyclopedia of Ecology, 5 vols. Elsevier, Oxford, pp. 2211–2217.

Chopin, T. (2006) Integrated multi-trophic aquaculture. Northern Aquacult. July/August 2006: 4, available at www.northernaquaculture.com

Chopin, T., Yarish, C., Wilkes, R., Belyea, E., Lu, S. and Mathieson, A. (1999) Developing *Porphyra*/salmon integrated aquaculture for bioremediation and diversification of the aquaculture industry. J. Appl. Phycol. **11**: 463–472.

Chopin, T., Buschmann, A.H., Halling, C., Troell, M., Kautsky, N., Neori, A., Kraemer, G.P., Zertuche-Gonzalez, J.A., Yarish, C. and Neefus, C. (2001) Integrating seaweeds into marine aquaculture systems: a key toward sustainability. J. Phycol. **37**: 975–986.

Chopin, T., Robinson, S.M.N., Troell, M., Neori, A., Bushmann, A.H. and Fang, J. (2008) Multitrophic integration for sustainable marine aquaculture, In: S.E. Jorgensen and B.D. Fath

(Editor-in-Chief) *Ecological Engineering*. Vol. 3 of Encyclopedia of Ecology, 5 vols. Elsevier, Oxford, pp. 2463–2475.
Chung, I.K. (2008) Seaweed coastal CO_2 removal belt in Korea. The United Nations Framework Convention on Climate Change, UNFCCC, and the Thirteenth Conference of the Parties, COP-13, Side event of seaweed coastal CO_2 removal belt in Korea & algal paper and biofuel, 6–10 December 2007, Bali, Indonesia.
Chung, I.K., Beardalla, J., Mehta, S., Sahoo, D. and Stojkovica, S. (2007) Using marine algae for carbon sequestration: a critical appraisal. The United Nations Framework Convention on Climate Change, UNFCCC, and the Fourteenth Conference of the Parties, COP-14, Side event of seaweed coastal CO_2 removal belt in Korea & algal paper and biofuel, 1–12 December 2008, Ponzan, Poland.
Cirik, S. and Cirik, S. (1999) Aquatic plants: biology, ecology and culture techniques of marine plants. Ege University, Fisheries Faculty Publications, No: 28, Bornova, Izmir, 188 p., ISBN 975-483-46-4 (In Turkish).
Cirik, S., Turan, G., Ak, I. and Koru, E. (2006) *Gracilaria verrucosa* (Rhodophyta) culture in Turkey. *International Conference on Coastal oceanography and Sustainable Marine Aquaculture: Confluence and Synergy*, 2–4 May 2006, Sabah, Malaysia.
Cohen, I. and Neori, A. (1991). *Ulva lactuca* biofilters for marine fishpond effluent. I. Ammonia uptake kinetics and nitrogen content. Bot. Mar. **34**: 475–482.
Critchley, A.T. and Ohno, M. (eds.) (1997) Cultivation and farming of marine plants, In: CD-ROM, Expert Centre for Taxonomic Identification (ETI), University of Amsterdam, Amsterdam, ISBN 3-540-14549-4. Springer, New York.
Critchley, A.T. and Ohno, M. (eds.) (1998) *Seaweed Resources of the World*. Japan International Cooperation Agency, Yokosuka, Japan, 429 pp.
Critchley, A.T. and Ohno, M. (eds.) (2001) *Cultivation and Farming of Marine Plants*. ETI World Biodiversity Database, CD-ROM Series. http://www.eti.uva.nl/products/catalogue/ cd_detail.php?id=177&referrer=search (accessed December 2007).
Critchley, A.T., Ohno, M. and Largo, D.B. (eds.) (2006) World seaweed resources, In: DVD-ROM, Expert Centre for Taxonomic Identification (ETI), University of Amsterdam, Amsterdam, ISBN: 90 75000 80 4. Springer, New York.
Dayton, P.K. and Tegner, M.J. (1984) Catastrophic storms, El Niño, and patch stability in a Southern California Kelp Community. Science **224**: 283–285.
De Poli, F., Nicolucci, C. and Monegato, A. (1994) Industrial production of paper obtained from prolific seaweeds in Venice Lagoon. Cellulosa e Carta (Italy) **45**(5–6): 41–47 (in Italian).
De Roeck-Holtzhauer, Y. (1991) Uses of seaweeds in cosmetics, In: M.D. Guiry and G. Blunden (eds.) *Seaweed Resources in Europe: Uses and Potential*. Wiley, Chichester, pp. 83–94.
Demirbas, A. (2008) Global renewable energy and biofuel scenarios, In: A. Demirbas (ed.) *Biodiesel: A Realistic Fuel Alternative for Diesel Engines*. Springer, London, ISBN: 978-1-84628-994-1 (print) 978-1-84628-995-8 (online), pp. 185–194.
Edwards, P. (2004) Traditional Chinese aquaculture and its impact outside China. World Aquacult. **35**:24–27.
FAO (2003) Guide to the seaweed industry (A). FAO Fisheries Technical Paper No. 441 Rome, 116 pp.
FAO (2006a) Aquaculture production 2004, *FAO Yearbook*, Fishery Statistics, Vol: 98/2. Rome.
FAO (2006b) *The State of World Fisheries and Aquaculture 2006*. Food and Agriculture Organization of the United Nations, Rome, Italy.
FAO (2009) *The State of World Fisheries and Aquaculture 2008*. Food and Agriculture Organization of the United Nations, Rome, Italy.
Fei, X. (2004) Solving the coastal eutrophication problem by large scale seaweed cultivation. Hydrobiologia **512**: 145–51.
Flowers, A.B. and Bird, K.T. (1987) Methane production from seaweeds, In: I. Akatsuka (ed.) *Introduction to Applied Phycology*. SPB Academic Publishing, The Hague, pp. 575–587.
Gao, K. and McKinley, K.R. (1994) Use of macroalgae for marine biomass production and CO_2 remediation: a review. J. Appl. Phycol. **6**: 45–60.

Gao, K., Aruga, Y., Asada, K., Ishihara, T., Akano, T. and Kiyohara, M. (1991) Enhanced growth of the red alga *Porphyra yezoensis* Ueda in high CO_2 concentrations. J. Appl. Phycol. **3**: 355–362.

Gao, K., Aruga, Y., Asada, K. and Kiyohara, M. (1993) Influence of enhanced CO_2 on growth and photosynthesis of the red algae *Gracilaria* sp. and *G. chilensis*. J. Appl. Phycol. **5**: 563–571.

Hanisak, M.D. and Ryther, J.H. (1986) The experimental cultivation of the red seaweed *Gracilaria tikvahie* as an 'energy crop': an overview, In: W.R. Barclay and R.P. McIntosh (eds.) *Algal Biomass Technologies, An Interdisciplinary Perspective*, Vol. 83. Nova Hedwigia, Berlin, pp. 212–217.

Issar, A.S. and Neori, A. (2010, this volume) Progressive development of new marine environments—IMTA (integrated multi-trophic aquaculture) production: a new policy proposed for the mitigation of impacts of climate change on coastal regions, particularly in arid and semi-arid climates.

Kain-Jones, J.M. (1991) Cultivation of attached seaweeds, In: M.D. Guiry and G. Blunden (eds.) *Seaweed Resources in Europe: Uses and Potential*. Wiley, Chichester, pp. 309–377.

Kang, Y.H., Shin, J.A., Kim, M.S. and Chung, I.K. (2008) A preliminary study of the bioremediation potential of *Codium fragile* applied to seaweed integrated multi-trophic aquaculture (IMTA) during the summer. J. Appl. Phycol. **20**(2): 183–190.

Kelly, M.S. and Dworjanyn, S. (2008) *The Potential of Marine Biomass for Anaerobic Biogas Production*. The Crown Estate, 103 pp., ISBN: 978-1-906410-05-6.

Krom, M.D. and Neori, A. (1989) A total nutrient budget for an experimental intensive fishpond with circularly moving seawater. Aquaculture **83**: 345–358.

Kubler, J.E., Johnston, A.M. and Raven, J.A. (1999) The effects of reduced and elevated CO_2 and O_2 on the seaweed *Lomentaria articulata*. Plant Cell Environ. **22**: 1303–1310.

Levavasseur, G., Edwards, G.E., Osmond, C.B. and Ramus, J. (1991) Inorganic carbon limitation of photosynthesis in *Ulva rotundata* (Chlorophyta). J. Phycol. **27**: 667–672.

Maberly, S.C. (1990) Exogenous sources of inorganic carbon for photosynthesis by marine macroalgae. J. Phycol. **26**: 439–449.

McHugh, D.J. (2003) A guide to the seaweed industry. FAO Fisheries Technical Paper 441, 107 pp. Food and Agriculture Organization of the United Nations, Rome, Italy.

Morand, P., Carpentier, B., Charlier, R.H., Maze, J., Orlandini, M., Plunkett, B.A. and de Waart, J. (1991) Bioconversion of seaweeds, In: M. Guiry and G. Blunden (eds.) *Seaweed Resources in Europe: Uses and Potential*. Wiley, Chichester, pp. 95–148.

Muraoka, D. (2004) Seaweed resources as a source of carbon fixation. Bull. Fish. Res. Agen. Suppl. **1**: 59–63.

Neori, A. (2007) Essential role of seaweed cultivation in integreted multi-trophic aquaculture farms for global expansion of mariculture: an analysis. J. Appl. Phycol. Online Publication.

Neori, A. (2008) Macro-algal (seaweed) biomass converts economically CO_2 into food, chemicals, and biofuel. The United Nations Framework Convention on Climate Change, UNFCCC, and the Fourteenth Conference of the Parties, COP-14, Side event of seaweed coastal CO_2 removal belt in Korea & algal paper and biofuel, 1–12 December 2008, Ponzan, Poland.

Neori, A. and Shpigel, M. (1999) Using algae to treat effluents and feed invertebrates in sustainable integrated mariculture. World Aquacult. **30**(2): 46–51.

Neori, A., Krom, M.D., Cohen, I. and Gordin, H. (1989) Water quality conditions and particulate chlorophyll *a* of new intensive seawater fishponds in Eilat, Israel: daily and diel variations. Aquaculture **80**: 63–78.

Neori, A., Krom, M.D., Ellner, S.P., Boyd, C.E., Popper, D., Rabinovitch, R., Davison, P.J., Dvir, O., Zuber, D., Ucko, M., Angel, D. and Gordin, H. (1996) Seaweed biofilters as regulators of water quality in integrated fish-seaweed culture units. Aquaculture **141**: 183–199.

Neori, A., Shpigel, M. and Ben-Ezra, D. (2000) A sustainable integrated system for culture of fish, seaweed and abalone. Aquaculture **186**: 279–291.

Neori, A., Chopin, T., Troell, M., Buschmann, A.H., Kraemer, G.P., Halling, C., Shpigel, M. and Yarish, C. (2004) Integrated aquaculture: rationale, evolution and state of the art, emphasizing seaweed biofiltration in modern mariculture. Aquaculture **231**: 361–391.

Nielsen, K.J. (2003) Nutrient loading and consumers: agents of change in opencoast macrophyte assemblages. Proc. Natl. Acad. Sci. **13**: 7660–7665.

Notoya, M. (2010, this volume). Production of biofuel by macro-alga with preservation of marine resources and environment.
Nunes, J.P., Ferreira, J.G., Gazeau, F., Lencart-Silva, J., Zhang, X.L., Zhu, M.Y. and Fang, J.G. (2003) A model for sustainable management of shellfish polyculture in coastal bays. Aquaculture **219**: 257–277.
Petrus, L. and Noordermeer, M. (2006) Biomass to biofuels, a chemical perspective. Green Chem. **8**: 861–867.
Renaud, S.M. and Luong-Van, J.T. (2006) Seasonal variation in the chemical composition of tropical Australia marine macroalgae. J. Appl. Phycol. **18**: 381–387.
Ritschard, R.L. (1992) Marine algae as a CO_2 sink. Water Air Soil Pollut. **64**: 289–303.
Ryther, J.H., Goldman, J.C., Gifford, J.E., Huguenin, J.E., Wing, A.S., Clarner, J.P., Williams, L.D. and Lapointe, B.E. (1975) Physical models of integrated waste recycling—marine polyculture systems. Aquaculture **5**: 63–177.
Sahoo, D. and Yarish, C. (2005) Mariculture of seaweeds, In: R.A. Anderson (ed.) *Algal Culturing Techniques*. Elsevier Academic Press, Burlington, MA, pp. 219–237.
Schramm, W. (1991a) Seaweeds for waste water treatment and recycling of nutrients, In: M.D. Guiry and G. Blunden (eds.) *Seaweed Resources in Europe: Uses and Potential*. Wiley, Chichester, pp. 149–168.
Schramm, W. (1991b) Cultivation of unattached seaweeds, In: M.D. Guiry and G. Blunden (eds.) *Seaweed Resources in Europe: Uses and Potential*. Wiley, Chichester, pp. 379–408.
Schuenhoff, A., Shpigel, M., Lupatsch, I., Ashkenazi, A., Msuya, F.E. and Neori, A. (2003) A semirecirculating, integrated system for the culture of fish and seaweed. Aquaculture **221**: 167–181.
Schuenhoff, A., Mata, L. and Santos, R. (2006) The tetrasporophyte of *Asparagopsis armata* as a novel seaweed biofilter. Aquaculture **252**: 3–11.
Shpigel, M. and Neori, A. (1996) The integrated culture of seaweed, abalone, fish, and clams in modular intensive land-based system. I. Proportion of size and projected revenue. Aquacult. Eng. **15**(5): 313–326.
Shpigel, M., Neori, A., Popper, D.M. and Gordin, H. (1993a) A proposed model for "clean" land based polyculture of fish, bivalves and seaweeds. Aquaculture **117**: 115–128.
Sinha, V.R.P., Fraley, L. and Chowdhy, B.S. (2001) Carbon dioxide utilization and seaweed production. *Proceedings of NETL: First National Conference on Carbon Sequestration* (http://www.netl.doe.gov/publications/proceedings/01/carbon_seq/p14.pdf).
Surif, M.B. and Raven, J.A. (1989) Exogenous inorganic carbon sources for photosynthesis in seawater by members of the Fucales and the Laminariales (Phaeophyta): ecological and taxonomic implications. Oecologia **78**: 97–103.
Troell, M., Halling, C., Nilsson, A., Buschmann, A.H., Kautsky, N. and Kautsky, L. (1997) Integrated marine culture of *Gracilaria chilensis* (Graciales, Rhodophyta). And salmon cages for reduced environmental impact and increased economic output. Aquaculture **156**: 45–61.
Troell, M., Rönnbäck, P., Halling, C., Kautsky, N. and Buschmann, A.H. (1999) Ecological engineering in aquaculture: use of seaweed for removing nutrients from intensive mariculture. J. Appl. Phycol. **11**: 89–97.
Troell, M., Halling, C., Neori, A., Chopin, T., Buschmann, A.H., Kautsky, N. and Yarish, C. (2003) Integrated mariculture: asking the right questions. Aquaculture **226**: 69–90.
Troell, M., Robertson-Andersson, D., Anderson, R.J., Bolton, J.J., Maneveldt, G., Halling, C. and Probyn, T. (2006) Abalone farming in South Africa: an overview with perspectives on kelp resources, abalone feed, potential for on-farm seaweed production and socio-economic importance. Aquaculture **257**: 266–281.
Turan, G. and Cirik, S. (2008) Culture, mineral-vitamin composition, and thalassotherapy application studies on seaweeds. *The 11th International Conference on Applied Phycology*, 22–27 June 2008, The National University of Ireland, Galway, Ireland.
Turan, G., Ak, I., Cirik, S., Koru, E. and Kaymakci-Basaran, A. (2006) *Gracilaria verrucosa* (Hudson) Papenfuss culture in intensive fish farm. Ege Univ. J. Fish. Faculty **23**(1/2): 305–309 (in Turkish).

Turan, G., Cirik, S., Tekogul, H., Koru, E., Seyhaneyildiz, S., Peker, O. and Can, E. (2007a) Seaweed Cultivation in Integrated Aquaculture Systems. XIV. National Fisheries Symposium, Mugla University, Fisheries Faculty, 4–7 September 2007, Mugla, Turkey (in Turkish).

Turan, G., Koru, E., Tekogul, H., Seyhaneyildiz, S., Can, E., Peker, O. and Cirik, S. (2007b) Seaweed Aquaculture in Turkey: Pilot Projects with Potential for Integrated Aquaculture Systems, Aquaculture Europe 07, October 24–27, 2007, İstanbul, Turkey.

Turan, G., Koru, E., Tekogul, H., Seyhaneyildiz, S., Peker, O. and Cirik, S. (2007c) Seaweed Resources of Turkey, Aquaculture Europe 07, October 24–27, 2007, İstanbul, Turkey.

Whitmarsh, D.J., Cook, E.J. and Black, K.D. (2006) Searching for sustainability in aquaculture: an investigation into the economic prospects for an integrated salmon-mussel production system. Mar. Policy **30**(3): 293–298.

Xu, Y., Fang, J., Tang, Q., Lin, J., Le, G. and Liao, L. (2008) Improvement of water quality by the macroalgae, *Gracilaria lemaneiformis* (Rhodophyta), near Aquaculture effluent outlets. J. World Aquacult. Soc. **39**(4): 549–556.

Yang, Y.F., Li, C.H., Nie, X.P., Tang, D.L. and Chung, I.K. (2004) Development of mariculture and its impacts in Chinese coastal waters. Rev. Fish. Biol. Fish. **14**: 1–10.

Yang, Y.F., Fei, X., Song, J., Hu, H., Wang, G. and Chung, I.K. (2006) Growth of *Gracilaria lemaneiformis* under different cultivation conditions and its effects on nutrient removal in Chinese coastal waters. Aquaculture **254**: 554–562.

Yarish, C. and Pereira, R. (2008) Mass production of marine macroalgae, In: S.E. Jørgensen and B.D. Fath (Editor-in-Chief) *Ecological Engineering*, Vol. 3 of Encyclopedia of Ecology, 5 vols. Elsevier, Oxford, pp. 2236–2247.

You, H.C. (2008) Innovative seaweed pulp, paper and biofuel. The United Nations Framework Convention on Climate Change, UNFCCC, and the Fourteenth Conference of the Parties, COP-14, Side event of seaweed coastal CO_2 removal belt in Korea & algal paper and biofuel, 1–12 December 2008, Ponzan, Poland.

Zou, D. (2005) Effects of elevated atmospheric CO_2 on growth, photosynthesis and nitrogen metabolism in the economic brown seaweed, *Hizikia fusiforme* (Sargassaceae, Phaeophyta). Aquaculture **250**: 726–735.

Biodata of **Werner Reisser**, author of *"The Future is Green: On the Biotechnological Potential of Green Algae"*

Professor Dr. Werner Reisser is teaching General and Applied Botany at the Botanical Institute of Leipzig University in Germany. He got his Ph.D. in 1977 from the University of Göttingen, Germany, by studies on endosymbiotic associations of ciliates and algae. His research interests center on taxonomy and ecophysiology of aeroterrestrial algae and molecular ecology of soil ecosystems.

E-mail: **reisser@rz.uni-leipzig.de**

THE FUTURE IS GREEN: ON THE BIOTECHNOLOGICAL POTENTIAL OF GREEN ALGAE

WERNER REISSER
Institute of Biology I, General and Applied Botany, University of Leipzig, D – 04103, Leipzig, Germany

1. Introduction

There are two main players that form the basis of nearly all global ecosystems in converting solar energy to biomass: algae and plants. While plants are omnipresent in public discussions dealing with such topics as climate change, bioreactors, biofuels and green biotechnology, the role and potential of algae is usually known only to experts. However, algae are present as primary producers in nearly all types of ecosystems, their versatile physiology allowing them an impressive range of adaption to aerial, terrestrial as well as aquatic habitats. As to its ecological impact, the most important group of algae is the phytoplankton, especially the nano- and picoplankton, which forms the basis of marine ecosystems. The phytoplankton produces about the same amount of oxygen as all land plants and is also involved in climatic processes by the production of volatile compounds and condensation nuclei for the formation of clouds.

Tapping the biotechnological potential of algae has a long tradition in human history (Spolaore et al., 2006). Algae, mainly kelps, are used as moisteners in soil and as fertilisers for human or animal food production. The ability of algae to absorb metals is used in biotreatment of contaminated soil. Microalgae are also working in self-supporting life systems as they are used in space travel. A plethora of algal products is on the market (Gantar and Svircev, 2008) obtained mainly from cyanobacterian genera *Aphanizomenon* and *Arthrospira* (*Spirulina*) and from chlorophycean genera *Chlorella*, *Dunaliella* and *Scenedesmus*. Algal products can be found in ice cream, puddings, dietary products and cosmetics. From algal cultures, polyunsaturated fatty acids are obtained along with antioxidants, suppressors of hypertension, vitamins and natural pigments such as carotenoids and phycobiliproteids. Microalgae also serve as food additives and are incorporated into the feed for aquacultures, farm animals and pets. Nonetheless, in comparison with plants, algae have played only a minor role in public awareness.

This may change now dramatically. The ever-increasing energy demand of world economy is recognised as a threat to the world climate owing to an increase in atmospheric CO_2 (greenhouse gas) released by burning of fossil fuels (coal, oil, natural gas). Therefore, there is a growing request for renewable energy sources that do not release CO_2 or – at least – do not emit additional CO_2 (Schiermeier et al., 2008). These are the classical sources of wind, water, solar, geothermal and nuclear energy, but also the hopeful newcomers, hydrogen technology and fusion power. However, although a lot of money has been and will be spent to exploit those alternative energy sources, it is obvious that in the near future they can cover only a small part of the world energy demand, either because they are more or less exploited (e.g., hydropower in Europe), their public acceptance is limited (e.g., nuclear fission power, wind plants) or financial resources are not as substantial as necessary (fusion power, hydrogen technology). What is more, it is obvious that in the near future most energy needed for transportation purposes will be used in combustion engines that require liquid fuels or gas. Therefore, it is desirable to try to replace fossil fuels by biofuels.

2. Biofuels

Biofuels (biogas, bioalcohol, biodiesel) are made from plant biomass, and are more or less carbon neutral, since in burning they do not increase the overall CO_2 content of the atmosphere but set free just the amount of CO_2 that was fixed before in photosynthesis of the plant (Schiermeier et al., 2008). Thus, in recent years, plant biomass has gained growing importance as an alternative energy source. Main current sources of bioalcohol (mostly bioethanol) are sugars of sugarcane and starch of corn and wheat. Today, about 20% of the US corn harvest is used to make bioethanol. This covers about 2% of the US demand for transportation fuels (Chisti, 2007).

Biodiesel is made from plant oil, mainly from rapeseed, palm oil and jatropha. Biogas is produced from different kinds of biomass. However, a serious flaw in the ecobalance of traditional biofuels becomes obvious when the complete CO_2 balance is calculated, i.e., when CO_2 costs for seed, fertilisers, herbicides, irrigation, harvest and processing are taken into account. There is also growing concern in public discussion about the fact that the use of edible plants as energy sources may raise the prices of food. It is also rather questionable to cut down tropical rainforests for planting oil palms or jatropha when the plant oil is not processed locally but transported a long way to the industrial countries. Therefore, for making biofuels it is desirable to use plants (energy plants) with the modest requirements in soil quality and water supply and that grow under conditions and at places not suitable for crop plants. Ideally, the whole plant biomass can be used for fuel production and not just the special parts of plants such as oil-containing fruits as in rapeseed or oil palm ('first-generation energy plants'). 'Second-generation energy plants' are already available, such as many prairie grasses and microalgae.

3. Microalgae as Second-Generation Energy Plants

The use of microalgae for the production of biofuels has many ecological and economical advantages. First of all, microalgae show a much higher efficiency in converting solar energy to biomass. From the biomass of corn grown on one hectare, about 2,000 m^3 biogas (methane) can be obtained; however, biomass of microalgae grown on the equivalent area produces about 200,000 m^3 (Solarbiofuels 2008). In microalgal biomass, the percentage of compounds suitable for the production of biofuels (e.g., starch, oil) is much higher than in crop plants, because there is no need to divert energy to the synthesis of fibre material, vascular and absorption tissues, etc. Microalgal cultures can be grown on a relatively small area that may not be appropriate for agriculture and – at least in the case of 'indoor systems' – they do not need irrigation and produce a high constant yield irrespective of outside environmental conditions such as temperature and draught. For producing a given amount of biomass, indoor cultures need about 1,000 times less water than crops. Preliminary data show that – on the same area – (Chisti, 2007) microalgal cultures produce about 15× more oil for biodiesel production than rapeseed does. To cover 25% of the US demand for transportation fuels by corn (Table 1), an area of about 4.6× the area that is currently used for US agriculture is needed. When oil from oil palms is used, about 12% of the agricultural area is required; however, when microalgal cultures are used only 2–5% of that area would be sufficient.

3.1. CULTURE SYSTEMS FOR MICROALGAE

For a large-scale culture of microalgae, two systems are used, the so-called 'indoor' and 'outdoor' systems (Figs. 1 and 2). Outdoor systems are sometimes also called 'open raceway ponds' and have a long tradition that can be traced back in history

Table 1. Crop efficiencies for the production of biodiesel. (Modified after Chisti, 2007.)

Crop	Yield of biodiesel (L × ha^{-1} × a^{-1})	Land area required currently (ha × 10^6)	(percentage of area covered by crops in the USA)[a]
Corn	157	843	463
Soybean	451	294	162
Rapeseed	1,206	110	60
Jatropha	1,892	70	38
Oil Palm	5,991	22	12
Microalgae[a]	58,700	9	5
Microalgae[c]	136,900	4	2

[a] To cover about 25% of all transportation fuels needed in the USA per year.
[b] Oil content in algal biomass (by weight): 30%.
[c] Oil content in algal biomass (by weight): 70%.

Figure 1. Outdoor system for the cultivation of microalgae: 'open raceway pond' Seambiotic (Israel) by permission.

Figure 2. Indoor system for the cultivation of microalgae: Photobioreactor IGV (Germany) by permission.

for centuries and were used already by the Mayas. Microalgae, usually cyanobacteria as, e.g., *Spirulina* spp., were grown in small lakes, ponds and ditches, harvested and spread nearby for drying and subsequently used as animal food. Modern systems consist of pools of different shapes, in which microalgae are grown in a shallow layer (20–40 cm) that is permanently agitated to guarantee optimal growth conditions. Most outdoor systems are run today in countries providing optimal natural conditions concerning temperature and sunshine such as Hawaii, Australia and Japan. The largest outdoor facilities spread on an area of about 440,000 m^2 (Spolaore et al., 2006) and produce about 8–12 g dryweight of algal biomass per m^2 and day (Ackermann, 2007).

Indoor systems are closed bioreactors, in which algae are grown under defined conditions of temperature, light, nutrient supply, etc. (Pulz, 2001). The largest commercially used indoor system has a volume of about 600 m^3 that is contained in glass tubes 500 km in length (Ackermann, 2007). Production of algal biomass amounts to 32 g m^{-2} and day (sunlight only).

3.2. COSTS

A comparison of pros and cons of both systems shows that open systems are more cost-efficient than closed systems when optimal and constant climatic conditions are given, e.g., optimum temperature and sufficient sunshine. However, it is also obvious that open systems are generally more prone to changes in environmental conditions, as to the input of spores, germs and particulate matter from the atmosphere and to extreme weather conditions such as thunderstorms, hurricanes, etc. Continuous production of algal biomass is easier to achieve by closed systems that, however, require higher financial investments. Various tests have shown that in closed systems, the maximum cell density that can be obtained under optimum conditions is about 30 times higher than that in open systems. Taking into account the economic advantage of a constant and predictable production of high-quality indoor systems presumably is more cost-efficient than in open systems. Estimates (Chisti, 2007) calculate the costs for the production of 1 kg of algal biomass to US$2.85 in closed systems, whereas in open systems it is about US$3.89. When algae are used for the production of biodiesel (30% of algal biomass is processable), costs of 1 L of biodiesel obtained from open systems would amount to US$1.81, from closed systems to US$1.40. The same author calculates for 1 L biodiesel made from palm oil US$0.66 and made from petroleum about US$0.49. However, it should be mentioned that calculations are a matter of discussion depending not only on the oil content of algal biomass but also on general operating costs of the facility that in part may be effected by environmental conditions such as ambient temperature and duration of sunshine. Other authors calculate for 1 L biodiesel made from microalgae costs ranging from about US$5.35 (Dimitov, 2007) to US$0.16 (Günzburg, 1993). At any rate, most calculations show that costs of biofuels made from microalgae are still higher than costs of petrofuels. However, in recent times, the market price of petrofuel is steadily pointing upwards and the break-even point may be reached sooner than anticipated.

3.3. HIGH-VALUE PRODUCTS

It is not astonishing at all that up to now any economic success of indoor and outdoor systems was achieved by the production of high-value products such as pharmaceuticals, cosmetics, products for healthcare, natural colours, unsaturated fatty acids, essential amino acids, etc. Those high-value substances allow a realistic competition of microalgal cultures with classical production methods such as isolation of linolic acids from herbs, etc. They allow microalgae cultures to bring in their specific advantages such as production under reliable sterile conditions, no risk of contamination with human viruses, prions, etc. Thus, it is conceivable to use microalgal cultures also for genetic engineering techniques to obtain, for example, specially designed antibodies, recombinant proteins, etc. Appropriate techniques are available for microalgae, e.g., in *Chlamydomonas* sp. (green algae) genetic manipulation has been successful (Patel-Predd, 2007). An interesting advantage of microalgae over crop plants concerning genetic manipulation might be that there should be no public concern about bringing 'manipulated' organisms into the ecosystem, because algal systems are not in contact with the environment.

4. Chances

4.1. INCREASE IN EFFICIENCY

As to the future role of microalgae as an alternative energy source, the key question is whether the production of biofuels by microalgal cultures will ever be able to compete on a large scale with petrofuels. The answer to this question is not easy. It depends on how its economical, ecological and political aspects are measured.

There are good reasons to assume that petrofuel will remain the most important energy source for transportation in the near future and that its price will keep going up. Therefore, it is a realistic goal to raise the efficiency of microalgal culture systems, mainly of indoor systems, in order to be able to compete with the price of petrofuels. Efficiency may be augmented by a lot of measures. The exploitation of light energy has to be made more efficient, e.g., by keeping algae in small layers as are tested in new types of bioreactors and by constructing fibre-optic devices that might increase the capacity of daylight. It is also important to avoid clumping of algae, adhesion to reactor walls, for example, by special coating techniques of inner tube surfaces and to establish an automatic control and harvesting system that guarantees a continuous operating. It will also be desirable to have nearby a cheap source of CO_2, such as power and cement plants, although it should also be made clear that by algal cultures their output of CO_2 never can be reduced to zero.

From the viewpoint of biology, it will be certainly promising to screen for further microalgal candidates since until now only a rather small part of natures fundus of about 30,000–40,000 algal species is used, mainly cyanobacteria (e.g. *Spirulina*) and green (chlorophycean) algae, such as coccoid (*Chlorella, Scenedemus*) and flagellate specimen (*Dunaliella, Chlamydomonas*). Other microalgal taxa are represented by a few xanthophycean (e.g., *Botryococcus*), eustigmatophycean (e.g., *Nannochloropsis*), prasinophycean (e.g., *Tetraselmis*) and bacillariophycean (diatom) taxa such as *Nitzschia* and *Phaeodactylum*.

Efforts should be aimed at finding species that do not clump, show a lower photosynthetic compensation point to increase yield at less input of light energy and perform optimum growth at elevated temperatures to reduce costs of cooling. It will also be desirable to obtain a higher output of 'interesting substances' under standard conditions. If, for example, the content of processable oil in the algal biomass could be raised from 30% to 50% of the dryweight, the price of biodiesel would become competitive. However, this requires much better understanding of the physiological mechanisms in microalgae responsible for the synthesis of value products such as lipids and oil. Most algae produce substantial amounts of triacylglycerols such as the production of components of storage lipids only under stress conditions, e.g., nutrient limitation and photooxidative stress (Hu et al., 2008). Thus, the challenge will be to find culture conditions that combine both optimum growth of microalgae, i.e., high yield of biomass and high lipid content in cells.

As to the production of bioethanol from algal biomass, it will be recommendable to look by screening or by genetic engineering for algae producing carbohydrates that are more suitable to fermentation than the standard starch basis is. The same consideration holds also for other cell parts, such as the cell wall. Ideally, it consists of material that can also be easily fermented or alternatively used, for example by the BTE-technique, to produce energy.

It could also be promising to think about the problem whether the end product of fermentation has to be ethanol. Ethanol shows a high solubility in water and therefore is not easy to separate from it; it is poisonous to cells and has relatively relatively low energy content. Butanol could be a good candidate for ersatz.

A further yet speculative energy source produced by microalgae is H_2. The photoproduction of H_2 is mainly studied in *Chlamydomonas* sp. and represents a rather promising field of microalgal technology (Patel-Predd, 2007). One of the current problems to tackle is that H_2 production is inhibited by oxygen. Various strategies to meet this obstacle are under study, for example, the insertion of leghemoglobin genes into algae. It still requires a lot of basic research and is far from being exploitable in practice.

4.2. BIOREFINERY

In general, the leitmotiv should be to raise the efficiency of biomass processing. This means to increase the output of energy from algal biomass, either as biofuels (biodiesel, bioethanol, biogas) or directly in the form of heat energy that may be used in power plants. Each form of energy has to deal with a different and already successful competitor in the market. Therefore, it will be important to increase the current revenue from biomass by combining energy yield with the production of additional high-value products. Every component of the biomass should be exploited to add value ('biorefinery', Chisti, 2007).

An economic advantage of biomass production by microalgae in closed systems in comparison with other energy plants is doubtless the fact that production is constant, predictable and not dependent on such hazards as drought, diseases, and hurricanes.

5. Outlook

Under ecological aspects, advantages of microalgal cultures in closed systems are obvious and have been already discussed here. They neither take up arable land nor potable water resources; they can be located on marginal land as well as in buildings downtown used for vertical farming. This means an additional unforeseen advantage of closed systems since they help avoiding the 'palmoil-paradoxon', i.e., tropical deforestation in the name of climate protection.

As second-generation energy plants, microalgae represent an alternative energy source not producing netto CO_2. However, this is not a unique qualification. The culture of other second-generation energy plants such as *Miscanthus* also does not produce netto CO_2 provided that cultivation is done without fertilisers and mechanical processing is limited. Nonetheless, assessment of hazards of continuous culture such as plant diseases and exhaustion of nutrients in soil is still unsatisfactory.

On the political level, a decision has to be made to what extent we want to stay dependent on petrofuels that are neither available at a constant price nor accessible to everyone. For some countries, the return to mining coal could be an alternative. As to experts, coal reservoirs exceed by far oil and gas supplies, but it will need a lot of money to (re)activate mining industry and also to establish effective methods not only to catch the CO_2 but also to store it away from the biogeochemical cycle.

Another problem that has to be decided by politics is how long we want to keep on using combustion engines for transportation. Is there a realistic chance to replace them by motors driven by electric power, for example? Power plants do not necessarily need petrofuels or biofuels. They can also be driven by biomass. In the context of microalgal cultures, this could mean that production of combustible biomass becomes more important and more demanded by the market than special components in algae. It is also conceivable that microalgal cultures will become part of the trade of CO_2 certificates.

The amount of public funding often depends on the number of jobs created and on the influence of pressure groups. Since in most countries, farmers associations are much more influential than the lobby of people growing microalgae, currently public money is primarily spent to support farmers who grow energy plants such as rapeseed or *Miscanthus*. This has also psychological reasons, because most people are more familiar with flowering plants than with 'slimy' algae. As to the acquisition of venture capital, the situation may be somewhat different.

The plain fact remains that microalgae culture facilities will become a success story only when they deliver their products – be it energy or chemicals – at a competitive price. However, it should also be kept in mind that current major players in alternative energy production such as solar and wind energy are still not competitive and supported by public money. Microalgal technology merits the same chance.

6. Summary

Microalgae provide a lot of value products for human nutrition and health care. In recent times, they have been propagated as second-generation energy plants producing biofuels, such as ethanol, biogas and hydrogen, in particular, biodiesel of oil-containing specimen. Cultivation of algae under controlled conditions in photobioreactors guarantees a high and constant yield of biomass independent from ambient climatic conditions, quality of soil and water supply. For producing a given amount of biomass, microalgal cultures need about 1,000 times less water than crops and produce on the same area about 15 times more oil for biodiesel than, for example, rapeseed. Costs of biodiesel from microalgal cultures mainly depend on oil content of algae and on algal by-products such as cellulose that can also be exploited. A breakthrough of algal biodiesel on the market will depend on both the price of petrodiesel and the political will to support a promising source of alternative energy.

7. References

Ackermann, U. (2007) Konzeptpapier: Zukunftsworkshop Mikrotechniken für eine effizientere Bioenergieerzeugung. http://www.mstonline.de/mikrosystemtechnik/mst-fuer-energie/medien/konzeptpapier

Chisti, Y. (2007) Biodiesel from microalgae. Biotechnol. Adv. **25**: 294–306.

Dimitov, K. (2007) Greenfuel technologies: a case study for industrial photosynthetic energy capture. http://www.nanostring.net/Algae/Casestudy.pdf

Gantar, M. and Svircev, Z. (2008) Microalgae and cyanobacteria: food for thought. J. Phycol. **44**: 260–268.

Günzburg, B.Z. (1993) Liquid fuel (oil) from halophilic algae: a renewable source of non-polluting energy. Renew. Energ. **3**: 249–252.

Hu, Q., Sommerfeld, M., Jarvis, E., Ghirardi, M., Posewitz, M., Seibert, M. and Darzins, A. (2008) Microalgal triacylglycerols as feedstocks for biofuel production: perspectives and advances. Plant J. **54**: 621–639.

Patel-Predd, P. (2007) Hydrogen from algae. http://www.technologyreview.com

Pulz, O. (2001) Photobioreactors: production systems for phototrophic biosystems. Appl. Microbiol. Biotechnol. **57**: 287–293.

Schiermeier, Q., Tollefson, J., Scully, T., Witze, A. and Morton, O. (2008) Electricity without carbon. Nature **454**: 816–823.

Solarbiofuels (2008) http://www.solarbiofuels.org

Spolaore, P., Joannis-Cassan, C., Duran, E. and Isambert, A. (2006) Commercial applications of microalgae. J. Biosci. Bioeng. **101**: 87–96.

Biodata of **Levent Cavas** and **Georg Pohnert**, authors of *"The Potential of Caulerpa spp. for Biotechnological and Pharmacological Applications"*

Associate Professor Dr. Levent Cavas is currently the member of the Faculty of Arts and Sciences, Department of Chemistry–Biochemistry in the University of Dokuz Eylül, Turkey. He obtained his Ph.D. from the University of Dokuz Eylül in 2005 and continued his studies and research at the University of Dokuz Eylül. Dr. Cavas's scientific interests are in the areas of biochemistry of invasive *Caulerpa* spp. in the Mediterranean Sea, specifically secondary metabolites from marine algae, antioxidants and lipid peroxidation.

E-mail: **lcavas@deu.edu.tr**

Professor Dr. Georg Pohnert pursued his doctoral studies in the group of Professor W. Boland. After receiving his Ph.D. in 1997, he moved to the Cornell University for a postdoc on the biochemical and biophysical characterisation of the *E. coli* P-protein with a focus on the phenylalanine receptor site. He moved in 1998 to the Max–Planck-Institute for Chemical Ecology, where he worked as group leader on chemical defence mechanisms of algae. In 2005, he was appointed as assistant professor at the Institute of Chemical Sciences and Engineering of the Ecole Polytechnique Fédérale de Lausanne, Switzerland. He moved to the Friedrich–Schiller University in Jena, Germany, in 2007, where he was appointed as chair in Instrumental Analytics. Professor Dr. Pohnert's research interests are in the field of plankton chemical ecology and in the elucidation of induced and activated defence mechanisms of macroalgae.

E-mail: **Georg.Pohnert@uni-jena.de**

Levent Cavas   **Georg Pohnert**

THE POTENTIAL OF *CAULERPA* SPP. FOR BIOTECHNOLOGICAL AND PHARMACOLOGICAL APPLICATIONS

LEVENT CAVAS[1] AND GEORG POHNERT[2]

[1]*Division of Biochemistry, Department of Chemistry, Faculty of Arts and Sciences, Dokuz Eylül University, 35160, İzmir, Turkey*
[2]*Institute for Inorganic and Analytical Chemistry, Friedrich Schiller University Jena, D-07743, Jena, Germany*

1. Introduction

The genus *Caulerpa* belonging to the Bryopsidophyceae consists of about 75 species, which are distributed worldwide in tropical and temperate regions (Fama et al., 2002). *Caulerpa* spp. are siphonous green algae with a unique cellular organisation. Despite the fact that members of this genus can reach several meters in length, the organisms are unicellular with giant differentiated cells (Menzel, 1988).

In recent years, this genus has attracted much attention because of *Caulerpa taxifolia*, which was termed 'killer alga' because of its invasive potential (Meinesz and Simberloff, 1999; Meinesz et al., 1995). *C. taxifolia* was introduced into the Mediterranean Sea in the late 1980s and, within a few years, it spread rapidly and started to affect coasts of at least six Mediterranean countries (Thibaut et al., 2001; Jousson et al., 1998). In the invaded areas, dense patches of *C. taxifolia* cover the sea floors, which affects massively the local flora and fauna. Invasive specimens from the same clone were also reported from California and Australia (Jousson et al., 2000; Anderson, 2005). Many eradication methods such as covering *C. taxifolia* meadows with dark-coloured plastic foils, treatment with algicides, heavy metals, dry ice or chlorine bleach have been applied so far (Williams and Schroeder, 2004; Uchimura et al., 2000). But these methods were only successful if small local patches of *C. taxifolia* were treated. However, large-scale applications of any technical solution in the Mediterranean Sea seem not to be feasible owing to the massive coverage of *C. taxifolia*. The specialised sea slug, *Lobiger serradifalci*, was considered to be used in the biological war against *C. taxifolia*. However, it was understood from laboratory experiments that *L. serradifalci* can divide *C. taxifolia* into small viable fragments that can re-grow, thereby even supporting the rapid proliferation of the alga (Zuljevic et al., 2001).

Another widely recognised member of the genus *Caulerpa* is *Caulerpa racemosa* var. *cylindracea*. This species was first observed in 1926 in the Mediterranean Sea at Sousse harbour of Tunisia (Hamel, 1926). At that time, it was considered as a lessepsien migrant. *C. racemosa* did not show any invasive properties up to 1991. However, thereafter this species spread with massive growth rates and now it has been observed at the coastlines of 13 Mediterranean countries (Albania, Algeria, Croatia, Cyprus, France, Greece, Italy, Libya, Malta, Monaco, Spain, Tunisia, Turkey) (Verlaque et al., 2003, M. Verlaque, 2003).

It is evident that an eradication of invasive *Caulerpa* spp. in the Mediterranean Sea or Australia is impossible, since large areas all along the coasts are affected. Nevertheless, *Caulerpa* spp. contain unique natural products and are sources for crude algal preparations, extracts and secondary metabolites with interesting activities. We believe that a commercially motivated harvesting paired with political management of this new resource might offer a chance to control these algae. In this chapter, we introduce the dominant secondary metabolites from the algae and highlight potential biotechnological and pharmaceutical applications of products derived from *Caulerpa* spp., thereby providing an outline of potential commercially interesting applications.

2. The Dominant Secondary Metabolite of *Caulerpa* Genus: Caulerpenyne (CYN)

Caulerpa spp. contain several linear terpenoid secondary metabolites and especially caulerpenyne (CYN, Fig. 1) has been associated with anti-cancer, anti-proliferative, anti-microbial, anti-herpetic and anti-viral properties in many reports. Both invasive *Caulerpa* spp. contain the dominant sesquiterpenoid metabolite CYN, which can be found in high concentrations up to 1.3% of the dry weight of the alga (Amade and Lemée, 1998; Dumay et al., 2002). CYN contents of invasive and non-invasive *Caulerpa* species of the Mediterranean are similar, suggesting that this molecule is not the key for the success of rapidly spreading species but rather a wider distributed metabolite of *Caulerpa* spp. in general (Jung et al., 2002).

Purified CYN is a feeding inhibitor against sea urchins (Erickson et al., 2006). But in the natural context, this compound might be considered to be rather a storage form for more reactive metabolites, which are formed enzymatically after wounding the giant algal cells (Weissflog et al., 2008). CYN contains an unusual bis-enoyl acetoxy functionalisation that is transformed rapidly by esterases once the algal cells are disrupted. This enzymatic transformation results in the formation of the highly reactive 1,4-bis-aldehyde oxytoxin-2 (Jung and Pohnert, 2001). The transformation is relevant for the activated chemical

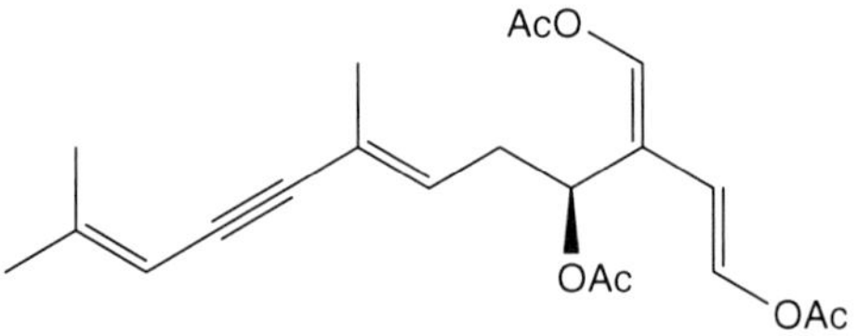

Figure 1. Caulerpenyne, the major terpenoid metabolite produced by *C. taxifolia* and *C. racemosa*.

defence and the survival of the unicellular macroalga after mechanical wounding (Adolph et al., 2005). Oxytoxin-2 is central for the rapid sealing of the giant cells after wounding, which enables survival and reproduction of *Caulerpa* spp. (Adolph et al., 2005). Once a cell is injured, a plug is formed around the wound by protein cross-linking with this activated secondary metabolite. The polymer material seals the cells and prevents leakage of the cytoplasm into the surrounding sea water (Dreher et al., 1982).

In general, terpenes from *Caulerpales* are known for their icthytoxic, anti-bacterial, anti-fungal and anti-proliferative properties (Paul and Fenical, 1986, 1987). Nevertheless, some specialised herbivores such as the sea slugs *Lobiger serradifalci* and *Oxynoe olivacea* have adapted to this defensive reaction and do not only consume *Caulerpa* spp. but also exploit their chemistry for their own chemical defence by sequestration (Gavagnin et al., 1994; Cimino et al., 1999).

3. Anti-cancer/Anti-tumour Effects

The first anti-cancer study using purified CYN was carried out by Fischell et al. (1995). They reported growth-inhibitory effects of CYN against eight cancer cell lines of human origin. Barbier et al. (2001) reported on the anti-proliferative activity of CYN on the tumour cell line SK-N-SH. They observed a loss of neuritis and a compaction of the microtubule network in response to low doses of CYN. IC_{50} values of CYN were similar to those of the well-established anti-cancer drug cisplatinum.

Anti-proliferative and apoptotic effects of both purified CYN and methanolic extracts of *C. racemosa* on the chemosensitive SH-SY-5Y and chemoresistant Kelly cell lines were reported by Cavas et al. (2006). IC_{50} values were obtained for SHSY5Y and Kelly cell lines as 0.59 ± 0.06, 1.06 ± 0.23 g wet weight alga/mL methanol and 5.64 ± 0.09, 6.02 ± 0.09 µM CYN, respectively.

Cytotoxic effect of a crude acetone extract of *C. racemosa* on a human melanoma cancer cell line was reported by Rocha et al. (2007). Ji et al. (2008) isolated polysaccharides from *C. racemosa* and showed their anti-tumour activity both *in vivo* and *in vitro* on K562 cells.

4. Enzyme Inhibition Studies

Kanagewa et al. (2000) reported telomerase inhibition by a methanolic extract of *C. sertularioides* on a MOLT-4 cell culture at a level of 1.25% (v/v).

A comprehensive screening for phospholipase A_2 (PLA_2) inhibitors from marine seaweeds revealed that CYN and structurally related metabolites from *C. prolifera*, *C. bikinensis* and *C. racemosa* are highly active (Mayer et al., 1993). CYN isolated from *C. prolifera* in 4.2 µM concentration lead to a 92% inhibition of PLA_2 activity. An esterase-based cleavage product (CYN lacking one acetyl group) isolated from *C. racemosa* showed 26% inhibition at 3.9 µM concentration for PLA_2. 12-Lipoxygenase inhibitory activity of *C. taxifolia* extracts was reported by Nimoniya et al. (1998).

Bitou et al. (1999) showed complete inhibition of lipase by purified CYN. According to their research, 50% inhibition was observed against triolein and 4-methylumbelliferyl oleate hydrolysis in 2 mM and 13 µM concentrations, respectively. Ben Rebah et al. (2008) showed that *C. prolifera* extract significantly decreased both dog gastric and human pancreatic lipase activities. Therefore, they proposed *C. prolifera* extract for the development of anti-obesity drugs.

Mao et al. (2006) discovered the two novel sesquiterpenoids, caulerpal A and B, from *C. taxifolia*. These two compounds have inhibitory property on the human protein tyrosine kinase. This enzyme is a target enzyme in the treatments of type-2 diabetes and obesity.

Inhibition of α-amylase by acetone extract of *C. racemosa* with an IC_{50} 0.09 mg/mL value has been shown by Teixeira et al. (2007).

5. Anti-viral/Anti-microbial Studies

A considerable inhibitory effect of acetone and ethanol extracts of *C. taxifolia* against the bacterium *Bacillus subtilis* and the yeast *Candida albicans* were observed by Crasta et al. (1997). Anti-bacterial activity in marine algae from the coast of Yucatan, Mexico, was investigated by Freile-Pelegrin and Morales (2004). After dissection of *Caulerpa* spp. in apical, basal and stolon regions and extraction, they observed that stolons of *Caulerpa* spp. exhibit the highest anti-bacterial activity. *In vitro* anti-bacterial activities of *C. racemosa* extracts were shown against both Gram-negative and Gram-positive pathogenic bacteria by Kandhasamy and Arunachalam (2008).

Anti-viral properties of a chloroform-methanolic extract of *C. taxifolia* were studied by Nicoletti et al. (1999). The inhibition of viral transcriptase enzyme activity and reduction of viral capsid protein p24 expression of feline immunodeficiency virus by the extracts were observed.

Potent anti-HSV-1 activities from *Caulerpa* spp. were reported by Lee et al. (2004). Ghosh et al. (2004) exhibited *in vitro* anti-herpetic activity of sulphated polysaccharide fractions from *C. racemosa* on TK- acyclovir-resistant strains of herpes simplex virus type I (HSV-1) and type 2 (HSV-2) in Vero cells without any cytotoxic effects. They also observed no anti-coagulant properties at the concentration of the IC_{50} value.

Anti-leishmanial activity in the crude extract of *C. racemosa* and *Caulerpa faridii* was reported by Sabina et al. (2005).

According to Wang et al. (2007), hot water extracts and the *n*-butanol fraction from *C. racemosa* exhibit strong inhibition against herpes simplex virus type 1 (HSV-1) and Coxsackie virus B3 (Cox B3). A sulphoquinovosyldiacylglycerol (SQDG) could be made responsible for this activity.

6. *Caulerpa* Species as Adsorbents for Pollutants

The high adsorption properties of preparations from *Caulerpa* species on dyes and heavy metals have recently attracted increasing interest. *Caulerpa* species are proposed as low-cost adsorbents in these studies. Several physicochemical characterisations such as determination of adsorption kinetics, equilibrium constants and thermodynamics of the adsorption process have been reported. Generally, pseudo-first-order and pseudo-second-order kinetic models are used to express adsorption kinetics. Langmuir, Freundlich and Dubinin–Radushkevich models are the most used isotherm models.

The first study on adsorptive properties of powdered *Caulerpa lentillifera* was carried out by Marungrueng and Pavasant (2006). The basic dye astrazon blue FGRL was adsorbed according to a pseudo-second-order model. The chemisorption occurs with a positive enthalpy value (Marungrueng and Pavasant, 2006). In a comparative investigation of adsorption properties of *C. lentillifera* and activated carbon, Marungrueng and Pavasant (2007) showed that the algal powder has about 1.75-fold higher adsorption capacity (417.19 mg/g) for methylene blue compared to activated carbon (238.12 mg/g).

Pavasant et al. (2006) investigated the affinity of dried and powdered *C. lentillifera* on metal ions such as Cu^{2+}, Cd^{2+}, Pb^{2+} and Zn^{2+}. FT-IR analysis allowed the identification of the functional groups which are associated with the metal adsorption (O–H bending, N–H bending, N–H stretching, C–N stretching, C–O stretching, S=O stretching and S–O stretching). A low equilibrium time of 10–20 min was required for the adsorption of Cu^{2+}, Cd^{2+}, Pb^{2+} and Zn^{2+} onto *C. lentillifera*. The adsorption process obeyed the Langmuir isotherm; the maximum sorption capacities were confirmed in the order $Pb^{2+} > Cu^{2+} > Cd^{2+} > Zn^{2+}$.

Sorption of copper, cadmium and lead ions in a binary component system prepared from the same macroalga was studied by Apiratikul and Pavasant (2006). The presence of a second ion reduces the adsorption capacity of the algal preparations. Pb^{2+} was the best sorbed species, followed by Cu^{2+} and Cd^{2+}.

Apiratikul and Pavasant (2008) investigated the batch and column biosorption of the above-mentioned heavy metals by using preparations of the same macroalga *C. lentillifera*. The authors gave column properties and performances (effects of flow rate and bed depth) related to metal ions biosorption. The release of Ca(II), Mg(II) and Mn(II) from algal residue during the sorption was observed and it was proposed that column systems could be employed for the remediation of wastewaters.

Aravindhan et al. (2007a) studied biosorption of basic blue (Sandocryl blue C-RL), which is used in leather industry, onto *Caulerpa scalpelliformis*. They developed a two-stage treatment with this seaweed to obtain complete decolorisation. Aravindhan et al. (2007b) in another study investigated the factors that affected sorption of yellow dye (Sandocryl golden yellow C-2G) onto *C. scalpelliformis*. They reported a maximum uptake of 27 mg of dye/g seaweed.

Cengiz and Cavas (2008) studied methylene blue adsorption onto invasive *C. racemosa* and proposed an alternative use of biomass of this alga after possible eradication efforts that could be based on manual uprooting.

Punjongharn et al. (2008) studied the influence of particle size and salinity on the biosorption of basic dyes by *C. lentillifera*. The authors prepared three powders from *C. lentillifera* with average grain sizes of *S* (0.1–0.84 mm), *M* (0.84–2.0 mm) and *L* (larger than 2.0 mm). They found that *S*- algal powder had the highest adsorption capacity and increased salinity caused the decreased adsorption capacity.

C. racemosa powder was also used to remove malachite green from aqueous solutions by a physisorption process (Bekci et al., 2009). In another study, the same group showed the affinity of *C. racemosa* powder on boron in the form of borate. They proposed the removal of boron with dried and powdered biomass of *C. racemosa* (Ant-Bursali et al., 2009).

7. Other Activities of *Caulerpa* Preparations

The non-specific agglutination of human red blood cells (haemagglutinating activity) was reported in extracts of *Caulerpa cupressoides* by Ainouz and Sampaio (1991). Thangam and Kathiresan (1991a) reported the mosquito larvicidal effect of *C. scalpelliformis* extracts against *Aedes aegypti* with an LC_{50} value of 54 mg/L. The same researchers also reported the synergetic effect of *C. scalpelliformis* extract with the synthetic insecticide 1,2,3,4,5,6-hexachlorocyclohexane (BHC) (Thangam and Kathiresan, 1991b).

Shen et al. (2008) investigated the immunomodulatory activity of a modified polysaccharide (obtained from *C. racemosa* var. *peltata*) called CrvpPS-nano-Se complex. After intragastric administration, it has been shown that CrvpPS-nano-Se induced the percentage of CD3+, CD3+CD4+, NK cells and the CD4+/CD8+ value.

Purification and partial characterisation of a lectin from *C. cupressoides* was pursued by Benevides et al. (2001). This lectin agglutinated trypsin-treated erythrocytes from humans and various animals. The haemagglutination activity was more effective for human blood group A erythrocytes compared with B and 0 erythrocytes and rabbit erythrocytes.

Sixteen species of Indian marine green algae were screened for their heparinoid-active sulphated polysaccharides by Shanmugam et al. (2001). *Caulerpa* species showed the highest activity, which is comparable with heparin. According to this report, *Caulerpales* contained 93–151 heparin units/mg.

Ara et al. (2002) showed that ethanolic extracts of *C. racemosa* exhibited hypolipidaemic activity and significantly decreased the serum total cholesterol and triglyceride levels in rats. Santoso et al. (2004) investigated the anti-oxidant activity of methanol extracts from Indonesian seaweeds. Among these seaweeds, it is reported that the extract from *Caulerpa sertularioides* had the strongest anti-oxidant activity.

Chemopreventive effect of *C. prolifera* extract against aflatoxin B1-initiated hepatotoxicity in female rats was investigated by Abdel-Wahhab et al. (2006).

Sivasankari et al. (2006) proposed *Caulerpa chemnitzia* as a seaweed liquid fertiliser, since its positive effects on growth and biochemical constituents of *Vigna sinensis* were observed. Cavas et al. (2007) confirmed that *C. racemosa* could also be used as a biostimulator for the growth of *Phaselus vulgaris* seedlings. It might be speculated that this effect is due to the pigment caulerpin, which has been previously identified from *Caulerpa* spp. (Maiti et al., 1978). Root growth assays with pure caulerpin gave essentially the same results as assays with the known plant growth promotor indole-3-acetic acid (IAA). The pigment resembles structurally this plant hormone and might thus mimic its activity (Raub et al., 1987).

In vitro nematicidal activities from *C. racemosa*, *C. scalpelliformis* and *C. taxifolia* were reported by Rizvi and Shameel (2006).

Paul and de Nys (2008) studied the possible use of *Caulerpa* species bioremediation in an integrated plant–animal tropical aquaculture. Marine macroalgae with high growth rates can effectively strip nutrients from marine aquaculture effluent. Because *Caulerpa* spp., as bloom-forming green tide algae, have high growth rates and are free floating, these seaweeds offer to be an excellent option for culture in settlement ponds, the most common bioremediation infrastructure in tropical aquaculture. Especially, since certain isolates of *C. racemosa* are consumed by humans as 'sea grape', it can be speculated that one aquaculture might provide two commercially interesting products – fish and edible algae. This would maximise profit and simultaneously reduce pollutants (Paul and de Nys, 2008).

The radical-scavenging and reducing ability of *C. lentillifera* and *C. racemosa* extracts were showed by Matanjun et al. (2008).

According to a recent study (Cengiz et al., 2008), a dried and powdered form of *C. racemosa* was proposed to be used as a low-cost immobilisation agent for bovine serum albumin, which is a model protein for protein immobilisation studies.

8. Patents

Several patents on *Caulerpa* preparations, extracts and purified metabolites have been approved, illustrating the high potential of these green seaweeds as resources for commercially attractive products. These include patents on cosmetics like hair treatment agents, plant growth regulators or medicinal applications. Here, for example, treatments of inflammation, diabetes or retardation of cardiovascular disorders based on *Caulerpa* spp.-derived products have been claimed.

9. Conclusion

Exotic *Caulerpa* species, *C. taxifolia* and *C. racemosa*, have spread rapidly in the Mediterranean Sea. So far, no valid eradication method has been developed and the amount of biomass currently found in the Mediterranean suggests that a mechanical eradication is out of reach. The extensive body of literature reviewed here makes it evident that these species could be exploited in many biotechnological and medical purposes. Exploitation of this resource might be interesting owing to the apparent nearly unlimited supply. A targeted economic development of these *Caulerpa* species from the Mediterranean might thus lead to a control of the spread.

10. Summary

The genus *Caulerpa* contains over 75 marine green algal species found in temperate and tropical waters. This genus has recently attracted much attention because of two invasive members, *Caulerpa taxifolia* and *C. racemosa* var. *cylindracea*, which occur in the Mediterranean Sea. These species have covered the sub-littoral habitats along vast coastal stretches of the Mediterranean Sea. Several often costly and unsuccessful efforts have been undertaken to control this invasion by mechanical or chemical eradication. In parallel, scientists have focused on exploitable traits that can be derived from these algae. Commercial products from these algae might motivate a control of the invasion by harvesting efforts. In this chapter, we focus on the properties of crude extracts as well as purified metabolites of *Caulerpa* sp. and their potential for commercial exploitation.

11. References

Abdel-Wahhab, M.A,, Ahmed, H.H. and Hagazi, M.M. (2006) Prevention of aflatoxin B-1-initiated hepatotoxicity in rat by marine algae extracts. J. Appl. Toxicol. **26**: 229–238.

Adolph, S., Jung, V., Rattke, J. and Pohnert, G. (2005) Wound closure in the invasive green alga *Caulerpa taxifolia* by enzymatic activation of a protein cross-linker. Angew. Chem. Int. Ed. **44**: 2806–2808.

Ainouz, I.L. and Sampaio, A.H. (1991) Screening of Brazilian marine-algae for Hemagglutinins. Bot. Mar. **34**: 211–214.

Amade, P. and Lemée, R. (1998) Chemical defence of the Mediterranean alga *Caulerpa taxifolia*: variations in Caulerpenyne production. Aquat. Toxicol. **43**: 287–300.

Anderson, L.W. (2005) California's reaction to *Caulerpa taxifolia*: a model for invasive species rapid response. Biol. Invasions **7**: 1003–1016.

Ant-Bursali, E., Cavas, L., Seki, Y., Seyhan, S. and Yurdakoc, M. (2009) Sorption of boron by invasive marine seaweed: *Caulerpa racemosa* var. *cylindracea*. Chem. Eng. J. **150**: 385–390.

Apiratikul, R. and Pavasant, P. (2006) Sorption isotherm model for binary component sorption of copper, cadmium, and lead ions using dried green macroalga, *Caulerpa lentillifera*. Chem. Eng. J. **119**: 135–145.

Apiratikul, R. and Pavasant, P. (2008) Batch and column studies of biosorption of heavy metals by *Caulerpa lentillifera*. Biores. Tech. **99**: 2766–2777.

Ara, J., Sultana, V., Qasim, R. and Ahmad, V.U. (2002) Hypolipidaemic activity of seaweed from Karachi coast. Phytother. Res. **16**: 479–483.

Aravindhan, R., Rao, J.R. and Nair, B.U. (2007a) Kinetic and equilibrium studies on biosorption of basic blue dye by green macro algae *Caulerpa scalpelliformis*. J. Environ. Sci. Health Part A-Toxic/Hazard. Subst. Environ. Eng. **42**: 621–631.

Aravindhan, R., Rao, J.R. and Nair, B.U. (2007b) Removal of basic yellow dye from aqueous solution by sorption on green alga *Caulerpa scalpelliformis*. J. Hazard. Mater. **142**: 68–76.
Barbier, P., Guise, S., Huitorel, P., Amade, P., Pesando, D., Briand, C. and Peyrot, V. (2001) Caulerpenyne from *Caulerpa taxifolia* has an antiproliferative activity on tumor cell line SK-N-SH and modifies the microtubule network. Life Sci. **70**: 415–429.
Benevides, N.M.B., Holanda, M.L., Melo, F.R., Pereira, M.G., Monteiro, A.C.O. and Freitas, A.L.P. (2001) Purification and partial characterization of the lectin from the marine green alga *Caulerpa cupressoides* (Vahl) C. Agardh. Bot. Mar. **44**: 17–22.
Bekci, Z., Seki, Y. and Cavas, L. (2009) Removal of malachite green by using an invasive marine alga *Caulerpa racemosa* var. *cylindracea*. J. Hazard. Mater. **161**: 1454–1460.
Ben Rebah, F., Smaoui, S., Frikha, F., Gargouri, Y. and Miled, N. (2008) Inhibitory effects of Tunisian marine algal extracts on digestive lipases. Appl. Biochem. Biotechnol. **151**: 71–79.
Bitou, N., Ninomiya, M., Tsujita, T. and Okuda, H. (1999) Screening of lipase inhibitors from marine algae. Lipids **34**: 441–445.
Cavas, L., Baskin, Y., Yurdakoc, K. and Olgun, N. (2006) Antiproliferative and newly-attributed apoptotic activities from a marina alga: *Caulerpa racemosa* var. *cylindracea*. J. Exp. Mar. Biol. Ecol. **339**: 111–119.
Cavas, L., Kandemir-Cavas, C. and Alyuruk, H. (2007) *Caulerpa racemosa* var.*cylindracea* ve *Dictyota dichotoma*'dan elde edilen sıvı gübrelerin *Phaselus vulgaris* üzerine etkileri. 11. Sualti Bilim ve Teknolojisi Toplantisi, Koç Üniversitesi, 3–4 Kasım 2007, Istanbul, Turkey.
Cengiz, S. and Cavas, L. (2008) Removal of methylene blue by invasive marine seaweed: *Caulerpa racemosa* var. *cylindracea*. Biores. Technol. **99**: 2357–2363.
Cengiz, S., Cavas, L. and Yurdakoc, K. (2008) A low cost immobilization agent from an invasive marine alga: *Caulerpa racemosa* var. *cylindracea* for bovine serum albumine. Turk. J. Biochem. **33**: 64–70.
Cimino, G., Fontana, A. and Gavagnin, M. (1999) Marine Opisthobranch Molluscs: chemistry and ecology in Sacoglossans and Dorids. Curr. Org. Chem. **3**: 327–372.
Crasta, P.J., Raviraja, N.S. and Sridhar, K.R. (1997) Antimicrobial activity of some marine algae of southwest coast of India. Ind. J. Mar. Sci. **26**: 201–205.
Dreher, T.W., Hawthorne, D.B. and Grant, B.R. (1982) The wound response of the siphonous green algal genus *Caulerpa* III: composition and origin of the wound plugs. Protoplasma **110**: 129–137.
Dumay, O., Pergent, G., Pergent-Martini, C. and Amade, P. (2002) Variations in Caulerpenyne contents in *Caulerpa taxifolia* and *Caulerpa racemosa*. J. Chem. Ecol. **28**: 343–352.
Erickson, A.A., Paul, V.J., Van Alstyne, K.L. and Kwiatkowski, L.M. (2006) Palatability of macroalgae that use different types of chemical defenses. J. Chem. Ecol. **32**: 1883–1895.
Fama, P., Wysor, B., Koosistra, W.H. and Zuccarello, G.C. (2002) Molecular phylogeny of the genus Caulerpa (Caulerpales Chlorophyta) inferred from chloroplast tufA gene. J. Phycol. **38**: 1040–1050.
Fischel, J.L., Lemee, R., Formento, P., Caldani, C., Moll, J.L., Pesando, D., Meinesz, A., Grelier, P., Pietra, F., Guerriero, A. and Milano, G. (1995) Cell growth inhibitory effects of caulerpenyne, a sesquiterpenoid from the marine algae *Caulerpa taxifolia*. Anticancer Res. **15**: 2155–2160.
Freile-Pelegrin, Y. and Morales, J.L. (2004) Antibacterial activity in marine algae from the coast of Yucatan, Mexico. Bot. Mar. **47**: 140–146.
Gavagnin, M., Marin, A., Castellucio, F., Villani, G. and Cimino, G. (1994) Defensive relationships between *Caulerpa prolifera* and its shelled sacoglossan predators. J. Exp. Mar. Biol. Ecol. **175**: 197–219.
Ghosh, P., Adhikari, U., Ghosal, P.K., Pujol, C.A., Carlucci, M.J., Damonte, E.B. and Ray, B. (2004) In vitro anti-herpetic activity of sulfated polysaccharide fractions from *Caulerpa racemosa*. Phytochemistry **65**: 3151–3157.
Hamel, G. (1926) Quelques algues rares ou nouvelles pour la floreme´diterrane´enne. Bull. Mus. Natl. Hist. Nat. **32**: 420.
Ji, H.W., Shao, H.Y., Zhang, C.H., Hong, P. and Xiong, H. (2008) Separation of the polysaccharides in *Caulerpa racemosa* and their chemical composition and antitumor activity. J. Appl. Polym. Sci. **110**: 1435–1440.

Jousson, O., Pawlowski, J., Zaninetti, L., Meinesz, A. and Boudouresque, C.F. (1998) Molecular evidence for the aquarium origin of the green alga *Caulerpa taxifolia* introduced to the Mediterranean Sea. Mar. Ecol. Prog. Ser. **172**: 275–280.

Jousson, O., Pawlowski, J., Zaninetti, L., Zechman, F.W., Dini, F., Di Guiseppe, G., Woodfield, R., Millar, A. and Meinesz, A. (2000) Invasive alga reaches California. Nature **408**: 157–158.

Jung, V., Thibaut, T., Meinesz, A. and Pohnert, G. (2002) Comparison of the wound-activated transformation of Caulerpenyne by invasive and noninvasive *Caulerpa* species of the Mediterranean. J. Chem. Ecol. **28**: 2091–2105.

Jung, V. and Pohnert, G. (2001) Rapid wound-activated transformation of the green algal defensive metabolite caulerpenyne. Tetrahedron **57**: 7169–7172.

Kandhasamy, M. and Arunachalam, K.D. (2008) Evaluation of in vitro antibacterial property of seaweeds of southeast coast of India. Afr. J. Biotechnol. **7**: 1958–1961.

Kanegawa, K., Harada, H., Myouga, H., Katakura, Y., Shirahata, S. and Kamei, Y. (2000) Telomerase inhibiting activity in vitro from natural resources, marine algae extracts. Cytotechnology **33**: 221–227.

Lee, J.B., Hayashi, K., Maeda, M. and Hayashi, T. (2004) Antiherpetic activities of sulfated polysaccharides from green algae. Planta Med. **70**: 813–817.

Maiti, B.C., Thomson, R.H. and Mahendran, M. (1978) Structure of caulerpin, a pigment from *Caulerpa* algae. J.Chem. Res.-S. **4**: 126–127.

Mao, S.C., Guo, Y.W. and Shen, X. (2006) Two novel aromatic valerenane-type sesquiterpenes from the Chinese green alga *Caulerpa taxifolia*. Bioorg. Med. Chem. Lett. **16**: 2947–2950.

Marungrueng, K. and Pavasant, P. (2006) Removal of basic dye (Astrazon Blue FGRL) using macroalga *Caulerpa lentillifera*. J. Environ. Manag. **78**: 268–274.

Marungrueng, K. and Pavasant, P. (2007) High performance biosorbent (*Caulerpa lentillifera*) for basic dye removal. Biores. Technol. **98**: 1567–1572.

Matanjun, P., Mohamed, S., Mustapha, N.M., Muhammad, K. and Ming, C.H. (2008) Antioxidant activities and phenolics content of eight species of seaweeds from north Borneo. J. Appl. Phycol. **20**: 367–373.

Mayer, A.M.S., Paul, V.J., Fenical, W., Norris, J.N., Carvalho, M.S. and Jacobs, R.S. (1993) Phospholipase A_2 inhibitors from marine algae. Hydrobiologia **260/261**: 521–529.

Meinesz, A., Benichou, L., Blachier, J., Komatsu, T., Lemée, R., Molenaar, H. and Mari, X. (1995) Variations in the structure, morphology and biomass of *Caulerpa taxifolia* in the Mediterranean Sea. Bot. Mar. **38**: 499–508.

Meinesz, A. and Simberloff, D. (1999) *Killer Algae,* University of Chicago Press, Chicago, IL.

Menzel, D. (1988) How do giant plant cells cope with injury? The wound response in siphonous green alga. Protoplasma **144**: 73–91.

Nicoletti, E., Della Pieta, F., Calderone, V., Bandecchi, P., Pistello, M., Morelli, I. and Cinelli, F. (1999) Antiviral properties of a crude extract from a green alga *Caulerpa taxifolia* (Vahl) C-Agardh. Phytother. Res. **13**: 245–247.

Ninomiya, M., Onishi, J. and Kusumi, T. (1998) 12-Lipoxygenase inhibitory activity of Japanese seaweeds and isolation of a Caulerpenyne derivative from the green alga *Caulerpa taxifolia* as an inhibitor. Fish. Sci. **64**: 346–347.

Paul, V.J. and Fenical, W. (1986) Chemical defense in tropical green algae, order Caulerpales. Mar. Ecol. Prog. Ser. **34**: 157–169.

Paul, V.J. and Fenical, W. (1987) Natural products chemistry and chemical defence in tropical marine algae of the phylum chlorophyta, In: P.J. Scheuer (ed.) *Bioorganic Marine Chemistry.* Springer-Verlag, Berlin, pp. 1–29.

Paul, N.A. and de Nys, R. (2008) Promise and pitfalls of locally abundant seaweeds as biofilters for integrated aquaculture. Aquaculture **281**: 49–55.

Pavasant, P., Apiratikul, R., Sungkhum, V., Suthiparinyanont, P., Wattanachira, S. and Marhaba, T.F. (2006) Biosorption of Cu^{2+}, Cd^{2+}, Pb^{2+}, and Zn^{2+} using dried marine green macroalga *Caulerpa lentillifera*. Biores. Technol. **97**: 2321–2329.

Punjongharn, P., Meevasana, K. and Pavasant, P. (2008) Influence of particle size and salinity on adsorption of basic dyes by agricultural waste: dried Seagrape (*Caulerpa lentillifera*). J. Environ. Sci.-China. **20**: 760–768.

Raub, M.F., Cardellina, J.H. and Schwede, J.G. (1987) The green algal pigment Caulerpin as a plant-growth regulator. Phytochemistry **26**: 619–620.

Rizvi, M.A. and Shameel, M. (2006) *In vitro* nematicidal activities of seaweed extracts from Karachi coast. Pak. J. Bot. **38**: 1245–1248.

Rocha, F.D., Soares, A.R., Houghton, P.J., Pereira, R.C., Kaplan, M.A.C. and Teixeira, V.L. (2007) Potential cytotoxic activity of some Brazilian seaweeds on human melanoma cells. Phytother. Res. **21**: 170–175.

Sabina, H., Tasneem, S., Sambreen Kausar, Y., Choudhary, M.I. and Aliya, R. (2005) Antileishmanial activity in the crude extract of various seaweed from the coast of Karachi, Pakistan. Pak. J. Bot. **37**: 163–168.

Santoso, J., Yoshie-Stark, Y. and Suzuki, T. (2004) Anti-oxidant activity of methanol extracts from Indonesian seaweeds in an oil emulsion model. Fish. Sci. **70**: 183–188.

Sivasankari, S., Venkatesalu, V., Anantharaj, M. and Chandrasekaran, M. (2006) Effect of seaweed extracts on the growth and biochemical constituents of *Vigna sinensis*. Biores. Technol. **97**: 1745–1751.

Shanmugam, M., Ramavat, B.K., Mody, K.H., Oza, R.M. and Tewari, A. (2001) Distribution of heparinoid-active sulphated polysaccharides in some Indian marine green algae. Ind. J. Mar. Sci. **30**: 222–227.

Shen, W.Z., Wang, H., Guo, G.Q. and Tuo, J. (2008) Immunomodulatory effects of *Caulerpa racemosa* var. *peltata* polysaccharide and its selenizing product on T lymphocytes and NK cells in mice. Sci. China Ser. C-Life Sci. **51**: 795–801.

Teixeira, V.L., Rocha, F.D., Houghton, P.J., Kaplan, M.A.C. and Pereira, R.C. (2007) Alpha-amylase inhibitors from Brazilian seaweeds and their hypoglycemic potential. Fitoterapia **78**: 35–36.

Thangam, T.S. and Kathiresan, K. (1991a) Mosquito larvicidal effect of seaweed extracts. Bot. Mar. **34**: 433–435.

Thangam, T.S. and Kathiresan, K. (1991b) Mosquito larvicidal activity of marine plant-extracts with synthetic insecticides. Bot. Mar. **34**: 537–539.

Thibaut, T., Meinesz, A., Amade, P., Charrier, S., Angelis, K.D., Ierardi, S., Mangialajo, L., Melnick, J. and Vidal, V. (2001) *Elysia subornata* (Mollusca) a potential control agent of the Alga *Caulerpa taxifolia* in the Mediterranean Sea. J. Mar. Biol. Assoc. UK **81**: 497–504.

Uchimura, M., Rival, A., Nato, A., Sandeaux, R., Sandeaux, J. and Baccou, J.C. (2000) Potential use of Cu^{2+}, K^+ and Na^+ for the destruction of *Caulerpa taxifolia*: differential effects on photosynthetic parameters. J. Appl. Phycol. **12**: 15–23.

Verlaque, M., Durand, C., Huisman, J.M., Boudouresque, C.F. and Parco, Y. (2003) On the identity and origin of the Mediterranean invasive *Caulerpa racemosa* (Caulerpales, Chlorophyta). Eur. J. Phycol. **38**: 325–339.

Wang, H., Li, Y.L., Shen, W.Z., Rui, W., Ma, X.J. and Cen, Y.Z. (2007) Antiviral activity of a sulfoquinovosyldiacylglycerol (SQDG) compound isolated from the green alga *Caulerpa racemosa*. Bot. Mar. **50**: 185–190.

Weissflog, J., Adolph, S., Wiesemeier, T. and Pohnert, G. (2008) Reduction of herbivory through wound-activated protein cross-linking by the invasive macroalga *Caulerpa taxifolia*. Chembiochemistry **4**: 29–32.

Williams, S.L. and Schroeder, S.L. (2004) Eradication of the invasive seaweed *Caulerpa taxifolia* by chlorine bleach. Mar. Ecol. Prog. Ser. **27**: 69–76.

Žuljevic, A., Thibaut, T., Elloukal, H. and Meinesz, A. (2001) Sea slug disperses the invasive *Caulerpa taxifolia*. J. Mar. Biol. Assoc. UK **81**: 343–344.

PART 8:
OTHER VIEWS TO GLOBAL CHANGE

**Klostermaier
Josef Roth
Glicksberg
Rozenson**

Biodata of **Klaus Konrad Klostermaier**, author of *"Ecology, Science and Religion"*

Klaus Konrad Klostermaier, F.R.S.C., is University Distinguished Professor Emeritus (1999) at the University of Manitoba. He obtained a Dr. Phil. in Philosophy in 1961 from the Gregorian University, Rome, and a Ph.D. in Ancient Indian History and Culture from the University of Bombay (now Mumbai) in 1969. He joined the Department of Religion at the University of Manitoba (Canada) in 1970 and held the headship from 1986 to 1997. His areas of research and teaching are history of religions, especially Indian religions, and science and religion. Among his major publications are *Mythologies and Philosophies of Salvation in the Theistic Traditions of India*, Wilfrid Laurier University Press (1984), *A Survey of Hinduism*, State University of New York Press (1989, 1994, and 2007), *Buddhism: A Short Introduction,* Oneworld Oxford (1999 and 2001), *Hinduism: A Short History*, Oneworld Oxford (2000), *The Nature of Nature: Explorations in Science, Philosophy and Religion*, Theosophical Publishing House Adyar, Madras 2004. He continues to write and to teach at the University of Manitoba.

E-mail: **kklostr@cc.umanitoba.ca**

ECOLOGY, SCIENCE, AND RELIGION

KLAUS K. KLOSTERMAIER
*Department of Religion, University of Manitoba, Winnipeg,
MB, R3T 2N2, Canada*

1. Religious Roots of the Ecological Crisis?

No less an authority than world historian Arnold J. Toynbee identified in a paper in the *International Journal of Environmental Studies* "monotheism" as the root cause of our environmental crisis. Referring specifically to *Genesis* I, 28 ["Be fruitful, multiply, fill the earth and conquer it. Be masters of the fish in the sea, the bird of heaven and all living animals on earth" (The Jerusalem Bible, p. 6)]. Toynbee not only questioned the right of humans to use and abuse the earth but also challenged the authority of the one who supposedly had given this command, asking: "Has nature no rights against this autocratic creator and against man, God's aggressive licensee?" Recalling ancient European traditions of nature worship, he concluded: "Monotheism, as enunciated in the Book of Genesis, has removed the age-old restraint that was once placed on man's greed by his awe. This primitive inhibition has been removed by the rise and spread of monotheism." Eastern religions, like the pre-Christian European, "counsel man even when he is applying his human science to coax nature into bestowing her bounty on man" (Toynbee, 1971, p. 141).

Toynbee's charge expanded and sharpened a thesis proposed in 1967 by Lynn White Jr., a historian of mediaeval technology. White had called Christianity "the most anthropocentric religion the world has ever seen," through whose influence "the old inhibitions to the exploitation of nature crumbled" and suggested – paradoxically – to name Francis of Assisi the patron saint of environmentalists (White, 1967).

In response to accusations like these, a great number of monographs and learned papers were published, not only defending "monotheism" by offering alternative interpretations of *Genesis* I, 28, but also highlighting the ecological potential of the world's religions. A series of major conferences were convened at Harvard University in the 1990s, whose proceedings appeared in a series of ten impressive volumes published by Harvard University Press under the title *Religions of the World and Ecology*, covering virtually every living religion from tribal traditions to Buddhism, Hinduism, Jainism, Taoism, Shinto, Confucianism, Judaism, Christianity, and Islam. Out of these conferences developed the *International Forum on Religion and Ecology*, co-chaired by John Grim and Mary Evelyn Tucker, editors of several of the volumes in the series.

Harold Coward, Director of the Center for Studies in Religion and Society at Victoria University, B.C., arranged in 1993 a high-powered workshop at Chateau Whistler, B.C., where religion scholars interacted with specialists from various sciences and economists to address questions of (over-)population, resource consumption, and the environment (Coward, 1995).

The environment was also a central topic at the 1993 meeting of the World Parliament of Religions in 1993, where Hans Küng tabled a document on Global Ethic and Responsibility (Küng, 1991).

As far as literature is concerned, the Internet offers numerous bibliographies on Religion and Ecology in general as well as more specific ones, such as *Hinduism and the Environment* (Noyce, 2002). *The Oxford Handbook of Religion and Ecology* (Gottlieb, 2006) is a useful one-volume reference work for the field.

If Toynbee and White had identified the Abrahamic religions as the root of the world's ecological crisis, Anil Agarwal, an Indian engineer turned journalist and ecological activist, made Hinduism, India's majority tradition, responsible for India's ecological malaise. "Hinduism," he says, "is a highly individualistic religion: the primary concern is to do one's own *dharma* for the sake of one's own well-being. Under the onslaught of modern-day secularism this has brought out the worst type of individualism in Hinduism" (Agarwal, 2000, p. 165).

Going one step further, W. Ophuls sees the ecological crisis as "primarily a moral crisis in which the ugliness and destruction outside in our environment simply mirror the spiritual wasteland within." And "the sickness of the earth reflects the sickness of the soul in modern industrial man, whose life is given over to gain, to the disease of endless getting and spending" (Ophuls, 1992).

2. The Ambiguity of Religions

David Kinsley's *Ecology and Religion* – a widely used text for university courses – juxtaposes chapters on "Christianity as Ecologically Harmful" and "Christianity as Ecologically Responsible," illustrating the ambiguous record of Christianity with regard to its attitude toward nature: on the one side extolling the greatness of nature as God's handiwork and on the other condemning nature as the source of humankind's downfall. A similar ambiguity can also be found in all other major traditions. None of the ancient religions directly addressed ecological issues or the need to protect the natural environment from human interference: all of them were built around other core concerns.

The "Abrahamic religions" are focused on God and salvation, sin and atonement. The Decalogue, the source of all Judeo–Christian ethics, is concerned with the majesty of God and interhuman relations: it does not contain any "ecological commandment." The Hebrew Bible condemns the worship of nature deities practiced by the people of Canaan. Paul, the most influential voice in early Christianity, held the whole of nature mortally afflicted with Adam's "original sin," and "groaning to be redeemed" by Christ. The writers of the Gospels saw the proof of Jesus' divinity in the performance of miracles that violated the general laws of nature. The Koran teaches the absolute Lordship of the Creator who has installed humans as "vice-gerents" of His creation.

Numerous historic examples can be found where followers of these religions tried to prove the superiority of their faith over that of "nonbelievers" through violent acts against nature, such as felling the sacred trees of "heathens" or by devastating the natural habitats of their spirits with impunity. *Agere contra naturam* became one of the principles of Christian moral teaching: obedience toward authorities was regarded as the highest virtue, proved through such tests as executing the commands of a religious superior to stick vegetable seedlings head first into the earth.

Eastern religions aim at the emancipation of the human spirit from bodily existence. To achieve this, the Jains developed a veritable science of *ahimsa*, a multitude of precepts for not injuring life in any form, down to minerals and metals, to liberate the spirit–soul from its burden of karma–matter. The Buddha discovered the chain of the conditions of co-origination that ties human consciousness to the world of nature, and taught his followers how to break it. The Upanishads emphasize the unreality of all things visible and tangible and urge to realize *brahman*, the transcendent all-embracing spirit that alone can be called real. Yogis become famous by defying the laws of nature: surviving for days buried underground, walking on water or sticking swords through their bodies without bleeding. If any of the teachings of the mainstream major religions had the added benefit of protecting nature, it was a side effect, not their main intention.

3. Discovering Ecological Wisdom in Ancient Texts

Stung by White's and Toynbee's criticism, the "monotheists" began to reread and reinterpret their scriptures from an ecological perspective. They discovered a great many texts that appear to provide support for contemporary ecological thinking. A recent edition of a popular English Bible translation marks all texts that refer to nature in green – all in all about a thousand.

Rabbi Sharon Joseph Levi sees an "ecological command" in the Talmud text: "God created Adam and led Adam through the trees in the garden. 'See my works, how fine and excellent they are. Know that all I have created is for your benefit. Reflect upon this and do not destroy my work, for if you do, no one will fix it after you'" (Levi, 1995, p. 94).

Walter Lowdermilk, a hydrologist with a Christian background, alarmed by the soil erosion that he had observed in China and in Mediterranean countries, suggested to add an 11th commandment to the Biblical Decalogue: "Thou shalt inherit the holy earth as a faithful steward, conserving its resources and productivity from generation to generation. Thou shalt safeguard thy fields from soil erosion, they living waters from drying up, they forests from desolation, and protect the hills from overgrazing by the herds, that they descendants may have abundance forever. If any shall fail in this stewardship of the land, they fruitful fields shall become sterile stony ground and wasting gullies, and thy descendants shall decrease and live in poverty or perish from off the face of the earth" (Quoted in Nash, 1989, p. 202).

Howard T. Odum, a professional ecologist, drew up "Ten Commandments of the Energy Ethic for Survival of Man in Nature" whose first rule is: "Thou shall not waste potential energy" ending with "Thou must find in thy religion, stability over growth, organization over competition, diversity over uniformity, system over self, and survival process over individual peace" (Odum, 1971, p. 244).

Christian theologians had for centuries left nature out from their considerations, concentrating instead on the "super-natural" as the proper field of their study. They too had to recognize the seriousness of the issue and began to reflect on how to "save the earth." The Catholic Hans Kung and the Protestant Wolfgang Pannenberg could be mentioned as pioneers in this field. Pope John-Paul's II 1990 message for the World Day of Peace calls the ecological crisis a moral crisis that demands the cooperation of all (Gottlieb, 1996, pp. 230–237). His successor, Benedict XVI, chose care for the environment as the focus of his address to the World Youth Festival in Sydney (Australia) in Summer 2008, and included "sins against nature" in a new catalogue of the traditional "seven deadly sins" – listed by TIME magazine among "The 50 Best Inventions of the Year" (TIME, November 10, 2008).

Conservative Protestants in general and Evangelicals in particular were for a long time leery of connecting their interpretation of Christianity with an interest in nature – they considered ecology-enthusiasts as pagans and rejected them together with Wiccas and similar new religions. However, a recent "Statement of the Evangelical Climate Initiative" includes "An Evangelical Call to Action" and an apology for past in-action in that field (http://www.christiansandclimate.org/statement).

Mawil Y. Izzi Deen, a Muslim scholar, interprets a Koran text that speaks of human vice-regency as divine entrustment of creation to humankind with the command to protect the earth: "The Prophet Muhammed (peace be upon him) considered all living creatures worthy of protection and kind treatment. He was once asked whether there will be a reward from God for charity towards animals. His reply was very explicit: 'For [charity shown to] each creature which has a wet heart there is a reward'" (Deen, 1990, p. 165).

Hindus, encouraged by Toynbee's praise, cite ancient Vedic prayers to Earth and Sun, Wind and Water as testimony to their tradition's ecological awareness. O. P. Dwivedi (1990), a Hindu scholar who has published extensively on this subject, quotes the ancient *Laws of Manu* prohibiting the pollution of lakes and rivers and threatening severe punishment to offenders. The planting of trees is commended as a religiously meritorious act in the *Puranas*, the "Bibles" of popular Hinduism. *Krishishastra,* India's traditional agricultural science contains practical advice on agriculture within a religious framework, linking sowing and harvesting with moon-phases and planetary movements as well as the popular Hindu festivals.

Buddhists refer to the *Jatakas*, stories about earlier incarnations of the Buddha in various animals, as evidence for the ecological sensitivity of their tradition.

L. G. Hewage, a prominent Buddhist scholar, criticized contemporary Western ecological thinking as too superficial. Ecology has also to address the "psycho-sphere" in which the roots of the ecological crisis are located: greed, hatred, and delusion: "The teachings of the Buddha contain adequate and appropriate information about the nature of this psycho-sphere and suggest empirically tested ways and means of understanding it and having full control over it at the final stage, by each individual who wishes to do so" (Hewage, 1982, p. 105).

Jains claim to be the oldest ecologists on record: their religion prescribes protection of all life. *Ahimsa*, not harming life, is their first and highest commandment. The Jain scholar Nathmal Tatia cites vegetarianism, practiced by all Jains, as a major contribution to the conservation of resources. Jains also established animal hospitals and shelters. Tatia's "Jain Guidelines to Meet the Ecological Crisis" include warnings against making the accumulation of wealth the aim of one's life (Tatia, 2002, p. 15).

While the majority of those for whom the Bible is scripture read Genesis I, 28, now as an entrustment of "stewardship" of the earth to humankind, there are still some who continue to see in it a license for large-scale human interference with nature. W. E. Fudpucker, an American Jesuit, sees in it a "technological imperative" to develop "technological Christianity" (Fudpucker, 1984). A former US interior secretary defended the destruction of large tracts of wilderness are as "Preparation for the Second Coming of Jesus." And an Editorial in The Christian Century reaffirmed a belief in the Biblical "Dominion Over Nature" by pleading for "Accepting the Risk" of further industrial development.

4. Scientific Roots of the Ecological Crisis?

Disagreement with Toynbee and White has not only been voiced by defenders of the "monotheisms" but also by respectable scientists from a variety of backgrounds. Far from blaming religion for everything from climate change to soil erosion, they point their fingers at modern Western science that, in their opinion, took a wrong direction somewhere between the seventeenth and the twentieth centuries.

Erwin Chargaff, one of the most successful biochemists of his generation, in later life considered modern science an "enemy of nature" and charged that today's science was "no longer searching for truth in nature but waging colonial warfare against nature" (Chargaff, 1979). When Hiroshima and Nagasaki were incinerated by atomic bombs, he felt "nauseating terror at the direction in which the natural sciences were going." And: "I saw the end of the essence of mankind, an end brought nearer, or even made possible, by the profession to which I belonged" (Chargaff, 1978, p. 40). Asked about alternatives to modern science, he suggested "to renounce it completely because of its devastating consequences" (Chargaff, 1989).

Rupert Sheldrake, a noted biologist, bemoaning the desacralization of nature in the Western world, puts the major blame on Francis Bacon, the prophet of a "new scientific priesthood," whose aim it was "to establish the power and dominion of the human race itself over the universe." Sheldrake holds that "the current devastation of the Amazon forests [was] made possible by technology and a Baconian faith in man's right to master nature." In his *Atlantis*, he continues, "[Bacon] describes a technocratic utopia in which a scientific priesthood made decisions for the good of the state as a whole and also decided which secrets of nature were to be kept secret." Vivisection and all kinds of experiments with animals were recommended. Bacon wanted nature to be "bound into service," "made a slave," and "put into constraint." Nature was to be "put on the rack" and "dissected," "forced out of her natural state and squeezed and molded." Bacon, after all, coined the slogan "knowledge is power," recommending the "mining of nature" for human profit. Nature, considered a living organism by premodern humanity, became for the moderns a machine to be manipulated (Sheldrake, 1991, pp. 41 ff).

Sayyed Hussein Nasr, a physicist by training and a renowned historian of science, holds "that although science is legitimate in itself, the role and function of science and its application have become illegitimate and even dangerous because of the lack of a higher form of knowledge into which science could be integrated and the destruction of the sacred and spiritual value of nature" (Nasr, 1976, p. 14). In his opinion, "man cannot save the natural environment except by rediscovering the nexus between Spirit and nature and becoming once again aware of the sacred quality of the works of the Supreme Artisan. And man cannot gain an awareness of the sacred aspect of nature without discovering the sacred within himself and ultimately the Sacred as such. The solution of the environmental crisis cannot come but from the cure of the spiritual malaise of modern man and the rediscovery of the world of the Spirit, which being compassionate always gives of Itself to those open and receptive to Its vivifying rays." Nasr evokes a vision of the cosmos as theophany, connecting with the symbolism of medieval philosophers of nature: "The cosmos reveals its meaning as a vast book whose pages are replete with the words of its author and possesses multiple levels of meaning like the revealed book of religion" (Nasr, 1982, p. 152).

5. The Greening of Science and of God

Sheldrake, even as spreading the blame for our present ecological crisis equally between religion and science, also sees signs for hope in both. He notices a "reanimation of the physical world" in recent science, which recognizes that nature is not tied down by external mechanical laws but is self-organizing from within: "Indeterminism, spontaneity and creativity have reemerged throughout the natural world. Immanent purposes or ends are now modeled in terms of attractors. And beneath everything, like a cosmic underworld, is the inscrutable realm of dark matter. All nature is evolutionary. The cosmos is like a great developing organism, and evolutionary creativity is inherent in nature herself"

(Sheldrake, 1991, pp. 41f). Some of today's physicists do not hesitate to ascribe consciousness to the universe (Kafatos and Nadeau, 1990). Ervin Laszlo's "Integral Theory of Everything" is based on the assumption that a cosmic consciousness informs everything, keeping an unbroken record of all that ever happened (Laszlo, 2006).

The discovery of an ecological dimension of religion has refocused the attention of the adherents of the various religious traditions on issues beyond the rumination over definitions of arcane elements of their creeds or the desire to narrowly demarcate their boundaries over against the "others." Reminding the world of the sacredness of nature, religious thinkers inject a personal dimension into a depersonalized science: reality is not exhausted by a description of its material components.

Seeing in the degradation of the environment not only the destruction of the basis of our physical lives but also an assault on the human spirit and the desecration of something inherently sacred, some are developing an ecology that includes science but that goes beyond quick technological fixes. David Suzuki, Canada's most prominent environmentalist, looking back on his own early career as a geneticist in the 1960s, remarked: "It is amazing, how much we accomplished and how little we understood" (Suzuki, 2002, Part I). Suzuki began to "understand" after coming into contact with the native Haida community, finding that the old creation myths had a better grasp of the "big picture" than modern science. James Lovelock, a biochemist by profession, struck by the interdependence of all components of our biosystem – the earth in conjunction with its creatures is actively creating and maintaining the conditions for the emergence and continuance of life – revived the ancient myth of the Greek Goddess Gaia to give a name to the "super-organism" that is earth.

Awe and wonderment, for Aristotle the source of philosophy, are returning to a science that believed that it had done away with all mysteries. Ernst Mayr, the great biologist-scholar confessed: "Virtually all biologists are religious in the deeper sense of the word, even though it may be a religion without revelation, as it was called by Julian Huxley. The unknown, and perhaps the unknowable, instills in us a sense of humility and awe" (Mayr, 1982, p. 81). Science is moving away from the mechanistic, materialistic, reductionist ideology that was adopted by nineteenth century theorists. Quantum Physics' principle of indeterminacy has overturned the old deterministic philosophy. Matter, once so confidently identified with reality as such, is dissolving into an ever-growing particle-zoo exhibiting short-lived bizarre phenomena. Though much more is known today about the chemistry of biological processes than a century ago, the belief in having solved the mystery of life by describing the physical–chemical processes in cells has been abandoned.

There is a new readiness among scientists and religionists to make common cause in the face of major challenges to the well-being of humankind: what could be a greater challenge that the impending climate-change? By providing a spiritual motivation to efforts to "save the earth," religious leaders can give a new dimension to the common endeavor and enlist a far greater number of people in the effort. Seeing something sacred in nature also widens the scope of environmentalism.

Increasingly, ecologists realize that our attitude to nature has much to do with our relationship to fellow-humans: living in peace with nature and maintaining peace among humans are strongly correlated! One of the documents that spells that out most clearly is the Club of Budapest's "Manifesto on Planetary Consciousness" that covers not only ecology in the more technical sense, but also world peace and the further evolution of humankind (Laszlo, 2008, pp. 133–139).

6. Eco-activism: Ecology as Religion?

Greenpeace is probably the world's best-known association of eco-activists. What is less known is its religious background: Robert Hunter, its founder intended his movement to represent "ecology as religion," describing its ethics as one of "personal responsibility and confrontation." Hunter also believed that whales are the highest developed living beings on earth – higher than humans: this is the background to the highly publicized "Save the Whales" campaigns. The members of Greenpeace also believe that they are fulfilling a prophecy made by a Cree woman, "Eyes of Fire," who predicted that at a time of universal ecological devastation "Warriors of the Rainbow" would arise to end the desecration of the earth (Hunter, 1979, p. 28). Importantly, Greenpeace also draws an essential connection between care for the environment and peace among humans: the two are intertwined! Though they consider confrontation an indispensable part of their strategy, they advocate nonviolence in its execution.

Another eco-activist group –"Earth First" – founded in 1980 by Dave Foreman, is openly resorting to violence and illegal activities like the spiking of trees, to ruin the chainsaws of the lumberjacks. It also considers itself a religion at whose center is Mother Earth: "All the Earth is Sacred" (Kinsley, 1995, p. 201). Earth-First holds that nonviolence is unnatural and that violence is necessary to defend Mother Earth against her enemies.

"People for the Ethical Treatment of Animals" (PETA) claim – rather boldly – that Jesus was a vegetarian and that meat-eaters commit a "holocaust on their plate" (http://www.peta.org/). Their spectacular "events" to persuade people from wearing animal furs often enlist pop-stars and society persons. Supporting at least some of their ideas, the *Anglican Society for the Welfare of Animals* issues information, composes prayers, and organizes animal blessings in churches on the feast of St. Francis of Assisi. A generation ago, Albert Schweitzer developed his theology of nonviolence on the foundation of the holiness of all forms of life.

Many of the over 800 contemporary grassroots ecological movements in India are religion-inspired, such as the Chipko, engaged in saving forests, the Mitti Bachao Abhiyan, a save-the-soil movement, the Vana Mahotsava, a tree planting organization, or the Braja Raksha Dal, engaged in protecting the hills connected with Krishna' early life from being mined for building materials – to mention only a few.

Interestingly, the religious movement that has done most for the Indian environment, Swadhyaya (literally "self-study"), founded by Pandurang Shastri Athavale in the 1940s, refuses to call itself an ecological movement and insists that whatever ecological benefits may follow from its activities, these are secondary. The primary aim had been from the very beginning a socioreligious reform of entire village communities: people have to learn to see their daily work as *seva* (religious service) and to regard nature as manifestation of the divine. Its, by now, several million members, living mainly in the states of Madhya Pradesh, Gujarat, and Maharashtra, have planted orchards and forests on hundreds of thousands of hectares of land earlier considered useless. They have transformed vast stretches of barren countryside into gardens that produce vegetables for home consumption and market, and they have performed wonders in the areas of water-preservation and village sanitation. Above all, they have given an example of voluntary cooperation on a large scale, overcoming divisions of village against village and religion against religion. All these "ecological successes" are considered "by-products" of the teaching that God is appearing to humans in and through nature: The earth is their Holy Mother, to be thanked for her gifts and to be cherished. *Vriksha mandiras*, tree-temples are the focal points for their worship. Athavale cited many ancient Hindu texts to support his "nature *dharma*" and taught people to practice "devotional farming" (Jain, 2008, pp. 10–63).

Athavale followed in some ways the example of the Bishnois, disciples of a fifteenth-century saint Jambeshvara, living in the forbidding Rajasthan desert. Among the 29 rules that the guru laid down, eight have to do with the protection of animals and trees. Jambheshavara also got several large tanks dug as water reservoirs for the dry season, still in existence today. Deepening these reservoirs is considered a meritorious act that still continues to be practiced. Planting and conserving trees was another of his commandments. Jambheshvara is even credited with having revived a dead tree. He taught people not to cut off branches from trees to feed their animals, as they had been doing during the dry hot summers, but to care for the trees. Soon, pockets of lush forests developed in the desert. An eighteenth-century Maharaja of Jodhpur, desirous of building a new palace, sent out his officers to scout for timber: they located a source in Khedajali, a nearby village. When they came to cut down the trees, Amrita Devi, a resident of the village, tried to stop them. The workmen, however, killed her and 362 other Bishnois who opposed the cutting of their trees. The Raja, on hearing about it, immediately stopped the killing and the cutting of trees. Today, a monument in Khedajali commemorates the 363 villagers who became martyrs for their trees. Recent instances have been documented where Bishnois sacrificed their lives protecting animals from being hunted. They also maintain nursing homes for cattle and consider it religiously meritorious to provide fodder and medical assistance to them.

Indian eco-activists also take inspiration from Mahatma Gandhi, who taught and practiced ecology long before it became fashionable in the West

(Torchia, 1997). India's traditions that never knew a hard and fast division between human life and other forms of lives make it easier for contemporary ecological movements to win adherents for their campaigns to save wildlife and protect the natural environment.

Among Thai Buddhists, a movement developed that performs the robing of trees to protect them from being felled: being robed, the trees become members of the Buddhist Sangha and thus inviolate.

Some Neo-Buddhist and Neo-Hindu communities in the West, such as the Green Gulch Farm and the Spirit Rock Meditation Centre in California, or ISKCON, present in many countries, also have a decidedly ecological agenda: they practice organic farming, take care to recycle as much as possible, and make reduction of consumption part of their religious routine.

There are also elements in the teachings and practices of older religions that are of ecological relevance: moderation in consumption, care in the use of resources, and sharing of possessions surely would help the environment if people followed these recommendations. Vegetarianism, enjoined by several of the major Asian religions, if worldwide adopted, would certainly be of immense environmental benefit. So would be the cultivation of charity, especially when extended to animals: it would result in a fairer distribution of resources, and also prevent wars, the source of much ecological devastation. Religions, by sensitizing people for dimensions of reality other than the grossly material, encourage "reverent thinking": teaching respect for nature rather than encourage its senseless exploitation (Skolimowski, 1989).

7. Deep Ecology, Eco-feminism, Eco-philosophy

Already the early nineteenth-century ecologists – such as Henry David Thoreau, John Muir, and Aldo Leopold – realized that "protecting nature" entailed much more than fencing in a parcel of land and preventing humans from exploiting its fauna and flora. They quite understood that the then incipient ecological crisis was at its heart a cultural and a moral crisis: the result of the cultivation of a modern Western ideal of human existence that saw its fulfillment in the amassing of material possessions and the reckless exploitation of the earth, praised as "progress." "Conspicuous consumption" conferred social status. A person's "net-worth" became expressed in dollar figures. Humans were identified as "consumers" – their main role being to keep the economy growing.

In contrast to this, traditional Asian religions saw the embodiment of the highest human ideal in the sage: the person who had gained a state of tranquil self-contentment. Enlightenment and wisdom went hand in hand with the restriction of wants and desires. Empathy with everything and everybody and sensitivity toward nature as well as toward fellow-humans were highly valued. Virtually all Asian traditions accept rebirth as a universal fact of life: Humans do not occupy a unique position and their fate, both present and future, depends on their relationships with all other forms of life. Everybody's life has gone through many transformations. Life as such deserves respect: not harming living beings was considered the highest moral principle.

Terms like (Chinese) *tao* and (Indian) *dharma* refer to a universal natural law that has humanistic as well as transcendental dimensions. *Tao/dharma* encompasses all reality: it is the "nature" of things as well as the "inborn law" of which humans become aware, when awakened. Implied in this notion is the idea that the universe is a meaningful, purposeful reality, within which humans find their personal fulfillment. The articulation of *tao/dharma* includes societal and historic aspects: it has to be rearticulated in every age and for each culture. The ecological crisis manifests a severe violation of *tao/dharma:* the only way to resolve it is to return to the principles of *tao/dharma*. Typically, Chinese and Indian cosmologies correlate material elements with socio-cultural and emotive-intellectual categories. In their view, the universe is not fully circumscribed by an inventory of its material content: it also contains life, feeling, and consciousness.

Influenced by the study of Eastern religious classics Arne Naess, a Norwegian philosopher, developed what he called "deep ecology." Abandoning the anthropo-centrism of Western ethic, he adopted the Eastern view that all life has intrinsic value and that the purpose of all life is self-realization. His "principle of identification" – reminiscent of the "exchange of the self and the other" recommended in the eighth century Buddhist classic *Bodhicaryavatara* – says that we humans, individually and collectively, are part of a whole that sustains us and that we, in turn, have a duty to sustain. Naess' ideas were taken up and partly transformed by other philosophers, such as J. Baird Callicott (1994) and Michael E. Zimmerman (1994).

"Eco-feminism" is a special branch of "deep ecology" (Adams, 1994). Feminists attribute much of the ecological crisis to the suppression of women in patriarchal culture, responsible for 'The Death of Nature' (Merchant, 1980). From its Western origins, critical of the male-dominated Christian churches, it has branched out also to Asia (Shiva, 1989). By empowering women, it expects to combat violence – including violence against nature – seen as a typical male attitude.

Henryk Skolimowski, a contemporary philosopher, thought that "deep ecology" was not deep enough because it was lacking ultimate ends. His own eco-philosophy, later expanded into "Ecological Humanism," has strong religious overtones: "There are aspects of traditional religions, which add something significant to man's substance, and if rejected or neglected seem to produce a crippling effect on man's life. Without worship, man shrinks. If you worship nothing, you are nothing." And: "The new theology underlying Ecological Humanism is that we are God-in-the-process-of-becoming. We are fragments of grace and spirituality in *statu nascendi*. We give testimony to our extraordinary (divine) potential by actualizing these fragments in us. By creating ourselves into radiant and spiritual beings, we are helping to create God-in-the-process-of-becoming. God and our divinity are the end of the road, at the end of time, at the end of mankind, in the finale. Our task is to become fully aware that, as the result of certain propensities of unfolding evolution, we possess the potential for making ourselves into spiritual beings, and thereby bring to fruition some of the seeds of God in the process of self-making. This ecological theology provides not only a new cosmological scheme; it also has an existential import: it brings into sharp focus the meaning of our individual life, which is redeemed insofar as we elicit from ourselves our potential godliness. This theology also justifies the emerging myth of the unity of the human family: we are all striving towards the actualization of something much greater than our individual selves, something that transcends the boundaries of all states and cultures" (Skolimowski, 1981, p. 115).

One of the most passionate voices crying in the ecological wilderness is that of Father Thomas Berry, whose *The Dream of the Earth* presents a program of spirituality based on ecology. It is not without significance that the book was first published by the Sierra Club, founded in 1892 by John Muir, one of the early American environmentalists.

8. A New Ecology for a Postnatural World?

Ecological theory reflects the reality picture that the sciences project. Physics, the basic natural science, has always exerted great influence on shaping the popular view of the natural world. Radical changes in physical world-views have a deep impact on the perception of nature by the general public. Newton's conception of the universe as mechanical clockwork, regulated by a few fairly simple mathematical laws, became the world-model also of the philosophy and theology of his age, as well as the backdrop to its literature and architecture. Planck's Quantum Theory and Einstein's Theory of Relativity had a similar effect. Today, it is Chaos Theory that is finding application in a wide range of areas, including ecology. A "new ecology" is emerging, that sees "flux" and "chaos" as the determinants of nature over against the traditional notion of "balance" (Lodge and Hamlin, 2006).

Flux is, no doubt, a universal fact. Evolution, too, is recognized as a general characteristic of our world. However, today's physics acknowledges unpredictability and indeterminacy at the subatomic level that does not translate directly into the macrocosmic world. In the "human-sized world," physical bodies obey statistically deterministic laws, and complex organisms develop in a largely predictable way. How else could we build houses to live in and dare to cross bridges, or why would we train doctors and engineers?

Order and spontaneous self-organization do not exclude each other and the "balance of nature" does not contradict flux and change: all of nature exists and evolves in time. Time is the key factor that must enter into all our reflection on nature. On the macro level, nature presents itself to us in a delicate but real balance: even relatively minor changes in the composition of the air we breathe or in the temperature of our atmosphere would make life impossible. Proponents of the "Anthropic Principle" point to the fine-tuning of the cosmic constants – within ratios of one to a billion! – making it possible for life and humanity to exist in a universe full of contingencies. Even minimal deviations from the existing balance of matter and anti-matter would have made the coming into existence of our world and of life impossible.

"Natural Balance" and "Flux" play out on two different levels of reality and coexist in the same real world: life-giving air is a finely balanced mix of oxygen and nitrogen and it is constantly in flux. A change in this "balance" would kill most of life – a cessation of "flux" would have the same effect.

When taking over scientific terminology into everyday language, we have to consider that scientists are often coining technical terms by giving a very specific meaning to commonly used words. "Relativity" in the physicists' vocabulary does not mean that "everything is relative," as some may think. Similarly, the word "chaos," as used by scientists, does not refer to the condition of a room strewn with all manner of objects, but it denotes the mathematically as yet undefined condition of a large number of elements before they are reaching a mathematically describable order. In an unforeseen and as yet unforeseeable way, from what seems a "chaotic" jumble of disparate elements, order, and symmetry arise. Not insignificantly, the book that has inspired much of the popular "chaos" discussion bears the title *Order Out of Chaos*: the nature in which we live and have our being is not "chaos" but is constituted by the order that arose from it! (Prigogine and Stengers, 1984) In the language of today's physics, nature is in a "dynamic equilibrium," i.e., the interaction of a number of factors generates a condition of existence that is self-supporting under certain given conditions. If the ratio of the components is changed beyond a calculable amount, the whole dynamic equilibrium collapses.

If the "new ecologists" state that "the balance of nature ... does not exist" (Lodge and Hamlin 2006, p. 306), we have to say that this is simply not true. There is a genuine and scientifically verifiable "balance of nature" without whose functioning we would not be here and without whose continued working we could not live: it is this really existing "balance of nature" that we have to preserve. It is again this "balance of nature" that tells us how far we can go in using/abusing our environment: in and through this "balance" nature "speaks to us." If we continue to disturb this "Sacred Balance" in a major way, the very existence of humankind and of a great many other forms of life will be in peril.

In its presumed 4.5 billion years of existence, the earth has gone through dramatic changes: from a lifeless mass of cosmic dust and a fiery ball of magma to a planet with oceans and continents teeming with life. Life itself underwent major catastrophes and cataclysmic changes. There was, there is, and there will be change and flux. The earth as a planet would continue to exist also without humanity, probably even without any life. However, the point of our "ecology" is precisely to make sure that humanity – together with other forms of life – can survive on our earth. For this purpose, we have to preserve the "balance of nature."

9. Reinventing Nature

The New Ecologists support their (false) claim of the nonexistence of the "balance of nature" not only by referring to a (misunderstood) contemporary branch of physics but also by the philosophical argument that "nature" is always "interpreted nature." That is: we do not know nature in and by itself, but we can speak of nature only as variously understood by humans, who always perceive nature through the medium of their specific cultural lenses. They argue that we reached "the end of nature" (Vogel, 2002) and that we now need a foundation outside nature to develop an environmental ethics "for a post-natural world" (Lodge and Hamlin, 2006).

The more secular minded "new ecologists" reach back to Kant's "autonomous self" as the source of such an ethics, or to some Heideggerian "authenticity." Apart from the difficulty, to communicate Kant's or Heidegger's thinking to ordinary people, it really does not address our problems. The more religious-minded "new ecologists" look to revealed religion, specifically to Christianity that is to offer the essential guidelines for a postmodern environmental ethic (Lodge and Hamlin, 2006, pp. 279–309). Contemporary Christianity, however, still has to learn how to deal with nature and it has to make amends for the many crimes against nature and humanity, committed in its name. There is further the fact that Christians are divided into so many different branches that hardly agree on anything at all, least of all on the interpretation of the Scripture to which they refer. And since the majority of humankind is not Christian, their scripture-based ecological ethic would hardly be universally acceptable, apart from its intrinsic shortcomings.

We need to work out an ethic from nature: from "nature interpreted" to be sure, even from "nature re-invented," but not unrelated to the reality that had been called nature throughout the ages. Nature is the source not only of science but also of religion. If we can legitimately connect meaning with ancient texts that can be interpreted in different ways, we can also accept a variety of interpretations of nature as the basis for a meaningful ecological ethic. Though there is always some latitude in interpretations, there are definite limits to it, both in literature and nature. As we must not interpret the text itself away if we try to understand a piece of literature, so we must not dispense with the reality of nature, when we attempt to understand it conceptually. As long as we live in a "natural" body, we depend on the realities of air and water, on plants and animals, and the rest of what we call nature. People in different parts of the world have different words for all that is included in "nature," but they do agree in their existence and the need for them. Nature and reality – one could make a strong case for their identity – are certainly widely overlapping terms: we neither can have individual existence nor culture – including ethics and religion – without nature.

The scientific interpretation of nature has changed dramatically in the last 100 years or so. The materialistic–mechanistic world-picture that had dominated an earlier atheistic and scientific age has all but been given up, except for a few "hardliners" like Jacques Monod or Richard Dawkins (Klostermaier, 2008). Over against the earlier reduction of nature to "atoms and the void," today we are convinced that everything in nature is interconnected at a very profound level. Rejecting the hubris of an age that believed to have uncovered everything there was to know, today's scientists admit that nature is unfathomable and mysterious

– some even ascribe consciousness to the cosmos as a whole (Kafatos and Nadeau, 1990). Most importantly of all, we begin to learn that we have to rejoin the subject (mind) to the object (nature): we have to learn to see ourselves as part of nature and nature as our larger self (Schroedinger, 1946). As Konrad Lorenz (Nobel Prize 1973 for Physiology and Medicine) said: "Since all moral responsibility of humans is determined by their value perceptions we must reject the endemically wrong belief that only what can be counted and measured is real. We must convincingly state that our subjective processes possess the same degree of reality as anything that can be expressed in the terminology of the exact natural sciences" (Lorenz, 1973).

Suggestions, that humankind has reached a point where it could "de-link" itself from nature, that cultural evolution has now replaced natural evolution, are becoming the less acceptable, the more we feel the effects of the humanly caused ecological crisis: we better try to "re-link" and "re-integrate" our species with nature! In our reinterpretation of nature, we must replace "dead matter" as central category by "consciousness." We must realize that it is not the vision and theory of a nature observed "through a dead man's eye" but "the vision and theory of value" that is "the starting point of understanding" (Kohak, 1987).

Nature is greater than science: all great scientists are agreed on that. "Physics," as Nobel Prize winning physicist Niels Bohr said, "is not telling us what nature *is*, but only what we can *say* about nature." Nature is not identical with an inventory of all the material objects that can be studied by the natural sciences. And *pace* Kepler: neither are quantities the only archetypes of nature, nor is mathematics – a useful tool for our present sciences – the only language of nature. Nature has real qualities that everybody can experience and that have been further explored by poets and seers. If we accept the anthropomorphism of a "language of nature," we can go one step further and say that nature – like humankind – speaks in many languages. Nature does indeed speak in several languages to those who know how to listen.

Bhikkhu Buddhadasa, a prominent twentieth-century Thai Buddhist, expressed it thus: "Trees, rocks, sand, even dirt and insects can speak. This does not mean, as some people believe, that they are spirits. Rather, if we reside in nature near trees and rocks we will discover feelings and thoughts arising that are truly out of the ordinary. ... The lessons nature teaches us lead us to a new birth beyond the suffering that results from attachment to self" (Swearer, 1979, p. 34). Buddhadasa has also learnt another lesson from nature: "The entire cosmos is co-operative. The sun, the moon, and the stars live together as cooperative. The same is true of humans, animals, trees and the earth. Our bodily parts function as a cooperative. When we realize that the world is a mutual, cooperative enterprise, that human beings are all mutual friends in the process of birth, old age, suffering and death, then we can build a noble, even a heavenly environment. If our lives are not based on this truth, then we'll all perish" (Swearer, 1997, p. 35).

The *Bhagavata Purana*, a popular Hindu scripture, tells a story about a young man by name of Dattatreya, who had chosen nature as his spiritual guide:

24 representatives of nature offer the equivalent of the guidance of an inspired guru. Nature teaches in metaphors. Thus, the earth teaches steadfastness and the wisdom to realize that all things, while apparently pursuing their own aims, follow the universal divine law. It also teaches that human existence is a being-for-others to be lived out in forbearance and humility. A tree teaches forbearance and patience. The story goes on and suggests that nature has a moral lesson to teach to those who look to it for guidance (Klostermaier, 2007, pp. 476–489).

The Swadhyaya movement, mentioned earlier, shows that the language of nature can still be understood today by people in intimate touch with nature who can see in nature a revelation of Reality. Vaishnavas see the world as the body of God, going so far as to identify mountain ranges as his bones and rivers as his veins. The sacredness of holy places is connected with nature itself, and does not depend on religion-specific beliefs that may be connected with them.

Nature is also greater than historic, man-made religions: by reestablishing an in-depth contact with nature organized religions must regenerate and reorient themselves in the service of humankind: all too often they appear as self-serving and tyrannical institutions that violate their own lofty principles. By reorienting themselves on the generosity and unity of nature, religions can overcome their endemic sectarianism and intolerance and their pettiness and arrogance.

There is genuine wisdom as well as true science in many old traditions that convey practical knowledge as well as provide spiritual satisfaction. David Suzuki's *The Sacred Balance* offers some striking contemporary evidence: For perhaps 2,000 years, the Balinese had cultivated rice in an area that was watered by springs from a mountain ridge, considered sacred. The rhythm of work was determined by the local water temple, where people congregated for worship as well as for discussion of practical matters. In the 1970s, the government of Indonesia, pushing the Green Revolution, forced the Balinese to adopt new varieties of rice that yielded three crops per year instead of the two traditional ones. This upset the entire traditional agricultural cycle. After a few years, pests destroyed much of the crops, rats became a major menace, and the use of increasing amounts of pesticides made people and animals sick. When the Balinese reverted to the old practice, governed by the traditional rules of the water temple, everything worked fine again: ducks took care of the pests, the simultaneous flooding of all fields, as determined by the water-temple, drove the rats away, the old varieties of rice yielded sufficient good-quality food and people took up their traditional rhythm of work and festivals, not forgetting to offer a share of their bounty to the gods at the water temple. Not only the economy but also the happiness of the people was restored.

There is no alternative to nature in our search for a basis for an ecological ethics: a nature that speaks to us in many ways and whose languages we have to learn if we wish to survive. It is clear, that neither the Baconian technological enslavement of nature nor post-modern cynicism nor the fundamentalist fideism of a "new ecology" is working. Nature is warning us today by various signs that it may no longer tolerate human abuse. She also let us know that so far we have not really understood her. Through its wiser offspring nature, she tells us what to do to save her and ourselves. But we must learn to listen!

10. Conclusions

Our world is facing simultaneously a multitude of crises: a worldwide financial and economic crisis "of historic proportions," a severe food crisis in many parts of Africa, wars and insurrections, corrupt and incompetent governments, not to speak of the multitudinous natural catastrophes of the past few years: tsunamis, hurricanes, floods, droughts, and earthquakes.

There are predictions of even more worldwide disaster impending: wars fought over scarce resources wiping out the majority of the world's population in the process; calculations, that the economic fallout from global warming would surpass the effect of all other crisis put together. No single person and no single government can solve these crises once for all: they will be with us forever, or – as long as humankind exists.

One good could arise from this predicament: the willingness of people to come together to address this global crisis globally. Instead of competing with each other, claiming superiority of race or religion, playing out ideological games or trying to trick each other out on world-markets, people may realize that it makes more sense to work together. The scientific acumen of the human race that has caused so much of our predicament could be employed to also solve the problems it has created. The religious emphasis on the sacredness of nature and the interdependence of all life could help to create an intellectual climate in which care for nature will be regarded a universal duty. Once a large enough part of humanity has undergone the necessary mind-change and has begun to act on these insights, a healing process will begin in nature for the benefit of all.

This essay has tried to show that adherents of many religions have become ecologically aware and active and that religiously motivated eco-activism is making a difference. It also made a case for a "return to nature" in the development of a viable theory of ecology, articulating features of a "re-invented nature" as the basis of an interpretation of nature as source of an ecological morality, finally suggesting that "peace with nature" is inseparably linked to "peace among people."

11. References

Adams, C.J. (ed.) (1994) *Ecofeminism and the Sacred*. Continuum, New York.
Agarwal, A. (2000) Can Hindu beliefs and values help India meet its ecological crisis? In: K.C. Chapple and M.E. Tucker (eds.) *Hinduism and Ecology*. Harvard University Press, Cambridge, MA, pp. 165–179.
Berry, Th. (1998) *The Dream of the Earth*. Sierra Club, San Francisco, CA.
Callicott, J.B. (1994) *Earth's Insights: A Survey of Ecological Ethics from the Mediterranean Basin to the Australian Outback*. University of California Press, Berkeley/Los Angeles, CA/London.

Chargaff, E. (1978) *Heraclitean Fire: Sketches from a Life Before Nature.* Rockefeller University Press, New York.
Chargaff, E. (1979) Hoehlt das Hirn, In: Spiegel, Nov. 19, 1979, pp. 240–243.
Chargaff, E. (1989) Gibt es Alternativen zu unserer gegenwaertigen Naturwissenschaft? In: Die Neue Rundschau Vol. 4/1989. Frankfurt am Main, 100, Jahrgang, pp. 5–15.
Coward, H. (1995) *Population, Consumption and the Environment.* SUNY, Albany.
Dawkins, R. (2006) *The God Delusion.* Houghton Mifflin, Boston, MA/New York.
Deen, M.Y.I. (1990) Islamic environmental ethics, law and society, In: R. Engel and J.G. Engel (eds.) *Ethics of Environment and Development.* Belhaven, London. Reprinted in Gottlieb, R.S. (ed.) (1996). *This Sacred Earth: Religion, Nature, Environment.* Routledge, New York/London, pp. 164–173.
Dwivedi, O.P. (1990) Satyagraha for conservation: awakening the spirit of Hinduism, In: Gottlieb, R.S. (ed.) (1996) *This Sacred Earth: Religion, Nature, Environment.* Routledge, New York/London, pp. 151–163.
Fudpucker, W.E. (1984) Through Christian technology to technological Christianity, In: C. Mitcham and J. Grote (eds.) *Theology and Technology: Essays in Christian Analysis and Exegesis.* University Press of America, Lanham, MD, pp. 53–69.
Gottlieb, R.S. (ed.) (1996) *This Sacred Earth: Religion, Nature, Environment.* Routledge, New York/London.
Gottlieb, R.S. (2006) *The Oxford Handbook of Religion and Ecology.* Oxford University Press, Oxford.
Hewage, L.G. (1982) Survival and development – towards a buddhist perspective, In: Mahabodhi April–June 1982, pp. 103–114.
Hunter, R. (1979) *Warriors of the Rainbow: A Chronicle of the Greenpeace Movement.* Holt, Rinehart & Winston, New York.
Jain, P. (2008) *Sustaining Dharma, Sustainable Ecology* [unpublished manuscript].
Kafatos, M. and Nadeau, N. (1990) *The Conscious Universe: Part and Whole in Modern Physical Theory.* Springer-Verlag, New York.
Kinsley, D. (1995) *Ecology and Religion: Ecological Spirituality in Cross-Cultural Perspective.* Prentice Hall, Upper Saddle River, NJ.
Klostermaier, K. (2007) Hinduism and ecology, In: *A Survey of Hinduism*, 3rd edn. State University of New York Press, Albany, NY, pp. 476–489.
Klostermaier, K. (2008) Reflections prompted by Richard Dawkins' The God Delusion. J. Ecumenical Stud. **43**(3): Fall 2008.
Kohak, E. (1984) *The Embers and the Stars: a philosophical enquiry into the moral sense of nature*, The University of Chicago Press, Chicago and London, 1984.
Kung, H. (1991) *Global Responsibility: In Search of a New World Ethic.* SCM Press, London.
Laszlo, E. (2006) *Science and the Reenchantment of the Cosmos: The Rise of the Integral Vision of Reality.* Inner Traditions, Rochester, New York.
Laszlo, E. (2008) *Quantum Shift in the Global Brain: How the New Scientific Reality Can Change Us and Our World, Inner Traditions.* Rochester, New York.
Levi, S.J. (1995) Judaism, population and the environment, In: H. Coward (ed.) *Population and the Environment: Religious and Secular Responses.* SUNY, Albany, NY, pp. 73–107.
Lodge, D.M. and Hamlin, C. (eds.) (2006) *Religion and the New Ecology: Environmental Responsibility in a World of Flux.* University of Notre Dame Press, Notre Dame, IN.
Lorenz, K. (1983) *Der Abbau des Menschlichen*, Piper: München 1983.
Mayr, E. (1982) *The Growth of Biological Thought: Diversity, Evolution and Inheritance.* The Belknap Press of Harvard University Press, Cambridge, MA.
Merchant, C. (1980) *The Death of Nature: Women Ecology and the Scientific Revolution.* Harper & Row, New York.
Naess A. (1985) Basic Principles of Deep Ecology, In: B. Dewall and G. Sessions (eds.) Deep Ecology: Living as If Nature Mattered. Gibbs Smith, Salt Lake City, UT.

Nash, R. (1989) *The Rights of Nature: A History of Environmental Ethics*. The University of Wisconsin Press, Madison, WI.
Nasr, S.H. (1968/1976) *Man and Nature: The Spiritual Crisis of Modern Man*. Unwin, London.
Nasr, S.H. (1982) *Knowledge and the Sacred*. Edinburgh University Press, Edinburgh.
Noyce, J. (2002) *Hinduism and the Environoment*, johnnoyce@hotmail.com
Odum, H.T. (1971) *Environment, Power and Society*. Wiley Inter-Science, New York.
Ophuls, W. (1992) *Ecology and the Politics of Scarcity*. W.H. Freeman, New York.
Pannenberg, W. (1993) *Toward a Theology of Nature*. Westminster/John Knox Press, Louisville, KY.
Prigogine, I. and Stengers, I. (1984) *Order Out of Chaos*. Bantam Books, Toronto.
Schroedinger, E. (1946) Der Geist der Naturwissenschaften, In: *Geist und Natur: Eranos Jahrbuch 1946*. Rhein Verlag, Zuerich.
Sheldrake, R. (1991) *The Rebirth of Nature: The Greening of Science and God.* Bantam Books, New York.
Shiva, V. (1989) *Staying Alive: Women, Ecology and Development.* Zed-Books, London.
Skolimowski, H. (1981) *Eco-Philosophy. Designing New Tactics for Living*. Boston-London.
Skolimowski, H. (1989) *The Theatre of the Mind: Evolution in the Sensitive Cosmos*. The Theosophical Publishing House, Wheaton, IL.
Suzuki, D. (1997, 2002, 2006) *The Sacred Balance, Greystone Books*. Vancouver, Toronto.
Suzuki, D. (2002) The Sacred Balance Video Series (four segments).
Swearer, D.K. (1979) The hermeneutics of Buddhist ecology in contemporary Thailand: Buddhadasa and Dhammapitaka, In: M.E. Tucker and D.R. Williams (eds.) *Buddhism and Ecology*. Harvard University Press, Cambridge, MA.
Tatia, N. (2002) "The Jain Worldview and Ecology" in: C.K. Chapple (ed.) *Jainism and Ecology*, Harvard University press.
The Christian Century, July 30–August 6, 1980, From an Editor at Large.
The Jerusalem Bible. Readers' Edition (1971), Doubleday & Co., Garden City, NY.
Torchia, A.D. (1997) Gandhi's Khadi spirit and deep ecology. Worldviews: Environ. Cult. Relig. 1: 231–247.
Toynbee, A.J. (1971) The religious background of the present environmental crisis. Int. J. Environ. Stud. 3: 141–146.
Vogel, S. (2002) Environmental philosophy after the end of nature. Environ. Ethics 24(I): 23–39.
White, L. Jr. (1967) The historical roots of our ecologic crisis. Science 155: 1203–1207.
Zimmernman, M.E. (1994) *Contesting Earth's Future Radical Ecology and Postmodernity*. University of California Press, Berkeley/Los Angeles, CA/London.

Biodata of **Hermann Josef Roth**, author of *"Nature and Resource Conservation as Value-Assessment Reflections on Theology and Ethics"*

Dr. Hermann Josef Roth is a Studiendirektor, biologist, Roman-Catholic Priest, and Cistercian-Monk; manager of the "Naturhistorischer Verein der Rheinlande und Westfalens" (NHV), who has a seat of the University at Bonn. Dr. Roth is member of the "Europainstitut für cisterciensische Geschichte, Spiritualität, Kunst und Liturgie" (EUCist) with seat at Päpstliche Hochschule Heiligenkreuz near Vienna. He obtained his Ph.D. from the University of Nijmegen (Netherlands) in 1990. Dr. Roth published many books and papers on dialogue between theology and natural sciences, on natural history of the Rhineland, and history of botany and zoology in Middle-Ages and during the eighteenth and nineteenth centuries. His main interest is in Monastic Medicine including ethnobotany in the Portuguese world. His worked on the project "Klostermedizin" at the University of Würzburg.

E-mail: **Hermannjroth@aol.com**

NATURE AND RESOURCE CONSERVATION AS VALUE-ASSESSMENT REFLECTIONS ON THEOLOGY AND ETHICS

HERMANN JOSEF ROTH
Naturhistorischer Verein, University at Bonn, Germany
Europainstitut für cisterciensische Geschichte, Päpstliche Hochschule Heiligenkreuz near Vienna, Austria

1. Introduction

The protection of natural resources results from rational hindsights. But which resources are indispensable? Coal, oil, or uranium? Presently, the preferences are being decided based on nonscientific criteria. The goals of natural and environmental conservation are not scientifically justifiable. They are being formulated according to ideological approaches. Politicians and decision-makers make use of economists and technicians without scientific fact repertoires for argumentation and implementation.

A concrete example (AWH, 1989) serves to illustrate the point: they argued in support of an expansion of the Cologne–Bonn airport citing a lack of capacity, related security problems, economic requirements, and provision of work. Nature conservationists protested vehemently claiming that the new take-off and landing runways would destroy the habitats of "Red List Species" (endangered flora and fauna) or would cause the disappearance of rare vegetation areas. Furthermore, the increased air traffic would increase the environmental damage. However, many of them did not express an understanding for the concern to protect nature. Some were attracted by the lure of new jobs. But it is basically accepted that in nature, creation and destruction are one-time events. After all, the dinosaurs went extinct.

What stirs the emotions here is not justified by the facts. There are basic differences between natural extinctions and those caused by human activities. The contrary position is equally weak: botanists or ornithologists, apart from citing the rareness of a species or biome, have no basis for species conservation. At least this controversy helped to hinder some contradictions and spurious arguments from entering the discourse. Hence, biologists and conservationists have abandoned the formerly popular but groundless difference between "useful" and "harmful."

The modern study of ecology focusing on the interdependence between organisms and the environment – justified by method as in other sciences – restricts itself to trying to rationally and objectively describe the world and its functional connections. Though this increases our knowledge, it does not relieve our conscience.

2. Foundations

The chapters in this book offer full historical, natural scientific, and other scientific data that, in total, offer an extensive picture of "The role of seaweeds in future global environments." The improved knowledge of scientific facts, however, will by no means lead to an avowal to one specific method of examination. Depending on the conflicting views of the interested parties, one of many options may be employed. The decisions are not based solely on science, but not without it either.

Ecology has played an important role in the refining of the scientific method. Along with other disciplines, it has sharpened the awareness of the interconnections of processes and their factors in nature, without minimizing the individual cause-and-effect analyses. It can and does clarify how each apparently insignificant disturbance affects the balance of the biome, and it also shows how it counter-reacts to the spread neglect of the unclear consequences of the apparent disturbances. Because ecology researches and clarifies the basic efficient principles of nature, it offers findings on which ethics (deontology) can also base its concepts.

Additionally, other sciences (i.e., social science) must account for facts and so ecological insights are brought into accord with recognizable human needs. In the long term, ecological action will only be possible when humans see themselves as a part of nature and recognize that ecological findings are also vital for their existence.

3. Ethical Aspects

Ethics begins fundamentally when one has determined to take rational findings into account. Everyone knows that smoking is dangerous, yet many still smoke.

Even environmental politicians and nature conservation workers take short flights from the Cologne–Bonn airport although there are available comfortable railways. Whosoever decides to base his own behavior according to ecological perspectives makes a considerable commitment in that moment of choice – as long as that person's determination holds. This means that for that person, for example, his actions are no longer determined solely by his own needs but the needs of the entire ecosystem are taken into account. The use of nature remains understood but keeps itself within reasonable bounds. The use becomes taboo when the bounds exceed reasonable exploitation or bring irretrievable destruction. An individual's orientation is easily hindered if his interests become unfortunately identical to those of the society.

Scientific insight alone then is disabled as a long-term basis and motivation for ethical behavior. We live and make decisions as much from the soul as from reason.

The story could begin differently as it did for Lynn White in 1926 in Ceylon (Krolzik and Knöpfel, 1986): Singhalese were building a new road under British rule. In the planning, they consistently omitted several sites. There were snake nests. The local people were not afraid of the animals but held the opinion that the animals had a right to live there as long as they wanted. As Buddhists, they were convinced that souls were reincarnated in snakes. The British Christians yielded to the people, as the building activity would certainly have dislodged the animals. The English did not value the snakes, but the Buddhists certainly did.

Animal conservation without environmental laws and "Red Lists" only holds meaning for those who find value from other sources. Two thousand years ago, the Rabbi from Nazareth referred to the Samaritans, and even barely today do we find reason to look over the fence of our Western tradition to find a few fleeting impressions of how others relate to nature. Perhaps this will help in providing motivation to study our own behavior. Finally, the impulse for an "ecological ecumenism" may result.

4. Environmental Thought in Different World Religions

4.1. BUDDHISM

The previous example provides more information about Buddhist sensibilities than any quote can. Each of the many ways of thinking within Buddhism has a very different understanding of reality and therefore each weighs the environmental problem differently. Many know of reincarnation as a Bodhisattva, one who remains on the last step before entering Nirvana out of compassion for creatures and denies himself entry to pave the way for others to reach enlightenment by teaching Buddhism. From there comes the belief: the necessity "to bring all of creation with us to enlightenment," to be able "to not leave them to their fate", or: "Grass, trees, earth – everything becomes Buddha" (Klöcker and Tworuschka, 1986). Respect is not only for nature in general, but for every individual life-form. An Indian doctrine (*Ahimsa*), not limited to Buddhism but also found in Hinduism (i.e., Mahatma Gandhi), prohibits any harm or killing of life. The parallel to Indian Jainism, which also arose at the same time as Buddhism, will also be clarified here.

4.2. HINDUISM

Hinduism (similar to Buddhism), the religion in India, where the monsoon and drought starkly present the changes of existence and disappearance, is more a way of life than a theory. Man experiences himself as part of the world. Between him and the rest of life there is no perceived basic difference as between animate and inanimate nature. The environment means "the entire surroundings consisting of the relationships among spiritual, small and large materials" (Klöcker and Tworuschka, 1986). Each and every thing is spun into the order of the world (*Dharma*), which on its part ensures itself through ritual acts, which also signify Dharma. Everything in nature then has multiple values: trees represent the power of vegetation and are worshipped and rivers are not only experienced as a water supply but also as holy, especially the Mother Ganges (*mata ganga*).

Traditional asceticism has brought man closer to the ideal self-restraint. Many have tried to occasionally learn from the ecology movement and practice frugality within certain lifestyles.

4.3. CHINESE RELIGION

Chinese religion has anticipated many of the ecological ideas of our time with its varieties of universal thinking that sense the dynamics of interdependence and aspires to balance and harmony. In Taoism, the self-worth of the world and nature are qualified if a rigorous asceticism is applied at the same time.

4.4. SHINTOISM

In modern Japanese Shintoism, as in Hinduism, there are traits of ancient nature religions. To experience a small piece of this mindset nearby, one can visit the parts of the Japanese Gardens in Leverkusen or the architecture exhibit in the East Asian Museum in Cologne. The intuition of how important nature can be as a religious symbol emerges. Only because of the proximity of these examples, which show us an otherwise strange religion, can this information be confirmed.

4.5. NATIVE AMERICAN RELIGIONS

Recently, the Native American tribal religions have gained popularity among the nature religions. The speech attributed to the Chief Seattle (1855) has become a cult text of the alternative, although it appears not to be authentic. He also used nature and hunted the buffalo, but only "to stay alive." And Archie Fire Lame Deer professed: "We never took more than we needed." Every time that man harvested plants or animals, the Native Americans begged forgiveness of the sacrificed life form and thanked Mother Earth who bestows all life (Klöcker and Tworuschka, 1986).

4.6. MIDDLE EASTERN (BIBLICAL) RELIGIONS

In strong contrast to all of those just mentioned are the "Biblical Religions" Judaism, Christianity, and Islam. But as these above-mentioned belief systems weakened their ideals throughout history and negated their principles, this historical change has also affected our culture. Century-long language difficulties and ideological intentions have brought about the misinterpretation of the (Hebrew) Bible and disseminated the translations, which sometimes were totally contrary to the original ideas.

Not only maintenance is implied, but also active management. The ecologically meaningful use of the earth for livelihood remained implicit, but the mandate was increased. For example, the Bible promotes preventative action against drought and erosion. Moreover, Jewish law (Torah) promotes many specific actions (Mitzvot), which have retained great significance until now and include steps relating to ecology (Num 2; Ex 22, 4), animals (Ex 23, 5; Lev 19, 26, and 22, 24; Dtn 25, 4), and nature conservation (Dtn 20, 19 and 22, 7), even as far as the modern topic of genetic modification (Lev 19, 19), in that it forbids the inappropriate use of animals and plants.

The Koran has clearly adopted the esteem for nature created by Allah. Man must never forget the power of control over creation bestowed upon him, that his authority is limited, and that he is constantly accountable to Allah. Islam is typical in that it clearly emphasizes man's responsibility. The appropriate behavior toward nature is not determined by one's own will but out of deference to Allah as the lord of heaven and earth. As in biblical thought, in Islam too "man is the decisive problem for existence, form, and use of Creation" (Steck, 1978). Both religions recognize the right to life of nonhuman life forms. One should ask at this point: Can nature conservation be more strongly justified than by God's ownership of nature?

Christianity goes one step further in the recognition that "in Jesus Christ, the creator himself becomes the created and goes into his creation" (Klöcker and Tworuschka, 1986). In so doing, the creator sanctified his work with his presence. It is tragic that the contradictory history of Christianity provokes critics to speak of its "merciless consequences." The author of this wording (Amery, 1973) blames Christianity in assisting the over-exploitation of nature and assigns it decisive culpability for the ecology crisis.

The eastern churches have remained especially true to the approach of early theologians. That approach tries to overcome the Greek creation myth in that it interprets the creature as God's self-revelation and as proof of his love for man. The function of the creature "consists of its usefulness to man's salvation. Because the creature is contained in man, it shares his identification to take part in holy life and may be deified through the same grace as man" (Panagopulos, 1987).

It was brought up earlier that life and doctrine often diverge in historical reality (and not only in Christianity). This led to the creation of doctrines that included contradictory ideas about the behavior toward nature. It is how deism clarified the absence of a relationship between god and the world. Straight materialism was, at times, able to fascinate, but apparently not satisfy, minds. Marxism–Leninism, among others, asked as a basic tenet of its philosophy the question about the relationship between awareness and nature and matter. Man is part of nature but puts himself above it by means of work. Through it he realizes his goals, changes nature, and molds it to his ends. He must not separate himself from nature, even as its master, but must help people practice "an ever-deeper understanding and application of its laws for his goals" (Klaus and Buhr, 1972).

5. Similarities

From the many ways of thought, several profound similarities are apparent. As a preliminary result of our cursory observation, perhaps we can establish here principally:

1. The esteem for nonhuman nature was and has been required at all times and in almost all cultures by individuals and groups and has also frequently been lived.
2. Nature is awarded its own right on multiple different grounds. In the monotheistic religions (Judaism, Christianity, Islam), the weight is easily shifted in favor of man for whom an extraordinarily special position is awarded.
3. Man's entitlement to use is considered self-evident, but is limited by nature's intrinsic organizing principles. Ecologically it must be formulated as follows: the limit lies where natural resources are depleted and are threatened to become irretrievable. However, as a basis, all environment-oriented thought and behavior should be determined with responsibility and care.

6. Practice

Sweeping consequences for nature and resource conservation based on traditional religious beliefs in today's world cannot be expected, especially regarding the sea. Protection of wells and other bodies of water in dry areas and deserts was a practice of survival. Not without reason does the bible directly or indirectly mention their value (Ps. 73, 15; Js 41, 18; Jac 3, 11; Rev 21, 6). The population increase in central Europe in the Christian Middle Ages (Leguay, 1999) forced more careful behavior toward water supplies. Indeed, all references point to – at least in the relatively well-supplied area of Central Europe – wells, rivers, and seas inland precisely in the area of settlement. Marine ecosystems shifted simply in connection to coastal protection and the foundation of port cities like Venice or those in the Netherlands (Radkau, 2000). The sea was regarded as mysterious and contradictorily valued as both dangerous and beneficent (Bechtold-Stäubli and Hoffmann-Krayer, 1935, 1987). In the "Book of Nature" (1547–1550) by Konrad von Megenberg, sea life was portrayed more fantastically than realistically (translated by Sollbach, 1990). Still, he acknowledged in support that the sea was the "Father of waters" and supplied the inland water (Bechtold-Stäubli and Hoffmann-Krayer, 1935, 1987).

Accordingly, water as a natural element enjoyed a special value and its use was always bound by moral responsibilities. In spite of scientific advances, the attentiveness of churchly moral teachings and practices of care are concentrated on areas of settlement. Therefore, the present-day concrete concerns for acute water scarcity is decisive in poor countries. This is showed by the organization of a conference of

"Ecumenical Water Networks" (EWN). As an initiative of churches, organizations, and movements, it pursues the goal of "protecting and guaranteeing of water supply for all people in the world, promoting community initiatives and projects to overcome the water crisis, and ensuring that the collective Christian voice will be heard in the debate on water problems."

The Pope sent a message on March 27, 2007, to the Director of the Food and Agriculture Organization (FAO), Jacques Diouf, that may be able to illustrate the way of thinking that comes from a religiously based ethic. It states among other things (quoted by the official announcement of the roman-catholic church in Germany):

> Water is an essential commodity for human life. The management of this valuable resource must be administered so that all people, especially the poor, have access and that both people living today and future generations are guaranteed a life on this planet fit for humans.

Access to water belongs to the inalienable rights of every human and it is a precondition for the realization of many other human rights like the rights to life, food, and health. Therefore, water can "not be treated as any other resource and its use must be rational and solidary. (...) The right to water is based on the dignity of man and not on a simply quantitative value, which regards water as an economic commodity. Without water life is endangered. Therefore, the right to water is an inalienable and universal right (...)

Towards this goal, the behavior towards water must be regarded as a socio-economic, ethical, and environmental challenge that concerns not only institutions but also the society as a whole. It is a challenge that must be faced with the principle of subsidiarity, namely by the participation by the private sector and mostly too, by local communities;

- With the principle of solidarity, the cornerstone of international cooperation, that bestows special attention to the poor
- With the principle of responsibility for both present and future generations that calls for the necessary revision of ... consumption and production models

This responsibility must be widely shared and become a moral and political imperative in a world that has access to extensive knowledge and technology (...) in order to achieve dramatic consequences. However, where a mutual hydrological dependence ground in a far-sighted attitude reigns that binds the users (...) in neighboring countries to a common system, then this responsibility can become the basis for inter-regional cooperation ...

We are all called upon to create a new way of living to restore value and care to this common resource of man that we must have for our society."

All of these reputable objectives should be discharged into the ambitions, which are aimed for in the different perspectives in this book.

7. Summary

Water and its natural resources stand at the center of the articles in this anthology. Water (Hock et al., 2001) is equally a material and spiritual element of religions and doctrines of salvation. Teachings and rituals of all religions mirror the basic importance of the commodity for man, animal, and plant. This view is prevalent in an immense number of liturgies, rituals, and uses. It is promulgated in typical fashions in a multitude of myths, mythologies, epics, legends, tales of miracles, or it is rationally explained in epistles and theological accounts. It is heard as well in hymns, mantras, psalms, litanies, and prayers of all kinds. Washing (Bowker, 1999) constitutes basic ceremony in many religions. "Water – Source of Life" was the motto of the United Nations International Decade. "Water is life" wrote Antoine de Saint-Exupéry. Water is practically "a culturally universal symbol." It stands not only for biological but also for spiritual existence.

It will be remembered in the "Year of Darwin" that the conception of the development of life belongs to age-old traditions of man. It took the "minority theory" of Anaximandros of Milet (first half of the sixth century BC) (Mayerhöfer, 1959–1970) a long time before it became a common belief. He believed that present oceans were the rest of an ancient ocean and speculated that modern land creatures had aquatic ancestors. Thus the basic assumptions of modern science were anticipated by an ancient philosopher.

Mythic-religious thought is mixed with real knowledge, particularly regarding water. Rivers and sources, therefore, have been and continue to be seen as holy places. Water is most frequently related in connection to the beginning of the world. In the Indian "Bhavishyotara-purana" (31,14), water is described as the source of all existence. According to the Babylonian "Enuma Elisch," the earth came from chaotic waters. Accordingly in the bible, Yahweh floats "above the waters" (Roth, 2008) before water and land were separated from one another.

This brings to mind, likewise, sources regarded as holy places (i.e., Lourdes) and holy rivers like the Ganges, the Jordan, and the Euphrates as one of the four paradise rivers of the bible (Gen 2, 14) and the four ruin-bringing angels (Rev. 9, 14).

Water is a basic element of liturgies and ceremonies. Baptismal water is used in all Christian denominations. Easter vigil is celebrated with water sanctification. On the twelfth night after morning mass, water of the three kings is accepted. There are similarities to other religions. Holy water is also a significant part of the Yasna ceremony of a Gahambar in Zoroastrianism (Arbeitspapier, 2008).

8. References

Amery, C. (1973) *Das Ende der Vorsehung*. Die gnadenlosen Folgen des Christentums, Hamburg.
Arbeitspapier "chrichten" (2008) Ökumenischer Rat der Kirchen (ed.) Bistum Würzburg.
AWH (ed.) (1989) *Die Wahner Heide*. Eine rheinische Landschaft im Spannungsfeld der Interessen, Natur- und Umweltschutz als ethische Verpflichtung, Köln, pp. 13–22.

Bechtold-Stäubli, H. and Hoffmann-Krayer, E. (eds.) (1935, 1987) In: Hünnerkopf. Meer. *Handwörterbuch des deutschen Aberglaubens.* Vol. 6. de Gruyter, Berlin, Sp. 65–69.

Bowker, J. (1999) *Das Oxford-Lexikon der Religionen.* Patmos Verlag, Düsseldorf.

Hock, K., Ernst, J., Kranemann, B. and Intorp, L. (2001) Wasser, In: *Lexikon für Theologie und Kirche.* Herder-Verlag, Freiburg, Basel, Rom, Wien, **10**, column 984–989.

Klaus, G. and Buhr, M. (1972) *Marxistisch-Leninistisches Wörterbuch der Philosophie.* Rowohlt Verlag, Hamburg.

Klöcker, M. and Tworuschka, U. (1986) *Ethik der Religionen,* In: Lehre und Leben, **5**, Umwelt.

Krolzik, U. and Knöpfel, E. (1986) *Verantwortung für die Schöpfung.* Materialien Nr. 7, Paderborn.

Leguay, J.-P. (1999) *La Pollution au Moyen Age – dans le Royaume de France et dans les Grands Fiefs.* Gisserot, Paris.

Mayerhöfer, J. (1959–1970) *Lexikon der Geschichte der Naturwissenschaften.* Brüder Hollinek, Wien.

Panagopulos, J. (1987) *Grenzen der christlichen Menschenlehre (Begegnungen mit der Orthodoxie),* Frankfurt.

Radkau, J. (2000) *Natur und Macht. Eine Weltgeschichte der Umwelt.* Beck Verlag, München, pp. 142–153.

Roth, H.J. (2008) Message of the Bible or Theory of Darwin? An interdisciplinary statement on the current controversy in Germany, In: J. Seckbach and R. Gordon (eds.) *Divine Action and Natural Selection. Science, Faith and Evolution.* World Scientific Publishing Company, Singapore, pp. 649–659.

Steck, O. (1978) *Welt und Umwelt. Biblische Konfrontationen,* p. 153.

Biodata of **Shlomo E. Glicksberg**, author of *"Global Warming According to Jewish Law: Three Circles of Reference"*

Dr. Shlomo E. Glicksberg obtained his second degree in Jewish History and his Ph.D. in Jewish law at the Bar-Ilan University in 2006. His doctoral thesis was on the subject of the preventing of environmental hazards in the Jewish law. He teaches at Efrata College, the Lander institute Jerusalem academic center, and Bar-Ilan University. Dr. Glicksberg is the rabbi of a large congregation in Jerusalem and the co-editor of "Siah Sadeh," which is an online journal on Judaism and the environment. His research interests are the historical, legal, and philosophical aspects of Jewish law with a focus on the meeting points of Judaism and the environment.

E-mail: **glicksberg@neto.net.il**

GLOBAL WARMING ACCORDING TO JEWISH LAW: THREE CIRCLES OF REFERENCE

SHLOMO E. GLICKSBERG
Efrata College of Education, Ben Yifuneh 17, Baka'a Jerusalem, 91102, Israel

1. Introduction

Judaism's literary sources are roughly divided into two main categories: Halakha and Aggadah. The Halakha consists of a system of practical rules and instructions, on two main levels: commandments between man and God, and between man and his fellow man. The Halakha, originating in biblical directives, was edited and formulated into a comprehensive codex around the year 200 CE, and called the Mishnah. The discussions on the Mishnah continued to branch out, and in 500 CE the Babylonian Talmud was compiled. The Halakhic body of literature continued to develop, and besides the Halakhic sources, one can mention – beyond the Talmudic commentative literature – also the codification literature of Jewish law, such as Maimonides' *Mishne Torah* (twelfth century) and Rabbi Yosef Karo's *Shulchan Aruch* (sixteenth century), as well as the responsa literature.

The Agaddah, which also originates in the study of biblical texts, is found within the Babylonian and Jerusalem Talmuds and collections of *Midrashim*,[1] and constitutes the primary basis for Jewish philosophy and theology. Books of diverse styles were written based on the Aggadah, to name a few: philosophical literature such as "Faiths and Beliefs" by Rabbi Saadia Gaon (ninth century), or Maimonides' "Guide of the Perplexed" (twelfth century); Mystical literature (Kabbalah) such as the "Zohar" attributed to Rabbi Shimon Bar Yochai (discovered in the thirteenth century), or Rabbi Haim Vital's *Etz Haim*, "Tree of Life" (sixteenth century); Hassidic literature such as *Toldot Yaakov Yosef*, "The History of Yaakov Yosef," by Rabbi Yaakov Yosef Katz of Polnoye (eighteenth century), or the *Tanya* by Rabbi Shneur Zalman of Liadi, the founder of Chabad (eighteenth to nineteenth century).

The issue of the attitude to nature and the environment is given wide expression in the above-mentioned sources. This survey will first bring a general description of the attitude toward the environment found in sources of Jewish thought, and

---

[1] *Midrashim* – Rabbinical commentary on the scriptures and oral law.

will then focus on the practical sources of Halakha, Jewish law. The aim here is to illustrate the ways in which the religious scholars from the first centuries of the first millennia dealt with environmental hazards of their time, and, in so doing, laid the Halakhic-legal foundations for later-day Halakhic scholars, when they came to address the issue of more modern hazards.

2. Three Different Approaches to Man–Environment Relations in Sources of Jewish Thought

Theoreticians and environmentalists generally refer to two contrasting perceptions of man's place in relation to the world surrounding him: the anthropocentric perception, which places man at the center, and perceives him as having stewardship or mastery of world resources[2]; and the contrasting biocentric perception, which places nature at the center, and views man as one of the species whose importance is no greater than those of other species in the world, as arises from the following description:

> The word anthropocentrism, whose roots are anthropo, or "human", and centrism, or "center", is a buzzword in environmental thinking. It expresses the notion that the world was made expressly for humanity. Many think that as long as we see ourselves as center (and master) of the universe, there will be no end to the environmental crisis. In today's environmental debate, the strongest arguments against the anthropocentrism come from the Deep Ecologists, who call for a "biocentric" world view.[3]

In forming an opinion on the issue of man's status in the world, one would assume that surely man is of central importance, since he was given the task to preserve nature, "to work it and guard it."[4] And yet, a review of various sources shows that besides

---

[2] Within the anthropocentric approach, two different approaches should be discerned: stewardship versus mastery. On this, see: Schwatz E Response (2002). Mastery and stewardship, wonder and connectedness: a typology of relations to nature in Jewish text and tradition. Judaism Ecol 93–108, Cambridge, MA.

[3] Bernstein E (ed) (1998) Ecology and the Jewish spirit. Jewish Lights Publishers, Woodstock, VT, pp. 230.

[4] Genesis 2:15. Notably, even this anthropocentric approach, which shows concern for man and his needs, could be viewed as having not only man's immediate needs in mind, but also the needs of mankind in generations to come. It seems that several sources illustrate this concern. One example for this is the discourse in the Babylonian Talmud in *Taanit*, 23a, which illustrates the importance of planting a carob tree, whose fruit a person might not enjoy himself, but which will provide fruit for future generations.

the anthropocentric approach in Judaism, the biocentric approach also exists.[5] See, for example, in the book of Job, when God addresses Job from out of a storm, His concern also included places uninhabited by people.[6] Later, too, God enumerates different species for which He is concerned, even though man has no need for them. Thus, God's concern for the world is direct, and is not necessarily related to man's welfare.[7] This also seems to be the understanding of Maimonides, that man, or humanity, is only one species of Creation amongst many.[8]

Of course, to these sources one must add many other sources of Jewish thought which express the biocentric approach. One should especially mention the perception of Beshtian Hassidism (pertaining to the Baal Shem Tov), which provides in its sources a clear illustration of this approach.[9]

However, for a more accurate understanding of some of the sources, one must refer to a third approach, one not mentioned so far: the theocentric approach, in which neither nature nor man is at the center, but rather God, or at least the consciousness of the presence of God.

The belief in the connection between the world and God, and in God's intervention in world events, consequently making demands on humanity in return, is one of Judaism's foundations, and can therefore determine man's attitude toward God's world.

This principle also constitutes Halakhic norms, the clearest example of which is found in the bible as explanation for the commandment on the jubilee year: "The land must not be sold permanently, because the land is mine."[10] The explanation accompanying the commandment expresses a clear theocentric perception, according to which the fact and acceptance of God's ownership of the land has practical consequences, and according to the famous Midrash, although "All I have created, I created for you," nevertheless, "Take care not to corrupt and destroy my world."[11]

---

[5] See expansion in Fink D Between dust and divinity: Maimonides and Jewish environmental ethics. In: Bernstein E (ed) (1998) Ecology and the Jewish spirit. Jewish Lights Publishers, Woodstock, VT, pp. 230–239.

[6] See, Job, 38: 1, 26–27.

[7] See, Ibid, 40:15–32.

[8] Maimonides, Guide for the perplexed, Part III, Ch. 13 as well as in Ch. 12.

[9] Studies have already been written on the subject, to mention just one: Manfred Gerstenfeld and Netanel Lederberg, "Nature and the Environment in Hasidic Sources", Jewish environmental perspectives, 5 (October 2002), pp. 1–11.

[10] Leviticus 25:23. Rashi's comments on this verse, that man should not be selfish about the land, since it does not belong to him.

[11] Ecclesiastes Rabbah (Vilna edition) 7:13. Yet, the main emphasis in the literature of the Sages and the diverse commentative literature is undoubtedly anthropocentric, placing man and his education at the center of importance, besides the emphasis on the responsibility of mankind toward nature and toward other generations who also deserve to enjoy that nature.

And yet, despite the existence of several biocentric sources, it seems that their main significance lies in their ability to create a spiritual climate that educates to viability and to an enhanced sensitivity toward the environment. However, these sources cannot serve as the infrastructure for an applicable legal doctrine containing clear instructions on what is allowed and forbidden, and how a society committed to these codes could apply and enforce their implementation.[12] For this, we must turn to the Halakhic sources and their analysis:

3. Man–Environment Relations in Halakhic Sources: Three Circles

Unlike the sources of Jewish thought, Jewish civil law, which contains the laws concerning nuisances, is fundamentally anthropocentric: The Halakha seeks to protect the individual and society from various hazards.

Generally, the Halakhic discussions can be described according to three circles of reference: local, public, and global. Regarding the first two circles, Halakha offers an organized legal doctrine. Indeed, in the absence of updated scientific knowledge available to the religious scholars of the relevant period in which this genre was created, no clear guidelines were formulated on the management of modern environmental issues, including global ones. However, within the framework of reference to the local and public circles, in the reality addressed by scholars of the Mishnah and Talmud periods, Halakhic foundations were laid, not only concerning modern hazards, but even the third circle of reference.

The Halakhic approach does not speak in terms of rules (conceptualization) but of cases (casuistics).[13] However, the modern *posek* (decisor) is accustomed in many instances to analyze situations referred to by earlier sources, to convert the cases to principles, which then could be applied to a changing reality.

The main corpus of these discussions is found in the Mishnah, in the second chapter of *Bava Batra* Tractate, which actually consists of the last section of the original *Nezikin* (damages) Tractate. As nuisances are the main focus of this chapter, consisting of 14 verses, or *Mishnayot*, its central principles will therefore be the focus of discussion here.

The novelty of this chapter, even prior to going into its details, is the very fact that it prohibits not only acts that harm another's body or property, but also

---

[12] Although, as we have shown elsewhere, it seems that the sources of Jewish thought did in fact influence Halakhic rulings, even if indirectly, by establishing a stricter norm, demanding a higher measure of piety and going beyond the letter of the law. See Glicksberg S (2005) Ecology in Jewish law: preventing personal environmental damage. Ph.D. Dissertation, Bar-Ilan University, Ramat Gan, pp. 273–276, 312–313, 315–317 (Hebrew).

[13] L., Moscovits, Talmudic reasoning: from casuistics to conceptualization, Tubingen 2000.

ones that detract from ones' quality of life or tranquility. Nuisance laws, therefore, introduce the concept that quality of life is also a "protected interest," which is legally protectable.

Here, we will first describe the two first circles and will then describe in brief some principles derived from them, and which could serve as a platform for an applicable Halakhic doctrine for the management of modern environmental hazards, including global ones, which go beyond the responsibility of the individual or the local public, and relate to the world at large.

3.1. THE LOCAL CIRCLE

Some examples: The Mishnah requires the distancing of hazardous materials from the neighbor's property: "A man should keep olive refuse, dung, salt, lime, and flint stones at least three handbreadths from his neighbor's wall or plaster it over."[14] The Mishnah makes similar demands on ones' conduct toward his upstairs neighbor: "An oven should not be fixed in a room unless there is above it an empty space of at least 4 cubits. If it is fixed in an upper chamber, there must be under it paved flooring at least three handbreadths thick, or for a small stove one handbreadth."[15]

And so also in the following Mishnah: "A man should not open a bakery or a dyer's workshop under his neighbor's storehouse, nor a cowshed. In point of fact the rabbis permitted [a bakery or dyer's workshop to be opened] under wine, but not a cowshed."[16] The Mishnah also requires the protection against noise: "One may protest against the opening of a shop within the courtyard and claim, 'I cannot sleep due to the noise of those entering and exiting'.[17]

3.2. THE PUBLIC CIRCLE

The Mishnah protects not only the individual's property and quality of life but also the public and its property. For reasons of protection from air pollution, the Mishnah requires the distancing of various hazards from inhabited settlements: "Animal carcasses, graves and tanneries must be distanced 50 cubits from a town. A tannery may be set up only to the east of a town. Rabbi Akiva says it may be set up on any side save the west, and it must be distanced 50 cubits [from the town]."[18] The Mishnah also prohibits air pollution caused by waste, and thus prohibits threshing grain

---

[14] Mishnah, Bava Batra, 2:1.
[15] Ibid., 2:2.
[16] Ibid., 2:3.
[17] Ibid.
[18] Ibid., 2:9.

close to town: "A permanent threshing floor may not be made within 50 cubits of the town. One may not make a permanent threshing floor within his own domain unless his ground extends 50 cubits in every direction, and he must distance it from his fellow's plants and ploughed land so that it will not cause damage."[19]

A further Mishnah prohibits the planting of a tree near an inhabited area, and even requires its uprooting once planted: "A tree may not be grown within a distance of 25 cubits from the town, or 50 cubits if it is a carob tree or a sycamore tree."[20] The Mishnah does not explain the measures taken or the nature of the nuisance. However, the discussion in the Babylonian Talmud on this issue quotes the Amora (Talmudic Sage) Ullah on this, that the preservation of the town's aesthetic beauty is the reason for distancing the trees, so that they will not obscure the view seen from the town, which would detract from the town's beauty. Later, *poskim* concluded from these sources that the town's aesthetic appearance should be nurtured, and that legal recourse could be pursued to remove any hindrance to this.[21]

According to the *Braita*[22] quoted in Bava Kama Tractate, on a town with a public standing such as Jerusalem, even stricter demands were made concerning quality of life and the environment: "Ten special regulations were applied to Jerusalem:... that neither beams nor balconies should be allowed to project there; that no dunghills should be made there; that no kilns should be kept there; that neither gardens nor orchards should be cultivated there."[23]

Why is the planting of gardens and orchards in Jerusalem forbidden? The Talmudic reason, due to "*sircha,*" is explained by the commentator Rashi as noxious odors, which originate either in the weeds thrown out from the gardens or in fertilizer. Thus, the cultivating of gardens and orchards is forbidden to prevent ecological nuisances. However, the modern-day Talmudist Jacob Nahum Epstein used philological comparisons to explain that the Talmudic term actually referred to a "Sandfliege" (sand fly) – a small fly found especially in gardens.[24]

The discussion in the Babylonian Talmud also explains the prohibition to light kilns, which is due to the smoke rising from them.[25]

In the *Tosefta*[26] to Bava Batra Tractate, a general rule was formulated by Rabbi Nathan stating that kilns must be kept 50 cubits from any town.[27]

---

[19] Ibid., 2:8.

[20] Bava Batra 2:7.

[21] On this Halakha and its evolvement, see, S. E. Glicksberg, "The tree as a nuisance, the evolution of a Halacha", *Mo'ed* 18 (2008), pp. 1–11.

[22] *Braita* – Tanaic teachings external to the Mishnah.

[23] Bava Kama 82:2.

[24] J.N. Epstein, "Talmudic Lexicon", *Tarbitz* 1 (1930), p. 124 (Hebrew).

[25] Bava Kama 82b. Rashi explains that the smoke would blacken the walls, which would disgrace the city.

[26] *Tosefta* – A collection of *Braitas*, constituting a supplement to the Mishnah.

[27] Tosefta, Bava Batra Tractate, 1:10.

3.3. THE GLOBAL CIRCLE

As stated, in the absence of updated scientific knowledge on the capability of mankind to cause real damage to the world and its resources,[28] the Sages did not provide precise guidelines as to what is allowed and forbidden. Nevertheless, analysis of the sources, part of which we brought here, raises several principles for implementing a Halakhic nuisance doctrine in modern times. These principles, defined and formulated by Halakhic scholars over centuries, referred originally to the two circles mentioned. However, it seems that they could be implemented also in cases of modern nuisances, even regarding those nuisances which harm not only a neighbor or residential area, but the whole world.

4. Principles for Removing Modern Nuisances

Following are several applicable principles:

4.1. URBAN ENVIRONMENTAL PLANNING[29]

Serious ecological nuisances will be removed from the town up to a distance from which they will no longer cause harm. Town planning should include the allocation of industrial areas located at a distance from town, to ensure that industrially related activities will not take place within city bounds.[30] Cottage industry can

---

[28] Nachmanides (Spain, thirteenth century), in his commentary on Deuteronomy 22:6, acknowledged the possibility that man could bring about significant changes to Creation: "This too is a commandment prohibiting the killing of an animal and its young on the same day. Alternatively, the text prohibits the extinction of an entire species, even if it permits slaughter of animals within a given species. One who kills the mother and her young on the same day or captivates them when they are free is considered as if he had eliminated an entire species." In this statement, Nachmanides undoubtedly was ahead of his time, since in the Middle Ages there was no technical scientific ability to foresee how mankind's actions could indeed harm nature. Many suppose that the notion that human activity could affect nature was first conceived of in the nineteenth century, with the publication of the essay by George Perkins Marsh, "Man and Nature," in 1864. See Botkin DB (1990) Discordant harmonies: a new ecology for the twenty-first century. Oxford University Press, Oxford, p. 32.

[29] Glicksberg S (2001) The attitude of Jewish law to public nuisances. M.A Thesis, Touro College, Jerusalem, pp. 9–52; Zichel M (ed) (1990) Ecology in Jewish sources, Bar Ilan University, Ramat Gan (Hebrew), pp. 87–104.

[30] Bava Batra, 2:7–9.

take place within town, but adequate distance must be maintained from residential homes.[31] These distances will be determined by experts, who will set regulations on this matter.

The aesthetic appearance of the town should be strictly kept,[32] including open spaces surrounding the town. Even stricter regulations will be maintained concerning towns with a public standing like Jerusalem.

However, prevention of existing nuisances is not enough. One should also plan the prevention of potential nuisances which could evolve in the future into serious hazards.[33]

4.2. SCIENTIFIC CRITERIA IN IDENTIFYING AND LOCATING A NUISANCE

The process of identification and assessment of hazards and their damage is a Halakhic-legal process, which can be learned from studying the various *Mishnayot*, the commentators on them and the decisors following them. This process can be incorporated into the management of modern-day hazards, including some "global" hazards.

A number of stages can determine whether the removal or prevention of a certain nuisance is effective or not. The process is as follows[34]: First, the nuisance must be identified as one that the Sages or *Poskim* would have acted upon to prevent. Even when it comes to modern nuisances which are not specifically mentioned in the texts, experts (using scientific criteria) can possibly characterize them as such.[35] Activity deemed by expert opinion as hazardous will be distanced or prohibited altogether.

A subsequent stage consists of the assessment of the severity and extent of damage, since not all activity levels or dosages are actually hazardous. This examination will also be made according to expert opinion.[36] The next stage will be to determine the distance to which the nuisance will be distanced, if at all, also according to expert opinion.[37]

---

[31] Ibid., 2:3.

[32] The definition of aesthetics will change according to time and place. See, Glicksberg, Moed, pp. 1–11.

[33] Bava Batra 17b; Maimonides, *Mishneh Torah*, Laws of Neighbors, Chapter 9, Halakha 10.

[34] For detailed discussion, See, Glicksberg, Ph.D. Dissertation, pp. 222–246. (n. 11).

[35] Thus, for e.g., in *Responsa Tziz Eliezer*, 15:39, which prohibits cigarette smoking, following his review of scientific findings, from which he learned that cigarette smoking indeed was a health hazard.

[36] *Responsa Ribash* (Rabbi Isaac bar Sheshet Barfat), 196; *Responsa Tashbetz* (Shimon Ben Tsemah Duran), 4:57.

[37] The Rama (Rabbi Moshe Isserles), *Choshen Mishpat* 155:20.

Occasionally, legal obstacles could be placed before the removal of a nuisance, as when its owners have legal rights due to precedence (*Chazaka*: legal presumption). However, when the nuisance is declared intolerable according to scientific criteria and norms, the owner could be forced to remove the nuisance, even in cases of lengthy possession, since in cases of severe damage, *Chazaka* does not apply.[38]

4.3. DISTANCING, MINIMIZING, OR PREVENTING A NUISANCE

A principle common to all the mishnah verses we reviewed was the obligation to distance the nuisance to a place where it is no longer harmful. As clarified in the previous section, the distances set by the sages in the Mishnah are not dogmatic, and can alter according to time and place, as determined by professionals.

The above refer to modern-day local and public nuisances. However, when referring to global hazards, expert assessment can determine that the hazard shall not only be distanced, but occasionally also have its activity minimized[39] and even prohibited altogether. This is due to the nature of a global perspective, which would naturally claim that the mere distancing of a nuisance is ineffective, since the issue is not damage to a local population but to world population and the world at large. Certain hazards cause the same degree of harm to the entire population. Therefore, the activity's damage versus its necessity should be weighed, and a decision made on the action needed, whether the activity should be removed, minimized, or prohibited entirely.

4.4. INDIRECT NUISANCES AND THE ABILITY TO ENFORCE THE MEASURE OF PIETY

The Babylonian Talmud discusses indirect nuisances, which although damaged a person's body or property, but in an indirect manner. The Talmudic discourse brings the opinion of Rabbi Yosef, that while one does not have to compensate another for

---

[38] The discussion in the Babylonian Talmud states that in cases of serious hazards such as the stench from lavatories or smoke, Chazaka does not hold.

[39] The style in the *Mishnayot*, which lay down not only prohibitions but also a proactive suggestion for a way of life for both sides by ways of distancing hazards, rather than prohibiting them entirely, can be understood as aiming at minimizing, but not preventing, damage. However, there are also cases where the sages totally and unequivocally prohibited the existence of hazards. Thus, for example, the *Mishnah* forbids the grazing of sheep and did not suffice with the distancing of the grazing areas from places of inhabitance or by allocating areas for this purpose. "It is not right to breed small cattle in the land of Israel" (Bava Kama 7:7).

creating a nuisance, it is prohibited to indirectly cause damage (*gerama*),[40] and one should prevent the creation even of an indirect nuisance.

Elsewhere, this author claimed that quite possibly some modern-day nuisances would be too indirect to be included even under Rabbi Yosef's prohibition of indirect damages. However, there exists an additional halakhic level, which is referred to as "beyond the letter of the law" (*lifnim mishurat hadin*), or the "measure of piety" (*midat chassidut*), and at times, the Sages applied to the nuisance owner to be more stringent with himself than required, and the decisors of each generation were even given the authority to impose such measures.[41]

4.5. LEGISLATION

The Sages of each generation made halakhic decisions on issues brought before them, but also had legislative authority. In many cases, the decisors did not suffice with analyzing and interpreting Halakhic sources to reach decisions, but also applied the decisions' spirit to create independent, new directives (*Takanot*). A similar authority was given to the public and its representatives. Over the generations, halakhic scholars and community leaders legislated many takanot for the protection of the environment from various hazards.[42] This course enables us to determine that even hazards which have not been proven to be scientifically harmful, if they were feared by the public, could be forced to be removed,[43] minimized, or prohibited.

4.6. REMOVING ONESELF FROM DAMAGE

Notwithstanding the unequivocal position of the *mishnayot* surveyed above, prohibiting the existence of hazards, the Halakhic ruling has been set according to the more lenient method of Tanaic (Mishnaic sage) Rabbi Yossi, who established that it is incumbent on the one who suffers the damage to distance himself from it.[44]

However, this approach was qualified by later-day Talmudic sages, who stated that in clear cases of hazards, as direct as a hit of an arrow, Rabbi Yossi concedes that the onus is on the perpetrator of the nuisance rather than the one subjected to it.[45]

---

[40] Bava Batra 22:2.
[41] See Glicksberg, Ph.D. Dissertation (n. 11).
[42] For a list of such *Takanot* (legislation), see Glicksberg, Ph.D. Dissertation, pp. 348–366 (n. 11).
[43] Formulators of *Takanot* could also update the precise distances, as determined by experts and as explained above.
[44] As is clarified by the discourse in the Babylonian Talmud on Bava Batra 18b.
[45] Bava Batra 25b.

In a parallel discussion regarding the distancing of a tree from a neighbor's pit, Rabbi Yossi's opinion, which stood in contradiction to the other sages, did not require the removal of the tree. The reason for this was given in a *Tosefta*: that "the sole purpose of having fields is for planting."[46] A similar tone is found in the Jerusalem Talmud: "The world is settled with trees".[47] Indeed, these sources reveal the position of the "nuisance owner," whose viewpoint was not considered by the Sages disagreeing with Rabbi Yossi.

Rabbi Yossi's point of reference, on which the Halakhic ruling is based, is that generally speaking, the legislation aims at striking a balance between different vital interests. On the one hand, the protection of the property of the one suffering from the damage, and on the other, the rights of the "owner of the nuisance," as long as the nuisances considered are not unreasonable. However, at a stage where one identifies and assesses serious hazards, the injured party is to be given legal recourse, and the authorities should enforce the removal or prevention of the nuisance.

5. Conclusion

This survey brought the environmentalist approach according to the sources of Jewish thought and Halakhic sources. Halakha and Agaddah are generally viewed as two distinct but mutually influential fields. It seems that on environmental issues too, the Agaddah and subsequent Jewish thought should be seen as providing the spiritual climate educating toward strict maintenance of the environment, while the Halakha provides the guidelines and tools. As shown, the literature of the Sages provides only a basis for the principles rather than particular guidelines on the management of modern hazards. And yet, it seems that modern environmentalist decisors and legislators could draw inspiration and motivation for their activity, both from the sources of Jewish thought, which provide guidelines for a positive connection with the environment, and from the Halakhic sources, which can illuminate how much the Sages' strived to nurture a local environment that was protected from hazards, aesthetic, and imbued with quality of life.

---

[46] Tosefta, Bava Batra Tractate (Lieberman Publishers), Chapter 1, Halakha 12.
[47] Jerusalem Talmud, Bava Batra, Ch. 2, Halakha 13, p 13:3.

Biodata of **Dr. Yisrael Rosenson**, author of *"Guarding the Globe: A Jewish Approach to Global Warming"*

Yisrael Rosenson, Ph.D., is President of the Efrata Teachers College in Jerusalem. For the past 26 years, he has been involved with teacher education and has researched the Bible and its relationship to relevant fields of inquiry, which include geography, ecology, history, and literature. He has published widely on various topics including midrash, Biblical criticism.

E-mail: **efrata@macam.ac.il**

GUARDING THE GLOBE: A JEWISH APPROACH TO GLOBAL WARMING

YISRAEL ROZENSON
*Efrata Teachers College, 17, Ben-Yefuneh St., PO Box 10263,
Jerusalem, 91102, Israel*

1. Introduction

For anyone conversant with the world of modern Judaism, it is almost pointless to emphasize the steadily rising preoccupation with questions regarding global ecology. If, in the recent past, the issue focused on the doctrines of a broad-minded few who realized the potential for a significant *environmental theory* that is rooted in various forms of Jewish sources over the generations, today, we may find *environmental ideology* in thematic literature and commentaries, as well as in popular literature, oral sermons, lectures, educational, didactic and pedagogical material, and other kinds of religious matter. This is true also for conferences on these subjects, institutions dedicated to them, and much more. The bibliography of materials that combine Jewish matters with environmental ones is thus expanding at a rapid pace, which means that herein, we will not be able to devote ourselves to the matter except in the most cursory manner. Nevertheless, one point bares clarification: Judaism has raised upon its banner the *Halacha* as its guiding element, and the question of ecology is beginning to take root even here. There are many *Halachic* anthologies and even entire sections devoted to the subject and its implications. Even if it is still difficult to talk about an abundance of material that enables the organization of Jewish life based on an ecological agenda, there are already nascent signs of this.[1]

---

[1] Determining this matter is not simple. There are problems in defining which groups in the religious world we are dealing with, and which form of socioreligious segmentation to adopt. For example, *Tehumin* is an important journal on *Halachic* matters that devotes much of its space to religion and science. In its first 20 volumes, only two papers were found on the topic: (1) Rabbi Yehuda Shaviv: *Beauty and Eternity – Chapters on Jewish Ecology*, in vol. 12, 5751; (2) Rabbi Yossef Gabriel Bekhoffer: *Recycling and the Halacha*, in vol. 16, 5756, pp. 302–296. Beginning 5760, the number of papers grew somewhat; (3) Rabbi Mordechai RALBAG: *Neighborly Damage – Noise, Visual Damage and Humidity*, in vol. 23, 5763, pp. 434–442; (4) Rabbi Shmuel David: *Cutting Down Fruit Trees to Settle Disputes Between Neighbors*, in vol. 26, 5766, pp. 286–292; (5) Prof. Gideon Sinai

The specific topic of concern in this chapter is *Global Warming* that is due to unbridled human activity and which does not take natural values into consideration. For our purpose, among the wealth of ecological problems that present themselves, global warming encapsulates a unique challenge, primarily because – in terms of topics that may be raised within the Jewish-ecological discourse – it is the most *distant*. What do we mean? First of all, we should state that it is the most distant from traditional subjects; awareness of the problem is recent, and only now is its entire import being realized. It is not surprising, therefore, that aside from Jewish-ecological slogans and utterances, there is very little material that delves into this matter.

A fundamental observation should be made on this matter. When dealing with new issues of environment based on earlier sources, the *time barrier* is a hindrance. Indeed, when dealing with ecology, there is little resemblance between our era and that of the Bible and the *Mishnah*, even if now and then we may recognize some similarities. Dealing with persistent problems that directly concerned our forefathers as they concern us is not like dealing with problems that have only recently arisen. Clearly, even if our topic of concern is not completely uniform, a matter that is not purely and simply technical has changed over the years, and the problems with which it challenged people living in earlier times are not the same as the problems that arise today. Stated more succinctly, dealing with a putrid pile of garbage outside one's front door can lean directly on earlier sources, even if yesterday's garbage is today's waste-material – although the latter admittedly contains a higher level of plastics than the former. On the other hand, when that waste resides in the atmosphere, different as it is in its form from traditional waste, dealing with it using traditional tools of thought becomes much more difficult.

To what do we refer when we use the term "to deal with"? Moral, philosophical and – to a certain extent – practical conclusions based on ancient texts in relation to new problems are of course possible, but these require hermeneutic insights and an intellectual approach that differs from that used when dealing with direct *Halachic* rulings. At the root of the matter lies the need to derive extremely general principles and only then the ability to examine the application of a principle to a new reality. To elucidate these relationships, the discourse encompasses several basic sources. For the following proposed analysis, a compilation of *Mishnaic* paragraphs in the *Baba Batra* tract is of special interest. These deal with concrete hazards and damage that threaten relatively small congregations – cities etc. Alternatively, it is no less important to examine those same Biblical sections that are wider in scope and which examine the world on a wider level. Indeed, at

---

and Avner Hazut: *Using Sewage Water – Halachic Thoughts*, in vol. 28, 5768, pp. 129–134; (6) Rabbi Moshe Shefter: *The Usage of Treated Water – a Reaction*, pp. 132–135. For the purpose of this paper, we may define these as environmental topics, but not really concerned with global matters. (The bibliographical material mentioned herein and in the annotations following is in Hebrew. Excerpts have been translated to indicate the scope and nature of subjects at hand.)

present, there is widespread interest in both kinds of sources, and this expresses itself extensively in religious literature,[2] especially since the matter at hand has not been independently and significantly dealt with.

Here is an attempt to derive conclusions from *Mishnaic* tracts whose essence deals with local ecology and to jump-start a sort of extrapolative process, to *stretch* some ecological qualities beyond the original intent of the texts. Concurrently, we will focus on that Biblical section dealing with what, to many readers, seems to deal with the globe as a complex of sorts – the Garden of Eden – and initiate an interpretation that will extract those conclusions from their localized context, adapting these to a global and universal viewpoint. Our fundamental claim is that dealing with these two forms of sources is directly pertinent to a more comprehensive and inclusive position.

2. Baba Batra: A Comment on the Order of the Tract

We will present a specific *Mishnaic* doctrine from the *Baba Batra* tract, which we feel may be considered a source for the discourse on preventing atmospheric pollution. Prior to this, however, we should delineate some guidelines for a simple and fundamental idea that arises from the organizing method of the tract in general. As we know, *Baba Batra* deals with the prevention of damages. The specific doctrine that will be presented here is part of a wider construct that has a basic value. Here is a brief presentation of this structure.

2.1. CHAPTER 1

Chapter 1 deals with *partnership*; the first word in the tract is "ha-shutafin" (lit: *the partnership*). The subject is the ability of a society's – relatively small though it may be – to perpetuate relationships of partnership and mutual responsibility. An important doctrine that relates indirectly to our topic in this chapter is: "Partners of a courtyard must share in the expense of building a gate or a door to it. ... An inhabitant of a city has to share in the building of a wall around the city, with the doors and the bolts." (*Baba Batra* I, 5). Theoretically, what is described here is a process that contradicts accepted environmental procedures,

---

[2] Two essays are significant, we believe: The third chapter in Nahum Rakover: *The Environment – Ideological and Legal Aspects in Jewish Sources*, Jerusalem, 5754 (pp. 47–76), deals with environmental pollution, and contains a sub-heading on air pollution (pp. 65–70), where the *Halachic* approach is clearly expressed. In Mansfred Gerstenfeld's book, *The Environment in Jewish Tradition – a Sustainable World*, Jerusalem, 2002, the emphasis is directly ideological, a development of ideas that may be relevant through maximal use of the Bible (one interesting example is the discussion of the *Manna* – the perfect food, pp. 81–82). The author also has tried to take the same approach in his essay, Y. Rozenson: *And It Was Very Good*, Jerusalem 5762.

the fortification of a society within its own defined and limiting boundary – those of the land within the land and those of the city within the city – alongside the demarcation ("a gate or a door," "a wall around the city, with the doors and the bolts") that protects them from the threatening exterior. Simplistically read, the exterior represents flesh-and-blood enemies; taken on a more symbolic level, it represents the opposite of civilization, which threatens to intrude upon and disrupt the daily life of an urban setting. In fact, there is no intention of being involved in this exterior or to change it, but rather to protect one's self from it in the manner in which a civilization normally protects itself. We should reiterate that protecting civilization from its outside enemies is a task undertaken by force of mutual responsibility. A society's self-organization, which is based on the model of protection and partnership to the point where the means of protection may be enforced upon the individual (*him*), may serve as the foundation for protection against a new enemy of society – the ecological hazard. The walls that are erected by this protective organization will, of course, be of a totally different kind, as we will clarify here.

2.2. CHAPTERS 2 AND 3

The tract deals with substantial hazards – one of which will be the topic of our next section – and the general collaboration for their prevention. In dealing with the general structure of these chapters, one should stress that the issue of *Hazaka* (lit: *occupancy*), that is, the determination of private ownership over assets, only arises later on, in Chapter 3: "The law of *hazakah* is, if one has occupied any property for three years from date to date ... and this applies to houses, pits, excavations, caves, pigeon-coops, bath-houses, press-houses, dry land, slaves, and the same is with all other articles which bring fruit frequently." (*Baba Batra* III, 1) Thus, the structure is as follows: (1) partnership and its expression in an area; (2) partnership as a way of dealing with various hazards; (3) occupancy.

We will not deal herein with the actual law, but with the implications of the arrangement, and we assume – of course – that the literary order of the tract is manifest and meaningful. The tract is not a coincidental conglomeration of doctrines! And we attach great ideological importance to the precedence of rules of partnership to those of occupancy – a form of purchase, the formation of private ownership: first comes the ability to generate a relationship based on responsibility toward one's fellow man and society, and only then the means of private purchase. Through methodical study of the tract from beginning to end, one internalizes first the idea of partnership as a prerequisite for ownership; within this framework and on its basis, communal ability to mobilize for the prevention of hazards is incorporated; the latter derives from partnership and precedes ownership, since responsibility precedes ownership.

These simple ideas, entrenched in the tracts formulation, do not directly bear upon the problems of atmospheric hazards; but they offer a wide ideological basis for discussing the topic.

3. The "Barn" Doctrine

The principles that may guide an educational discussion on the problem of global warming can be derived from specific doctrines. Here, we will deal with one of them "A barn must not be placed within 50 ells of the town; the same is the case if one wishes to make a barn on his own property – he may do so, provided he has 50 ells of space on each side of it. One must also remove a barn from the plants and from the newly ploughed field of his neighbor ... to prevent any harm to the plants or the field" (*Baba Batra* II 8).

This doctrine allows for the following association. On the left, we have disassembled the doctrine to its elements, and on the right we provide a basic ecological interpretation to the elements: Thus, the city, which in the previous chapter – by force of partnership – was enclosed by a protective wall (*Baba Batra* I 5), is now protected through entirely different means: not using a physical wall that may prevent manifestations of external violence, but through the legislative prevention of a trivial hazard deriving from simple activities – a hazard that is theoretically unpreventable: who could imagine life without a barn?![3] A barn, indeed, bears little relationship to global warming; but it presents us with an ideological basis for comparison, if only thanks to the atmospheric association, but mainly because the heat is a result of essential circumstances that are basic benefits to humans, and which – however – need to be distanced from others (Table 1).

Table 1. Old terms and their meanings.

A barn	A sustainable hazard related to the air
must not be placed	An acceptable ecological reality is created
within fifty ells of	Quantification/explanation of that reality
the town	An inhabited area/civilization
If one wishes to make a barn on his own property, he may do so,	Creating an acceptable ecological reality is the responsibility of the individual
provided he has fifty ells of space on each side	The terms
One must also remove a barn from the ... field of his neighbor	The responsibility is on the individual
to prevent any harm to the plants or the field	The rationale

---

[3] A similar interpretation may also be widely applied to the following doctrine: "Carcasses, cemeteries, and tanneries must be removed to a distance of fifty ells. A tannery must not be established except on the east side of the city; R. Akiva, however, maintains that it may be established on every side except the west, and a space of fifty ells is to be left" (*Baba Batra* II 9). However, this is beyond the scope of this discussion.

The structure of this specific doctrine is also noteworthy: in brief, the use of indicative expressions – "must not be placed," "if one wishes," "he has (left) 50 ells of space." There is a movement from the general to the specific. At first, reference is to the general (public): it "must be placed." Then, moving on to the private individual: "if one wishes." Finally, symmetrically, we return to the distancing, but in singular: "he has left." The kernel of the structure is the limitation: "must not be place;" but this is enveloped in the distancing that indicates the individual's ability to restrain and concede.

The movement in the doctrine from the general to the individual ("it must be" – "he has") is the same movement that reflects the situation of the individual within a communal society. Here, we have, not only the obligation that falls upon some form of communal authority, but also – and perhaps primarily – upon each of the community members to avoid harming a neighbor or a friend. This is an individual task and not just a public duty. The structure is conditional upon limiting personal attributes and psychologically fostering a structure that generates feelings of closeness and empathy. Under these circumstances, a technical ecological ruling becomes a means of augmenting membership and communal fraternity.

It is possible to consider the ramifications of the above on global warming. On the basis of the presentation of the above doctrine, we can imagine a metaphorical development from a small example to a greater one. The barn is a super-structure that represents damage to space and air – the atmosphere. But it is also important to demonstrate that this is a personal problem: each one of us is instructed to distance – not merely the collective or the central authority. We shall now take a sample hazard/nuisance and – based on this example – examine the wider implications.

Theoretically, a clearly utilitarian approach is being demonstrated – ecology for humans, the need to rally around the ecological banner; otherwise human benefit within the community will not be achieved. We must demonstrate restraint through partnership, relinquish rights so that others – and consequently ourselves – will be able to continue existing. Can we, through the power of such ecology, protect the planet in general? Can the barn represent a different kind of hazard than that in which its activities are traditionally rooted? This may be the case and – as in the barn whose basic danger (the chaff) is inherent – the basic fear of global warming may be raised. Indeed, the inherent danger can be presented as greater than that of the barn itself. Nevertheless, such greater problems may require another kind of argumentation.

Here, therefore, we have presented a utilitarian ecology,[4] but it is of greater educational import. It may even be said to have added *educational* value: through maintaining the ecological dimension in reality, one may strengthen values of

---

[4] Regarding related questions, see Shwartz E Response (2002), mastery and stewardship, wonder and connectedness: a typology to nature in the Jewish texts and tradition. Judaism Ecol 93–108, Cambridge, MA.

partnership; upon the ecological platform, communal unity is strengthened. However, this simplistic relationship between education and ecology is insufficient and it lacks another moral dimension, one which is presented in the story of the Garden of Eden, and which may serve as a lever for making the world an entirely better place.

4. The Garden of Eden

The primal goal of the edict – "to cultivate it and keep it" (Genesis II 15) – has become a symbol for any discussion on Jewish ecology.[5] This goal is defined in relationship to an environment whose delineated borders are quite clear. This is a *Garden of Eden* that was quite prevalent in the castles of kings, and its literary representations appear in various mythologies. Needless to say, such gardens were inaccessible to regular people; nor were they permitted to be present therein. In Genesis, on the other hand, even if a tree or trees within were divinely forbidden as a food source, this specific Garden of Eden was not holy; it is not the exclusive abode of a god or a king, but – rather – an effective habitat that offers harmonious coexistence with nature. This basic situation may lead to varying commentaries and different educational conclusions. Here, we will adopt the approach that sees the Garden of Eden as a model for fostering global responsibility, *global* since – according to the borders of the Garden as delineated in Genesis by "great rivers" – it was huge, in a sense covering the entire known world. And thus, a garden that in earlier traditions was considered holy or royal, bordered off and differentiated from regular mortals, in Exodus becomes a human habitat.

In reference to this, we should mention that the Garden is clearly defined botanically as one that contains trees. As stated, the Garden may be presented as an allegory – not merely as the location of the first human couple, but as an actual habitat. Nevertheless, in this allegory, the emphasis is on trees, which represent the environment, in spite of the fact that, ostensibly, one could have referred to other natural phenomena. For our purpose, the crucial verses are, "The LORD God planted a garden toward the east, in Eden; and there He placed the man whom He had formed. Out of the ground the LORD God caused to grow every tree that is pleasing to the sight and good for food" (Exodus II 8–9). Consider: "The LORD God planted" stresses God's action, nature. God creates a natural environment. And yet, the chosen verb, *planted* refers specifically to trees. "There He placed the man": nature, which God has planted, is a habitat. "The LORD God caused to grow" is an additional emphasis on the natural disposition. "Every tree" once again emphasizes trees. Why is it so important to highlight as a protagonist in the Garden of Eden parable? It seems that there is a cultural

---

[5] As mentioned in previous notes, we cannot expand here; but any perusal of this expression in any search engine easily demonstrates the extent to which this has become a motto for nature preservation in the spirit of Judaism.

convention at play. The parable is saturated with the premise that is apparent at every stage of the Bible – the realization that man and trees are comparable. The presumption that "man is a tree in the field," which has so many expressions in early Biblical culture, guides this parable, and from an environmental aspect, the conclusion is clear. Man and the environment are very close to one another: man is as a tree, and the tree is likened to a man; harming the tree is akin to harming man; an impressive harmony spreads between the natural habitat and those that inhabit it. True, this basic idea can be developed in different directions, especially providing it with a psychologistic explanation; however, here too, the idea of harmony with one's environment holds true.[6]

In regard to preservation, this is a fundamental matter, not just applying to basic hazards[7]; it also includes the idea of *preservation of the species*. Man lives in the Garden and maintains it; thus, consequentially, man-the-guard may also utilize nature to manufacture clothes from fig leaves: "and they sewed fig leaves together and made themselves loin coverings" (Genesis III 7) – to make, but not to damage the fig tree itself. This represents proportional usage in accordance with need and no more; it is somewhat reminiscent of the aforementioned barn, which is required for maintaining life. Moreover, since the topic is trees, one must surmise that not only that specific tree must be preserved, but the complex in its entirety; i.e., to preserve the specific species and avoid the formation of circumstances that will harm it. Anyone who desires to consider the Garden as a symbol can identify here the warning against harming plants, which are a basis for life; since the garden is not related to the local community, but rather represents the global environment, what may harm it is a change in atmospheric conditions or other global conditions. Each plant requires specific conditions that are related to specific atmospheric and ground characteristics; any change in these may harm specific plants no less than direct activities, such as uprooting or fires.[8] The simple conclusion is that maintaining the Garden is equal to maintaining global conditions on all their elements and types.

---

[6]See also Rozenson (2002b).

[7]Ibn Ezra is often quoted: "to guard – against all animals and prevent them from entering there and polluting." The idea is an attractive one of battling pollution and aspiring to aesthetics, but the separation between plants and animals is not consistent with profound environmental notions.

[8]Although we are dealing here with Genesis, the Book of Ezekiel and certainly other early cultures recognize the cedar as a central element the Garden of Eden: "The cedars in God's garden could not match it; the cypresses could not compare with its boughs, And the plane trees could not match its branches. No tree in God's garden could compare with it in its beauty." (Ezekiel XXXI 8) It is noteworthy that the Cedar's sensitivity to climatic changes, a tree that is primarily endemic to the Lebanese mountains, is widely resonant throughout our sources.

But the most fundamental point, to our mind, is the destiny of the Garden; after all, the Garden no longer serves as a concrete environment. At this point, a different sort of interpretational step must be taken. The banishment was ostensibly decreed because of the Tree of Life: "and now, he might stretch out his hand, and take also from the tree of life, and eat, and live forever; therefore, the LORD God sent him out from the Garden of Eden, to cultivate the ground from which he was taken. So He drove the man out; and at the east of the Garden of Eden, He stationed the cherubim and the flaming sword which turned every direction to guard the way to the tree of life" (Genesis III 22–24). A psychologistic interpretation would present this as a symbol for the need to mature, meaning to understand that death cannot be beaten, and that God must not be forced to provide everlasting life; in accordance, human life must be planned to contain meaning. For our purpose, however, it is noteworthy that, not only man's access to that specific tree has been blocked, but also his admission to the entire complex. The Garden is inaccessible; a strip of land is withheld from mankind; he cannot cultivate it because he has been directed toward another – the soil. The situation is reversed: God guards this piece of land from man. There is a portion that is inaccessible to us because God, who created the world, is not only concerned with man's welfare; the world exists independently. Apparently, the fear that is expressed here is of man, who in his pursuit of the Tree of Life will destroy anything, including God's Garden. Indeed, the Garden has a double meaning: on the one hand, it was dedicated to human life, and on the other hand, it is inaccessible. He who reads the story in a superficial manner may settle this contradiction within the context of time – first the Garden was accessible, then it was not. Even those who read it as an allegory may do this, though in a different manner: the Garden was expropriated from man, who resided within, for the sake of the Garden's protection. This is not the literal meaning; but an idea seems to be presented, that the desire for the Tree of Life involves pride and a broadened heart, which are – after all – the motivating force behind man's continual destruction of God's earth and not only his Tree of Life. Man, who aspires to eternal life, is not a danger to God but to his *world*.

5. Two Forms of Preservation

The motto that derives from the above is preservation. The *preservation* motif is fundamental in the organization of a society and a community. These first "guard themselves from," and only then "act for." In a wider sense, a community will guard itself from the outside, but also the garden within must be maintained. Both forms of preservation have an environmental significance: guarding the community includes guarding in the face of various ecological hazards; the barn – important as it may be – threatens the community. Maintaining the garden has a totally different meaning: God guards nature; man was extracted from it; it must be guarded from man, who cannot be trusted and needs to be made aware of this.

More importantly, man needs to join with God to help guard the garden from mankind. The first concept of preservation – the urban/communal – is important because it places the idea of partnership before private ownership, a basic tenet for any environmental organization. However, if there is an educational aspect to this, it is especially salient in its utilitarian aspect. On the other hand, the Garden, which is protected from man, provides a challenge of an entirely different sort: it is not merely the guarding but the avoidance of entry, which is important – man's self-restraint in the face of the temptation to divert energy toward battling God's guardianship.

The topic of global warming provides a challenge quite simply, because it deals with something distant. Indeed, it may draw inspiration from sources that deal with more proximate matters; nevertheless, thanks to its distance that it directly impinges upon those same ancient matters that involve Man in the Garden of Eden.

6. References

Bekhoffer, Y.G. (1996) Recycling and the Halacha. Tehumin **16**: 296–302.
David, S. (2006) Cutting down fruit trees to settle disputes between neighbors. Tehumin **26**: 286–292.
Ferstenfeld, M. (2002) The environment in Jewish "tradition – a sustainable world, Jerusalem. Moreshet Hamishpat B-Isrel (in Hebrew).
Rakover, N. (1994) The environment – ideological and legal aspects, Jersualem. Moreshet Hamishpat B-Isrel (in Hebrew).
Ralbag, M. (2003) Neighborly damage – noise, visual damage and humidity. Tehumin **23**: 434–442.
Rozenson, Y. (2002) And it was very good. Jerusalem. Moreshet Hamishpat B-Isrel (in Hebrew).
Rozenson, Y. (2002) What was that tree? About Midrash, commentary and education concepts. Reflect Jewish Educ. **4–3**, 51–73.
Sefler, M. (2008) The usage of treated water – a reflection. Tehumin **28**: 132–135.
Shaviv, Y. (1991) Beauty and eternity – chapters on Jewish ecology. Tehumim **12**: 296–302.
Shwartz, E. (1990) Response, mastery and stewardship, wonder and connectedness: a typology to nature in the Jewish texts and tradition, In: H. Tirosh-Samuelson (ed.) *Judaism and Ecology: Created World and Revealed World.* Center for the Study of World Religions, Harvard Divinity School, Cambridge, pp. 102–193.
Sinai, G. and Hazut, A. (2008) Using sewage water – Halachic thoughts. Tehumin **28**: 129–134.

ORGANISM INDEX

A

Acanthophora dimorpha, 267
Acanthophora fragilissima, 267
Acanthophora muscoides, 267
Acanthophora nayadiformis, 37
Acanthophora, 267
Acanthophora spicifera, 267
Acrochaete operculata, 144, 145
Acrochaetium codicola, 37
Acrochaetium robustum, 37
Acrochaetium spathoglossi, 37
Acrochaetium subseriatum, 37
Acropora, 272
Acrothamnion preissii, 37
Acrothrix fragilis, 39
Actinotrichia fragilis, 267
Adenocystis, 202
Aedes aegypti, 392
Agardhiella subulata, 37
Aglaothamnion feldmanniae, 37
Ahnfeltia plicata, 143
Ahnfeltiopsis devoniensis, 170
Ahnfeltiopsis flabelliformis, 37
Alteromonas, 268
Amphiroa foliacea, 267
Ampithoe valida, 211, 212
Anabaena cylindrica, 237
Anotrichium okamurae, 37
Antithamnion amphigeneum, 37
Antithamnionella boergesenii, 37
Antithamnionella elegans, 37
Antithamnionella spirographidis, 37
Antithamnionella sublittoralis, 37
Antithamnionella ternifolia, 37
Antithamnion nipponicum, 37
Aphanizomenon, 375
Apoglossum gregarium, 37
Arthrospira, 375

Arthrospira (Spirulina) platensis, 185
Ascophyllum, 76, 77, 80, 85–87
Ascophyllum nodosum, 74, 76–81, 83–86, 143, 190
Asparagopsis armata, 37, 43, 348
Asparagopsis/Falkenbergia, 362, 363
Asparagopsis taxiformis, 37, 42, 43
Astroides calycularis, 42

B

Bachelotia antillarum, 162
Bacillus, 289
Bacillus algicola, 297
Bacillus subtilis, 390
Bangia, 342
Bifurcaria bifurcata, 163
Blidingia minima, 202
Bonnemaisonia hamifera, 37
Boodlea composita, 268
Boodlea composite, 266
Bossiela orbigniana, 122
Bostrychia, 202
Botryocladia madagascarensis, 37
Botryococcus, 380
Botrytella parva, 39
Brassica campestris, 301
Bryopsis, 202

C

Caepidium, 202
Callithamnion gaudichaudii, 205, 210
Candida albicans, 390
Caulacanthus ustulatus, 162, 163
Caulerpa, 362, 363, 387–394
Caulerpa bikinensis, 389
Caulerpa chemnitzia, 393
Caulerpa cupressoides, 392
Caulerpa faridii, 390

Caulerpa lentillifera, 391–393
Caulerpales, 389, 392
Caulerpa mexicana, 38
Caulerpa prolifera, 389, 390, 393
Caulerpa racemosa, 42, 43, 387–390, 392, 393
Caulerpa racemosa var. *cylindracea*, 38, 42, 43, 45, 387, 394
Caulerpa racemosa var. *lamourouxii*, 38
Caulerpa racemosa var.*peltata*, 392
Caulerpa racemosa var. *turbinata*, 38, 42
Caulerpa scalpelliformis, 39, 391–393
Caulerpa sertularioides, 389, 392
Caulerpa taxifolia, 39, 42, 43, 45, 387, 388, 390, 393, 394
Centroceras, 267
Centrostephanus rodgersii, 57
Ceramium, 205, 206, 210, 266, 267
Ceramium bisporum, 37
Ceramium strobiliforme, 37
Chaetomorpha crassa, 268
Champia, 268
Chanos chanos, 276
Chlamydomonas, 379–381
Chlamydomonas rheinhardii, 237
Chlorella, 245, 375, 380
Chlorella pyrenoidosa, 237
Chlorella vulgaris, 237
Chondracanthus acicularis, 163
Chondria, 202
Chondria coerulescens, 37
Chondria curvilineata, 37
Chondria pygmaea, 37
Chondrophycus papillosus, 268
Chondrus, 77, 324, 362, 363
Chondrus crispus, 77, 144, 145, 190, 191, 332
Chondrus giganteus, 37
Chorda filum, 39, 148
Chrysymenia wrightii, 37
Chytridium polysiphoniae, 144
Cladophora, 202
Cladophora falklandica, 202
Cladophora herpestica, 39
Cladophora patentiramea, 39
Cladophoropsis, 202
Cladosiphon zosterae, 36
Codium, 202, 211

Codium arabicum, 108
Codium decorticatum, 163, 187
Codium fragile, 43, 86, 186
Codium fragile subsp. *tomentosoides*, 39
Codium taylori, 39
Colpomenia, 202
Colpomenia peregrina, 39
Corallina, 162, 205, 208, 210
Corallina elongata, 163
Corallina officinalis, 202, 205, 207–210
Corallina pilulifera, 122
Crassostrea gigas, 41
Cystoseira, 162, 362, 363
Cystoseira baccata, 163
Cystoseira tamariscifolia, 163
Cytophaga-Flavobacterium, 265

D

Dasya sessilis, 37
Dasycladus vermicularis, 166
Dasysiphonia, 37
D. cervicornis, 268
Derbesia boergesenii, 39
Derbesia rhizophora, 39
Desmarestia viridis, 36
Devaleraea ramentacea, 191
Dictyota, 206, 210
Dictyota dichotoma, 186, 187, 202, 206
Dictyota divaricata, 268
Dumontia, 329
Dunaliella, 375, 380
Dunaliella bioculata, 237
Dunaliella salina, 237
Dusidicus gigas, 56

E

Ecklonia, 221, 224, 362, 363
Ecklonia cava, 223
Ecklonia stolonifera, 220, 364
Ectocarpus, 144
Ectocarpus rhodochondroides, 186
Ectocarpus siliculosus, 143
Ectocarpus siliculosus var. *hiemalis*, 36
Eisenia, 224
Emiliania huxleyi, 122, 143
Endophyton ramosum, 144
Enteromopha linza, 121, 205, 210
Enteromorpha, 202, 211, 212

Enteromorpha intestinalis, 348
Enteromorpha linza, 205, 210
Eucheuma, 256, 264, 270–273, 275, 324, 362
Eucheuma denticulatum, 255, 256, 263, 271, 272, 279, 362
Eucheuma/Kappaphycus, 343, 361, 363
Eucheuma striatum, 270
Euglena gracilis, 237

F

Feldmannophycus okamurae, 37
Flabellatus, 37
Flavobacterium, 268
Fucus, 85
Fucus/Ascophyllum, 84, 85
Fucus evanescens, 77, 293
Fucus gardneri, 188, 190
Fucus serratus, 77, 119, 165, 171, 186, 189
Fucus spiralis, 39
Fucus vesiculosus, 76, 77, 81, 83–86, 143

G

Galaxaura rugosa, 37
Ganonema farinosa, 37
Gelidiella acerosa, 187
Gelidium, 163, 324, 326
Gelidium corneum, 170, 171
Gelidium Gigartina, 362
Gelidium pusillum, 162
Gelidium sesquipedale, 163
Gigartina, 363
Gigartina intermedia, 121
Gigartina skottsberdii, 332
Gloiopeltis furcata, 121
Goniotrichopsis sublittoralis, 38
Graciaria tenuisitipitata, 122
Gracilaria, 118, 120, 122, 186, 188, 266, 321–333, 343, 361–364
Gracilaria arcuata, 38, 267
Gracilaria arcuata var. arcuata, 325
Gracilaria bursa-pastoris, 325, 326
Gracilaria caudata, 327
Gracilaria changii, 330
Gracilaria chilensis, 118, 122, 324, 326, 330, 364
Gracilaria conferta, 143
Gracilaria corda, 327
Gracilaria cornea, 169, 191, 325, 329
Gracilaria corticata, 327–332
Gracilaria corticata var. cylindrica, 325
Gracilaria debilis, 324
Gracilaria dura, 323, 325, 331
Gracilaria edulis, 324, 325, 328, 331
Gracilaria foliifera, 187
Gracilaria gaditana, 122
Gracilaria gracilis, 202, 324–326, 328, 329, 331, 348
Gracilaria henriquesiana, 327
Gracilaria heteroclada, 325
Gracilaria lemaneiformis, 186–188, 190, 287, 288, 326, 327, 332
Gracilaria mammillaris, 327
Gracilaria pacifica, 332
Gracilaria salicornia, 333
Gracilaria sjoestedtii, 328, 331
Gracilaria symmetrica, 327
Gracilaria tenuistipitata, 119, 120, 122, 169, 190
Gracilaria textorii, 327, 328, 331
Gracilaria tikvahia, 324
Gracilaria verrucosa, 324–331
Gracilibacillus, 297
Gracilibacillus halotolerans, 297
Grateloupia asiatica, 38, 325
Grateloupia dichotoma, 169
Grateloupia lanceola, 169
Grateloupia lanceolata, 38, 364
Grateloupia patens, 38
Grateloupia subpectinata, 38
Grateloupia turuturu, 38, 86
Griffithsia corallinoides, 38

H

Halimeda, 112
Halomonas, 291, 293
Halomonas alkantarctica, 293
Halomonas boliviensis, 291, 293
Halomonas campisalis, 293
Halomonas hydrothermalis, 291
Halomonas magadiensis, 291, 293
Halomonas marina, 293
Halomonas muralis, 293
Halomonas variabilis, 291, 293
Halomonas venusta, 291
Halophila stipulacea, 43
Halothrix lumbricalis, 39

Harveyella mirabilis, 144
Herposiphonia parca, 38
Hildenbrandtia, 202
Himanthalia elongata, 142
Hizikia fusiforme, 118, 122, 364
Hizikia fusiformis, 121
Hizikia fusimorme, 120
Hydroclathrus clathratus, 268
Hydropuntia edulis, 268
Hypnea, 362, 363
Hypnea cornuta, 38
Hypnea flagelliformis, 38
Hypnea musciformis, 268
Hypnea pannosa, 268
Hypnea spinella, 38, 268
Hypnea valentiae, 38, 268

I
Idothea baltica, 211, 212
Iridaea, 202
Ishige okamura, 121

J
Jania rubens, 163

K
Kappaphycus, 255–279, 324, 361, 363
Kappaphycus alvarezii, 255, 256, 258, 259, 261, 263–273, 275–279, 362, 363
Kappaphycus alvarezii var. tambalang, 277, 278
Kappaphycus striatum, 258, 265, 270, 275, 278
Kappaphycus striatum var. sacol, 259

L
Laminaria, 144–147, 186, 189, 221, 223, 224, 343, 361–365
Laminaria angutata, 223
Laminaria digitata, 141, 146–149
Laminaria hyperborea, 142–148, 189
Laminaria japonica, 144, 362, 364
Laminaria ochroleuca, 148, 187
Laminaria pallida, 148
Laminaria religiosa, 144
Laminaria/Saccharina, 141, 144, 147, 148
Laminaria sensu lato, 141
Laminaria tomentosoides, 145

Laminariocolax aecidioides, 145, 146, 148
Laminariocolax macrocystis, 147
Laminariocolax tomentosoides, 146, 147
Laminariocolax tomentosoides sub deformans, 146
Laminarionema elsbetiae, 145, 146, 148
Laurencia caduciramulosa, 38
Laurencia obtusa, 163
Laurencia okamurae, 38
Leathesia, 202
Leathesia difformis, 39
Lessonia nigrescens, 148
Lichina pygmaea, 170
Lithophyllum incrustans, 163
Lithophyllum yessoense, 38
Litopeneaus vannamei, 276
Lobiger serradifalci, 387, 389
Lomentaria articulata, 118, 364
Lomentaria hakodatensis, 38
Lophocladia lallemandii, 38, 43, 44

M
Macoma, 56
Macoma balthica, 55
Macrocystis, 362–364
Macrocystis pyrifera, 147, 148, 187, 202, 332
Mastocarpus stellatus, 77, 191, 332
Mazzaella laminarioides, 144, 332
Metapenaeus monoceros, 42
Miscanthus, 382
Monostroma, 362
Mytilus edulis, 84, 85

N
Nannochloropsis, 380
Nemalion vermiculare, 38
Neomeris annulata, 39
Neosiphonia, 266–268, 278
Neosiphonia apiculata, 267–269
Neosiphonia savatieri, 267
Nitophyllum stellato-corticatum, 38
Nitzschia, 380

O
Ocullina patagonica, 59
Oxynoe olivacea, 389

ORGANISM INDEX

P

Pachymeniopsis lanceolata, 364
Padina australis, 268
Padina boergesenii, 39
Padina boryana, 39
Padina santae-crucis, 268
Palmaria, 362, 363
Palmaria palmata, 77, 142, 322
Pandalus Borealis, 56
Pelvetia compressa, 74
Petalonia fascia, 187
Peyssonelia, 163
Phaeodactylum, 380
Phaselus vulgaris, 393
Pilayella littoralis, 77
Pinctada martensi, 276
Pisaster ochraceus, 22, 56
Pleonosporium caribaeum, 38
Pleurocapsa, 144
Plocamium secundatum, 38
Polysiphonia, 211, 212, 266–268, 322, 324
Polysiphonia atlantica, 38
Polysiphonia brodiaei, 202
Polysiphonia fucoides, 38
Polysiphonia harveyi, 38
Polysiphonia lanosa, 77
Polysiphonia morrowii, 38
Polysiphonia paniculata, 38
Porites, 59
Porphyra, 144, 188–190, 202, 206, 224, 321, 341–350, 361–363, 365
Porphyra abbottae, 346
Porphyra amplissima, 346, 349
Porphyra columbina, 169, 190, 202, 205, 208, 210
Porphyra dioica, 345–350
Porphyra haitanensis, 121, 189, 191, 343, 346–348
Porphyra katadai, 346
Porphyra leucosticta, 119, 120, 122, 169, 189, 347
Porphyra linearis, 119, 345–347
Porphyra moriensis, 345
Porphyra pseudolinearis, 345, 346
Porphyra purpurea, 344, 346–348
Porphyra rosengurttii, 170
Porphyra seriata, 364
Porphyra tenera, 344, 362
Porphyra torta, 344, 346
Porphyra umbilicalis, 169, 189, 191, 321, 346, 347
Porphyra yezoensis, 38, 108, 118–120, 188, 189, 287, 342–344, 346, 347, 349, 362, 364
Porphyridium cruentum, 237
Posidonia oceanica, 165
Prymnesium parvum, 237
Pseudobryopsis/Trichosolen, 107
Pterocladia, 362
Pterosiphonia complanata, 163
Pterosiphonia tanakae, 38
Punctaria tenuissima, 39
Pylaiella littoralis, 36, 144

R

Rama, 202
Rhizoclonium, 202
Rhodomela confervoides, 144
Rhodophysema georgii, 38
Rhodymenia erythraea, 38
Rugulopterix okamurae, 39

S

Saccharina japonica, 39
Saccharina latissima, 141, 142, 144–148
Saccorhiza polyschides, 148
Sarconema filiforme, 38
Sarconema scinaioides, 38
Sardinella, 23
Sargassum, 187, 221, 223, 224, 362, 363
Sargassum hemiphyllum, 121
Sargassum horneri, 187
Sargassum macrocarpum, 222, 223
Sargassum muticum, 39, 44
Sargassum patensi, 223
Sargassum yezoensis, 223
Scenedemus, 380
Scenedesmus (Spirulina), 375
Scenedesmus dimorphus, 237
Scenedesmus obliquus, 237
Scenedesmus quadricauda, 237
Scytosiphon dotyi, 39
Siganus luridus, 42
Siganus rivulatus, 42
Solieria dura, 38
Solieria filiformis, 38

Sparisoma cretense, 42
Spathoglossum variabile, 39
Sphaerotrichia firma, 39
Spirogyra, 237
Spirulina, 375, 380
Spirulina maxima, 237
Spirulina platensis, 237
Spongomorpha, 202
Stomolophus nomurai, 224
Stypocaulon scoparium, 163
Stypopodium schimperi, 39, 43, 44
Symphyocladia marchantioides, 38
Synechoccus, 237

T
Tegula, 55
Tetraselmis, 380
Tetraselmis maculata, 237
Thalassoma pavo, 42
Trachinotus carolinus, 276

U
Ulva, 120, 162, 202, 206, 210, 226, 266, 278, 312, 348, 362, 363
Ulva clathrata, 268
Ulva compressa, 268
Ulva curvata, 187
Ulva expansa, 186
Ulva fasciata, 39, 43, 268
Ulva intestinalis, 348
Ulva lactuca, 121, 122, 186, 348
Ulva media, 268
Ulva pertusa, 39, 166, 190, 268, 348, 364
Ulva pinnatifida, 202, 288
Ulva reticulata, 268
Ulvaria obscura, 39
Ulva rigida, 118, 122, 186, 202, 203, 205, 210
Ulva rotundata, 187, 348
Undaria, 224, 343, 361–363, 365
Undaria pinnatifida, 39, 44, 202, 287, 289, 362, 364
Urospora, 202

V
Vibrio, 23
Vibrio-Aeromonas, 265
Vibrio fischeri, 302
Vibrio, 268, 297
Vigna sinensis, 393

W
Womersleyella setacea, 38, 43

SUBJECT INDEX

A

Abrahamic religions, 404
Acadian Seaplants Limited (ASL), 78, 82, 83, 85, 87, 253
Accessory pigments, 207
Acclimation, 55, 118, 120, 123, 164–171, 173, 187
Aeolian dust, 309
Agamospores, 344
Agar, 247, 275, 291, 296, 312, 322, 326, 347, 361
Age-old insights, 403, 432
Aggadah, 437
Algae, 23, 43, 55, 74, 118, 129, 141, 161, 185, 202, 219, 231, 265, 289, 308, 321, 341, 360, 375, 387
　abundance, 56, 60, 74, 79, 130, 141, 159, 161, 208, 331
　biomass, 360, 377–381
　blooms, 287
　distribution, 77, 78, 80, 105, 108, 111, 146, 187, 191, 331
　oil, 236, 237, 239–243
　turf, 23, 24, 26, 141
Algaebase, 96, 108, 109
Algae-to-fuel conversion, 240, 243, 246
Algal-pathogen interactions, 144
Alginate, 77, 79, 147, 148, 289, 290, 361
Alginate-degrading bacteria, 148, 296–301
Alternative energy, xvii, xviii, 376, 380, 382, 383
Amazon rain forest, xvi, 408
Ambient stressors, 207
Ambiguity of religions, 404–405
Annelida, 132
Antarctica, xv, xvi, xviii, 13
Antarctic ice sheets, 14, 310
Antenna pigments, 190
Anthropic principle, 414

Anthropocentric, 403, 438–440
　perception, 438
Anthropogenic activities, 160, 161, 330, 346
Anthropogenic CO_2, 121, 360, 367
Anthropogenic disturbance, 129
Anthropogenic influences, 141
Anthropogenic stressors, 26
Anticancer, 388, 389
Antileishmanial, 390
Antimicrobial, 388, 390
Antioxidants, 149, 166, 170–172, 190, 207, 343, 375
Antitumor, 389
Antiviral, xvii, 388, 390
Aquaculture, 35, 36, 40, 227, 255, 272–277, 307–316, 322, 333, 343, 344, 346–350, 359–368, 375, 393
Aquatic biota, 203, 310
Aquatic crops, 360
Aquatic ecosystems, 159–165, 167–171, 173, 174, 185, 201, 202, 330
Aquatic plants, 310, 347, 359, 360
Aquatic productivity, 201, 315
Aral lake, xvi
Archeological evidence, xv, 8
Argentina, 201–212, 288
Argumentation, 425, 456
Arid climates, 307, 308, 315
Artificial lagoons, 312–314
Asexual propagules, 325, 326
Asexual reproduction, 324
Assemblages, 61, 100, 101, 109, 111, 112, 130
Astronomical tides, 6
Atmosphere, xv, xvi–xviii, 62, 85
Atmospheric gas, 203

B

Baba Batra, 452–455
Bacterial degradation

Ballast water, 35
Bangiales, 321, 341
Bay of Fundy, 75, 76, 85, 86
Benthic, xvii, 53, 57, 62, 105, 111, 129, 135, 160, 333
　communities, 56, 59, 63, 97, 104, 105, 130, 163
　marine, 100, 108, 142, 143, 162
Bible, 403–407, 428–430, 432, 439, 452, 453, 458
Biblical religion, 428–429
Bicarbonate utilization, 123
BIMP-EAGA region. See Brunei–Indonesia–Malaysia–Philippines (East Association of Southeast Asian Nations (ASEAN) Growth Area region
Biocentric, 438–440
　perception, 438
Biochemical factories, 360
Biodiesel, 231, 232, 234–236, 238, 240, 242–244, 359, 360, 366, 367, 376, 377, 379–383
Biodiversity, 24, 25, 54, 57, 60, 62, 63, 74, 109, 112, 134, 163–165, 168, 171, 173, 273
Bio-energy, 219–220, 227
Bioethanol, 231, 234, 240, 359, 360, 376, 381
Biofilter(s), 259, 276, 277, 346–349, 361
Biofuel, xvii, 219–228, 231–247, 311–313, 316, 359–361, 365–367, 375–377, 379–382
Biogenic calcification, 59
Biogeochemical feedback, 149
Biogeographical origin, 33, 41
Biogeographical shifts, 57
Biogeography, 54, 60, 109
Bioinvasion, 58, 270, 271, 273
Biological activity, 361
Biological elements, 160
Biological energy, 227
Biological interactions, 22
Biological invasions, 33, 36, 45, 46
Biomass, 76, 78–81, 83, 85, 87, 97, 141, 160, 161, 163, 168, 186, 228, 240–242, 259, 262, 267, 275, 287, 288, 295, 313, 314, 321, 325, 347–350, 359, 360, 364, 366, 367, 375–383, 392, 393
Biorefinery, 381

Bioremediation, 117, 275, 346, 349, 393
Bioresources, 219, 221–222
Biosorption, 391, 392
Biotechnology, 361, 367, 375
Biotic interactions, 143–147, 150, 151, 333
Bioturbation, 129
Bivalves, 130, 313
Bloom-forming patents, 393
Book of Nature, 430
Brazil, 234, 256, 259, 269, 270
Brunei–Indonesia–Malaysia–Philippines (East Association of Southeast Asian Nations (ASEAN) Growth Area (BIMP-EAGA) region, 256, 257
Buddhism, 403, 427
Buddhists, 406, 412, 427
Butanol, 381, 390

C
C3, 310
Calcification, 23, 58, 59, 121–123, 205
Canada, 74, 75, 82, 87, 409
Canadian maritimes, 73–87
Cape Cod, 73
Carbonaceous skeletons, 308
Carbonate sediment, 129
Carbonate system, 117, 121
Carbon concentrating mechanisms (CCM), 118, 119, 123
Carbon credits, 255
Carbon cycle, 12, 14, 117, 365–367
Carbon dioxide (CO_2), 22–24, 33, 54, 56, 60, 73, 80, 117–123, 166–168, 220, 223, 232, 236, 239, 241, 242, 245, 255, 294, 308–314, 316, 333, 349, 359, 365, 380, 382
　balance, 21, 310, 376
　enrichment, 117–123, 316, 364
　fixation, 172, 188, 219, 224
　uptake, 58, 224, 347, 360, 367
　utilization, 59, 293, 360, 363–364
Carbon fixation, 80
Carbonic anhydrase (CA), 118, 120, 166, 347
Carbon limitation, 59, 118, 121
Carbon sequestration, 308, 311, 360
Carbon sink, 360
Carpospores, 323–325, 328, 329, 331, 332

SUBJECT INDEX

Carrageenan, 255, 257, 259, 262, 263, 265, 268, 269, 275–277, 279, 321, 347, 361
Caulerpenyne (CYN), 388–390
CCD sensors, 97
CCM. *See* Carbon concentrating mechanisms
Cellular oxidation, 172
CF_3, xv
CFC, xv, xvi, 186
Chad lake, xvi
Chazaka, 445
Cheap oil, 231, 244
Chemical composition, 236–237, 288–290, 294, 295, 309
Chemical elements, 160
Chinese religion, 428
Chlorophyll, 103, 210, 236
Chlorophyll fluorescence, 164, 171
Chubut Province, 202
Cisplatinum, 389
Climate change, 5, 21–26, 34–35, 53–63, 73–87, 111, 112, 141–151, 159–175, 208, 279, 307, 309, 310, 314, 315, 321–333, 367, 375, 407, 409, 458
Cloud cover, 149, 311
CMOS sensors, 97
C:N ratio, 122–123, 295
CO_3^{2-}, 58, 117, 121
Coal, xvii, xviii, 235, 236, 314, 367, 376, 382, 425
Coastal communities, 61, 109, 262, 277, 279
Coastal ecosystems, 43, 61, 141–143, 149, 160, 163, 167, 222, 288, 363
Coastal lagoons, 312
Coastal mangroves, 26, 61, 108
Coastal region, 5, 309
Coastal shelves, 314–315
Coastal waters, 8, 22, 101, 105, 109, 117, 121, 159–164, 167, 171, 175, 185, 189, 275, 313, 346, 347, 367
Coastal wetlands, 26
Coastal zone, 22–23
Coccolithophorids, 59
Cold-temperate kelp, 142
Community structure, 55, 56, 58, 60, 63, 74, 130, 167
Composting, 287–295, 298–302
Computer-designed maps, 97

Conchocelis, 188, 189, 321, 343–346
Conchospores, 189, 343–346
Conservation, 21–26, 60, 63, 130, 135, 220, 222, 407, 425–432
Coral bleaching, 24, 53, 55
Coralline algae, 59, 60, 135, 222, 330
Coral reefs, 22–24, 26, 53, 61, 109, 110, 270, 271, 273
Corn, 220, 234, 238, 376, 377
Corn ethanol, 234
Cosquer Cave, 7, 14
Creation, 55, 95, 96, 405, 406, 409, 425, 427, 429, 439, 446
Crop production, 236, 243, 244
Crops, xvii, 79, 220, 223, 235, 236, 238, 241, 242, 263, 267–269, 342, 345, 362, 376, 377, 380
Crude oil, xvii, 231, 234, 240
Crustacea, 132, 134, 135
Cryptic introductions, 36, 44
Cryptofauna, 130–132, 134, 135
Cultivation, 223–225, 243, 256–259, 262–264, 267, 269, 272–276, 278, 279, 287, 321, 332, 345, 349, 362, 363, 366, 378, 382, 383, 412
Culture seaweeds, 313–315, 360–361, 363, 365–367
Cultures polyunsaturated, 375
Culture systems, 122, 223, 275, 277, 312, 341–350, 361, 377–378, 380
Culture techniques, 223, 260, 261
CYN. *See* Caulerpenyne

D

Database, 10, 96–100, 109, 111
Dead Sea, xvi, 293
Deep ecology, 412–413
Deep waters, 61, 147, 191, 270, 277
Defense responses, 147–150
Deforestation, xvi, xvii, 220, 382
Dehydration, 118, 191
Deontology, 426
Desert dust, 307–309
Desiccation, 77, 121, 170, 172, 191, 207, 208, 328
Dharma, 404, 411, 413, 427
Digitized maps, 97
Dikes, 311–316

Dimethylsulphide (DMS), 149
Dinosaurs, 425
Dioecious species, 322, 341, 344
Diploid phase, 322, 323, 326, 329, 343
Diseases, 23, 25, 143, 144, 146–148, 150, 264–270, 381, 382, 404
Dissolved inorganic carbon (DIC), 117, 118, 120, 122, 123
Dissolved organic carbon, 308
Distribution, 22, 23, 53, 54, 57, 60, 62, 63, 73, 74, 76–78, 80, 85, 104–111, 142, 146, 163, 165, 167, 171, 187, 191, 259, 270, 325, 326, 331, 341, 344, 412
  boundary, 143
  shifts, 95, 143
Distromatic, 342
DNA
  analysis, 291, 297, 330
  damage, 165, 167, 185, 189, 191
Domestic pollution, 33
Dust bowl, xvi

E
Earth, ix, xv–xix, 21, 33, 34, 73, 96, 104, 159, 186, 201, 202, 220, 241, 247, 332, 403–407, 409–412, 414, 415, 417–419, 427–429, 432, 459
Eco-feminism, 412–414
Eco-friendly, 228
Ecological crisis, 403–404, 406–408, 412, 413, 417
Ecological ecumenism, 427
Ecologically harmful, 404
Ecological toxicity, 300–302
Ecology, xv, 59, 61, 109, 117, 164, 175, 403–419, 425, 426, 428, 429, 443, 451–453, 456, 457
  and religion, 403–419
Economic importance, 77, 342–343
Eco-philosophy, 109, 159–175
Ecophysiological responses, ix
Ecosystems, ix, xvii, xviii, 43, 45, 53, 54, 56, 58–63, 74, 141, 142, 149, 159–171, 173, 174, 185, 201, 202, 219, 221, 222, 224, 227, 288, 312, 315, 330, 348, 349, 363, 375, 426, 430
Edible seaweeds, xvii, 288, 321
Educational value, 456
Elevated CO_2 level, 59, 117–123

Emersed photosynthesis, 120–121
Endophyte species, 141–150
Endophytic algae, 144, 150
Endospores, 344
Energy
  balance, 360
  crops, 359
  demand, 359, 375, 376
  plants, 376, 377, 381, 382
  policy, 359
Environment, 5, 21, 33, 53, 73, 95, 118, 129, 142, 159, 186, 201, 219, 231, 269, 287, 307, 321, 343, 359, 377, 403, 425, 437, 451
  analysis, 95–112
  challenge, 279, 431
  factors, 105, 143, 166, 171, 187, 202
  hazards, 438, 440, 441, 444–447, 452, 454–456, 458, 459
  ideology, 451
  preservation, 219–228, 288, 459, 460
  theory, 451
Environmentalists, 403, 414, 438
Enzymatic extraction, 239
Enzymatic transformation, 388–389
Enzyme inhibition, 389–390
Epiphytes, 77, 86, 149, 265–270, 277, 278
  infestation, 266–269
Epiphytic, 145, 265–267
Episodic ozone, 201
Erosion, 5, 24, 61, 164, 273, 405, 407, 429
Ethical behavior, 426
Ethics, 404, 410, 416, 418, 425–432
Eucheumatoids species, 273, 279
Eulittoral, 187, 191, 201, 205–209
European Geostationary Navigation Overlay Service (EGNOS), 99
Eutrophication, 23, 141, 143, 161, 166, 169, 219, 222, 224, 225, 273, 278, 287, 288
Evangelical Climate Initiative, 406
Exotic species, 270
Extinctions, xvi, 22, 23, 63, 74, 163, 165, 425, 443

F
FAO, 274, 312, 349
Feeding inhibitor, 388
Feedstocks, 231, 235, 242, 243, 366, 367

SUBJECT INDEX 471

Fertilization, 275–278, 308, 322, 323, 328–329
Fertilizer, xvii, 79, 227, 245, 247, 288, 300, 301, 309, 315, 360, 362, 363, 442
Fisheries, 227, 309
Fisheries management, 25
Fishing catch, 22
Fixed oils, 238
Floods, xv, xvi, xviii, 24, 111, 141, 311, 314, 315, 418, 419
Fluorometry, 164
Food, xvii, 55, 56, 61, 62, 142, 211–212, 219, 227, 232, 234, 244–247, 259, 262, 273, 275, 288, 308, 309, 311, 313, 316, 327, 347, 359–363, 367, 375, 376, 378, 418, 419, 431, 457
  crisis, 220, 234, 419
  crops, 431, 434, 436
  preservatives, 361
  web, 54, 59, 142, 167, 173, 174
Fossil fuel, xv, 219, 220, 232, 313–315, 359, 376
France, 7, 14, 35, 95, 111, 141, 160, 325, 326, 388
Fresh-water aquifers, 5
Fucoids, 73–87
Fucus, 76, 77, 81, 83–85, 119, 142, 165, 171, 186, 188–190, 293
Functional group, 56, 58, 62, 162, 391
Future, ix, x, xix, 5–14, 24, 46, 54, 55, 57, 63, 73–87, 99, 102–104, 106, 108–112, 142, 149, 165, 173–175, 219, 220, 233, 234, 243, 245, 269, 276, 278–279, 307, 310, 312, 313, 364, 375–383, 412, 426, 431, 438, 444

G

Gametophyte, 144, 148, 322–324, 326–328, 341, 343–346
Garden of Eden, 453, 457–460
Gas emissions, xviii, 21, 33, 360, 367
Gaza Strip, 315
Gel strength, 259, 265, 268, 276
GenBank, 99, 109
Genesis, 403, 407, 457–459
Genetic diversity, 326
Genetic markers, 327
Geographical distribution, 76
Geographical shift, 57

Geographic information systems (GIS), 95–112, 162
Geological time-scale, 5
Georeferencing, 98–100, 108, 109
Glacial cycle(s), 7, 13, 34
Glacio-hydro-isostatic change, 5
Global aquaculture, 312, 365–367
Global change, 23, 43, 95, 100, 105, 109–112, 174, 330
Global Positioning System, 6, 98, 99
  database, 99
Global responsibility, 457
Global warming, 5, 7, 33–46, 55–58, 73, 142, 144, 219, 255, 307–309, 311, 315, 331, 359–368, 419, 437–447, 451–460
God, 404–406, 408–411, 413, 414, 418, 429, 437, 439, 457, 459, 460
God planted, 457
Government regulations, 368
GPS. *See* Global Positioning System
Grazing, xvi, 77, 78, 143, 243, 272, 405
Green algae, 166, 167, 202, 375–383, 387, 392
Greenhouse effect, xvi, xviii, 21, 309
Greenhouse gas emissions, xviii, 21, 33, 360, 367
Greenland, 13, 14, 310
Growth, 54, 56, 58–60, 77, 80, 85, 117–119, 122, 123, 129, 130, 132, 144, 146, 147, 166, 167, 170, 171, 173, 185–188, 205, 223, 224, 231–233, 235, 238, 243, 245–247, 256, 259, 261, 272, 275–278, 287–289, 297, 299–301, 310–313, 324–326, 328, 330, 332, 333, 346, 348–350, 360, 363, 364, 366, 368, 378, 380, 381, 388, 389, 393, 406
Growth rate, 58, 77, 119, 122, 144, 146, 147, 186, 232, 259, 275–278, 287, 291, 297, 333, 346, 348, 362, 364, 388, 393
Guide of the Perplexed, 437
Gulf of California, 129–136
Gulf of Maine, 75
Gulf of St. Lawrence, 75, 76, 81, 82, 87

H

Hadera, 11, 22
Halacha, 451, 452
Halakha, 437, 438, 440, 447
Halotolerant bacteria, 287–301

Haploid phase, 322–324, 329, 343, 344
Hard-copy maps, 97
Harmful, xvi, xviii, xix, 146, 148, 185, 191, 266, 310, 332, 404, 425, 445, 446
Harvest, 78–85, 87, 129, 223, 224, 227, 257–262, 268, 287, 288, 313, 321, 322, 345, 348, 349, 360, 363, 365, 366, 376, 378, 380, 388, 394, 406, 428
Hassidic literature, 437
Hazaka, 454
HCO_3^-, 58, 117–121, 123, 347
HDVB. *See* High density vertical bioreactor
Health foods, 227, 275
Heat balance, 21
Hebrew, 404, 428, 440, 442, 443, 452
Heliophany, 201
Hemagglutinating activity, 392
Herbivores, 26, 389
High density vertical bioreactor (HDVB), 243, 244
Hinduism, 403, 404, 406, 427–428
Host-parasite interaction, 144
Human activities, xv–xvi, 35, 58, 73, 160, 163, 168, 225, 425, 443, 452
Human effects, 35, 53, 58, 63, 73, 111, 160, 161, 163, 168, 222, 225, 332, 417, 425
Human life, ix, 219, 412, 431, 459
Human mediated, 53
Hydromorphologic elements, 160

I

Ice caps, 9
Ice damage, 81–82, 84
Ice-ice, 264, 265, 267, 269, 270, 277
Ice sheets, 9, 12, 13, 14, 34, 310
IMTA. *See* Integrated multi-trophic marine aquaculture
IMTA farms, 275, 313, 315, 363
Indoor system, 377–380
Integrated multi-trophic marine aquaculture (IMTA), 274–277, 307–316, 347–350, 361, 363, 367
Intensive aquaculture, 274, 350
Intergovernmental Panel on Climate Change (IPCC), 9, 12, 13, 53, 73, 159, 310

Intertidal, 22, 23, 56, 57, 61, 63, 103, 104, 117, 160, 162, 172, 173, 187, 188, 191, 202, 264, 330, 331, 333
  seaweeds, 74–76, 120–121
  species, 57, 121, 123, 191, 346
Intracellular nitrogen storage, 190
Invasion, 33, 36, 45, 46, 58, 63, 86, 111, 112, 145, 222, 235, 270–273, 394
Invasive, 42–44, 86, 145, 202, 270–273, 333, 387, 388, 392, 394
  organism, 270–273
  species, 86, 273
Iota-carrageenan, 255, 259, 263, 265, 279
IPCC. *See* Intergovernmental Panel on Climate Change
Irradiance, 55, 120, 163–166, 171, 173, 174, 186–188, 191, 202, 203, 205, 210, 328, 332, 333
Islam, 403, 428–430
Isoyake, 222
Israel, 7, 8, 22, 61, 119, 120, 274, 308, 312, 314, 315, 445
Israel Oceanographic and Limnological Research Institute, 22

J

Japan, 41, 44, 86, 96, 99, 220–228, 333, 360, 362, 378
Jesus Christ, 404, 407, 410, 429
Jewish approach, 451–460
Jewish-ecological slogans, 452
Jewish law, 429, 437–447
Jewish thought, 437–440, 447

K

Kappa-carrageenan, 255, 257, 263, 279
Kelp, 56, 79, 106, 141–150, 189, 331, 364, 375
Key species, 25, 54, 142, 208
Killer alga, 387
Kingdom Plantae, 34
Kingdom Stramenopiles, 34
K_m, 118
Konrad von Megenberg, 430
Koran, 405, 406, 429

L

Lagoons, 43, 44, 159, 160, 167, 171, 175, 278, 311–314, 316, 325

SUBJECT INDEX

Lakes, 40, 159, 167, 170, 171, 174, 175, 220, 293, 313, 378, 406
Laminaria saccharina, field surveys, 144–145
Land-based tank cultivation, 349
Latitudinal shifts, 142
Laver, 342, 362
Laws of Manu, 406
Lessepsian migration, 23
Lessepsian species, 40, 43
Life cycle, 105, 133, 166, 222, 322–329, 333, 343–346
Life history, 321, 322, 324–326, 328, 329, 333, 342
Life stages, 77, 111, 132, 185, 188–189, 191, 332, 333
Linolic acids, 379
Local extinctions, 22, 74
Long-line technique, 258–260
Low cost technologies, 363, 367
Low-intertidal subtidal, 162, 346

M

MAA. *See* Mycosporine-like aminoacids
Macroalgae, xvii, 23–24, 26, 59, 74, 77, 95, 100, 101, 103, 105, 108–110, 112, 117, 119, 122, 141–143, 159–167, 169, 170, 173, 185–191, 201–212, 219–228, 236, 330, 332, 347, 359, 364, 389, 391, 393
   communities, 95, 161, 202, 221
   diversity, 202
   shift, 23, 105, 142, 164, 273
Man–Environment Relations, 438–443
Mapping, 95, 97, 100–104, 106, 108–110
Marine bacteria, 291, 292, 296, 297, 299
Marine biodiversity, 63, 74
Marine biota, xvi, 142, 309
Marine communities, 22, 25, 54, 108, 162, 163
Marine conservation, 21–26
Marine environment(s), 21, 26, 54, 55, 60, 111, 112, 159, 164, 165, 221, 287–289, 307–316, 321
Marine habitats, xviii, 134
Marine Isotope Stage (MIS), 7, 14
Marine reserves, 134
Marine resources preservation, 219–228
Marxism-Leninism, 429
Medicinal applications control of the invasion, 393

Mediterranean, 8–14, 22, 25, 33, 57, 61, 310, 312, 314, 387, 388, 393, 405
Mediterranean Sea, 5–14, 22, 23, 33–46, 57, 61, 63, 164, 387, 388, 393, 394
Melting ice, 21, 309, 310
Merciless consequences, 429
Mesocosm experiments, 54, 59
Mesocosms, 173–175, 205
Methodical study, 454
Microalgae, 166, 236, 311, 313, 347, 375–382
Microsatellite DNA markers, 322, 326
Middle Pleistocene, 6–8
Mishnah, 437, 440–442, 445, 452
Mitigation, 25, 275, 363
Mitzvot, 429
Moba, 221–224, 226
Moisture content, 262, 293
Molecular markers, 326–327
Molecular screens, 342
Mollusk, 59, 132–135
Monitoring, 57, 100–104, 110, 135, 166, 173, 269
Monocious, 324
Monoculture, 279, 312, 367
Monoecious species, 341, 344
Monospecies aquaculture, 313
Monostromatic, 342
Monotheism, 403, 407
Moral, 404–406, 412, 413, 417, 418, 430, 431, 452, 457
Motivation, 409, 426, 427, 447
Multicellular photosynthetic organisms (MPOs), 34, 36, 41
Mycosporine-like aminoacids (MAA), 166, 169, 170, 189–191, 207–209, 212

N

Native American, 428
Natural balance, 415
Nature conservation, 425, 426, 429
Nature management, 315
Nature protection, 405, 425, 429, 430, 441, 459
N concentrations (NO_3-N + NH_4-N), 287
New ecology, 414–415, 418, 443
NH_4-N, 349
Niche modeling, 95–112

Nile Delta, 310, 315
Nirvana, 427
Nitrate reductase (NR), 122, 166, 364
Nitrogen assimilation, 119, 122, 364
Nitrogen metabolism, 122, 123
NOAA. *See* United States National Oceanic and Atmospheric Administration
NO$_3$-N, 349
Nori, 95–98, 108, 342, 349, 362, xvii
Nova Scotia, 74, 75, 78, 79, 83, 87
NO$_x$, 359
Nutrients, 24, 26, 61, 80, 97, 101, 111, 117, 129, 159–175, 185, 190, 224, 225, 235, 240, 243, 245, 247, 255, 272, 275–277, 287, 297, 302, 307–313, 315, 331, 345–350, 359, 361, 363, 367, 378, 381, 382, 393
 credits, 255
 removal, 275, 346, 348–350
 supply, 169, 235, 307–309, 311, 378
 uptake capacity, 349
Nutrification, 367

O

Occupancy, 454
Ocean(s), 5, 22, 24, 53, 54, 57, 62, 73, 100, 121, 149, 159, 219, 221, 223–225, 232, 308, 309, 365, 432
 acidification, 58–60, 121
 nutrification, 367
 warming, 57, 142
Ocean Regional Circulation, 22
Oil, 220, 231–24, 275, 314, 360, 366, 376, 377, 379, 381–383, 425
 extraction, 236, 238, 239, 247
 seeds, 237, 238
 spills, 33
 yield(s), 232, 235–237, 241–243
Ontogeny, 54
Open sea, 224, 274–277, 349
Optical density, 211, 212
Organic pollution, 287
Osmosis regulation, 191
Osmotic regulators, 207
Outdoor system, 377–379
Overfishing, 23, 25
Over-population, 404

Oxidative burst, 147, 148, 172
Ozone, 149, 159–167, 186, 201–204, 212, 332
 depletion, 159–167, 186, 201
 hole, xvi, 165, 203, 212

P

Palm, 220, 232, 238, 242, 366, 376, 379
Palythine, 170, 208, 209
PAM fluorometry, 164, 205
PAR. *See* Photosynthetically active radiation
Partnership, 368, 453–457, 460
Patagonia, 168, 201–212, 288
Pathogenicity, 150
P concentrations, 347
Peak oil, 231–247
Perennating stage, 343
PFD. *See* Photon flux density
pH, 22, 23, 25, 54, 58–60, 111, 117, 121–123, 166, 170, 278, 289, 291, 293–295, 297, 299
Pharmaceuticals, 231, 361, 367, 379, 388
Pharmacology, 387–394
Phenology, 53, 54, 325–326, 331
Philippines, 255–257, 262–266, 268–271, 275, 325, 362
Philosophical, 416, 437, 452
Philosophical literature, 437
Philosophy, 314, 409, 412–414, 429, 437
Phlorotannins, 189, 190
Phosphate production, 245–247
Photic zone, xvii
Photobioreactor, 378, 383
Photochemical efficiency, 120
Photodamage, 166, 191, 206
Photoinhibition, 23, 163, 165, 166, 171, 188–190, 201, 206
Photon flux density (PFD), 345, 346
Photooxidative, 381
Photorespiration, 118, 310
Photosynthesis, xvii, 117–121, 123, 164, 166, 167, 171, 173, 185, 187, 188, 191, 203, 205–207, 236, 241, 242, 245, 310, 332, 333, 347, 363, 364, 376
Photosynthetic acclimation, 120, 123, 167

SUBJECT INDEX

Photosynthetically active radiation (PAR), 106, 147, 166, 171, 185, 187–191, 201, 203–206, 208–210, 242, 332
Photosynthetic efficiency (PSE), 187, 188, 240–242, 244, 360
Photosynthetic electron transport, 212
Phycobiliprotein, 120, 122, 169
Phycocolloid(s), 227, 275, 321, 361
Phycocolloid industry, 321
Phylogenetic analysis, 341
Physico-chemical elements, 160
Physiology, 23, 43, 54, 55, 60, 117, 173, 341–350, 375, 417
Phytoplankton, 55, 56, 59, 62, 122, 159–161, 170, 172, 174, 185, 187, 188, 191, 201, 223, 308, 310, 311, 375
Planet Earth, ix, xix, xv, xvi, xvii, xviii, 21, 33, 34, 73, 96, 104, 159, 186, 201, 202, 220, 241, 247, 332, 403–407, 409–412, 414, 415, 417–419, 427–429, 432, 459
Polar vortex, 203
Political management, 388
Pollution, xv, xvi, xviii, 23, 26, 33, 63, 160, 162, 170, 219, 222, 287, 288, 359, 360, 406, 441, 453, 458
Polyculture, 275, 279, 313, 361
Polyembryony, 323
Population dynamics, 57, 203, 328
Population genetic structure, 329
Postglacial rebound, 9
Post-harvest management, 257–262
Post-natural, 416
Pre-Christian, 403
Pre-industrial age, 13
Preservation of the species, 458
Primary production, 22, 117, 167, 170, 172, 185, 201, 308–310, 312, 360
Production, 22, 60, 78–85, 111, 117, 121, 149, 160, 163, 167, 170–173, 185, 188, 201, 203, 219–228, 231–236, 242–247, 255–257, 264, 273–277, 279, 294–296, 307–316, 322, 323, 325, 328–329, 332, 343, 345, 347–349, 359–368, 375–382, 431
Productivity, xv, xvii, 78–85, 145, 149, 186, 188, 201, 222–223, 259, 265, 275–277, 308, 310, 315, 316, 360, 363, 364, 405
Progressive development, 307–316

Public-private partnerships, 368
Pulse amplitude modulated (PAM), 164, 172, 205

Q
Quantum yield, 120, 164, 171, 189, 191, 206
Quarantine, 41, 270, 273

R
RAdio Detecting And Ranging (RADAR), 97
RAPD, 327
Reactors, 240, 243–244, 380
Recruitment, 56, 58, 61, 77, 81, 84–85, 129–136, 188, 322, 326, 328, 332, 333
Red list species, 425, 427
Red Sea, xv, 36, 39, 40, 45
Reef areas, 258
Re-inventing nature, 415–419
Religion, 403–419, 427–430, 432, 451
Remote sensing, 95–112
Renewable energy, 359, 376
Reproductive cells, 341, 344
Resource conservation, 425–432
Rhodobionta, 34, 43
Rhodolith beds, 129–136
Ribosomal RNA gene, 342
Ribulose-1, 5-bisphosphate carboxylase-oxygenase large (rbcL) subunit gene, 342
Rising sea level, 6–14, 21, 22, 24, 25, 26, 54, 61, 63, 309, 311, 312
Rising temperature, 24, 54, 56
Rocky intertidal, 55, 56, 60, 61, 73–87, 100
Rocky shore, 57, 60, 61, 76, 161, 164, 203, 329
Roman-catholic church, 431
rRNA, 291, 293, 297, 298, 341
Rubisco, 118, 120, 122, 123, 166

S
Salinity, 22, 23, 53, 61, 73, 97, 160, 171, 246, 264, 267, 278, 289–291, 310, 328, 333, 345, 346, 392
  tolerance, 293
Salt intrusion, 5
Sampling design, 131, 171
Santa Cruz Province, 202

Sardinia, 7
Satellite altimetry, 8, 9
Save the earth, 406, 409
Scallops larvae, 134
Science, 269, 271, 311, 342, 367, 403–419, 425, 426, 432
Scientific Committee on Food (SCF), 259
Scytonemin, 189
Sea beds, 202, 224, 225
Sea currents, 310
Sea floor, 5, 134, 387
Sea grasses, 23, 24, 221
Sea level, xvi, xviii, 5–14, 21, 22, 24–26, 53, 54, 61, 63, 159, 309–312, 315
Seashore environment, 221, 287
Seasonal variation, 130, 134, 135, 364
Sea vegetables, 359
Seawater temperature, xvii, 74, 77, 82, 85, 143, 186, 264, 269, 278, 279, 325, 331, 332
Seaweed(s), xvii–xviii, 33, 61, 73, 95, 117, 149, 185, 202, 219, 236, 255, 287, 308, 321, 342, 359, 389, 426
  aquaculture, 343, 361–363, 367
  assemblages, 61, 100, 101, 112
  biomass, 259, 262, 275, 287, 288, 313, 360, 366
  composting, 287–289, 291–293, 298–300
  crop, 269, 342
  cultivation system, 276
  ecophysiology, 109, 339–340
  farming, 262–265, 279, 363
  farms, 264, 312, 366–367
  industry, 74, 79, 339, 362, 367
  recycling, 287–289
  reproduction, 321, 331
Secondary metabolite, 170, 388–389
Sedimentation, 61, 63, 130, 328
Semi-intensive IMTA, 312
Shade plants, 212
Shellfish, 221–225, 227, 275, 308, 311, 313, 363
Shelter, xvii, 76, 77, 141, 211, 256, 407
Shinorine, 170, 208, 209
Shintoism, 428
Sicily, 7, 42
Small subunit (SSU), 341, 342
SO_2, 359

Sodium alginate, 79, 297, 298
Soil amendment, 129
Solar radiation, 25, 166, 185, 186, 188–191, 201–206, 208, 210–212, 241, 242
Solar visible radiation, 185
Solar zenith angle, 203
SONAR, 97
Soybean, 235, 238, 242, 366
Spatial data, 95–98, 108–109
Spatial information, 95–97
Species composition, 59, 74, 143, 161, 162, 171
Species introduction, 33, 36, 270
Species richness, 130, 161–164, 171, 173
Spore coalescence, 330
Spore dispersal, 328, 329
Spore germination, 167, 189, 329–330
Spore production, 323, 328–329
Sporophyte, 141–151, 322, 324, 343, 346
SQDG. See Sulfoquinovosyldiacylglycerol
Squeeze effects, 22
Stalagmites, 7
Substantial hazards, 454
Subtidal, 56, 61, 62, 86, 101, 111, 117, 163, 167, 171, 172, 191, 202, 325, 330, 331, 346
Subtidal seaweeds, 101
Sugarcane, 220, 234, 376
Sulfoquinovosyldiacylglycerol (SQDG), 390
Super-natural, 406
Supralittoral, 164, 167, 201, 203, 205
Surface seawater temperature (SST), 22, 35, 36, 42–44, 74, 75, 77, 81, 82, 85, 86, 105, 108, 264, 269, 278, 279
Sustainable culture, 341–350
Sustainable development, 288, 307, 350
Synthesis de novo, 122

T

Talmud, 405, 437, 438, 440, 442, 445–447
Tanzania, 255, 262–265, 268, 271, 276, 278
Tao, 413
Tectonics, 5–7

SUBJECT INDEX

Temperature, xv–xviii, 13, 21–26, 35, 42, 53–60, 62, 73–75, 77, 81, 82, 85–87, 97, 101, 105, 142–144, 150, 159, 163–168, 171–174, 186, 221, 231, 264, 267, 269, 272, 278, 279, 289–291, 293–295, 297, 299, 309, 310, 324, 325, 328, 330–333, 345, 363, 377–380, 414
  increase, 54–58, 142, 163, 363
  limitation, 142
Tetraspores, 323, 324, 326–329, 332
The LORD, 429, 457, 459
Theocentric approach, 439
Theology, 410, 413, 414, 425–432, 437
Thermal expansion, 9, 13, 21
Thermophilic species, 41–43, 45
The Tree of Life, 437, 459
Three circles, 437–447
Tidal zone, 57, 61, 76–77, 81, 84–86, 117, 160, 173, 187, 191, 202
Tide-gauging, 6, 8–11
Tide-gauging station, 8, 11
"To cultivate it and keep it," 457
Torah, 429, 437
To stay alive, 428
Toxicity, 148, 300–302
Tradition, 258, 313, 375–377, 403, 404, 406, 409, 412–414, 418, 427, 428, 430, 432, 438, 452, 453, 456, 457
Trophic interactions, 55
Tropical, xv, 24, 40–45, 55, 58, 61, 108, 122, 164, 168, 256, 270, 322, 331, 364, 376, 382, 387, 393, 394
Tropical species, 40–42, 45, 58, 331
Turf algae, 23, 24
Turkey, 95, 388

U

Ubatuba, 259
Ultrasonic extraction, 240
Ultraviolet (UV) radiation, ix, xvi–xviii, 53, 147, 149, 150, 159–175, 185–191, 201–212, 311, 326, 332, 343
  absorbing compounds, 189–191, 206–212
  UV-A, 185–189, 191, 201, 203–206, 209
  UV-B, 165, 174, 185–191, 201, 203–205, 212

United States National Oceanic and Atmospheric Administration (NOAA), 21, 111
Universal symbol, 432
Uranium, xix, 7, 425
Uranium–Thorium (U–Th) dating, 7
Urchins, 57, 59, 60, 388
Useful, 103, 110, 163, 186, 219, 221–225, 227, 242, 339, 342, 349, 366, 404, 417, 425
Utilitarian ecology, 456

V

Vertical mixing, 310
Vice-gerents, 405
Vulnerability, 24, 25, 77, 141, 142, 159–175, 269, 333

W

Wakame, 287–302
Warming, xv–xviii, 5, 7, 13, 21, 33–46, 55–58, 73, 142, 144, 159, 164, 174, 219, 255, 307–315, 331, 359–368, 419, 437–447, 451–460
Warming trend, 142
Warm-water species, 35, 42, 58, 164, 331
Water is life, 432
Water motion, 129
Water quality, 96, 129, 159, 160, 162, 163, 219, 222, 256
Water temperatures, 22, 24, 56 58, 73, 75, 85, 159, 164, 221, 264, 272, 331
Wave action, 77, 80, 129, 144, 146, 164, 272
Weather changes, xv, 23, 164, 379
Wide Area Augmentation System (WAAS), 99

Z

Zanzibar, 255, 263, 264, 267, 272, 276, 278
Zohar, 437
Zonation, 23, 121, 167, 191, 212
Zygotospores, 344

AUTHOR INDEX

B
Bleicher-Lhonneur, Genevieve, xxi, 252, 255–283
Boudouresque, Charles F., xxi, 31, 33–50

C
Cavas, Levent, xxi, 385, 387–397
Craigie, James S., xxi, 70, 73–90
Critchley, Alan T., xxi, 71, 73–90, 253, 255–283

D
De Clerck, Olivier, xxi, 93, 95–114

E
Eggert, Anja, xxi, 139, 141–154
Einav, Rachel, ix, xi, xxi

F
Figueroa, Félix L., xxii, 157, 159–182

G
Gao, Kunshan, xxii, 115, 117–126, 183, 185–198, 347
Glicksberg, Shlomo E., xxii, 435, 437–447

H
Häder, Donat-P., xxii, 200–214
Hayashi, Leila, xxii, 251, 255–283
Helbling, E. Walter, xxii, 199, 201–214
Hurtado, Anicia Q., xxii, 251, 255–283

I
Israel, Alvaro, ix, xi, xxii
Issar, Arie S., xxii, 305, 307–318

J
Jha, Bhavanath, xxiii, 320–338

K
Klein, Micha, xxiii, 4–17, 341
Klostermaier, Klaus K., xxiii, 401, 403–421
Korbee, Nathalie, xxiii, 157, 159–182
Küpper, Frithjof C., xxiii, 140–154

L
Lichter, Michal, xxiii, 3, 5–17

M
Mantri, Vaibhav A., xxiii, 319, 321–338
Medina-López, Marco A., xxiii, 127, 129–138
Msuya, Flower E., xxiv, 252, 255–283

N
Nagata, Shinichi, xxiv, 286–304
Neori, Amir, xxiv, 274, 277, 305, 307–318, 357, 359–372
Notoya, Masahiro, xxiv, 217, 219–228

O
Olsvig-Whittaker, Linda, xxiv, 19, 21–28

P
Pauly, Klaas, xxiv, 93, 95–114
Pereira, Rui, xxiv, 339–354
Peters, Akira F., xxiv, 139, 141–159
Pohnert, Georg, xxiv, 385, 387–397

R
Reddy, C.R.K., xxiv, 319, 321–338
Reisser, Werner, xxv, 373, 375–383
Rhodes, Chris J., xxv, 229, 231–248
Rilov, Gil, xxv, 51, 53–68
Riosmena-Rodriguez, Rafael, xxv, 127, 129–138
Roth, Hermann Josef, xxv, 423, 425–433
Rozenson, Yisrael, xxv, 449, 451–460

S
Seckbach, Joseph, xv–xx, xxv
Sivan, Dorit, xxv, 4–17

T
Tang, Jing-C., xxv, 285, 287–304
Taniguchi, Hideji, xxvi, 285, 287–304
Treves, Haim, xxvi, 51, 53–68
Turan, Gamze, xxvi, 357, 359–372

U
Ugarte, Raul A., xxvi, 69, 73–90

V
Verlaque, Marc, xxvi, 31, 33–50
Villafañe, Virginia E., xxvi, 199, 201–214

X
Xu, Juntian, xxvi, 183, 185–198

Y
Yarish, Charles, xxvi, 339–354

Z
Zhou, Qixing, xxvi, 286–304
Zou, Dinghui, xxvii, 115, 117–126
Zviely, Dov, xxvii, 3, 5–17